G000149122

MECHANICS OF SANDWICH STRUCTURES

Mechanics of Sandwich Structures

Proceedings of the EUROMECH 360 Colloquium held in Saint-Étienne, France, 13-15 May 1997

Edited by

A. Vautrin
Department of Mechanical and Materials Engineering,
École Nationale Supérieure des Mines,
Saint-Étienne, France

KLUWER ACADEMIC PUBLISHERS
DORDRECHT / BOSTON / LONDON

Library of Congress Cataloging-in-Publication Data

ISBN 0-7923-5050-2

Published by Kluwer Academic Publishers,
P.O. Box 17, 3300 AA Dordrecht, The Netherlands.

Sold and distributed in North, Central and South America
by Kluwer Academic Publishers,
101 Philip Drive, Norwell, MA 02061, U.S.A.

In all other countries, sold and distributed
by Kluwer Academic Publishers,
P.O. Box 322, 3300 AH Dordrecht, The Netherlands.

Printed on acid-free paper

Printed in the Netherlands

EUROPEAN MECHANICS COLLOQUIUM 360

MECHANICS OF SANDWICH STRUCTURES

MODELLING

NUMERICAL SIMULATION

EXPERIMENTAL IDENTIFICATION

SAINT-ÉTIENNE, 13 - 15 MAY 1997

Chairmen: Prof. A. VAUTRIN, Prof. A. TORRES MARQUES

SCIENTIFIC COMMITTEE

Prof. H. G. ALLEN, University of Southampton, Southampton, United Kingdom

Dr. P. DAVIES, IFREMER, Brest, France

Dr. A. J. M. FERREIRA, DEMEGI, Universidade do Porto, Porto, Portugal

Prof. M. GRÉDIAC, École des Mines de Saint-Étienne, Saint-Étienne, France

Prof. M. MAIER, Institut für Verbundwerkstoffe, Kaiserslautern, Germany

Dr. F. MARTINS DE BRITO, Laboratorio Nacional Engenharia Civil, Lisboa, Portugal

Prof. H. -R. MEYER-PIENING, ETH Zürich, Zürich, Switzerland

Prof. C. MOTA SOARES, Instituto de Engenharia Mecânica, IST, Lisboa, Portugal

Prof. K. -A. OLSSON, Royal Institute of Technology, Stockholm, Sweden

Prof. R. TETI, University of Naples, Naples, Italy

Prof. A. TORRES MARQUES, DEMEGI, Universidade do Porto, Porto, Portugal

Prof. M. TOURATIER, École Nationale Supérieure des Arts et Métiers, Paris, France

Prof. A. VAUTRIN, École des Mines de Saint-Étienne, Saint-Étienne, France

Prof. I. VERPOEST, Katholieke Universiteit Leuven, Leuven, Belgium

LOCAL ORGANIZING COMMITTEE

Prof. M. GRÉDIAC, Chairman
As. Prof. W. S. HAN
Mr. V. MANET
As. Prof. F. PIERRON
Prof. Y. SURREL
Prof. A. VAUTRIN

CONTENTS

BUCKLING AND SINGULARITIES

DYNAMIC AND IMPACT BEHAVIOUR

SANDWICH DESIGN

INTRODUCTION

The main advantages of sandwiches as structural components are now well-known and well-established. Due to the progress in polymer science and engineering and advances in manufacturing processes, sandwich structures can blend various functional and structural properties and therefore lead to highly innovating systems. The current difficulty to overcome is to provide designers with proper methodologies and tools that could enable them to design improved sandwich structures. Such dedicated design tools should be efficient, reliable, flexible and user-friendly. They should be based on advanced knowledge of sandwich behaviour at global and local scales. Such approach relies on our capability to test, identify, control and model structure performances.

The impressive variety of core and face materials and the rapid developments in forming processes give new opportunities to design components which have more complex shapes and higher integrated functional and structural properties. Interest in sandwiches is permanently growing in industry and refined testing and modelling approaches should be encouraged to set up relevant guidelines to design reliable advanced structures.

The European Society for Mechanics sponsored the EUROMECH 360 Colloquium on the 'Mechanics of Sandwich Structures' in Saint-Étienne, France, on 13 - 15 May 1997. The main purpose of EUROMECH 360 was to go into the most recent progresses in sandwich analysis and design, including mechanical modelling and testing. It was expected that the Colloquium should contribute to define new research directions to support development of advanced applications in strategic industrial sectors such as ground transportations or building and civil engineering.

5 keynote lectures, 28 oral presentations and 19 posters have been selected by the scientific committee for presentation at EUROMECH 360. 74 participants from 19 different countries attended the sessions among which 33 were French scientists. The extended abstracts of the papers (a two - three page abstract per paper) have been provided to the participants.

The keynote lectures have highlighted different theoretical or applied problems. They have served as introductions to the specialized oral sessions. They focused on the classification of sandwich structures (Prof. H.G. Allen, University of Southampton, U.K.), the development of advanced sandwich structures for naval vessels (Prof. K.A. Olsson et al., Royal Institute of Technology of Stockholm, S), the damaged tolerance of aeronautical sandwich structures (Dr. Guedra-Degeorges et al., Aérospatiale, F), the numerical simulation of the crash behaviour of sandwich structures with fibre reinforced polymer-faces (Prof. M. Maier and Dr. S. Kerth, Institut für Verbundwerkstoffe, Kaiserlautern, D) and the numerical models and optimisation of sandwich structures (Dr. J. César de Sà et al., University of Porto, P).

The main themes covered by the Colloquium are:

* Modelling of sandwich structure behaviour;
* Dynamic properties of sandwich structures;
* Identification of sandwich and core material mechanical properties;
* Industrial applications and manufacturing.

A few papers tackled more specific topics such as sandwich joining and manufacturing, or local and global sandwich panel stability.

The present Proceedings comprise 49 extended papers which have been reviewed by the scientific committee of EUROMECH 360. Two experts have been in charge of the review of each paper after the Colloquium took place. Many changes have been suggested by the reviewers and introduced by the authors before publication.

The papers have been arranged in seven thematic sections. Most of the papers could belong to several sections, therefore the classification here should be considered as a simple way to help the reader to retrieve any scientific information easily.

I- Finite Element Modelling
II- Modelling
III Buckling and singularities
IV - Dynamic and impact behaviour
V - Experimental testing and control
VI - Sandwich and constituents properties
VII- Sandwich design

The main conclusions coming up from the round table that concluded the Colloquium are summarized below :

* there is an actual difficulty to find out complete sets of *core material* properties suited to structure design; development of rational testing routes is absolutely necessary to provide reliable static and dynamic properties of foams and honeycombs to be used for designing;
* new testing methodologies to get comprehensive view of sandwich panel behaviour under service loadings, including environmental effects, should be strongly encouraged; in particular there is a specific demand for panel impact strength characterization and prediction;
* particular advances in panel local buckling analysis are needed when large structures are in view; it requires progress in numerical modelling and testing, improvements in boundary condition control and strain field measurements are necessary;
* advanced mechanical modelling of joints and inserts is required; new approaches to optimize the joining design should be appreciated when dynamic or impact loadings are likely to occur, and safety conditions have to be guaranteed such as in transportation applications;
* new materials for sandwich skins and cores, such as anisotropic polymer composites or knitted materials, are highly promising, improved physical

and mechanical characterization and modelling are to be supported to point out the specific advantages they enable to achieve, including the forming process and joining technics;

* improved methodologies to check and compare finite element efficiency have to be set up; a great deal of sandwich dedicated finite elements have been developed in the past few years, however their validity ranges have not been determined and their advantages over often existing finite elements have not been pointed out; two routes should be introduced :
    * comparisons between finite elements and analytical solutions, developement of new analytical solutions is still relevant since they usually give better insight into the assumptions that are made;
    * comparisons between finite elements solutions and experimental results, special attention has to be paid to the boundary conditions and strain measurements when panels are tested;
* computer aided sandwich design tools should be supported; in particular, rational materials selection methods are needed since sandwich panels are supposed to integrate different functions and cut down production costs;

A benchmarking test to compare the reliability and efficiency of different sandwich beam analyses has been set up. 10 European research groups are involved in the process which intends to calculate the stress and strain fields in beams subjected to four point bending. Special attention is paid to the stress concentration near the loading points and between skin and core. Both linear and non linear finite element calculations are performed and compared. Comparisons with analytical analyses are carried out as well. The results of the benchmarking test will be published and extensions to other geometries and loading cases will be contemplated.

This volume provides the reader with a current view of methods of mechanical modelling, materials and structures testing approaches and overall conclusions dealing with the mechanical properties of the sandwiches, including impact performances and instabilities. It can be valuable to engineers and scientists in industry, as well as to professors and students at universities, to keep them up to date in this rapidly growing field.

ALAIN VAUTRIN
Saint-Étienne, November 18, 1997.

ACKNOWLEDGMENTS

The European Mechanics Colloquium "MECHANICS OF SANDWICH STRUCTURES, Modelling, Numerical Simulation, Experimental Identification" has been held under the auspices of the European Society for Mechanics and organised at École des Mines de Saint-Étienne from 13 to 15 May 1997.

Financial support and patronage have been received from national and local French organizations:

- the French Ministry of National Education, Research and Technology *Ministère de l'Éducation Nationale, de la Recherche et de la Technologie (MENRT)*,
- the French University Mechanics Society *Association Universitaire de Mécanique (AUM)*,
- the Rhône-Alpes Regional Council *Conseil Régional de Rhône-Alpes*,
- the Saint-Étienne City Council *Conseil Municipal de Saint-Étienne*.

The selection of the scientific programme was possible thanks to the International Scientific Committee, composed of : Prof. A. TORRES MARQUES (PT) and Prof. A. VAUTRIN (F), Chairmen of the Colloquium, Prof. H.G. ALLEN (U.K.), Dr. P. DAVIES (F), Dr. A.J.M. FERREIRA (PT), Prof. R. TETI (IT), Prof. M. GRÉDIAC (F), Prof. M. MAIER (DE), Dr. F. MARTINS DE BRITO (PT), Prof. H.-R. MEYER-PIENING (CH), Prof. C. MOTA SOARES (PT), Prof. K.-A. OLSSON (SE), Prof. M. TOURATIER (F) and Prof. I. VERPOEST (BE).

The French Local Organizing Committee was chaired by Prof. A. VAUTRIN and composed of Prof. M. GRÉDIAC, Dr. W. S. HAN, Dr. F. PIERRON, Prof. Y. SURREL and Prof. A. TORRES MARQUES. I am very obliged to my colleagues for their constant and efficient involvement in the Colloquium organization and I wish to express all my thanks to them.

Constant assistance of the École des Mines de Saint-Étienne in the conference holding was appreciated, in particular the participation of the International Affairs Department was highly helpful.

I am very grateful to all the members of the Mechanics and Materials Engineering Department, and more especially to Mr V. MANET, PhD student, and Mrs L. SANTANGELO and N. VIAL-BONACCI, in charge of the secretariat of the Colloquium, for their valuable contribution.

ALAIN VAUTRIN
Saint-Étienne, May 15, 1997.

CLASSIFICATION OF STRUCTURAL SANDWICH PANEL BEHAVIOUR

Howard G. Allen* and Zhengnong Feng†
*Emeritus Professor of Structural Engineering; †Research student
Department of Civil and Environmental Engineering,
University of Southampton, Southampton SO17 1BJ, UK

1. Primary and Secondary deformation

1.1 INTRODUCTION

The structural behaviour of a sandwich panel depends critically on the geometrical relationships between face and core thicknesses and the span, and on the relative stiffnesses of the face and core materials. The number of combinations possible, and the complexity of the equations that sometimes emerge from analyses of sandwich behaviour, can make it difficult for the designer to decide which aspects of sandwich behaviour are important, and which can be neglected safely.

This paper describes a method which enables the designer to decide, without lengthy calculations, how a given sandwich panel is likely to behave. The designer can then decide which kind of analysis will be most appropriate. The discussion leads on to some other aspects of sandwich behaviour that can be important, but which are often overlooked, such as the nature of the boundary conditions for sandwich panels and the effects of discontinuities in the core. The paper concludes with suggestions for the approximate analysis of sandwich plates carrying high lateral pressures.

1.2 DEFINITIONS AND NOTATION

It is convenient to defines some terms for use later.
(a) *Composite beam theory* (**CBT**) The sandwich is treated as an ordinary composite beam. Plane sections remain plane. There is no shear deformation.
(b) *Elementary sandwich theory* (**EST**). Stresses and ('bending') deflections are calculated by composite beam theory. But there is an additional ('shear')deflection associated with shear strains in the core.
(c) *Advanced sandwich theory* (**AST**). The faces must bend locally in order to follow the shear deformation of the core. Thus the additional shear deflections in (b) are reduced by the local bending stiffness of the faces.

Note. In (a) and (b), but not (c), it can be appropriate to treat the faces as 'thin'.
Note. In (b) and (c) it is common to ignore the contribution of the core to the overall flexural rigidity, D. This implies that the shear stress is uniform throughout the depth of the core. If the core is stiff enough to make a significant contribution to D, the analysis can be modified accordingly, but this is rarely necessary in practice.

This paper deals mainly with AST. Fig. 1 shows a conventional representation of the deformation of a sandwich beam, viewed from one side. The deformation is defined in terms of the total displacement, w, and the corresponding gradient w'. In a simple beam this gradient would also be the rotation of the plane cross-section of the beam. In a sandwich, however, there is an additional displacement associated with the shear strain, γ, in the core. What was a plane cross-section in the unloaded sandwich is now a zig-zag line aceg. It is perfectly possible to perform the analysis of the sandwich using w and γ as

A. Vautrin (ed.), Mechanics of Sandwich Structures, 1–12.

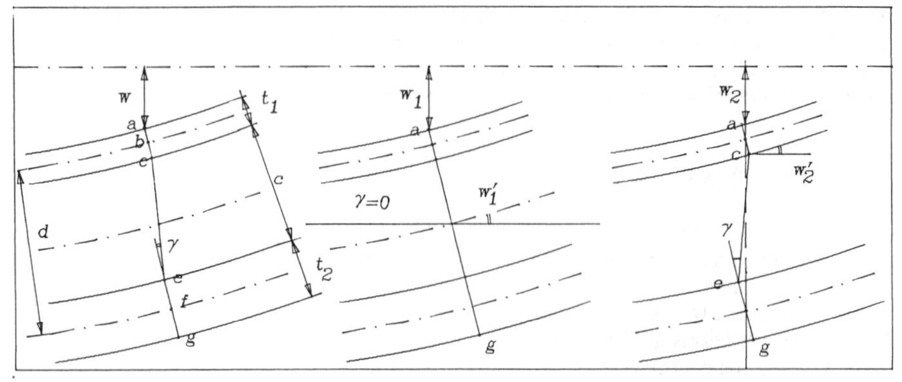

Fig.1 Total displacement (w) and shear strain (γ)

Fig.2a Primary displacement (w_1)

Fig. 2b Secondary displacement (w_2)

the variables in the equations, and this procedure is sometimes called 'the zig-zag method'.

In this paper a different notation is preferred, in which the total transverse displacement w is split into two quite independent parts, the primary deformation, w_1, and the secondary deformation, w_2. These are illustrated in Fig. 2. In the *primary deformation* plane sections (ag) remain plane and the sandwich functions as a simple composite beam, without shear deformation of the core. In the *secondary deformation* the faces bend about their own separate neutral axes and the core undergoes shear deformation. The primary and secondary deformations can occur quite independently of each other; they can be superimposed to obtain the total deformation. The task of the analyst is to determine the relative proportions of primary and secondary deformation present in a given situation.

From the definition of primary and secondary deformation, and from the geometry of Fig. 2b:

(1) $$w = w_1 + w_2$$

(2) $$\gamma = w_2'\left(\frac{d}{c}\right) \qquad where \quad w_2' = \frac{dw}{dx}$$

1.3 EXTREME LIMITS OF SANDWICH BEHAVIOUR.

When the core is *very stiff* in shear, the shear strain is negligible, the secondary deformation vanishes, and the CBT can be used. When the core is *very flexible* in shear, the faces are no longer coupled together effectively. Provided the core still maintains the faces at the correct distance apart, they bend as two independent beams, but sharing the same deformation. Ordinary beam theory can be applied, using only the secondary deformation.

1.4 INTERMEDIATE SANDWICH BEHAVIOUR

Between the extremes just described, sandwich theory must be applied. When the core is *fairly rigid* in shear, the EST is sufficient. When the core is *fairly flexible* in shear, the AST is necessary. and the stress pattern no longer corresponds to elementary beam theory. The problem is this: how is one to define a *'fairly stiff'* core and a *'fairly flexible'* core ? This question can be answered very easily in any particular situation, without recourse to difficult analysis. A single procedure is applicable to beams, columns, transversely-loaded plates and plates buckling under edge load.

1.5 BASIS OF THE METHOD

The differential equations for a sandwich beam, column or panel can be set up with the primary and secondary deformations, w_1 and w_2 as variables. These can be constructed directly from the requirements of equilibrium and compatibility, or they can be obtained from the requirement that an expression for the total energy must have a stationary value with respect to variations in the primary and secondary deformations. Detailed derivations using the energy method were given by ALLEN (1972, 1993); an abbreviated derivation will be available as a supplement to this paper.

These equations can be used to obtain exact solutions for particular problems, such as beams with central point load or distributed load, ALLEN (1969,1973). For the present purpose it is more useful to consider sinusoidal loading and deformation.

In the case of a beam of span L it will be found that a sinusoidal loading of intensity:

$$q = q_0 \sin\left(\frac{\pi x}{L}\right) \tag{3}$$

produces sinusoidal primary and secondary deformations:

$$w_1 = w_{10} \sin\left(\frac{\pi x}{L}\right); \qquad w_2 = w_{20} \sin\left(\frac{\pi x}{L}\right); \tag{4}$$

where the amplitudes w_{10} and w_{20} can be obtained from q_0.

In a similar way, a sinusoidally distributed load on a rectangular panel ($x = 0,a;$ $y = 0,b$) produces sinusoidal primary and secondary deformations.

For a pin-ended column of length L, the buckled configuration is a combination of sinusoidal primary and secondary deformations.

For a rectangular panel with an in-plane edge load P_x per unit length, any selected bucking mode with (m,n) half-waves is a combination of sinusoidal primary and secondary deformations with the same number of half-waves.

Because of the choice of variables, the solutions to all of these problems have the same form. The most useful aspect of the solutions can be summarised as follows.

For *simply-supported beams:*

$$w = \frac{q_0}{D}\left(\frac{L}{\pi}\right)^4 r \tag{5}$$

where w is the total deflection at the centre, D is the flexural rigidity of the entire beam (EI), q_0 is the amplitude of the sinusoidal transverse load, L is the span, and r is a coefficient to be discussed below.

For *pin-ended columns,*

$$P_{cr} = \frac{\pi^2 D}{L^2} \frac{1}{r} \tag{6}$$

where P_{cr} is the critical load and L is the effective length.

For a *rectangular panel* with *sinusoidal transverse load,*

$$w = \frac{q_0}{D}\left(\frac{L}{\pi}\right)^4 r \tag{7}$$

where w is the central displacement due to a sinusoidal load of amplitude q_0, and D is the flexural rigidity per unit width. In this case L is a notional length defined as

$$\frac{1}{L^2} = \frac{1}{a^2} + \frac{1}{b^2} \tag{8}$$

For a *rectangular panel* with a *compressive edge load,* P_x per unit length, buckling in a single half-wave in each direction, the critical load is

$$P_{xcr} = \frac{D\,a^2\pi^2}{L^4}\,\frac{1}{r} \tag{9}$$

and L is given by (8).

In all of these equations the coefficient r is given by

$$r = \frac{1 + \left(1 - \dfrac{D_f}{D}\right)\left(\dfrac{\pi^2 D}{L^2 D_Q}\right)}{1 + \left(1 - \dfrac{D_f}{D}\right)\left(\dfrac{\pi^2 D}{L^2 D_Q}\right)\left(\dfrac{D_f}{D}\right)} \tag{10}$$

Here D is the flexural rigidity of the complete cross-section of a beam or column, D_f is the sum of the separate flexural rigidities of the faces, D_Q is the shear stiffness of the core, and L is either the length of a beam or column or it is given by (8). In a simple sandwich of width b, with faces of equal thickness, t, a core thickness c, and $d = c+t$,

$$D_f = \frac{E_f b t^3}{6}; \qquad D = D_f + \frac{E_f b t d^2}{2}; \qquad D_Q = \frac{G_c b d^2}{c} \tag{11}$$

In a panel, b=1 and D, D_f and D_Q are *per unit width*.

The quantity D_f/D is the ratio of the bending stiffness of the faces to the stiffness of the sandwich as a whole. The quantity $D/(D_Q L^2)$ is the ratio of the bending stiffness to the shear stiffness, made non-dimensional by the inclusion of L^2. These two non-dimensional *parameters* define the way in which the sandwich behaves.

1.6 THE SANDWICH MASTER DIAGRAM

The coefficient r has a very special significance. In bending problems it represents the ratio of the total deflection of the sandwich to the ordinary bending deflection (from CBT). In buckling problems, the critical load of the sandwich is $(1/r)$ times the critical load when there is no shear deformation of the core. Fig. 3 shows the ratio r plotted against $D/D_Q L^2$ for various values of D_f/D. For any given sandwich the designer can easily calculate these ratios and, hence, obtain the value of r. The location of the point on the diagram shows what kind of analysis will be needed.

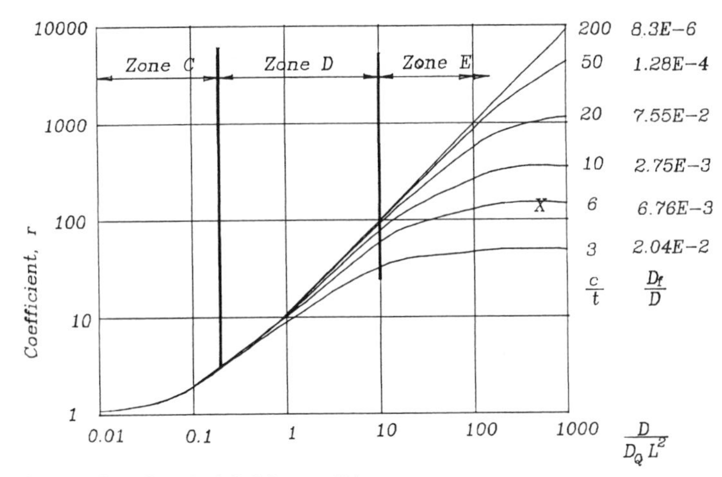

Fig. 3 The Sandwich Master Diagram

Case A, $D/(D_QL^2) = 0$ This happens when D_Q is infinite. There is no shear deformation and the structure behaves as an ordinary composite beam, column or panel and r = 1. CBT can be used.

Zone B, $0 < D/(D_QL^2) < 0.01$ The sandwich acts as a composite structure, with not more than 10% extra deflection (due to shear) under transverse load, and not more than 10% reduction of critical end load. In this case $r <= 1.1$.

Zone C, $0.01 < D/(D_QL^2) < 0.2$ Substantial shear deflection occurs, but the sandwich still acts as a composite structure. In this case $1.1 < r < 3$.

In zones B and C, the ratio D_f/D does not have much influence on the value of *r*. The shear deformation can be calculated by EST without much error, but there may be significant local bending stresses in the faces. These can only be calculated by using AST, as described above.

Zone D, $0.2 < D/(D_QL^2) < 10$ Shear deformation of the core is quite large and there is much local bending of the faces, causing high local bending stresses. Critical loads are much reduced. The value of *r* lies between 3 and 81 and the structure does not perform very effectively as a sandwich. AST is essential.

Zone E $10 < D/(D_QL^2)$ Depending on the value of D_f/D (or c/t) this is the region where the curves tend to become horizontal. The significance of a point such as *X* in the diagram is that the faces now act as two independent beams (columns or panels) and all sandwich action is lost. The value of *r* taken from the diagram actually provides the correct solution for the case in which the two faces bend independently. Clearly, this case should be avoided in the design of real sandwich structures !

The non-dimensional parameters can also be written as follows:

$$\frac{D_f}{D} = \left(1 + 3\left(\frac{d}{t}\right)^2\right)^{-1}; \qquad \frac{D}{D_Q L^2} = \frac{1}{6}\left(\frac{c}{L}\right)^2 \frac{E_f}{G_c}\frac{t}{c}\left(3 + \left(\frac{t}{d}\right)^2\right) \tag{12}$$

1.7 EXAMPLE

A simply-supported sandwich beam has a span (L) of 500mm, with an overhang of 50mm at each end, making the total length equal to 600mm. It carries a central point load, $W = 500$N. Properties of the cross-section are:

Face thickness, t	2.5mm	
Core thickness, c	25mm	
Distance between centroids of faces, $d = c + t =$	27.5mm	$c/t = 10$; $L/c = 20$
Width, b	1.0mm	
Young's modulus of faces, E_f	20,000 N/mm^2	
Shear modulus of core, G_c	25.07 N/mm^2	

From equations (11): $D_f = 52{,}083$ Nmm^2; $D = 18.958 \times 10^6$ Nmm^2; $D_Q = 758.368$ N

The non-dimensional sandwich parameters are therefore $D_f/D = 0.0027475$; $D/(D_Q L^2) = 0.10$. These parameters can also be obtained directly from equations 12.

The sandwich master diagram (Fig. 3) shows that, for $D/(D_Q L^2) = 0.10$, the value of r is about 2, and virtually independent of D_f/D. The beam is firmly in Zone C and one may conclude that it has a total deflection about twice the normal bending deflection, and that the faces are not thick enough to modify the shear deformation very much. The same conclusion applies to a laterally-loaded panel with a shorter span equal to 500mm. A column of the same length would suffer a reduction of about 50% (i.e. by a factor of $1/r$) in the critical stress, as would a panel 500mm wide under a compressive edge load. Equation (10) gives the value of r as 1.932.

Strictly speaking these values of r are for sinusoidal loading, but they are not very sensitive to the pattern of loading. Therefore they allow the designer to see what kind of behaviour to expect from a sandwich, whether it is a beam, a column or a panel, irrespective of the pattern of loading.

The 'exact' equations of AST for type of beam were given by ALLEN(1969). The solution for this example is:

Primary deflection	68.467mm
Secondary deflection	78.107mm
Total deflection	146.574mm

An analysis based on EST gives the following result:

Bending deflection, $WL^3/(48D)$	= 68.681mm
Shear deflection, $WL/(4D_Q)$	= 82.414mm
Total deflection	= 151.095mm

The true value of r for this beam is therefore 146.574/68.681 = 2.134, which is about 10% greater than the value from equation (10). Considering the generalisations on which equation 10 and Fig. 3 are based, and the fact that the 'exact' AST analysis is not a trivial calculation, this is quite acceptable.

The numerical deflections confirm what was said about the beam being in Zone C. The bending deflection (EST) is almost identical with the primary deflection (AST). The AST secondary deflection (78.107mm) is slightly less than the EST 'shear' deflection (82.414mm) because the local bending of the faces interferes with the latter. Nevertheless, as will be shown later, the local bending of the faces can lead to quite high local stresses in the faces and it may cause significant changes in the shear stress in the core.

2. Boundary Conditions

Many (though not all) practical sandwiches lie in Zone B, where the core is so stiff in shear that the secondary deformation is not much modified by the local bending of the faces. Why, then, should it be necessary to take the trouble to consider AST at all ? The answer lies in the boundary conditions, which cannot always be satisfactorily explained without recourse to AST.

In order to illustrate this point, consider a beam in three-point bending, with a central load W and a span L. Fig. 4 shows the support at the right-hand end of the beam.

EST (using thin faces) suggests that the forces in the faces and the core are as shown in the 'exploded' diagram in Fig. 5. The shear stress in the core is everywhere equal to $W/(2bd)$, ($d=c$ in this case) and this postulates the existence of an internal stiffener (as shown) to transfer the shear to the support. Of course, this internal stiffener is rarely provided in practice. But before we look at the effect of removing it, consider the implication of AST.

Fig. 6 shows the end of the same beam, but this time with two supports, one under each face. These divide the load between the two faces in proportion to their local bending stiffnesses. This corresponds very closely with the AST, which shows that the primary and secondary deformations appear as in Fig. 7. The secondary deformation must satisfy certain conditions:

Well to the left of the support (at a) the shear strain in the core is high.
At the free end of the beam (d) the shear strain is zero.
The shear strain must be the same on both sides of the support (at b & c).
The faces must bend locally to match the shear deformation of the core.

All of these conditions are achieved automatically by the AST, but the consequences are interesting. Fig. 8 shows that the shear force in the core (Q_1) at the support is $W/4$ (not $W/2$) and that it decreases smoothly to zero at the free end. Fig. 8 also shows that the shear force in the two faces in local bending (Q_2) is $W/4$ to the left of the supports and $-W/4$ just to the right. Combining the two graphs in Fig. 8 gives a total shear force of $W/2$ on the left and zero on the right, as required.

The important consequence of this is that even if supports are provided as in Fig. 6, the faces jointly carry half the shear force at the support. This is true even if the faces are thin, and even in the case shown in Fig. 5. Also, this shear force in the faces is associated with high local bending stresses. In the numerical example given above, the key results at the support are:

Primary bending moment	+1034.5 Nmm
Secondary bending moment	-1034.5 Nmm
Secondary bending stress	± 1191 N/mm^2

For comparison, at midspan the bending moment is $WL/4 = 62{,}500$ Nmm and the nominal bending stress is $M/(bdt) = 909.1$ N/mm^2. In other words, the *local bending stress at the support is greater than the nominal bending stress at mid-span.* The stress calculations given here are broadly confirmed by finite element analysis.

In practice a sandwich beam will usually be supported at the lower face, as in Fig. 9. Analysis based on AST cannot deal with this case explicitly, but it is reasonable to expect that the distribution of the shear in the core will be much the same as described above, that the local bending in the upper face will be small, and that nearly all the secondary bending moment will be concentrated in the lower face.

Another interesting case is shown in Fig. 10, where a rigid block has been inserted. At first sight, it appears that the face '*aa*' of the block can transmit the shear in the core directly to the support. However, the faces at points '*a*' must be at right-angles to the line '*aa*', and the shear strain in the core must be zero adjacent to the block. This implies that the shear stress in the core is zero at points close to the line '*aa*', so the shear force must be carried wholly by the faces in this locality. In other words, the block does not help to reduce the high local bending in the faces at the support. This analysis also implies that it does not make much difference if there is no bond between the core and the block !

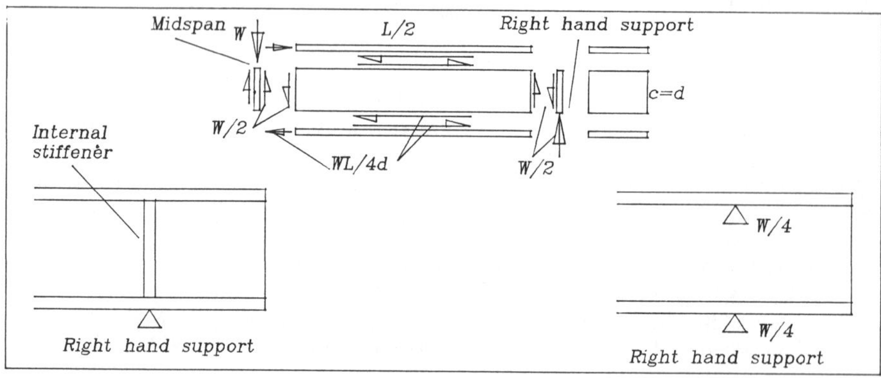

Fig. 4 Support with internal stiffener

Fig. 5 Exploded view of half the beam

Fig. 6 Double support

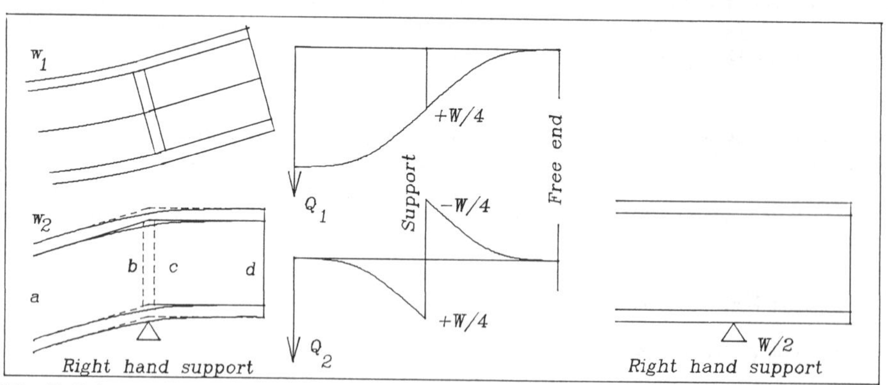

Fig. 7 Primary and secondary deflections

Fig. 8 Primary and secondary shear forces

Fig. 9 Single support, no stiffener

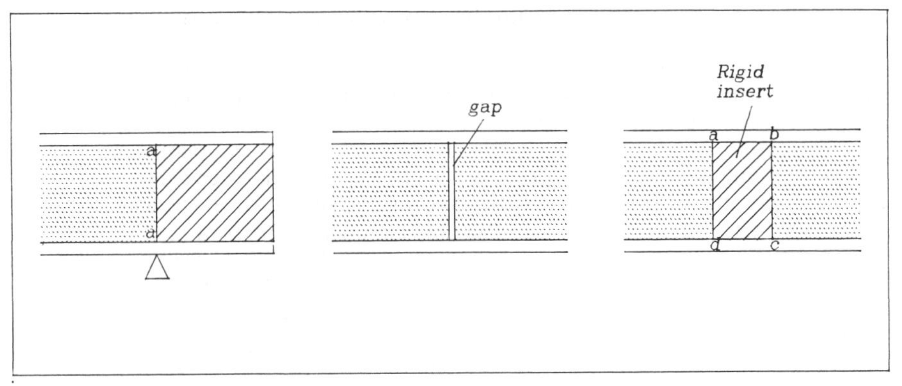

Fig. 10 Rigid end block **Fig. 11 Gap in core** **Fig. 12 Rigid insert**

3. Defects and Inserts

It sometimes happens that the core of a beam or a panel is made up from contiguous blocks of material. This might be the case with a balsa core, or with a foam core that is made to fit a curved mould. Elementary theory suggests that, if adjacent blocks of core material are not properly bonded together, the structure cannot function as a sandwich, because there is no continuous path along which the shear forces can be transmitted. In practice the bond between adjacent blocks may not always be maintained, and tests suggest that the absence of bond may not be fatal. Fig. 11 shows a small gap in the core of a sandwich beam. Obviously the shear stress at the free faces must be zero, so all of the shear force in the region of the gap must be transmitted to the faces, which must undergo intense local bending.

Another kind of irregularity is created by the insertion of rigid blocks of material, as in Fig. 12, perhaps to enable objects to be firmly attached to the sandwich. As in the case of the rigid end-block, the shear strain in the core must be zero at the points marked *a, b, c, d*. This implies that the shear stress in the core is zero adjacent to the block, and that all of the shear force is transmitted through local bending of the faces in this region.

The existence of gaps and inserts may have a fairly insignificant effect on the overall deflection of the sandwich structure, and on the overall pattern of bending and shear stresses. But this fortunate result is achieved at the expense of high local bending and shear stresses in the faces in the immediate vicinity of the gap or insert. This might be important if the faces are liable to delamination, or if the interfacial bond is poor, especially under dynamic loads.

4. Panels under High Transverse Pressures

4.1 BASIC EQUATIONS

In the past the classical analyses of thin plates under transverse pressure have been extended to allow for the increased deflection associated with shear deformation. Graphs have been published for design purposes. It is well known that the validity of these analyses is limited to 'small deflections'. This is a very severe limitation for single-skin plates and it is unclear what the limit should be placed on small-deflection theory when it is applied to sandwich panels. The question arises: 'How is it possible to *estimate* the behaviour of sandwich panels under large transverse pressures?' A rather speculative answer to this question is given below.

Tests on single-skin rectangular panels with clamped edges show very clearly that the behaviour under high transverse pressures is closely approximated by taking a strip of material in the short span direction and treating it as a catenary; i.e., as a tension member with no flexural stiffness. It is suggested that this concept could be extended to sandwich panels. The tests also show that the results are very sensitive to small unintentional in-plane movements of the boundaries, which must therefore be taken into account.

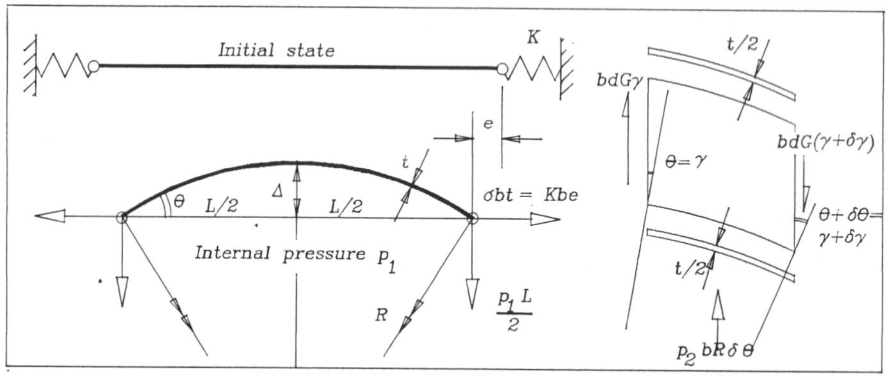

Fig. 13 Section through single-skin panel under transverse pressure p_1.

Fig. 14 Additional pressure p_2 resisted by shear of the core.

Starting with a single-skin panel, Fig. 13 shows a cross-section taken in the short-span direction, span L. Under a pressure p_1 the panel is assumed to deform in a circular arc of radius R, and each end is attached to a spring of stiffness K (force per unit length of edge, per unit displacement). The thickness of the panel is t, the deflection is Δ and the angle θ is as shown. The uniform tensile stress is σ and the corresponding strain is ϵ. The following equations are easily established.

For equilibrium,

$$\sigma = \frac{p_1 R}{t} \tag{13}$$

For an elastic material,

$$\sigma = E \epsilon \tag{14}$$

Geometry:

$$\Delta = R\,(1 - \cos\theta) \tag{15}$$

$$\frac{L}{2} = R \sin\theta \tag{16}$$

The force in the arc (per width b) is equal to the force in the spring, which is K times the extension of the spring. The extension of the spring is the geometrical shortening of the arc minus the extension of the arch due to the strain:

(17) $\sigma\, b\, t \cos\theta = K\, b\, R\,(\theta - \sin\theta - \epsilon \sin\theta)$

Values of K may be chosen to represent the stiffness of the structure to which the panel is attached. Only the case of a rigid boundary ($K = \infty$) will be considered here, in which case equation (17) reduces to

(18) $\theta - \sin\theta - \epsilon\sin\theta = 0$

Now suppose that the single thickness t is split into the two faces of a sandwich, and that under extreme loading conditions the core of the sandwich is forced to adopt the same pattern of deformation as the faces. The approximate shear strain in the core is shown in Fig. 14, and it is easy to show that the shear stress in the core supports an additional pressure p_2 where

(19) $$p_2 = \frac{2\, d\, G\, \theta}{L}$$

The total pressure resisted by the sandwich panel is $p = p_1 + p_2$. All of these equations enable the quantities R/L, ϵ, p/E and Δ/L to be expressed in terms of the angle θ. Furthermore the data for any problem can be expressed in terms of two ratios:

(20) $$R_1 = \frac{L}{t} \qquad R_2 = \frac{L}{d}\,\frac{E_f}{G_c}$$

Note that t here represents the *sum* of the face thicknesses and d is the distance between the centroids of the faces.

4.2 EXAMPLE

The data used in the previous example give:

(21) $$R_1 = \frac{500}{5.0} = 100 \qquad R_2 = \frac{500}{27.5}\,\frac{20000}{25.07} = 14505$$

It is easy to tabulate the various results in terms of θ and, hence, to cross-plot any one variable against any other. Fig. 15 shows the deflection plotted against the pressure, all for $R_1 = 100$. Curves are plotted also to show the effect of factoring R_2 by 100 or by 0.2.

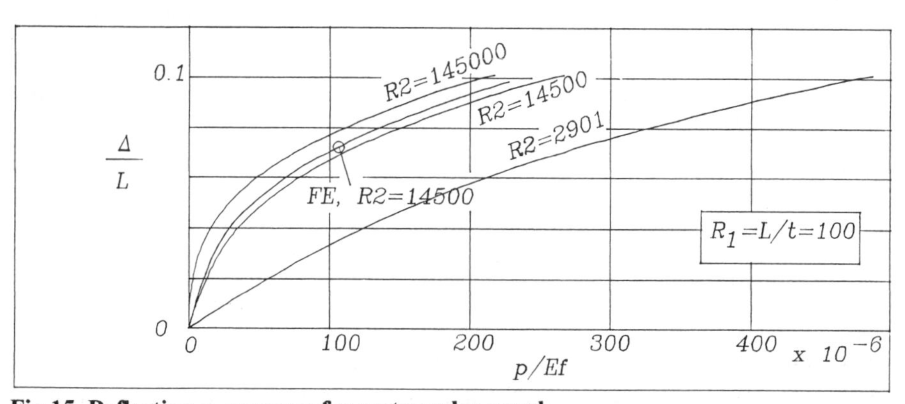

Fig 15 Deflection v. pressure for rectangular panel

The curve for $R_2 = 145000$ represents a very flexible core; the curve for $R_2 = 2901$ illustrates the extra load-carrying capacity of a panel with a stiff core. The curve marked FE has been obtained from a large-deflection finite element analysis of the sandwich panel with sides 1000 x 500 mm. It can be seen that the difference between this and the approximate curve for $R_2 = 14500$ is quite small.

Note that small-deflection plate analyses provide only the gradients of the curves at the origin, which is not generally very useful.

Similar graphs can be plotted to show how the strain (ϵ) or the angle θ vary with the pressure. The relationships between deflection, strain, and the angle θ do not depend on the ratios R_1 and R_2 because they are determined solely by geometry. The angle θ is useful because it is equal to the maximum shear strain in the core.

4.3 IMPLICATIONS

The analysis given here is speculative and it needs to be tested thoroughly before it can be applied in practice. However, it is both surprising and encouraging that such a simple approach can provide results which are so close to those of a complete FE analysis. By extending the analysis to incorporate different values of the spring stiffness, K, it would seem possible to provide quick estimates for the behaviour of complete systems of panels, continuous across ribs or bulkheads. In reality, failure is likely to occur by local bending of the faces at the edges of the panel, but it seems possible that approximate solutions can be found for this also. The method is limited at present to panels in which the long side is at least twice as long as the short side.

5. Conclusion

It is hoped that this paper will have provided some insight into

(a) The application of AST (Advanced Sandwich Theory), using primary and secondary deformations as the unknowns.

(b) The use of the Sandwich Master Diagram to quickly judge the way in which any given sandwich panel is likely to behave.

(c) The proper assessment of boundary conditions, defects and inserts in sandwich panels.

(d) Ways in which simple methods might be used to estimate large-deflection behaviour of sandwich panels under high transverse pressure.

References

ALLEN, H.G. (1969) *Analysis and Design of Structural Sandwich Panels*, Pergamon, Oxford, pp. 21-33.

ALLEN, H.G. (1972) Sandwich Construction. *Civil Engineering Departmental Report* CE/1/72, Southampton University.

ALLEN, H.G. (1973) Sandwich Panels with Thick or Flexurally Stiff Faces. In: *Sheet Steel in Building*, The Iron & Steel Institute, London, pp.10-18.

ALLEN, H.G. (1993) Theory of Sandwich Beams and Plates. In: R.A.Shenoi and J.F.Wellicome (eds.), *Composites and Materials in Maritime Structures*, CUP. Vol 1, pp 205-235.

THE DEVELOPMENT OF SANDWICH STRUCTURES FOR NAVAL VESSELS DURING 25 YEARS

K.-E. MÄKINEN
and
S.-E. HELLBRATT
Karlskronavarvet AB
S-371 82 Karlskrona
Sweden

K.-A. OLSSON
Royal Institute of Technology
Department of Aeronautic
Institute of Lightweight Structures
S-100 44 Stockholm
Sweden

During the last 25 years the FRP-sandwich technique has been used as a building technique for large ship hulls at Karlskronavarvet AB. The development of the FRP-sandwich technique started in the late-60's in close co-operation between Karlskronavarvet (KKRV), Royal Institute of Technology (KTH) and the Swedish Defence Materials Administration (FMV). An extensive research and development work has been performed since then to increase the structural performance and reduce the structural weight of the sandwich. Both experimental and numerical work has been performed to accomplish this and the structural weight have been decreased by more than 50% from the first designs built at Karlskronavarvet. Although the structural weight has been reduced, the structural performance has not decreased but rather increased in most areas.

1. Introduction

Glass fibre Reinforced Plastic (GRP) structures have been used in Marine applications for more than 30 years. The construction technique using the GRP-sandwich design has almost the same age. In the beginning small pleasure boats, up to 10 m in length were built. Sizes have continuously been increased, both for single skin and sandwich design. Today large vessels of glass fibre reinforced plastic with length over 50 m are built at several places around the world, both for military and commercial applications.

The demands to withstand dynamic loads caused by non-contact explosions have been increased for military vessels. For commercial vessel the dynamic slamming loads are increasing due to higher speed requirements than earlier. Speed requirements over 40 knots are now commonly both for military and commercial vessels and speeds of more than 60 knots for vessels with length of 40 m or more is not unrealistic. This means that the demands for low structural weight are increasing. More advanced materials are therefore of more and more interest despite their higher cost.

A. Vautrin (ed.), Mechanics of Sandwich Structures, 13–28.

Increasing length will also lead to increasing demands in hull girder deflections caused by the longitudinal bending moment and torsion of the vessel, which was a minor problem earlier due to the shorter length. The hull girder beam deflection is mostly depending on the bending stiffness (EI) for the actual vessel. Comparative analysis of a hull of a large patrol craft, with a length of more than 50 m, designed in steel or alternatively in GRP-sandwich, showed that the hull girder deflection increased from about 50 mm for the steel hull to 150 mm for the sandwich hull. The shear deformation, which can be neglected for the steel hull, contributes about 20% of the total deformation for the GRP-sandwich design. By optimising the fibre orientation for the whole design and using unidirectional fibres in certain hull members, the difference can be reduced. An extensive research and development work with both theoretical and experimental investigations has been performed during the last 25 years, which have now resulted in the latest project for the Swedish Navy, the "Visby-class", which is a stealth optimised Fibre Reinforced Plastic sandwich (FRP-sandwich) vessel with a length of 72 m built entirely of carbon fibre.

The object of this paper is to give an overview of the design and manufacturing methods, with emphasise on the structural weight of the different vessels manufactured by Karlskronavarvet.

2. Historical Background

The research and development of the GRP-sandwich building technique began in Sweden in the middle of the 60's when the old wooden mine sweepers of the Swedish Navy had to be replaced with new modern mine hunters made of non-magnetic material with high shock resistance.

The Swedish Defence Materials Administration (FMV) in close co-operation with the Royal Institute of Technology (KTH) and Karlskronavarvet (KKRV), started an intense research and development program in order to find a new hull concept that could meet the shock requirements. Various types of hull panels, such as panels made of wood, single skin GRP and GRP-sandwich, were produced at Karlskronavarvet and shock tested, since the shock resistance was considered to be of prime interest.

The tested wooden panels showed to have either too low shock resistance or too high structural weight. The panels that fulfilled the shock requirements had too high structural weight to be of interest. The framed GRP panels, both single skin and sandwich panels suffered delamination from problems. The stiffeners peeled off the laminate at too low shock levels.

The results of the tests showed that the frameless GRP-sandwich design, consisting of rigid PVC foam core with two skins of glass fibre reinforced polyester plastic, was very competitive both technically and economically. In order to find the optimum frameless sandwich design a second test program with 11 different sandwich designs was carried out.

The panels were mounted on a steel box which was submerged to a depth of 4 m, figure 1. Reduced size charges were placed 8.25 m from the centre of the panel. Charges of about 100 kg TNT were also used, but at larger distance from the panel.

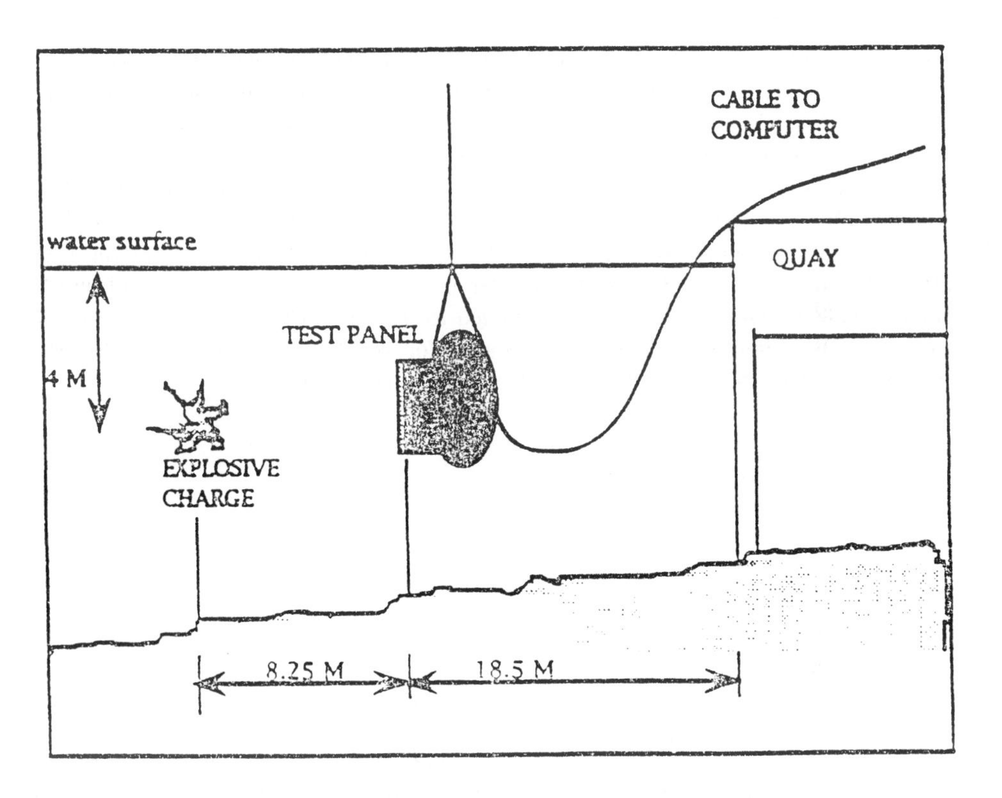

Figure 1. Test arrangement for the first panel test

After these first tests had been performed, a research and development group, with members from Karlskronavarvet, the Swedish Defence Materials Administration and the Royal Institute of Technology, was established, in order to find an optimal solution from both technical and production point of view. This group still exists and has now members from Sweden, Denmark and Finland. The group initiates a substantial amount of research programs, in different areas, in order to improve the sandwich design.

An area that has recently been investigated is the design of T-joints in sandwich structures, which usually contribute significantly to the total structural weight. A substantial amount of work has been carried out to optimise the strength properties and facilitate production methods in order to minimize the production costs. This can illustrate a typical research program that has been initiated by the group.

3. Sandwich Structures

3.1. SANDWICH VESSELS

From the experience gained during the early development work the semi mine-sweeper "VIKSTEN" was built in 1974, figure 2. Limited experience was available to use the sandwich concept for larger boats and the design turned out the be very robust with a typical structural weight for the bottom panels of about 56 kg/m ^{2}

The sandwich design was built up with a PVC foam core and laminates of woven Roving (WR) in combination with Chopped Strand Mats (CSM)

The vessel was built on a male wooden frame onto which the core material was attached and joined together with polyester putty. The outer surface were sanded smooth and the outer laminate was laminated to the core. The hull was then turned over and the same procedure was repeated for the inside of the hull. Finally prefabricated girders, bulkheads and decks were laminated into the hull.

Figure 2. "VIKSTEN"

For a Mine Counter Measure Vessel (MCMV) shock is of prime interest and an attempt was therefore made to analyse a section of a vessel with numerical tools. Only approximate methods were used at that time since the implementation of today's theory for underwater explosions was not available, but test data was available to compare the results with. The only available commercial program in those days, that had sandwich elements and the possibility to simulate transient loads was STARDYNE 3. The cavitation effect of the water close to the hull surface caused by the reflection of the shock wave was known but too complex to analyse at the time. In order to establish loads that simulate this behaviour a simplified analytical procedure was used. A pressure pulse with the same maximum peak (p_0) pressure as in equation 1, but with a negative distribution after 0.5 ms, was used in order to simulate this cavitation behaviour.

The pressure history at a fixed location starts with an instantaneous pressure increase to a pressure peak followed by a decay which, in its initial portion can be approximated by an exponential function.

$$p = p_0 e^{-t/\theta} \qquad \text{(MPa)} \qquad (1)$$

where, p_0 = peak pressure (MPa), t = time (s), θ = decay constant (s)

The maximum pressure, as well as, the decay constant, depends on the size and type of charges and distance to charge. For charges made from HBX the maximum pressure and the decay constant can be expressed as

$$p_0 = 56.6 \cdot (Q^{0.33} / R)^{1.15} \qquad \text{(MPa)}$$
$$\theta = 0.084 \cdot Q^{0.33} \cdot (Q^{0.33} / R)^{-0.23} \qquad \text{(ms)}$$

where, Q = charge size (kg), R = distance between charge and panel (m)

The equation above does not take into account the interference effect caused by the water surface, bottom conditions or any other parameter that might effect the shock wave on the way to the structure. When a shock wave reaches the water surface it is reflected as a tension wave. The water surface receives a vertical velocity and the water just underneath the surface, as a result of the reflected wave, cavitates and forms a cavitated region. This cavitated region is usually ignored even with the methods that are used today.

The pressure pulse used in the analysis had the equation

$$p = p_0 \cdot (1 - a/2 \cdot t) \cdot e^{at} \qquad (2)$$

where $a = 1/t_0$ and $t_0 = 0.0001$ sec

The results of the Finite Element Analysis (FEA) showed in general a fairly good conformity. The calculated shear stresses in the core material showed to be of both the same magnitude and location as the measured stresses in the full scale section. The location of the highest shear stresses was predicted from the FEA and strain gauges were glued in these areas in order to verify the validity of the analysis. The laminate stress level also showed a fairly good agreement with the measured values in the test section, but the calculated acceleration levels showed to be too low and the displacement level was in general found to be too high.

The experience gained from the manufacturing of VIKSTEN and from two 45 m coast guard vessels was implemented into the design of "Landsort", figure 3. At this time the classifications societies had started to come up with design rules for sandwich design and these were used to some extent in the design of ”Landsort”. However many of the restrictions in the rules where taken from the design of steel or aluminium vessels and were not really applicable to sandwich vessels. During the years the design rules have changed and are more intended for the composite material that is used in a sandwich design, but there are still areas that need to be further investigated in order to utilise the sandwich concept in full. The design rules still contain limitations on structural deformations and laminate thickness that make the sandwich structure very conservative if designed according to the design rules.

Figure 3. MCMV class "LANDSORT"

The structure of the "Landsort-class" was built up in a way similar to VIKSTEN and had a weight of the bottom panels of about 45 kg/m^2. This reduction could be accomplished thanks to decreasing the laminate thickness.

Over the years a number of ships where built with the same material configuration as in the "Landsort-class" of vessels. In the late 80’s Karlskronavarvet designed the experimental vessel "SMYGE", figure 4. SMYGE is a Surface Effective Ship (SES) and is stealth optimised to minimize the radar signature for the vessel. Since this vessel was intended as a test platform for new technology it was the first time Karlskronavarvet used more advanced fibres in the hull laminates. In the design structural optimisation was used for the first time and the optimisation package OASIS-ALLADIN developed at KTH was used for this purpose.

Figure 4. "SMYGE"

In the optimisation, the laminate thickness, fibre orientation, core density and core thickness were used as design variables and the objective was to minimise the weight. One result from the optimisation was that increasing the core thickness and reducing the density showed to be an advantage from a weight point of view. However this is not practical from a manufacturing point of view so the core properties and dimensions were kept the same as in the previous designs.

By changing fibre-type and fibre-orientation in the direction of the forces of the different members of the structure, the weight of a typical bottom panel could be reduced to about 35 kg/m^2 for this vessel. SMYGE was manufactured from woven roving with either glass or aramid fibres or a combination of both together with polyester resin and a PVC core.

At this time, Karlskronavarvet also started to use knitted fabrics, figure 5b, in addition to the woven fabrics, figure 5a, that had been earlier used in the laminates.

The knitted fabric makes the reinforcement more compact and less resin rich which is needed to give a higher fibre content and a lower weight of the laminate. The laminate will not only be lighter, it will also be stiffer since the fibres are already flat and therefore better carry the loads without distortions, figure 5.

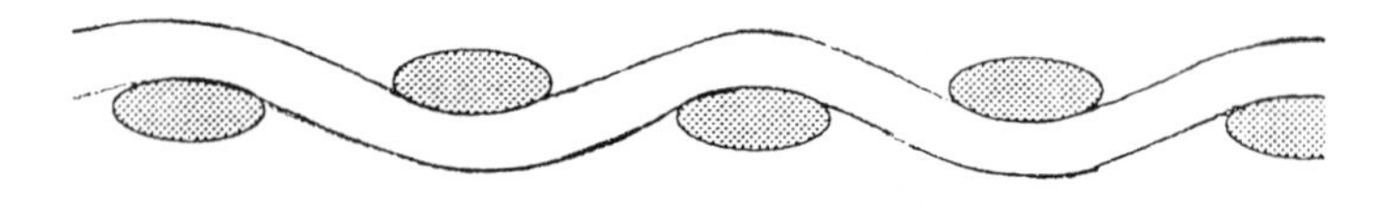

Figure 5a Woven fabrics

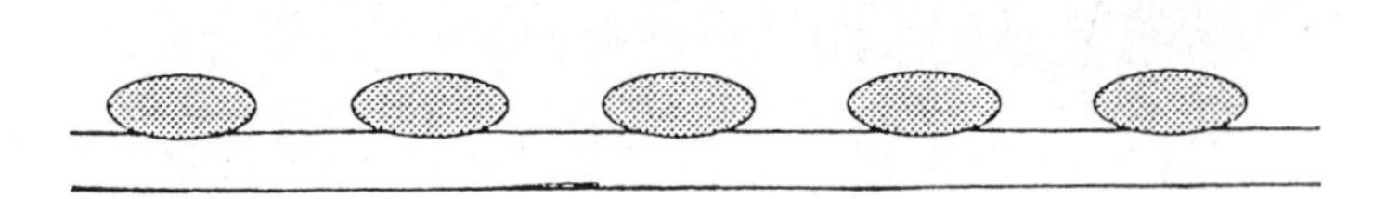

Figure 5b Knitted fabrics

The properties of laminates and the whole sandwich using this new type of fabrics were verified through testing and numerical calculations. Karlskronavarvet, has over the years, built up a database with laminate properties for laminates with different fibre types, fibre lay-ups and resin types. The laminate properties in the database have been verified through both testing and calculations. As a result of the use of knitted fabrics the weight of a typical sandwich panel could be decreased to about 26 kg/m^2 , which is used in the Styrsö-class, figure 6.

As the vessels become larger and larger, the weight becomes a critical factor for the speed of the vessel. The structure contributes significantly to the total weight of the vessel. The development of the high speed stealth vessel YS 2000 or "Visby-class", figure 7, has therefore lead Karlskronavarvet to look into other alternative materials, manufacturing methods and design methods in order to improve the performance of the vessel.

In the early development stage of the project a numerical model of the vessel was developed. The intention was to perform a structural optimisation to minimize the weight of the vessel, with core densities, core thickness, laminate thickness and lay-up as design variables. After the first optimisation with weight as the objective function, a second optimisation was made with cost as the objective function and the results from both runs were the same.

Figure 6. "Styrsö-class" of vessels

Figure 7. "Visby-class" of vessels

The result showed that the core densities and thickness could be decreased further in several areas compared to the conceptual design and the traditional glass fibre laminates were changed to much thinner carbon fibre laminates with orthotropic properties in large areas. The laminates were allowed to be very thin in areas where a new design criterion for damage and impact determined the thickness instead the old minimum thickness criterion together with the more traditional tension, compression or buckling criteria. The optimisation was made on a model of the whole vessel but several Finite Element runs were made on different parts of the vessel in order to investigate local effects. For example the structure for the gun foundation was investigated carefully.

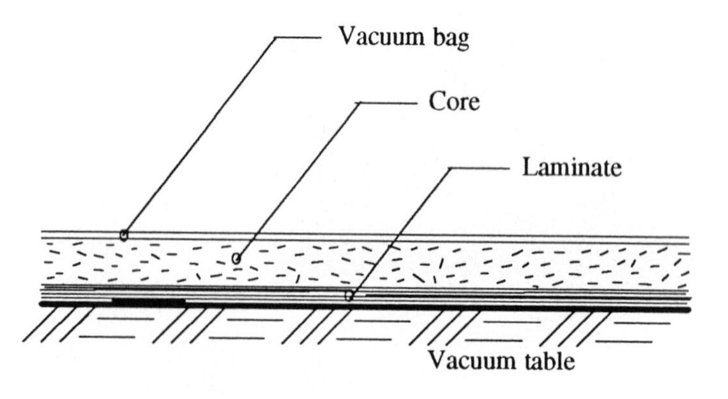

Figure 8. Vacuum injected half sandwich panel

Even though the carbon fibres are much more expensive than glass fibres the number of layers were reduced to that extent that the reduced labour cost compensated the higher material price of the carbon fibre.

In order to take full advantage of the better properties of the carbon fibres compared to glass fibres other manufacturing methods than the wooden male frame mould, have to be used. For the "Visby-class" of vessels Karlskronavarvet combines the wooden female frame mould method and vacuum infusion. Full and half sandwich panels are produced with special infusion technique developed at the yard.

The term "half sandwich", figure 8, means that one laminate and the core is laid up on a vacuum table and the resin is injected into the laminate under vacuum. These panels are then attached to the wooden female mould and the inner laminate of the sandwich is applied in the traditional way, figure 9. These changes in design, material and manufacturing method mean that the structural weight is reduced to about 20 kg/m^2 for typical bottom panels that in the first design of "VIKSTEN" had a weight of 56 kg/m^2.

The double curved surfaces are manufactured in the traditional way. Prefabricated vacuum infused elements for girders, bulkheads and deck are then laminated into the hull structure. Improvements of the joints for attaching these elements to the hull structure have been made both concerning the weight and the manufacturing cost.

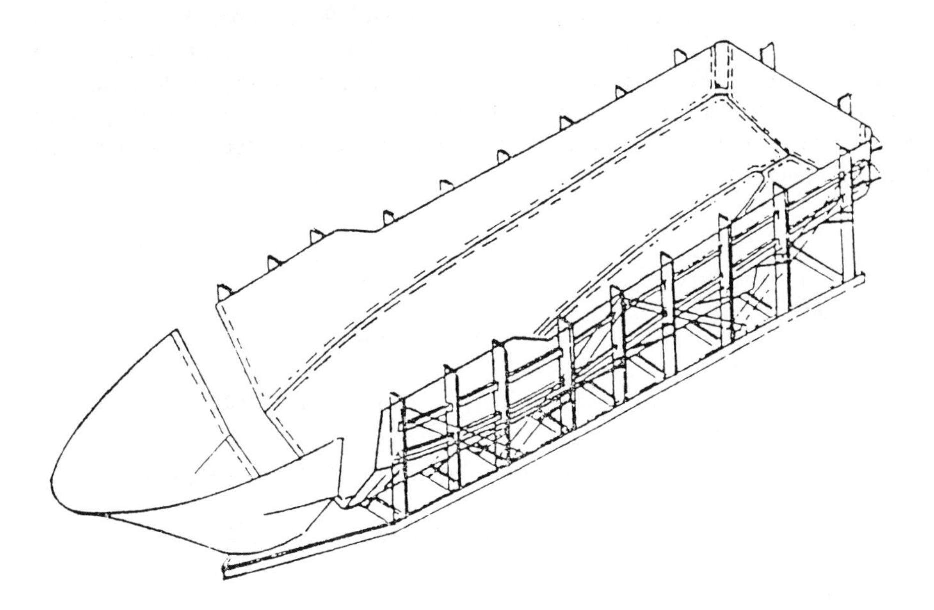

Figure 9. Wooden female frame for attachment of half sandwich panels

3.2. T-JOINTS IN SANDWICH VESSELS

In the design process the weight of the vessel is carefully checked throughout the whole process and it is found that a significant contribution to the total weight comes from the attachments of the prefabricated bulkheads, girders etc. The weight of these joints can be as much as 10-15% of the total structural weight.

For the first vessels that were built, the same type of attachment was used in all the joints in the vessel. Later Karlskronavarvet has started to use different types of attachments depending on how much the actual joint is loaded and the consequence of failure.

During the research of sandwich structures these joints have been investigated both theoretically and experimentally at the Royal Institute of Technology. The fillet shape, the core density and the putty type in the filler are all essential for the strength of the attachment, figure 10. The first joints had a triangular fillet but early work showed that a fillet with a radius has advantages both in static and fatigue strength.

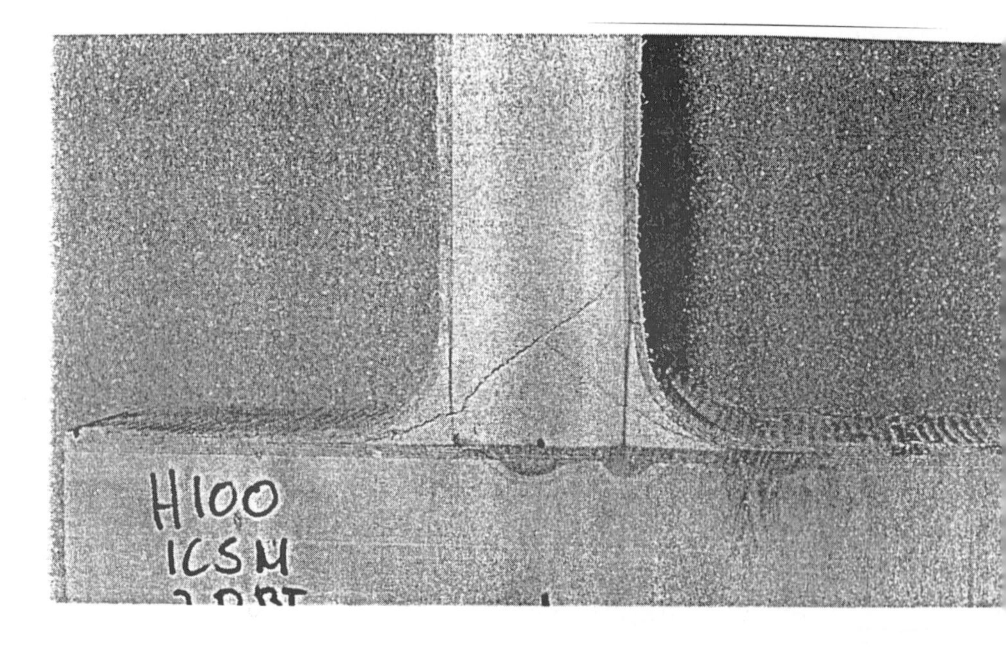

Figure 10a. T-joint with radius fillet

Figure 10b. T-joint without fillet

Figure 10c. T-joint in structural putty

From the experience of later research and development work, Karlskronavarvet has been using different types of attachments for the joints. The most loaded areas in the bottom structure, exposed to loads from slamming and underwater explosions, are furnished with larger radius fillets in conjunction with an elastic structural putty. However the laminate thickness in the attachment is reduced compared to the traditional T-joints. Interior elements with lower loads have a weaker joint where the members are just attached with structural putty, figure 10c.

3.3. INTEGRATED FUNCTIONS

One of the advantages in the design of a sandwich structure is that functions can be integrated into the structure. As mentioned earlier the stealth technology is a function that may not only depend on flat panels but also on different kinds of absorbing tissues built into the structure (also may be load carrying). This can be done if the faces are laminated with GRP for example.

Another typical example, when using PVC foam core, is the thermal insulation that this material posses. The core has in this case three functions: a) to separate the skins in order for these to obtain "normal" forces, b) withstand the shear forces in the structure, c) provide sufficient thermal insulation

Since the surface of a laminated surface is smooth and chemically resistance to water, oil etc., tanks for fresh water and fuel can be integrated into the structure and no surface treatment is necessary inside the tanks.

3.4. MAINTENANCE AND REPAIR

Maintenance of the structure involves inspection and painting. Inspection of a FRP-sandwich structure is more difficult than that of a steel or aluminium structure. Repair of FRP-sandwich structures requires good environmental conditions, but the procedures are relatively straight-forward. There are a number of guide-lines for repairing FRP and FRP-sandwich structures, but they are not standardised in the same way as for steel and aluminium, due to the wide variety of material combinations available and recent development of the materials. Small field repairs can be handled by trained personnel quickly and easily. Major repairs require shore based facilities and experienced personnel dealing with composite materials.

A FRP-sandwich vessel often has very good damage tolerance. The most common damages are caused by running aground or from sharp quays. After such damages the ship will retain floatability and can in some cases be allowed to continue operations for several weeks before dry docking and repair.

A FRP-sandwich also has good reparability. This was demonstrated by a panel, which was damaged on purpose all the way through the laminate into the core. The panel was submerged in the sea for 13 years before repair. The water absorption of the damaged panel was less than 1%. After being allowed to dry for one week the panel was repaired. The mechanical strength of the repaired panel was 95% of its original strength.

4. Other Design Aspects

4.1. FIRE RESISTANCE

Sandwich designs today used for boat and shipbuilding have cores of e.g. balsa, PVC or polyurethane foam. The skins consist, almost without exception, of fibreglass reinforced polyester. All combinations of these materials have in common that they mainly consist of organic material which burn or generate combustible vapour and frequently emit heavy smoke when exposed to heating. Accordingly, a sandwich structure appears to be a difficult problem from a fire technical point of view.

In the development work of the sandwich vessels, sandwich panels were tested according to the SOLAS standard fire test which requires the structure to:

- be constructed of steel or equivalent material
- be stiffened in a suitable way
- prevent penetration of smoke and flames for a maximum of 60 minutes in case the standard fire test
- be constructed in such a way that the rise of temperature on the non-exposed side shall be below 139 °C

A tested panel with PVC core and glass fibre reinforced polyester faces and intumecent paint could withstand this test. The exposed side of the panel was ignited after 12 minutes and burned during the remaining test time, no smoke penetrated the panel and the maximum temperature rise on the non-exposed side of the panel was measured to be 11 °C. The panel passed the test except for the first point where is states that the material should be steel or equivalent. The stability of the panel is not satisfied when one of the faces is burned away, however stiffeners were added to both sides of the bulkheads which then provide sufficient stiffness together with one skin.

During another test of the whole compartments in a vessel, it was found that the best way to limit a fire is to use smoke detectors connected to automatic door closers that will seal the compartment in question and extinguish the fire due to lack of oxygen.

4.2. NON-DESTRUCTIVE TEST METHODS (NDT)

During the past years, several research and development projects regarding NDT methods have been carried out. One of the most promising method, the infrared thermography has been tested at Karlskronavarvet. The results of these tests are in some respect depressing.

Face-to-core debonds can be relatively exactly located, but if the debonds or cavities are located deeper in the core material or behind a corner fillet, it is very difficult to obtain the required reliability. Though this method is quite difficult to implement, it is one of the few methods that has the advantage of being global.

The most reliable method to detect debonds is still the "coin-tapping" method. Debonds can be located with the above mentioned and some other methods. Reduced bond can never be detected with NDT methods and destructive test methods still have to be used in the future to ensure the strength of the bond.

4.3. COST CONSTRAINTS

It has been a general belief that a ship built in GRP-sandwich is considerably more expensive then a corresponding vessel built in steel or aluminium. However, experiences gained from building similar ships for more than 20 years clearly shows that this is not the case. It is true that the material cost for the hull itself is higher, but the total cost, including labour hours for hull production, outfitting and maintenance is lower.

A comparative weight and cost analysis for a 20 m long weight optimised patrol craft built in aluminium versus GRP-sandwich showed that both the weight and total cost should be about 10% lower for the GRP vessel. With the latest achievements in the field of manufacturing, where vacuum assisted infusion technique is used, the production cost is even more favourable for the FRP-sandwich design.

Service life of a carefully designed and built ship is potentially long if necessary maintenance and repair of smaller damages is carried out and the UV-degradation is prohibited by maintaining the paint system. The maintenance cost for the hulls of the Swedish Minesweeper "Viksten" with more than 20 years in service, as well as for the Mine Hunter class "Landsort" with more 10 years in operation, are limited to the cost for repainting.

Experiences from well documented ships with more than 20 years in service, shows no degradation of the hull structures or absorption of water where test samples are taken. This and the economic aspects gained during these years indicate that the sandwich concept is very competitive and we will certainly see more use of the sandwich concept in the future.

5. Refernces

Allen, H. G. (1969) *Analysis and Design of Structural Sandwich Panels* , Pergamon Press

Esping, B., Holm D. & Rommel O. (1989) Structural Optimization of a Surface Effective Ship, First International Conference on Sandwich Construction, June 19-22, Stockholm, Sweden, Proceedings Printed by Engineering Materials Advisory Services Ltd (EMAS), UK

Gullberg, O. & Olsson, K.-A. (1990) Design and Construction of GRP-Sandwich Ship Hulls, *Marine Structures*, Vol 3, pp 93-109

Hellbratt, S.-E. (1993) Design and Production of GRP-sandwich vessels, Report No. 93-5, Department of Lightweight Structures, Royal Institute of Technology, Stockholm, Sweden

Mäkinen, K.-E. (1995) Dynamically Loaded Sandwich Structures, Report No. 95-20, Department of Lightweight Structures, Royal Institute of Technology, Stockholm, Sweden

Olsson, K.-A. (1987) GRP-Sandwich Design and Production in Sweden, International Conference on Polymers in Defence Bristol, England

Plantema, F. J. (1966) *Sandwich Construction* , John Wiley & Sons, New York

Sjögren, J., Helsing, C.-G., Olsson,K.-A., Levander C.-G. & S-E. Hellbratt, S.-E. (1984) Swedish Development of MCMV-hull, Design and Production, Paper presented at the International Symposium on Mine Warefare Vessel and Systems, June 12-15, London

Zenkert, D. (1995) *An Introduction to Sandwich Construction*, Engineering Materials Advisory Services Ltd (EMAS, Chameleon Press Ltd., 1995

DAMAGE TOLERANCE OF AERONAUTICAL SANDWICH STRUCTURES

D. GUEDRA-DEGEORGES, P. THEVENET, S. MAISON
Aérospatiale - Centre Commun de Recherches Louis Blériot
12, rue Pasteur
92152 Suresnes
France

1. Introduction

The aim of this paper is to give an overview of the present aeronautical applications of sandwich materials and to describe the present damage tolerance approaches versus type of defects like debonding or impact. Some research orientations are proposed in order to improve, on the one hand, experimental identification of defect behaviour and, on the other hand, numerical simulation of aeronautical sandwich structures including such defects.

2. Overview of sandwich material applications in aeronautical industry

The use of sandwich materials allows to respect weight savings and stiffness design criteria. Such materials can be applied to primary and secondary structures. Primary structures are these ones from moderately to strongly loaded. Secondary structures are these ones weakly loaded or not.

The core of sandwich materials is mainly made of aluminium, Nomex honeycomb or organic foam, with a maximum thickness of 50 mm. The skins are made of glass, carbon, kevlar unidirectional plies or fabrics with thermoset organic matrices. The thickness is often lower or equal to 2 mm.

There are three aeronautical application fields of these materials : helicopter, aircraft and space. Concerning helicopter construction, Figure 1 shows various structural components (secondary structure) made of sandwich. For civil aircraft, some parts of wing (leading edge, wing flap) are made of sandwich (Fig. 2). Regarding space structures, Figure 3 shows a structural component of a launcher. This is the interface tube between the satellite and the launcher, made of high modulus carbon fibres for the skins and aluminium honeycomb for the core.

A. Vautrin (ed.), Mechanics of Sandwich Structures, 29–36.

Figure 1. Structural sandwich components of the Dauphin helicopter

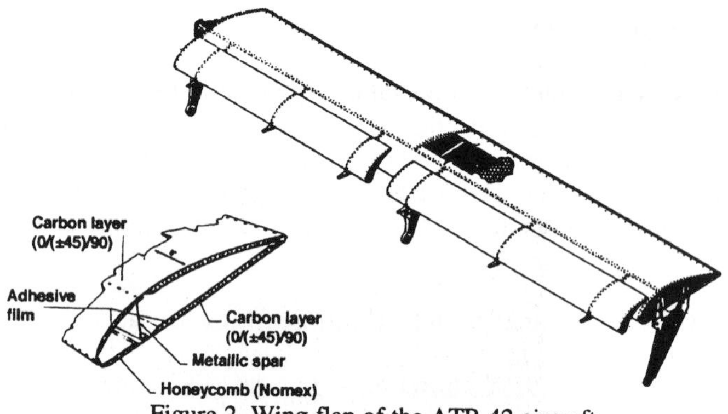

Figure 2. Wing flap of the ATR 42 aircraft

Figure 3. Sandwich tube between satellite and launcher

3. Concept of damage tolerance

Two specific defects in aeronautical sandwich materials are considered :

- the debonding between skin and core, due to lack of pressure and/or adhesive during curing, or due to the presence of a release film,
- low velocity impacts, occuring during maintenance or mounting (tool drop), or occuring in service (stone projections during landing or takeoff).

A better knowledge of their behaviour under loading from basic tests allows to take them into account in design of structural components and to make decision for repairs.

3.1. BEHAVIOUR OF SKIN-CORE DEBONDING DEFECT

Debonding can be critical under combined shear-compression loading. The typical behaviour is experimentally characterized by :

- at first local buckling (blistering) of the debonded skin,
- then failure by damage propagation in pristine areas or failure at the blister top.

For example, a typical failure of a sandwich coupon after propagation of a Ø50 mm central debonding defect (artificial) under a compressive loading (skin : high modulus carbon/epoxy, core : aluminium honeycomb) is presented hereafter (Fig. 4) :

Figure 4. Sandwich coupon (with a circular debonding) after failure

Such a behaviour can be numerically simulated. The following non-linear geometrical calculation based on finite element method allows to simulate the buckling and post-buckling behaviour of a sandwich specimen (Boeing type 150x100) with a circular debonding Ø50 mm under a uniaxial compressive loading. The skins are made of carbon fabrics G803/914 (1.2 and 1.8 mm) and the core is made of Nomex 15 mm. The F.E. model is drawn in Figure 5.

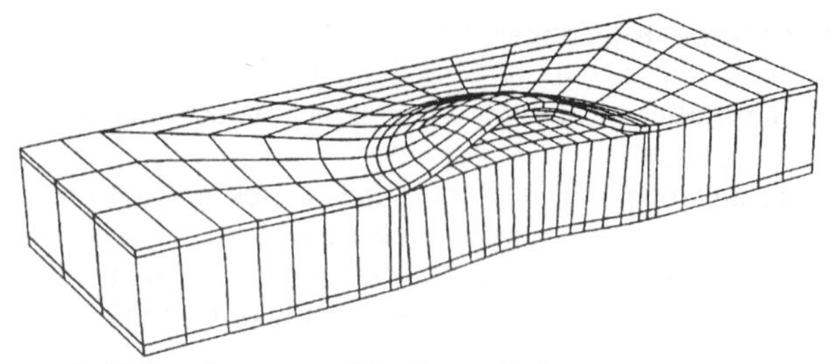

Figure 5. Finite element model of a sandwich coupon : post-buckling

The numerical longitudinal strains at the blister centre (j6) and close to the debonding front (j7) are compared with the experimental values (Fig. 6). From a threshold of the compressive load (about 30.0 kN) the numerical behaviour is differing in comparison with the experimental one where the debonding is spreading. This propagation can be predicted by the use of fracture mechanics adapted to composite materials. The Aérospatiale approach consists of computation of the global strain energy release rate (G) and the mode partition (G_i, with i = 1, 2, 3) on the debonding front for each load increment (Fig. 7). Then the use of a mixed mode propagation criterion ($c = f(\Sigma(G_i/G_{ic}))$) allows to predict the criticity of this debonding front.

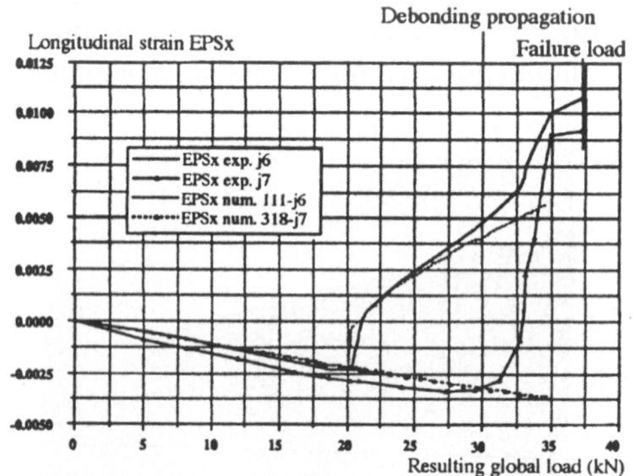

Figure 6. Comparison of the numerical/experimental strains

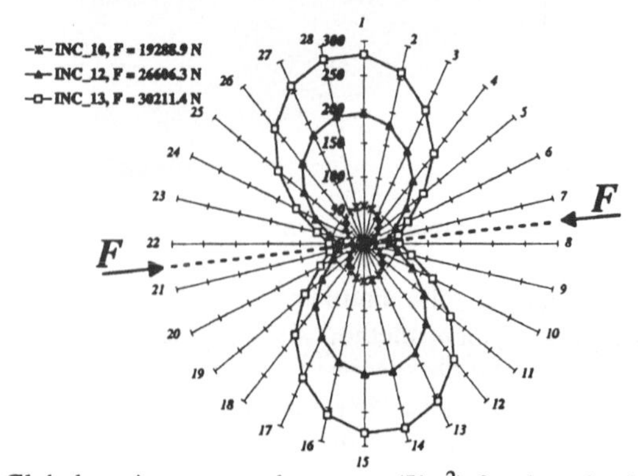

Figure 7. Global strain energy release rate (J/m^2) for three load increments

The main difficulty is the assessment of the threshold values G_{ic} for sandwich materials by classical tests like for the monolithic ones. So a new research orientation is taken for the development of reliable test procedures for the assessment of G_{1c}, G_{2c} and the determining of mixed mode propagation criteria. Another orientation is the study of the effects of fatigue, ageing and temperature on the debonding propagation.

3.2. LOW VELOCITY IMPACT

3.2.1. *Behaviour during impact*

The prediction of the mechanical behaviour of a sandwich structure subjected to low velocity impact is a key factor to improve the efficiency (cost, safety margins) of the damage tolerance concept on primary parts of aerospace applications.

Contrary to the debonding defect, the impact damage is visible. Some perpendicular impact tests performed on several coupons at different energy levels with a falling weight (impactor Ø16 mm) give the following results (Fig. 8) :

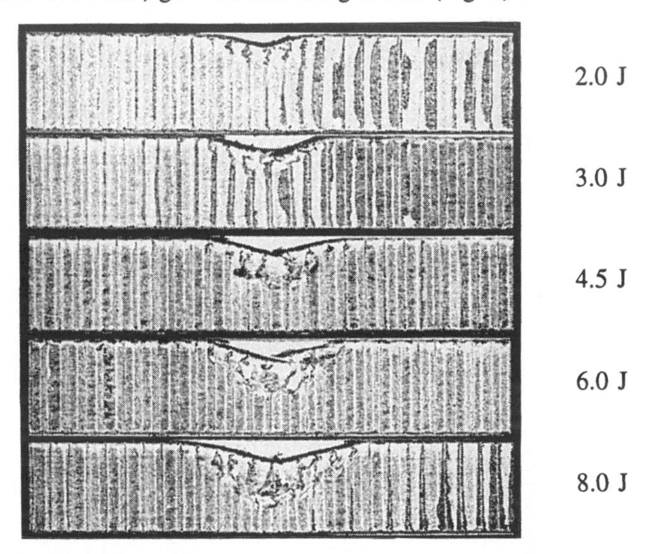

Figure 8. Impact damage on a sandwich for several energy levels

Residual indentation depth is greater when the impact energy increases. A simple analytical model allows to predict this indentation depth and the damaged area. Four assumptions are made : the impactor is hemispheric (radius R), the contact surface length between the impactor and the coupon is smaller than the diameter of the impactor, the impact is perpendicular to the coupon, stiffness of the impactor is much higher than the coupon stiffness. From the contact law of Hertz, one can write equation (1) between the residual indentation depth p and the impact energy :

$$p = z - z^* = \left(\frac{15.W_{impact}}{8.E_c.\sqrt{R}} \right)^{2/5} - z^* \qquad (1)$$

where z is the vertical displacement of the impactor, E_c is the transverse core modulus and z^* is a constant value depending on the sandwich material and corresponds to an impact energy threshold for which the indentation appears. So this residual indentation depth is defined as follows :

- when $z < z^*$, then $p = 0$
- when $z \geq z^*$, then $p = z - z^*$

A simple geometric calculation gives the damaged area (area of the core crush) S_d as a function of the residual indentation depth p :

$$S_d = \alpha.\pi.(2.R.p - p^2) \tag{2}$$

where α is a proportionality coefficient determined by some experiments. The two equations (1) and (2) can be combined to assess the damaged area as a function of the impact energy. Some calibration and verification tests are performed on the following configuration (Fig. 9) for example :

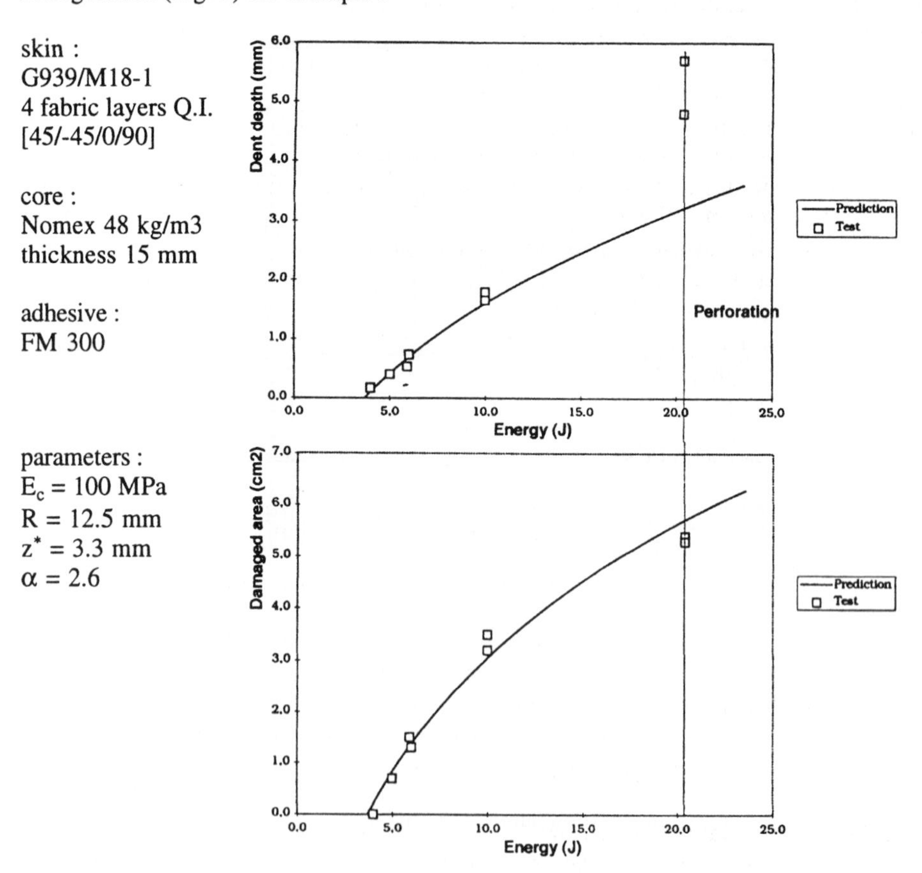

Figure 9. Calibration and verification of the analytical model of prediction

Other tests show that z^* depends on the skin thickness and that there is no typical law for α.

3.2.2. *Behaviour under loading after impact*

After impact, the prediction of residual strength is another research direction. From residual four-point bending tests performed on sandwich coupons, the mechanism of damage propagation can be described as follows : initial core crush induced by impact propagates during the residual loading till final buckling of the impact skin. A similar behaviour is observed in the case of compressive loading where the honeycomb crush spreads up to the specimen sides perpendicularly to the compression direction. The residual strength decreases with the increase of the impact energy level.

Such behaviour can be simulate by a finite element model with an original strategy computation. The chosen sandwich specimen is made of two G939/M18 skins (carbon fabric/thermosetting resin, lay-up [45,0]) and one honeycomb core (32 kg/m^3, thickness 15 mm). Three-dimensional multi-layer elements are used. This model takes into account the main parameters like the size of the damaged area, the indentation depth and the local stiffness loss for impacted skin (modulus E_1) and honeycomb (moduli E_3 and G_{13}). A first calculation shows that a compression over-stress appears in the core and on the crush propagation line determined by tests. Then a parametric study allows to quantify, in accordance with the experimental results, the stiffness losses and to investigate the influence of the model parameters on the calculation results. It is deduced that the transverse modulus E_3 of the honeycomb is the most important parameter. The skin modulus E_1, as well as the core shear modulus G_{13}, can be neglected. The calculation strategy consists of creating a zone following the line observed during test, in which the core stiffness (modulus E_3) is decreased in the same way as this one of the impacted zone. The first studied geometry consists of a narrow strip from center of impact (fig. 10). The length of this zone is one parameter of the model.

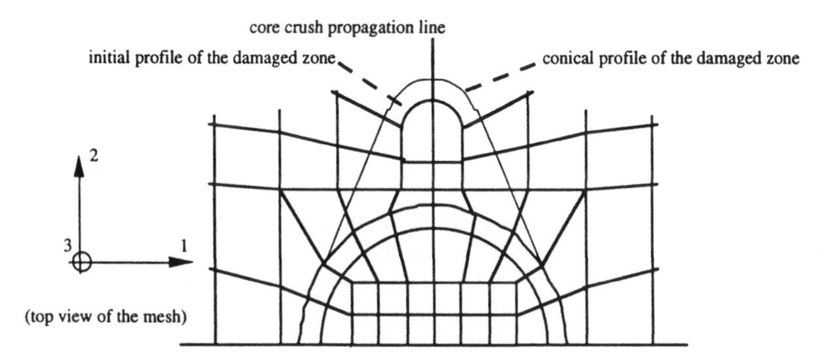

Figure 10. Local meshing of the damaged area

For the first calculation the propagation length is taken very short (crush propagation initiation). Stress analysis in the core shows that the maximum compression is not located at the tip of this zone but on both sides of this one. It means that the width of the damaged zone has to be increased, leading to a conical shape (fig. 10). When performing a calculation on this updated mesh, the compression over-stress appears at the tip of the damaged area and on the propagation line observed experimentally. The new parameters of this mesh are the radius of the cone top, the cone length and the initial crush propagation depth. A new parametric study allows to investigate the influence of these parameters on the numerical results and to determine appropriate values in accordance with the test results. It also leads to choose a failure criterion. The idea is that, for a certain core crush length, the impacted skin is not enough stabilized for preventing buckling of specimen. Therefore the strategy consists of determining, for several growing core crush length, both the load to be applied for extending the propagation and the buckling load of the structure. The structure failure is reached when the core crush propagation load is greater than the buckling load of the upper skin. During the propagation the crush area is always conical and its length is increased at each step. Finally the different steps of the simulation strategy are the followings (Fig. 11) :

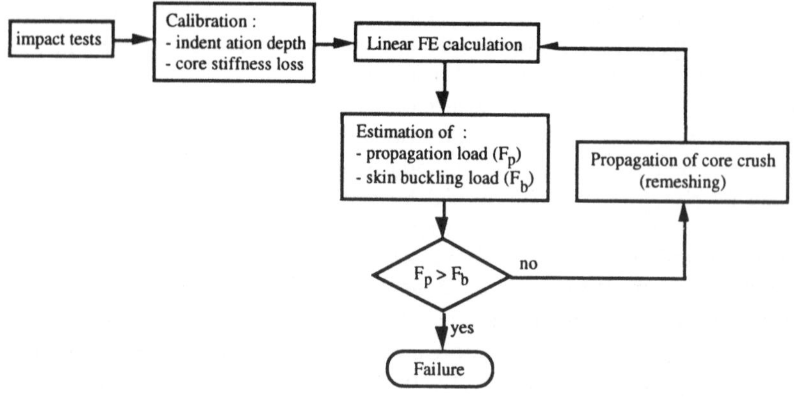

Figure 11. Calculation strategy

This modelling based on a finite element discretization and including physics of damage mechanisms should be helpful for designers to make decisions when a defect is observed on a real structure. This work also allows to identify the main parameters governing the failure of impacted sandwich specimens under bending load.

4. Conclusions

Sandwich materials in aeronautical application are used in order to respect objectives in terms of weigth and cost reductions, generally for stiffness design parts. Damage tolerance design requires the knowledge of the mechanical behaviour of critical damages such as debonding and impact. To achieve this goal, Aérospatiale has developped specific modelling approaches based on fracture mechanics. Specific test procedures have been designed to identify damage propagation criteria for debonding and core crush.

MODELING OF THE CRASH BEHAVIOUR OF EDGE LOADED SANDWICH STRUCTURES WITH FIBRE REINFORCED POLYMER FACES

S. KERTH AND M. MAIER
Institut fuer Verbundwerkstoffe GmbH
67663 Kaiserslautern, Germany

AND

M. NOHR
DaimlerBenz AG, Forschung und Technik, Ulm

Abstract. This paper focuses on the crash behaviour of sandwich structural parts with foam cores and FRP-faces which are loaded normally to their thickness direction. A basic methodology for numerical simulation of sandwich structures with foam cores is presented. The material properties of PMI foam are derived from material tests and the material cards for numerical crash simulation with LS-DYNA3D are generated semi-automatically. It is focused on the modeling of the adhesive joints between core and faces complex sandwich geometries the simulation of core-face separation.

1. Introduction

It has been found that the energy absorption capacity per unit weight of composite structural parts under uniaxial crash load is much better compared to those of steel or aluminium. For that reason they are increasingly used in crash loaded structures of any kind of vehicles, especially in automobiles. Beside special composite crash elements (e.g. crash boxes), complex composite structures like longitudinal girder, mud guards, underbodies, and even complete car bodies are subjects of research. A lot of research has been done on crash elements with simple geometry like e.g. tubes with cylindric and quadratic cross sections and cones ((Maier, 1990), (Hull, 1991), (Thornton, 1986)). Fundamental studies and crash tests on specimens with a more complex geometry are necessary. The sandwich structures have been chosen in order to introduce double belt press (Ostgathe, 1995) and stamp forming technique (Breuer, 1995) which is both available at IVW for in-

A. Vautrin (ed.), Mechanics of Sandwich Structures, 37–44.

dustrial scale manufacturing. Apart from the good structural capabilities of sandwich structures, little knowledge has been gathered concerning the crashworthiness. In the frame of this paper the experimental and numerical results for the crash behaviour of sine wave sandwich panels are presented. The Main goal of crashworthiness calculations is to reduce the number of expensive crash tests and thus, to shorten developing time.

2. Modeling of sandwich structural parts

2.1. MODELING OF ADHESIVE JOINTS BETWEEN FOAM CORE AND FRP FACES

In all cases the FRP-sandwich faces were modeled by shell elements while the core was modeled by volume elements. The concepts differ concerning the modeling of the adhesive joint between the faces and the core and can be divided in two groups shown in figure 1. Either discrete elements are used (Schweizerhof, 1995) or the adhesive joint is modeled with solid elements.

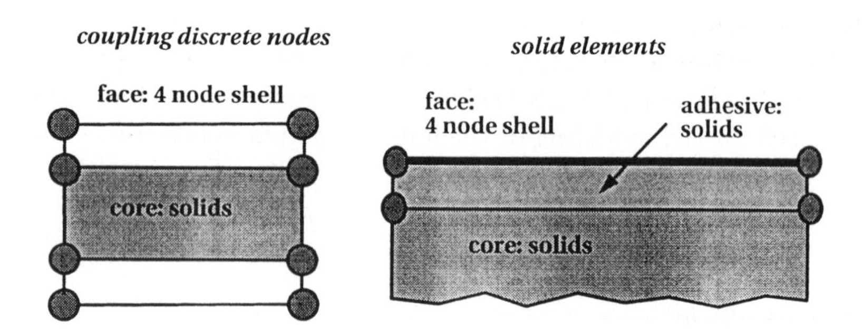

Figure 1. Concepts for adhesive joint modeling

2.1.1. *Nodal Constraints and Spotwelds*

The nodes of different parts are coupled with respect to one or more degrees of freedom only. There are different suboptions which will just be mentioned without further comment.

- Coupling only translational degrees of freedom - no failure option
- Coupling via a truss element - carrying only axial forces - failure optionally
- Coupling translational and rotational degrees of freedom - failure optionally

2.1.2. *Node-Tied-to-Surface*

This option, controlled by the sliding interface definition, is similar to the option described in chapter 2.1.1 with the advantage that it is not necessary to have adjacent nodes in the mesh. It is only necessary to define a node which is tied to the surface of an element (with no offset to the defined middle-plane of the element). The motion of this node is coupled with the kinematical motion of the surface at the point of contact. Only translational degrees of freedom can be coupled thus, only axial forces and in some cases shear forces can be transmitted. The *node-tied-to-surface* option can also be used in combination with the discrete beam elements, which makes it a very general option.

2.1.3. *Discrete Beam Elements*

The *Discrete Beam Elements* can be compared with a 3-dimensional non-linear spring/damper with the capability to act translations and rotations. The discrete beam element option (beam element type 6) is controlled by the material types 66, 67 and 68 in LS-DYNA3D. Because of the necessity of modeling nonlinear behaviour with failure, the recommended material model is type 68. It allows the definition of translational and rotational stiffness properties as well as the definition of the yield force versus plastic displacement or the moment limit versus rotation. The last option allows the user to calibrate the numerical model with an experimental result. A quadratic failure criterion can be defined including forces and moments or displacements and rotations in an arbitrary fashion. A local coordinate system has to be given by the user, with the x-axis in the direction of the two nodes. This is necessary to get proper information about the results in the post-processing. The discrete beam does not require an orientation node, in contrast to real beams. The discrete beam option allows the most versatile modeling for the problem of adhesive joint. It contains the other possibilities mentioned above as subsets. In addition one has the possibility to add arbitrarily nonlinear behaviour, which allows to calibrate the numerical model with an experiment in an arbitrary fashion.

2.1.4. *Solid elements*

Solid elements with an *elastic-plastic with failure* material behaviour can also be used to model the adhesive joint between core and face. The advantage is the visibility of failed solid elements in the post processing as well as homogeneous modeling of the whole structure compared to the discrete elements. A disadvantage can bee seen in an increase of calculation time.

2.2. MODELING OF FIBRE REINFORCED POLYMER SANDWICH FACES

The numerical crash simulation was performed using LS-DYNA3D (Hallquist, 1993).The material model for the crash simulation of composites was developed in the frame of an EUREKA research project (N.N., 1991). It is implemented in a shell element allowing economic simulation of complex structures. The material model is based on the continuum mechanic and the Classical Laminate Theory. Each layer of the laminate is considered to be homogeneous and behaves orthotropic in stiffness and strength. The material properties are determined from standard material tests. The material model can be subdivided in 3 sections:

- linear-elastic behaviour (CLT) in the under-critical region
- onset of failure with respect to special failure criteria (Hashin, 1980)
- post-failure behaviour and damage growth

After failure occurred, the material properties stiffness and strength are modified by the so-called degradation model, depending on the failure mode. The failure mode *delamination* and *fibre pull-out* can not explicitly be taken into account as a result of shell theory restrictions. Delamination is represented by reduction of bending stiffness with a global factor. Predamage of the laminate due to crack propagation in the crash front is represented by linear degradation of all strength values for elements beyond the crash front. [1]

The here used material combination (glass fabric reinforced polyamide 12 (GF-PA12)) has been chosen in order to test the capacity of the material model which was originally developed for unidirectional reinforced thermoset composites. In this case, a bi-directional reinforcement is combined with a ductile matrix system. The ductile matrix behaviour is taken into account by setting the parameter for predamage very small, with respect to experimental observations. The specimens have been triggered by a 45^0 chamfer in order to assure a progressive failure with a defined onset location. The chamfer is represented by stepwise reduction of the wall thickness. The initial velocity and the mass of the striker have been taken from the experiment. The material properties for GF-PA12 have been measured and, where necessary, calculated with micromechanical relations.

[1] The interested reader is advised to (Maier *et. al.*, 1992), (Maier *et. al.*, 1992-2), (N.N., 1991) to get more detailed information.

2.3. SANDWICH CORE FROM PMI FOAM

Basic tests have been performed to evaluate the material behaviour of PMI [2] foam. The uniaxial compression behaviour which is mostly interesting for crash simulation has been determined at a 100 x 100 mm cube. Several material models for foams in LS-DYNA3D have been evaluated. Finally, material model 63 (*isotropic crushable foam*) has been chosen because of its good match to the experimental results and its stability. The pressure versus relative deformation of the foam cube is represented by terms of a load curve in LS-DYNA3D. This load curve as well as Young's modulus and Poisson's ratio are generated automatically by a tool programme from the experimental data. It is assumed - and experimentally proven - that Poisson's ratio is very small ($\nu = 0.01$). So the volumetric strain is proportional to the deformed length of the cube. With this assumption experiment and simulation can be compared.

3. Results

3.1. CRASH TEST

To be able to compare the numerical results, crash tests have been performed on the IVW crash test rig. It was designed for experiments on larger specimens and structural parts. A photograph of the layout is shown in Figure 2.

A striker which can attain a mass scalable between 42 kg and 200 kg is accelerated horizontally from 1 m/s up to 18 m/s and impacts the specimen which is clamped to the force measurement plate equipped with four piezoelectric discs (sampling rate 300 kHz). Other measured parameters are the acceleration and the displacement of the carriage as well as the initial velocity at the onset of the crash event. The crash event can be filmed with up to 10,000 pictures per second.

All crash tests were performed with a striker mass of 106.55 kg at an initial velocity of 5.7 m/s (20 km/h). The kinetic energy was 1.73 kNm. The specimens were clamped to the force measurement plate. The impactor hit the chamfered side of the specimen.

The sine wave sandwich part has been manufactured by stamp forming of a flat sandwich panel with GF-PA12 faces and PMI foam core. A PMI film has been added in order to assure good adhesive bonding between core

[2]Polymethacrylimid

Figure 2. Crash test rig with main components

and face.

3.2. NUMERICAL SIMULATION

For the simulation, the sandwich has been modeled with 3168 solid and 2176 shell elements. The calculation on a PowerIndigo took 11 CPU hours for the model with the nodal constraint option and 18 CPU hours if the adhesive joint is modeled with solid elements. For the simulation of the sine wave sandwich part no significant influence of the modeling on the maximum force and the energy absorption behaviour has been found. The reason is the immediate destruction of the adhesive joint due to the material sound wave. Core face separation takes place at the whole structure within 1 ms.

3.3. COMPARISON OF NUMERICAL SIMULATION AND CRASH TEST

Figure 3 compares the force vs. displacement and the absorbed energy vs. displacement diagrams. The over estimation of the peak force may be a result of the course mesh and the trigger modelation. For performance reasons, the reduction of the wall thickness - representing the chamfer trigger - is done in only two steps. After 10 mm buckling takes place and the force decreases in simulation and experiment. After 40 mm of displacement, the sandwich is totally destroyed in the experiment. In the simulation the middle force is calculated on a higher level because the total destruction of the FRP faces is not represented. Figure 4 shows a sequence of the high speed

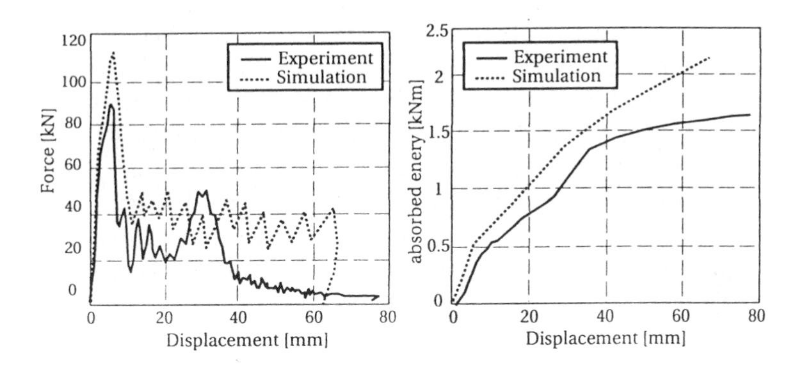

Figure 3. Comparison of experiment and simulation

film and snapshots from the simulation.

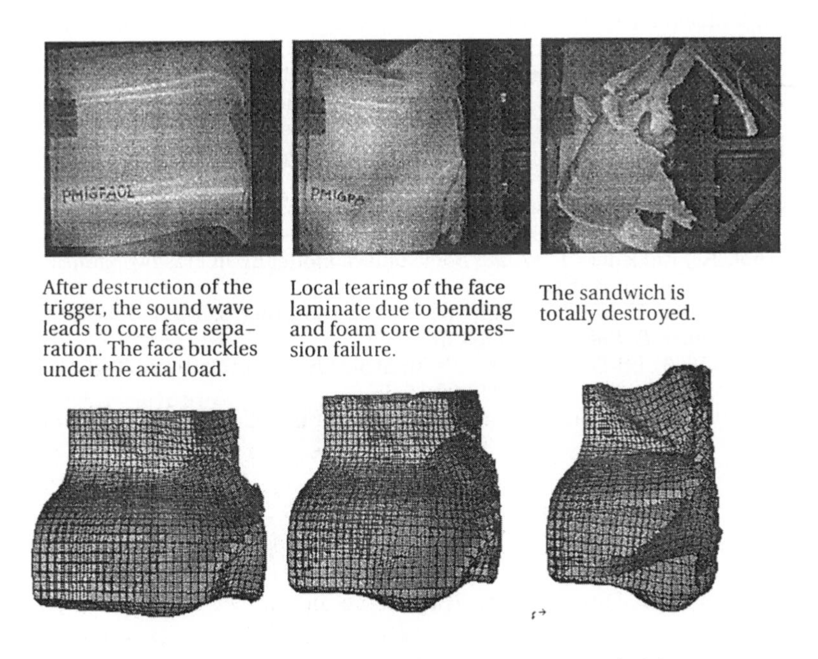

After destruction of the trigger, the sound wave leads to core face separation. The face buckles under the axial load.
Local tearing of the face laminate due to bending and foam core compression failure.
The sandwich is totally destroyed.

Figure 4. Deformation of the sine wave sandwich

4. Summary and Conclusions

Fundamental investigations have been made to introduce sandwich structures with foam core and FRP faces in numerical crashworthiness analysis. Several possibilities have been evaluated to model the adhesive joint between core and face of a sandwich part. The adhesive joint can be modeled automatically with a so-called *closest neighbor algorithm* which decreases the model generation effort. The existing material model for the FRP faces originally developed for unidirectional reinforced thermoset composites was used for the first approach. Extension to bi-directional fibre reinforcement and thermoplastic matrix systems is now subject of research. The industrial scale manufacturing techniques for stamp formed sandwich panels together with the good energy absorption capabilities of the structures enable them to have an application class in structural an crashworthiness structures.

References

Maier, M. (1990) Experimentelle Untersuchung und numerische Simulation des Crashverhaltens von Faserverbundwerkstoffen, Ph.D. Thesis, University of Kaiserslautern

Hull, D. (1991) A Unified Approach to Progressive Crushing of Fibre-Reinforced Composite Tubes, *Composite Science and Technology* **40**, 377-421.

Thornton P.H. (1986) The Crashbehaviour of Glass Fiber reinforced Plastic Sections, *Composite Sciences and Technology*, 247-262.

Maier, M., Matzenmiller, A. Schweizerhof, K. (1992) Numerische Simulation des Schaedigungsverhaltens von Faserverbundwerkstoffen, Abschlussbericht zum CARMAT-Projekt, BASF AG, Ludwigshafen, CAD-FEM GmbH, Ebersberg.

Maier, M., Vinckier, D., Thoma, K., Altstaett, V., Mueller, R. (1992) Numerische Simulation zur Untersuchung des Crashverhaltens von Faserverbundwerkstoffen, Abschlussbericht zum CARMAT-Projekt, BASF AG, Ludwigshafen, CONDAT GmbH, Scheyern/Fernhag.

Ostgathe, M. (1995) Faserverstaerkte Thermoplasthalbzeuge fuer die Umformung organischer Bleche, SAMPE Proceedings: Symposium Neue Werkstoffe, Kaiserslautern

Schweizerhof, K., Engleder, T. (1995) Joint failure modeling in crashworthiness analysis, Study, University of Karlsruhe.

Breuer, U., Neitzel, M., (1995) High speed stamp forming of thermoplastic composite sheets, *Polymer & Polymer Composites*

Hallquist, O., (1993) LS-DYNA3D User's Manual, Livermore Software Technology Corp.

CARMAT, BMFT Projekt Nr. 03 M 1021 A 9

Hashin, Z., (1980) Failure Criteria for Unidirectional Fiber Composites *Journal of Applied Mechanics* **47**, 239-334.

A COMPARATIVE ANALYSIS OF SOME THEORIES AND FINITE ELEMENTS FOR SANDWICH PLATES AND SHELLS

P. Vannucci
Dipartimento di Ingegneria Strutturale
Università di Pisa. Via Diotisalvi, 2 - 56126 Pisa - Italy

S. Aivazzadeh, G. Verchery
ISAT (Institut Superieur de l'Automobile et des Transports)
Université de Bourgogne
49, Rue Mademoiselle Burgeois - 58000 Nevers -France

1. INTRODUCTION

In this paper we give a comparative analysis of the performance of some theories and finite elements for the calculation of composite sandwich plates and shells

In the first part of the paper, two distinct analytical theories are considered. The former is the well known three-dimensional solution of Pagano, and the second one is the Pham-Dang solution, in its two versions. In the second part, we will consider the case of some finite elements, used by well known finite elements codes. Namely these are: the element DSQ (Discret Shear Quadrilateral), a four nodes element based on the theory of Mindlin Reissner for the thick plates, the eight nodes quadrilateral element proposed by Ahmad and a eight nodes solid element.

Number of numerical tests have been performed on two distinct cases: a square plate and a rectangular plate with a base to height ratio equal to two. In both cases, the plates are simply supported, and subjected to a uniform load equal to 1. We have then considered three different ratios base/thickness, and for each one of these cases we have taken into account three different ratios of the skin thickness to total thickness.

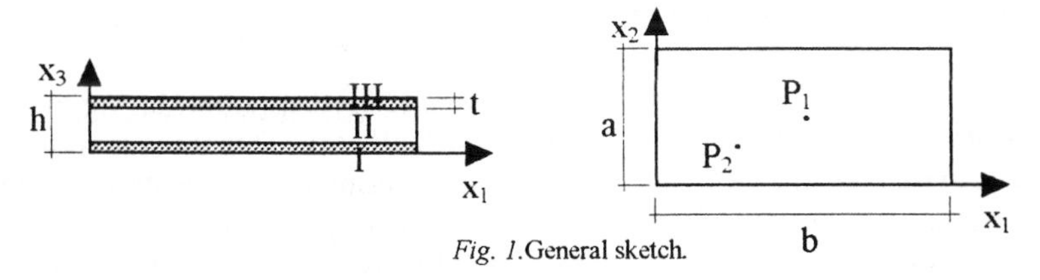

*Fig. 1.*General sketch.

2. THE SOLUTION OF PAGANO

The solution of Pagano [1], 1970, is obtained via a three dimensional approach in the theoretical framework of the linear elastic theory. It is not a sandwich-dedicated solution, in the sense that no kind of simplifying hypothesis is made about the stress and displacement fields, and it can be used also for stratified plates with more than

A. Vautrin (ed.), Mechanics of Sandwich Structures, 45–52.

three layers. Referring us to Fig. 1, the solution considers only rectangular plates and orthotropic materials, that have their orthotropy axes parallel to the sides of the plate. There are no body forces and there is only a distributed load on the upper surface. Here we have done a slight modification to the original solution, which allow us to consider also loads on the lower surface. The boundary conditions correspond to a simply supported plate on the four edges.

The method for obtaining the solution is quite straightforward: the equilibrium equations are written in terms of the displacements components, which are assumed to be of the following type:

$$\begin{aligned} u_1 &= U_1(x_3)\ \cos p\, x_1\ \sin q\, x_2, \\ u_2 &= U_2(x_3)\ \sin p\, x_1\ \cos q\, x_2,\ p=\frac{n\pi}{a},\ q=\frac{m\pi}{b}. \\ u_3 &= U_3(x_3)\ \sin p\, x_1\ \sin q\, x_2, \end{aligned} \tag{1}$$

These displacement fields respect the given boundary conditions; substituting the above relations in the equilibrium equations of each layer, an eigenvalue problem is found, which gives 6 unknown constants for each layer, to be determined imposing proper boundary and interface conditions, namely:

$$\begin{aligned} \sigma_{33}(-h/2) &= q^{-}\sin p\, x_1\ \sin q\, x_2, \\ \sigma_{33}(+h/2) &= q^{+}\sin p\, x_1\ \sin q\, x_2, \\ \sigma_{\alpha 3}(\pm h/2) &= 0,\quad \alpha=1,2 \end{aligned} \tag{2}$$

for the upper and lower surfaces, and

$$\mathbf{u}_k = \mathbf{u}_{k-1},\ (\sigma_{i3})_k = (\sigma_{i3})_{k-1},\ k=2,3\ , \tag{3}$$

at each interface. As apparent from the above developments, the load is decomposed in a Fourier-like series, where for each harmonic the chargement is of the type indicated by eqns (2). So, for a sandwich plate, 18 linear equations must be solved for each harmonic to find the numerical constants. As no kind of simplifying hypothesis is made in Pagano's solution, it is often considered as a reference solution in elasticity.

3. THE SOLUTION OF PHAM DANG

The solution of Pham Dang [2], 1976, has been studied by the author during his PhD thesis, under the guide of G. Verchery, co-author of the theory. This time, it is the case of a sandwich plate theory, that is a solution which is valid only for a stratified panel with three layers, with the skin plies thinner than the core ply, and symmetric respect to the middle plane. The basic assumptions of the theory are the following ones: the material of the layers is of monoclinic type, thus more general than the Pagano case; the boundary conditions at the upper and lower surfaces are the following ones:

$$\begin{aligned} \sigma_{\alpha 3}(-h/2) &= 0,\ \sigma_{33}(-h/2) = -q^{-}(x_1,x_2), \\ \sigma_{\alpha 3}(+h/2) &= 0,\ \sigma_{33}(+h/2) = +q^{+}(x_1,x_2), \end{aligned} \tag{4}$$

and the interface conditions are

$$(u_\alpha)_k = (u_\alpha)_{k-1},\ \alpha=1,2;\ (\sigma_{i3})_k = (\sigma_{i3})_{k-1},\ i=1,2,3 \text{ and } k=2,3. \tag{5}$$

Considering the characteristics of a sandwich plate, it is assumed that the skin plies have a membrane-like behavior, and that the shear stresses $\sigma_{\alpha 3}$ do not depend upon x_3 in the core ply. For what concern the distribution of the shear stresses $\sigma_{\alpha 3}$ in the skins, two distinct assumptions are made, that give rise to two different solutions. In the first one, it is assumed a linear variation of $\sigma_{\alpha 3}$ in the skin layers, while in the second one the variation is assumed to be parabolic. There are six unknowns to be determined, and the solution is obtained by means of a variational approach: the following mixed functional, introduced by the authors, is considered

$$P = \int_V \mathrm{P}dv - \int_{\partial t} t_\alpha^* u_\alpha ds + \int_{\partial u} t_3 u_3^* ds, \quad \mathrm{P} = \frac{1}{2}\sigma_{ij}\varepsilon_{ij} - \sigma_{3\alpha}u_{3,\alpha} - \sigma_{33}u_{3,3} \tag{6}$$

where ∂t is the boundary where the stress t_α^* is given, whilst ∂u is the boundary where the displacement u_3^* is assigned. The assumed principal variables are u_α, $\sigma_{\alpha 3}$ and σ_{33}, and using the relations that give the other unknowns as function of the principal variables, the general expression of P for a monoclinic material is found for each layer.

An analytical solution has been obtained by the authors in the case of a simply supported rectangular plate, made by orthotropic materials, with axes of orthotropy parallel to the edges of the plate, that is, in the same hypotheses as in the Pagano's solution. Also in this case, the load is expressed by the same Fourier series as in eqns (2), and for each harmonic a linear system of seven equations must be solved. This means that the Pham Dang solution is much more rapid than the Pagano's one.

It is worth noting that in both Pagano and Pham Dang cases an analytical solution, in the form of Fourier-like series, can be obtained only in the hypotheses given above. In particular, it is important that the axes of orthotropy be parallel to the edges of the plate. If it is not so, it is impossible to obtain solutions like those briefly exposed.

4. THE FINITE ELEMENTS USED

As mentioned above, we have used three different finite elements. The first one is the element DSQ, that means Discrete Shear Quadrilateral. It is a four nodes element, and like its similar triangular element DST has been conceived by Lardeur and Batoz [3], [4], to model thin to thick plates. It is based upon the well known theory of Mindlin-Reissner for the analysis of thick plates. The original version of this element has been adapted in the code MEF COMPOSIC 2.8 [5] to model layered plates, and it is also automatically considered the elimination of shear locking that may occur in the case of thin plates. Each node of one element has six degrees of freedom, the three translations and the three rotations.

The second one is the so-called element of Ahmad [6]. It is a eight nodes quadrilateral element (here we have used the version SHELL91 of the numerical code ANSYS 5.3 [7]), able to account for shear deformations. A coefficient depending on plate stiffness is introduced to avoid shear locking. The transversal shear stresses computed with the constitutive law are adjusted to account for their transversal variation, which is assumed to be parabolic.

The last element considered is a solid element , named SOLID46, Taylor *et alii* [8],of the element set of the code ANSYS 5.3 [7]. It is a layered version of a eight nodes solid element designed to model layered thick shells or solids, and it has three degrees of freedom at each node, the three displacements. It allows to model also layered solids with plies of variable thickness.

5. THE NUMERICAL SIMULATIONS

Two distinct cases have been considered, a square and a rectangular plate, with a ratio base to height of two. In both cases, the plates are simply supported at the four edges, and the load is uniform and unitary. For the theories of Pagano and Pham Dang 1 and 2 and also for the case of the solid finite element, the load is applied on both the faces, one half on each one, while for the shell finite elements, it is entirely applied on the middle surface. The load has been decomposed into 2500 harmonics for the Pagano solution, and into 10000 for the Pham Dang theories, whose computation is quickest. For the Pagano case, due to high sensibility of the solution to numerical errors, an algorythm with a quadruple precision has been used. In both cases, a uniform mesh of 400 elements has been used for the finite elements calculations. All the tests have been carried out on a DEC ALPHA 500 and on a SUN SPARCSTATION LX OS 5.5.

For each plate, three different ratios a/h of the side width to total thickness have been considered (for the case of the rectangular plate, a is the shortest side): 5, 10 and 100. Moreon, for each one of the above cases, three different ratios t/h of the skin layers thickness to total thickness have been considered: 0.01, 0.05 and 0.10. On the whole, 9 different cases have been taken into account for each plate and for each case, for a total of 108 different calculations.

The material of the skin layers is orthotropic, with the following elastic moduli:

$$\begin{gathered} E_{11}=17500,\ E_{22}=E_{33}=7000,\ G_{12}=G_{13}=3500,\ G_{23}=1400, \\ \nu_{12}=\nu_{13}=0.01,\ \nu_{23}=0.25. \end{gathered} \tag{7}$$

In the case of the core, the material is transversely orthotropic, with elastic constants

$$\begin{gathered} E_{11}=E_{22}=280,\ E_{33}=3000,\ G_{12}=112,\ G_{13}=G_{23}=420, \\ \nu_{12}=0.25,\ \nu_{13}=\nu_{23}=0.02. \end{gathered} \tag{8}$$

The results have been analised for two distinct points in the plates. The first one, P_1, is the center of the plate, while P_2 is located on the diagonal, at middle way between the center and a corner. In the first point, we have considered the values of the vertical displacement u_3 and of the normal stresses σ_{11} and σ_{22}, evaluated at the surface. In the second point we have considered the shear stresses σ_{12}, at the surface, σ_{13} and σ_{23}, at the interface All the results in the tables that follows are given in a non-dimensional form, multiplying each result by:

$$c_1=10\frac{E_{22}h^3}{q\,a^4}\ \text{for}\ u_3;\ c_2=\frac{100\ h}{E_{22}\ a}\ \text{for}\ \sigma_{\alpha\beta};\ c_3=1000c_2\ \text{for}\ \sigma_{\alpha 3}. \tag{9}$$

Hereon, q is the unitary load, a=10 is the side width, h is the total plate thickness and E_{22} is the Young's modulus in the direction x_2 for the skin layers.

6. RESULTS AND DISCUSSION

The results for the square plate are shown in tables 1 to 3, and those for the rectangular plate are written in tables 4 to 6. We analize each theory comparing it with the solution of Pagano.

The Pham Dang 1 solution gives very good results in terms of displacement; the same for normal stresses, that anyway can show differences up to 40%, mainly for higher values of t/h . For what concern the shear stresses, the response for σ_{12} is rather good (differences less then 18%). The transversal shear stresses are well evaluated, but in some cases (rectangular plate, t/h=0.01) they show some great differences with the Pagano's value, up to 75%. The Pham Dang 2 solution gives results that are in very great accordance with the other formulation of the same theory, no kind of really important differences can be observed (only a certain tendency to underestimate the stiffness mainly for thick plates and for higher values of t/h).

TABLE 1. Square plate, a/h=5

t/h	Computation method	Point P_1				Point P_2	
		u_3	σ_{11}	σ_{22}	σ_{12}	σ_{13}	σ_{23}
0.01	PAGANO	1.852	0.855	0.047	0.022	2.628	0.094
	PHAM DANG 1	1.835	0.834	0.047	0.021	2.782	0.170
	PHAM DANG 2	1.841	0.834	0.047	0.021	2.782	0.170
	ANSYS - SHELL	1.865	0.861	0.055	0.021	2.826	0.146
	ANSYS - SOLID	1.792	0.825	0.055	0.021	2.734	0.188
	MEF - COMPOSIC	1.882	0.868	0.047	0.021	2.651	0.071
0.05	PAGANO	1.010	0.196	0.023	0.009	2.675	0.342
	PHAM DANG 1	1.003	0.163	0.023	0.008	2.684	0.386
	PHAM DANG 2	1.031	0.163	0.024	0.008	2.680	0.391
	ANSYS - SHELL	1.097	0.178	0.028	0.008	2.822	0.360
	ANSYS - SOLID	0.896	0.173	0.025	0.008	2.568	0.348
	MEF - COMPOSIC	1.036	0.182	0.025	0.008	2.680	0.271
0.10	PAGANO	0.844	0.137	0.019	0.006	2.559	0.564
	PHAM DANG 1	0.851	0.079	0.017	0.005	2.626	0.614
	PHAM DANG 2	0.910	0.078	0.017	0.005	2.617	0.618
	ANSYS - SHELL	1.030	0.091	0.023	0.005	2.882	0.663
	ANSYS - SOLID	0.728	0.090	0.018	0.005	4.565	0.482
	MEF - COMPOSIC	0.879	0.094	0.019	0.005	2.648	0.468

TABLE 2. Square plate, a/h=10

t/h	Computation method	Point P_1				Point P_2	
		u_3	σ_{11}	σ_{22}	σ_{12}	σ_{13}	σ_{23}
0.01	PAGANO	1.264	1.792	0.062	0.036	3.155	0.033
	PHAM DANG 1	1.254	1.756	0.061	0.035	2.932	0.057
	PHAM DANG 2	1.256	1.756	0.061	0.035	2.932	0.057
	ANSYS - SHELL	1.251	1.784	0.076	0.035	2.967	0.050
	ANSYS - SOLID	1.220	1.747	0.079	0.034	2.874	0.087
	MEF - COMPOSIC	1.267	1.799	0.060	0.035	2.806	0.027
0.05	PAGANO	0.456	0.416	0.022	0.010	3.057	0.041
	PHAM DANG 1	0.453	0.378	0.021	0.010	3.061	0.049
	PHAM DANG 2	0.460	0.377	0.021	0.010	3.056	0.053
	ANSYS - SHELL	0.474	0.400	0.026	0.010	3.221	0.048
	ANSYS - SOLID	0.424	0.393	0.025	0.009	3.028	0.066
	MEF - COMPOSIC	0.460	0.406	0.021	0.010	3.056	0.046
0.10	PAGANO	0.341	0.245	0.017	0.007	3.130	0.121
	PHAM DANG 1	0.341	0.194	0.015	0.006	3.131	0.133
	PHAM DANG 2	0.357	0.192	0.016	0.006	3.107	0.149
	ANSYS - SHELL	0.391	0.216	0.021	0.007	3.467	0.167
	ANSYS - SOLID	0.301	0.213	0.017	0.004	2.838	0.024
	MEF - COMPOSIC	0.346	0.221	0.017	0.007	3.138	0.083

The 8 nodes element of Ahmad has a slight tendency to diminish the stiffness for the cases of thick plate (greatest difference in the displacements about 20%) and, on the contrary, to increase the same for very thin plates. The evaluations of σ_{11}, σ_{22} and σ_{12} are quite good (usually, differences less than 20%), but in the case of σ_{22} for the square plate there can be also differences up to 40%, see for instance the case with a/h=100 and t/h= 0.01. For what concern the transversal shear stresses, the evaluation is rather good, with maximum differences usually less than 20%; only in the case of σ_{23} for the rectangular plate there are large differences (up to 75% and more) for plates with t/h=0.01.

TABLE 3. Square plate, a/h=100

t/h	Computation method	Point P_1				Point P_2	
		u_3	σ_{11}	σ_{22}	σ_{12}	σ_{13}	σ_{23}
	PAGANO	1.062	18.179	0.496	0.325	2.843	0.014
	PHAM DANG 1	1.054	17.854	0.489	0.319	3.004	0.025
0.01	PHAM DANG 2	1.054	17.854	0.489	0.319	3.004	0.025
	ANSYS - SHELL	1.032	17.873	0.653	0.319	3.126	0.012
	ANSYS - SOLID	1.023	17.695	0.697	0.316	2.863	0.109
	MEF - COMPOSIC	1.052	18.048	0.487	0.322	2.810	0.030
	PAGANO	0.247	4.198	0.096	0.071	3.289	0.066
	PHAM DANG 1	0.245	3.959	0.091	0.067	3.303	0.071
0.05	PHAM DANG 2	0.245	3.956	0.091	0.067	3.300	0.071
	ANSYS - SHELL	0.240	4.120	0.133	0.069	3.476	0.046
	ANSYS - SOLID	0.238	4.087	0.142	0.069	3.079	0.012
	MEF - COMPOSIC	0.244	4.167	0.094	0.070	3.265	0.106
	PAGANO	0.139	2.350	0.052	0.040	3.509	0.086
	PHAM DANG 1	0.138	2.104	0.047	0.036	3.511	0.087
0.10	PHAM DANG 2	0.137	2.096	0.047	0.035	3.497	0.087
	ANSYS - SHELL	0.136	2.305	0.073	0.038	3.926	0.074
	ANSYS - SOLID	0.134	2.286	0.077	0.039	3.111	0.030
	MEF - COMPOSIC	0.138	2.330	0.051	0.039	3.487	0.124

TABLE 4. Rectangular plate, a/h=5

t/h	Computation method	Point P_1				Point P_2	
		u_3	σ_{11}	σ_{22}	σ_{12}	σ_{13}	σ_{23}
	PAGANO	9.328	1.546	0.295	0.068	2.949	0.893
	PHAM DANG 1	9.215	1.507	0.288	0.067	2.506	1.578
0.01	PHAM DANG 2	9.228	1.507	0.288	0.067	2.507	1.578
	ANSYS - SHELL	9.329	1.605	0.309	0.067	2.577	1.566
	ANSYS - SOLID	8.736	1.536	0.302	0.064	4.468	1.560
	MEF - COMPOSIC	9.632	1.592	0.301	0.069	2.300	0.908
	PAGANO	3.509	0.349	0.100	0.019	2.579	1.543
	PHAM DANG 1	3.469	0.326	0.094	0.018	2.604	1.712
0.05	PHAM DANG 2	3.529	0.325	0.094	0.018	2.597	1.714
	ANSYS - SHELL	3.651	0.359	0.105	0.018	2.700	1.817
	ANSYS - SOLID	3.248	0.349	0.098	0.018	4.474	0.226
	MEF - COMPOSIC	3.640	0.365	0.103	0.019	2.566	1.511
	PAGANO	2.473	0.193	0.065	0.012	2.493	1.904
	PHAM DANG 1	2.459	0.153	0.057	0.010	2.455	2.003
0.10	PHAM DANG 2	2.584	0.151	0.057	0.010	2.441	1.999
	ANSYS - SHELL	2.749	0.172	0.069	0.010	2.563	2.340
	ANSYS - SOLID	2.184	0.182	0.061	0.003	4.163	0.251
	MEF - COMPOSIC	2.559	0.182	0.066	0.011	2.425	1.888

In the case of the 4 nodes solid element, there is a little overestimation of the stiffness, mainly for the case of thick plates and for thin skin plies. The evaluation of σ_{11} has the same type of not accentuated discrepancies with respect to Pagano's solution, except in the case of the square plate with a/h= 5 and t/h= 0.10, when a difference of about 40% is registered. For what concern the evaluation of σ_{22} the same remarks as for the 8 nodes shell elements can be done. The answer to this strange behavior must be found in the fact that the distribution of the stress σ_{22} at the surface can be sensibly influenced by the geometry and by the elastic moduli of the materials; in some cases, such distribution is like in Fig. 2., and the difference between the maximum value and that in the center can amount also to about 100%. All the methods have shown these behavior, and the differences in the value of σ_{22} at the center are likely to derive from such a behavior, differently evaluated by the theories. Normally, there are no important discrepancies in the values of σ_{12}, except in two cases (square plate, a/h= 5, t/h=0.10 and rectangular plate, a/h= 10 and t/h=0.10), when the differences are very large, up to 75% and quite inexplicable. For what concern the stress σ_{13}, the differences are usually small, but in some cases, namely for thick plates, they can reach also 75%.

TABLE 5. Rectangular plate, a/h=10

t/h	Computation method	Point P_1				Point P_2	
		u_3	σ_{11}	σ_{22}	σ_{12}	σ_{13}	σ_{23}
0.01	PAGANO	8.474	3.311	0.558	0.139	2.314	0.875
	PHAM DANG 1	8.376	3.237	0.546	0.136	2.644	1.468
	PHAM DANG 2	8.378	3.236	0.546	0.136	2.644	1.468
	ANSYS - SHELL	8.302	3.380	0.578	0.136	2.714	1.458
	ANSYS - SOLID	8.022	3.282	0.577	0.131	2.624	1.456
	MEF - COMPOSIC	8.561	3.347	0.561	0.139	2.421	0.838
0.05	PAGANO	2.612	0.874	0.166	0.039	3.077	1.210
	PHAM DANG 1	2.582	0.811	0.156	0.036	3.155	1.351
	PHAM DANG 2	2.595	0.809	0.156	0.036	3.148	1.353
	ANSYS - SHELL	2.611	0.881	0.173	0.037	3.365	1.410
	ANSYS - SOLID	2.436	0.847	0.165	0.036	3.014	1.281
	MEF - COMPOSIC	2.667	0.886	0.168	0.038	3.170	1.170
0.10	PAGANO	1.647	0.483	0.102	0.022	3.195	1.420
	PHAM DANG 1	1.635	0.418	0.091	0.020	3.249	1.492
	PHAM DANG 2	1.660	0.413	0.091	0.020	3.219	1.498
	ANSYS - SHELL	1.722	0.471	0.108	0.021	3.596	1.711
	ANSYS - SOLID	1.512	0.473	0.100	0.021	2.990	1.308
	MEF - COMPOSIC	1.694	0.484	0.104	0.022	3.282	1.384

TABLE 6. Rectangular plate, a/h=100

t/h	Computation method	Point P_1				Point P_2	
		u_3	σ_{11}	σ_{22}	σ_{12}	σ_{13}	σ_{23}
0.01	PAGANO	8.181	33.957	5.457	1.398	2.350	0.850
	PHAM DANG 1	8.087	33.217	5.341	1.368	2.688	1.426
	PHAM DANG 2	8.086	33.216	5.341	1.368	2.688	1.426
	ANSYS - SHELL	7.855	33.990	5.572	1.357	3.077	1.844
	ANSYS - SOLID	7.624	33.073	5.574	1.298	2.540	1.199
	MEF - COMPOSIC	7.573	31.521	5.053	1.290	2.263	0.758
0.05	PAGANO	2.264	9.496	1.491	0.386	3.350	1.054
	PHAM DANG 1	2.237	8.912	1.401	0.363	3.398	1.177
	PHAM DANG 2	2.233	8.903	1.400	0.362	3.394	1.176
	ANSYS - SHELL	2.160	9.433	1.513	0.371	3.631	1.427
	ANSYS - SOLID	2.115	9.253	1.520	0.360	3.165	1.091
	MEF - COMPOSIC	2.245	9.434	1.481	0.382	3.215	0.936
0.10	PAGANO	1.303	5.463	0.856	0.222	3.681	1.133
	PHAM DANG 1	1.292	4.873	0.764	0.198	3.689	1.191
	PHAM DANG 2	1.282	4.852	0.761	0.197	3.674	1.186
	ANSYS - SHELL	1.244	5.423	0.868	0.212	4.156	1.432
	ANSYS - SOLID	1.220	5.331	0.869	0.207	3.276	1.064
	MEF - COMPOSIC	1.292	5.429	0.850	0.219	3.572	1.020

In the case of σ_{23}, there are often great differences (even an abnormous value of about 750% has been found), mainly for thick plates, in both senses of an over- or under-estimation.

Finally, the DSQ element, which is the one that perhaps gives on the whole the best responses: it gives results within 20% of approximation with respect to Pagano's solution, for all the characteristics and the cases considered. Only two times these discrepancies reach higher values: the stress σ_{11} for the case of the square plate with a/h= 5 and t/h= 0.10, with a difference of about 35%, and σ_{23} for a/h= 100, when the differences can reach also 100%.

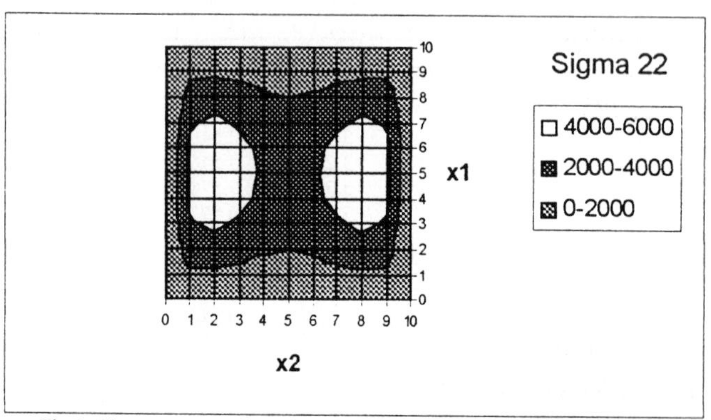

Fig 2. *The variation of σ_{22} for a/h=100, t/h=0.10 (Pagano's solution)*

On the whole, we can say that the evaluation of the displacements is rather well done by all the considered methods; some greater difficulties can be encountered in the evaluation of the shear stresses, mainly of the transversal shears, where the methods show still great differences among them. In this sense, the best performance seems to be that of the four nodes shell element.

7. REFERENCES

1. Pagano, N. J. (1970) Exact Solutions for Rectangular Bidirectional Composites and Sandwich Plates, *J Composite Materials*, **4**, 20-34.
2. Pham Dang, T and Verchery, G. (1976) Théorie des plaques sandwich assurant la continuité du déplacement et de la contrainte aux interfaces. *C. R. Acad. Sc. Paris*, **2 A**
3. Batoz, J. L. and Lardeur, P. (1989) A Discrete Shear Triangular Nine DoF Element for the Analysis of Thick to very Thin Plates, *Int J Num Meth Engng*, **28**.
4. Lardeur, P. and Batoz, J. L. (1989) Composite Plate Analysis Using a New Discrete Shear Triangular Finite Element, *Int J Num Meth Engng*, **27**.
5. MEF - COMPOSIC Manuel de référence. Framasoft CSI, Paris (1995).
6. Ahmad, S.Irons, B. M. and Zienkiewicz, O. C. (1970) Analysis of Thick and Thin Shell Structures by Curved Finite Elements, *Int J Num Meth Engng*, **2**, 419-451.
7. ANSYS 5.3 Theory Manual. USA (1996).
8. Taylor, R. L., Beresford, P. J. and Wilson, E. L. (1976) A Non-Conforming Element for Stress Analysis, *Int J Num Meth Engng*, **10**, 1211-1219.
9. Zienkiewicz, O. C. (1977) *The Finite Element Method*, Mc Graw-Hill, London.

STATIC ANALYSIS OF SANDWICH PLATES BY FINITE ELEMENTS

V. MANET, W.-S. HAN AND A. VAUTRIN
École des Mines de Saint-Etienne
Materials and Mechanical Engineering Department
158, cours Fauriel
42023 Saint-Etienne Cedex 2

Notations

{ }	vector (column)
⟨ ⟩	transposed vector (row)
[ ]	matrix
Ω	element's interior (volume or surface)
Γ	element's boundary (surface or curve) $\Gamma = \partial\Omega = \Gamma_u \cup \Gamma_\sigma ; \Gamma_u \cap \Gamma_\sigma = \varnothing$
Γ_u	part of Γ with prescribed displacements
Γ_σ	part of Γ with imposed forces
Γ_I	interface
U	(small) displacements
σ	Cauchy's stresses
ε	(small) strains
λ	Lagrange multipliers
D_{ijkl}	Hooke's operator $\{\sigma\} = [D]\{\varepsilon\}$
S_{ijkl}	$[S] = [D]^{-1}$
$[\mathcal{L}]$	differential operator $\{\varepsilon\} = [\mathcal{L}]\{U\}$
$\overline{f}_\Omega$	applied body forces in Ω
$\overline{T}$	applied forces on Γ_σ
$\overline{U}$	prescribed displacements on Γ_u

1. Introduction

Sandwich plates being more and more involved in structural components, it becomes essential to develop analysis tools taking their specificities into account.

In this paper, we present some computational methods developed in order to describe the mechanical behaviour of sandwich plates in a more realistic way. Emphasis has been put on the determination of stresses at

A. Vautrin (ed.), Mechanics of Sandwich Structures, 53–60.

the interfaces between the skins and the core which have in general very heterogeneous mechanical properties.

A sandwich plate is a 3-layer laminate, whose layers have very different mechanical properties: the skins are generally very stiff and work in membrane, whereas the core, which has low stiffness and density, is subjected to shear efforts. It yields quite particular behaviour and failure modes (crack modes, instability modes, local denting...), as explained in (Teti and Caprino, 1989).

The main points of our sandwich modelling are the following:

- transverse shear effects are taken into account: this point is extremely important since the core is only subjected to shear loading;
- the continuity of displacements and the equilibrium state of stresses at each interface of layers are fulfilled: these conditions ensure the physical interface condition;
- large differences of the geometrical and mechanical properties between layers are correctly modelized;
- the relative magnitudes of stresses: the plane stress assumption used in the classical theory of laminated plates is not acceptable for sandwich plates.

Computation of stresses at interfaces is of particular importance for sandwich panels submitted to flexural bending. Nevertheless, it remains difficult to calculate these quantities in a accurate way: at the interfaces, only some components of stresses are continuous. We also face the problem of working with "reduced" stress fields ("reduced" means "which does not contain all the components"). For this purpose several approaches have been developed and are presented in this paper.

2. Hybrid sandwich finite elements

In this section, we present the development of different special finite elements taking the aforementioned points into account.

Classical finite elements in displacements yield a correct displacement field, but cannot fulfill the equilibrium state of stresses, which are derived from displacements. On the other hand, mixed formulations, such as Reissner's one, which use both displacement and stress fields as variables, lead to the continuity of all components of both fields: these models induce too high number of degrees of freedom and non-required continuities.

Therefore, it is necessary to develop methods leading to the continuity of only required components. For this purpose, we introduced a reduced stress field as unknown. It can be done by insuring the displacements continuity at the interfaces by Lagrange multipliers; these multipliers are easily identified as the reduced stress field. The developed hybrid sandwich elements are initially composed of three sub-elements through the thickness, exactly as a sandwich plate. The displacement field is interpolated quadratically in each layer to take flexural effects into account in a better way.

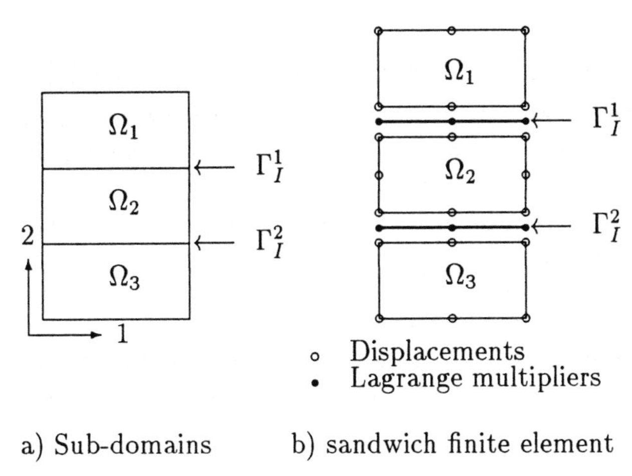

Figure 1. Sub-domains and finite element

We developed two families of hybrid sandwich elements. In the first type, we use the functional of the potential energy in each sub-layer and ensure the displacement continuity at interfaces with Lagrange multipliers representing the reduced stress field. In the second case, we use the condensed Pian and Tong functional in each sub-domain and Lagrange multipliers to ensure the displacements continuity. These functionnals can be found in (Washizu, 1982).

The first hybrid sandwich element leads to a non-symmetrical and non definite-positive stiffness matrix, the second one to a symmetrical but non definite-positive stiffness matrix, which is typical of mixed systems. Variational principles used by these elements are given in (Manet and Han, 1997).

We use the term "hybrid" because the displacement field in approximated in the volume and Lagrange multipliers only along interfaces. We do not use the term "mixed hybrid" even if we have two types of unknowns (displacements and Lagrange multipliers) to point out that Lagrange multipliers are not a physical field (although they have a physical meaning).

3. Post-processing

Specials finite elements lead to "non-usual" stiffness matrices, in particular non definite-positive matrices. So, their implementation in commercialized FEM softwares is sometimes difficult.

3.1. STRESS PROJECTION

As an alternative solution, we developed a "stress-projection" method in order to improve the results obtained by displacement finite elements.

This method, initially proposed by (Zienkiewicz and Taylor, 1994), permits to improve the accuracy of nodal stresses extrapolated from stresses derived from the displacements at Gauss points.

The principle of the method is to minimize the difference between the stresses obtained from displacements and the stresses issued from a mixed formulation, in each element.

For sandwich structures, problems arise from the high heterogeneity of mechanical properties between the skins and the core. The computation of stresses at the interfaces cannot be done so easily, otherwise stresses would not fulfill the equilibrium state. We use the stress projection method, not on one element, but on two elements on each side of an interface, so that the integration volume includes the interface. In such a case, the calculated stress field fulfills the equilibrium state.

In the displacements formulation, stresses are obtained from nodal displacements $\{q\}$ by: $\{\sigma_u\} = [D]\,[\mathcal{L}]\,[N_u]\,\{q\}$, where $[D]$ is the generalized Hooke's matrix and $[N_u]$ the matrix of shape functions. In a mixed formulation, stresses are obtained directly from nodal stresses $\{\tau\}$ by: $\{\sigma_m\} = [N_\sigma]\,\{\tau\}$. If we now minimize the difference $\int_\Omega \{\sigma_m\} - \{\sigma_u\}\ d\Omega$ after having premultiplied by $[N_\sigma]^T$, we reach:

$$[M_\sigma]\,\{\tau\} = \{P_u\}$$

with:

$$[M_\sigma] = \int_\Omega [N_\sigma]^T\,[N_\sigma]\ d\Omega \quad \text{and} \quad \{P_u\} = \int_\Omega [N_\sigma]^T\,[D]\,[\mathcal{L}]\,[N_u]\ d\Omega\ \{q\}$$

Hence:

$$\{\tau\} = [M_\sigma]^{-1}\,\{P_u\}$$

Contrary to (Zienkiewicz and Taylor, 1994), we perform the minimization process on two elements on each side of an interface, so that the result will yield a stress field which fulfills the force equilibrium state.

3.2. LOCAL REISSNER

Reissner's mixed formulation, as that of Pian and Tong, leads to a stiffness matrix in which the stress field can be condensed: the stress field does no more belong to the set on global unknowns, and is therefore discontinuous. This condensed method leads to the displacement field and is equivalent to the classical displacement method.

A post-processing method (in fact a decondensation method) can then be applied to retrieve the stress field, but once again on two elements:

$$\{\tau\} = [A]^{-1} [B] \{q\}$$

in order to fulfill force equilibrium state at interfaces, with:

$$[A] = \int_{\Omega} [N_\sigma]^T [S] [N_\sigma] \; d\Omega \quad \text{and} \quad [B] = \int_{\Omega} [N_\sigma]^T [\mathcal{L}] [N_u] \; d\Omega$$

This method is mathematically equivalent to the previous one.

4. Applications

In this section, we only present two simple examples. More results concerning hybrid sandwich elements can be found in (Manet and Han, 1997).

In the following, "Sandwich" denote solutions obtained with our hybrid sandwich elements (for legibility, these two elements are merged because their results are very close); "Disp." is the solution obtained with classical finite elements in displacements without post-processing; "P-P" denotes results obtained with the post-processing methods (the stress projection and local Reissner methods yield very close results).

4.1. A SANDWICH BEAM

We consider a simply supported beam, with total length L and total height H, under uniform pressure as shown in figure 2. The analytical solution is given in (Lerooy, 1983). The core represents 80% of the total thickness, and $L/H = 12$.

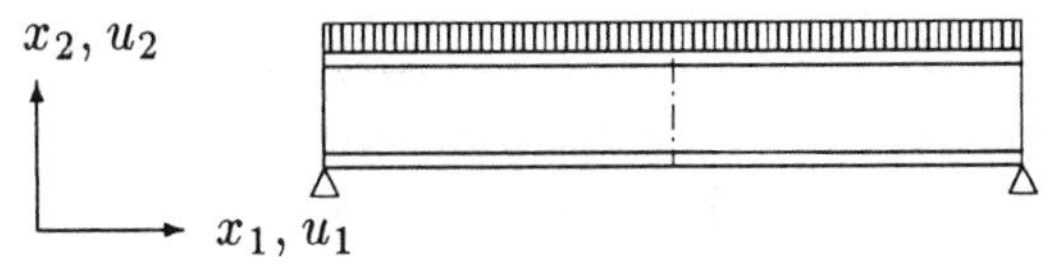

Figure 2. Simply supported sandwich beam

Convergence of the mid-section deflection is given: i) in figure 3.a as a function of the number of degrees of freedom; and ii) in figure 3.b as

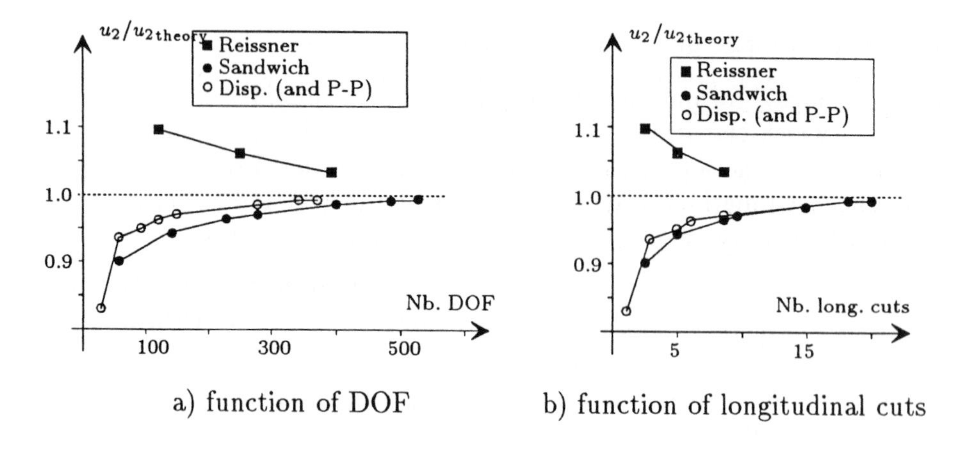

a) function of DOF b) function of longitudinal cuts

Figure 3. Convergence of central deflection

a function of the number of longitudinal cuts on the half beam. Hybrid sandwich elements have the same convergence speed as the displacements method in terms of the number of longitudinal cuts; the fact that these models have more degrees of freedom (Lagrange multipliers) explains the shift of figure 3.a.

The transverse shear stress distribution through the thickness is shown in figure 4.a. The result for the classical finite element method is not plotted, because this component is not continuous.

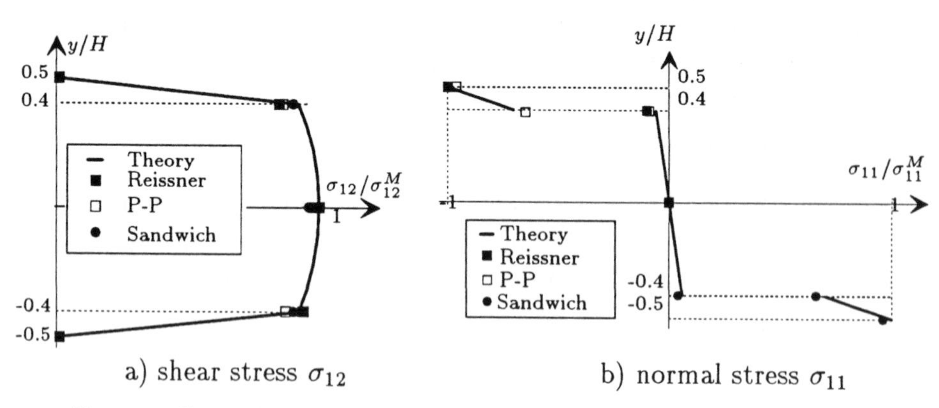

a) shear stress σ_{12} b) normal stress σ_{11}

Figure 4. Comparison of stress distribution through the thickness at $x = L/4$

Figure 4.b shows the distribution of σ_{11} through the thickness. Results obtained by Reissner's method and post-processing are only plotted for $\sigma_{11}/\sigma_{11}^M < 0$. For hybrid sandwich elements, they are plotted for $\sigma_{11}/\sigma_{11}^M > 0$ for better legibility. This component does not have to be continuous at the interfaces. This figure shows that Reissner's method leads to a exceeding continuity. Our elements and post-processing methods give an appropriate shape of this discontinuous component.

From figures 3 and 4 it is clear that the presented methods yield results in agreement with the theory, for displacements as well as for the stress distribution through the thickness, and especially at interfaces.

4.2. A SANDWICH PLATE

We now extend the previous example to 3D. Let us consider a simply supported square plate under uniform pressure, as shown in figure 5. The core represents 80% of the total height H, and $L/H = 12$, with L the length of the side of the plate. The reference solution is given in (Pagano, 1970).

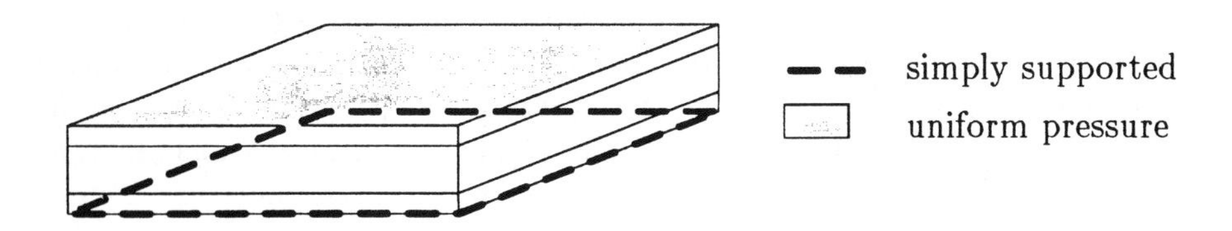

Figure 5. Symply supported square sandwich plate

Keeping the previous example in mind, we limit our study to the distribution of σ_{23} and σ_{33} through the thickness of the plate. These results are shown in figure 6.

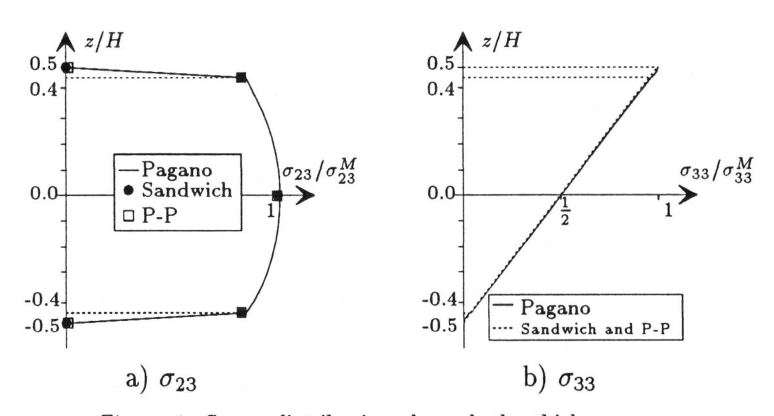

Figure 6. Stress distribution through the thickness

5. Conclusion

The two different approaches, the hybrid sandwich finite elements and the post-processing, have been presented and assessed. From that, we can draw the following conclusion:

- both methods give good results for displacements as well as for stresses: *i)* only required stress components are continuous at the interfaces and the force equilibrium state is fulfilled; *ii)* at the free edge of structures, the normal stress vector, $\sigma_{ij}n_j$, converges to zero.
  The hybrid sandwich finite elements are formulated without assumption of the theory of laminated plates. Even though they have a little higher number of degrees of freedom, they represent mechanical behaviours of the sandwich plates in a more realistic way.
- the post-processing methods permit to use directly the displacement results obtained by any existing FEM softwares.
  The computing time of the present post-processing methods is a little higher than the direct computation, which uses only shapes functions ($\{\sigma\} = [D][\mathcal{L}][N_u]\{q\}$) but has been shown to be numerically the worst, as explained in (Hinton, 1974). Most of FEM programs use more sophisticated methods, generally a local least square method (ANSYS for example) to compute nodal stresses. The present post-processing methods have the same computation time as these methods, but obtains the same accuracy with fewer elements.

We can emphasize that the two present approaches permit to calculate displacements and stresses in an accurate way, and that they can be easily implemented in existing FEM softwares.

References

Hinton, E. and Campbell, J. S. (1974) Local and global smoothing of discontinuous finite element functions using a least squares method, *Int. J. for Num. Meth. in Eng.* **8**, pp. 461–480.

Lerooy, J.-F. (1983) *Calcul des contraintes de cisaillement transversales dans les structures modérement épaisses*, PhD thesis, Institut National Polytechnique de Lorraine.

Manet, V. and Han, W.-S. (1997) La modélisation des plaques sandwich par éléments finis hybrides et ses applications, *Actes du troisième colloque national en calcul des structures* **2**, Presses académiques de l'Ouest, Nantes, pp. 657–663.

Pagano, N.J. (1970) Exact solutions for rectangular bidirectional composites and sandwich plates, *J. Composite Materials* **4**, pp. 20–34.

Teti, R. and Caprino, G. (1989) Mechanical behavior of structural sandwiches, *Sandwich Construction*, pp. 53–67.

Washizu, K. (1982) *Variational methods in elasticity and plasticity*, Pergamon Press.

Zienkiewicz, O.C. and Taylor, R.L. (1994) *The Finite Element Method* **1**, MacGraw-Hill, London.

A REFINED MULTILAYERED FEM MODEL APPLIED TO SANDWICH STRUCTURES

E. CARRERA AND F. NIGLIA
DIAS, Politecnico di Torino,
Corso Duca degli Abruzzi 24, 10129 Torino, Italy

Abstract. This paper presents a computationally efficient and mechanically accurate finite element formulation for linear and nonlinear analysis of sandwich flat panels. C^0-continuity as in Reissner-Mindlin model is preserved, zig-zag form for the in-plane displacement fields is allowed and interlaminar equilibrium at the interface between core and skins for the transverse shear stress components is accounted for. The numerical investigation assesses the proposed model by comparing it to Classical Lamination Theory, First order Shear Deformation Theory and other available results. The efficiency and accuracy, of the presented model is then demonstered by solving few problems dealing with linear and nonlinear bending, natural frequencies, in-plane buckling and postbuckling analysis.

1. Introduction

Sandwich panels have played and continue to play a considerable impact to design traditional as well as advanced ship, automotive and aerospace vehicles. They are multilayered structures consisting of one or more high-strength, stiff layers (faces), bonded to one or more low-density, flexible layers (core). The interest on the analysis of their static, dynamic, linear and nonlinear behavior and failure mechanisms is displayed by the enormous and relevant available literature concerning such a topic, few references are quoted at the attached list (Hallen, 1969–Noor *et al.*, 1996). As a very recent example in the fields of sandwich panels and shells we mention the review article by Noor, Burton and Bert (1996), where about 1300 papers were listed as references. To this paper the readers are addressed to provide more exhaustive overview and literature on the used classical and recent ap-

A. Vautrin (ed.), Mechanics of Sandwich Structures, 61–69.

proaches to analyze sandwich structures as well as numerical techniques to solve their related governing equations. The possibility of finding numerical solutions of the intricate equations governing their elastodynamic behavior is of prime interest to the structural analyst. Unfortunately closed form solutions can be obtained in very few cases mainly related to linear analysis, simple geometries, isotropic materials and particular boundary conditions (Srinivas and Rao, 1970). In the most general case the use of approximated solutions technique becomes mandatory. In this context the progress made in the last three decades by the computational mechanics, in particular by the finite element method FEM, helps very much to subjugate the above mentioned difficulties (Zienkiewicz, 1986).
In a recent research Carrera (1996) on the topics of efficient finite element formulation for accurate analysis of multilayered structures, the author has proposed a family of finite plate elements which preserve computational efficiency, include the so called *zig-zag form* in the thickness coordinate for the in-plane displacements, and fulfill *interlaminar equilibrium* of the transverse shear stresses. Two independent fields for the displacements (of Reissner-Mindlin type plus two zig-zag functions) and for the transverse shear stresses (parabolic in each layer) were assumed, see fig.1. C^0-continuity was retained as in classical Reissner-Mindlin preserving well know computational requirements (Zienkiewicz, 1986). Stress unknowns were eliminated *a priori* by writing a so called weak form of Hooke's law between the introduced stress and displacements unknowns. This model was denoted by the acronym RMZC (Reissner-Mindlin, Zig-zag, interlaminar Continuity). The developed finite elements were successfully applied to static and dynamic linear and geometrically nonlinear analysis of multilayered plates by Carrera and Kröplin (1997) and Carrera and Krause (1997). As further advantage, RMZC model leads to CLT (Classical Lamination Theory) and FSDT (First Shear Deformation Theory) as particular cases. The present work is intended to apply the RMZC model to sandwich structures in both field of linear and nonlinear analysis. The RMZC theory and related FEM formulation are described in Sec.2 and 3. Few problems concerning the linearized stability (free vibration and buckling), large deflection and postbuckling response are treated in Sec.4.

2. Description of the RMZC model

N_l denotes the number of layers, the integer k indicates the layer number, starting from the bottom of the plate. The letters x and y denote the plate middle surface coordinates. Ω is the x, y plate domain. The lamina are assumed to be homogeneous and orthotropic and the material is supposed to work in linear elastic field. For convenience, several reference coordinates are introduced along the plate thickness. z is the global coordinate along the plate thickness h. z_k ($z_k = z - z_{0k}$, z_{0k} denoting the distance of the middle surface of the k-layer from the $x-y$ plane) is the local coordinate along the k-layer thickness h_k. $\zeta = \frac{2z}{h}$ and $\zeta_k = \frac{2z_k}{h_k}$ are the non-dimensional global plate-coordinate and local k-layer-coordinate, respectively. A_k denotes the z_k-domain at the k-layer and θ_k the orientation with respect to the global x-axis. Stress, strain and displacement components along the global tri-orthogonal Cartesian system are indifferently denoted by subscripts $1, 2, 3$ or x, y, z. Non linearities of von Kármán type are retained in the present investigations and the normal strain component ϵ_{33} is discarded (Carrera and Kröplin, 1997). In order to meet the proposal of the mechanical model presented in the next subsections, both stiffness coefficient Q_{ij} $(i, j = 1, 2, 6)$ and compliance coefficient S_{ij} $(i, j = 4, 5)$ are used in the following form of Hooke's law:

$$\{\sigma^p\}_k = [Q_{pp}]_k\{\epsilon^p\}_k, \quad \{\epsilon^n\}_k = [S_{nn}]_k\{\sigma^n\}_k \tag{1}$$

$$[Q_{pp}]_k = \begin{bmatrix} Q_{11} & Q_{12} & Q_{16} \\ Q_{12} & Q_{22} & Q_{66} \\ Q_{16} & Q_{26} & Q_{66} \end{bmatrix}_k ; \quad [S_{nn}]_k = \begin{bmatrix} S_{44} & S_{45} \\ S_{45} & S_{55} \end{bmatrix}_k$$

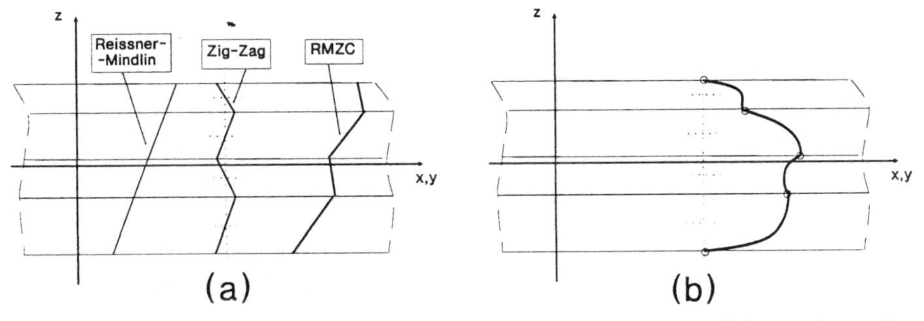

Figure 1. Assumed models along the thickness coordinate. (a) In-plane displacements u_1 and u_2; (b) transverse shear stresses σ_{13} and σ_{23}.

The compliance coefficients are: $S_{44}=\frac{Q_{55}}{\Delta}$; $S_{55}=\frac{Q_{44}}{\Delta}$; $S_{45}=-\frac{Q_{45}}{\Delta}$; $\Delta=Q_{44}Q_{55}-Q_{45}^2$. Further, $\{\sigma^p\}_k^T=\{\sigma_{11}^k,\sigma_{22}^k,\sigma_{12}^k\}$, $\{\epsilon^p\}^T=\{\epsilon_{11},\epsilon_{22},\epsilon_{12}\}$, $\{\sigma^n\}_k^T=\{\sigma_{13}^k,\sigma_{23}^k\}$, $\{\epsilon^n\}^T=\{\epsilon_{13},\epsilon_{23}\}$ are the in-plane and out-of-plane stress and strain components, respectively. In order to include the zig-zag effects, two zig-zag terms are added to the standard Reissner-Mindlin displacement model (see Fig.1).

$$\begin{aligned} u_i^k(x,y,z) &= U_i^0(x,y) + \tfrac{h}{2}\zeta U_i^1(x,y) + \zeta_k(-1)^k D_i(x,y) \\ u_3^k(x,y,z) &= U_3^0(x,y) \qquad i = 1,2 \end{aligned} \tag{2}$$

U_1^0, U_2^0 and U_3^0 are the displacement components of a point on the reference surface Ω of the plate. U_1^1 and U_2^1 denote the rotations of the normal to the reference surface in the planes $x-z$ and $y-z$, respectively. $\zeta_k(-1)^k D_1$ and $\zeta_k(-1)^k D_2$ are the zig-zag terms; these terms have the goal of reproducing the discontinuity of the first derivative along z. In order to guarantee transverse shear stress continuity at the interface between two adjacent layers, an independent stress field is introduced as in the following. The order of the z expansion for the transverse stresses $\{\sigma_n\}$ is established to be quadratic at each k-layer (see Fig.1).

$$\sigma_{i3}^k(x,y,z) = \sigma_{i3}^{kt}(x,y)F_0(z_k) + F_1(z_k)R_{i3}^k(x,y) + \sigma_{i3}^{kb}(x,y)F_2(z_k) \qquad (3)$$

Six k-dependent functions are used: the two stress resultants $R_{i3}^k(x,y) = \int_{A_k} \sigma_{i3}^k(x,y)dz$, and the four interface transverse stress values at the top and the bottom of the k-layer (denoted by σ_{13}^{kt}, σ_{23}^{kt} and σ_{13}^{kb}, σ_{23}^{kb}, respectively), with $F_0 = -\frac{1}{4} + \frac{\zeta_k}{2} + \frac{3}{4}\zeta_k^2$, $F_1 = \frac{3(1-\zeta_k^2)}{2h_k}$, $F_2 = -\frac{1}{4} - \frac{\zeta_k}{2} + \frac{3}{4}\zeta_k^2$. If the equilibria conditions at each interface have to be fulfilled, then the following set of boundary conditions must be linked to the introduced stress unknowns: $\sigma_{i3}^{(k+1)b} = \sigma_{i3}^{kt}$, $\sigma_{i3}^{1b} = \bar{\sigma}_{i3}^t$, $\sigma_{i3}^{N_l t} = \bar{\sigma}_{i3}^t$ Bars denote imposed transverse shear stresses. Thus *the assumed transverse stress model is capable of fulfilling both interlaminar equilibria and top/bottom-plate imposed transverse stress conditions.*

In order to eliminate the stress unknowns, reference is made to the method described by the first author in Carrera (1995) where the Mixed Reissner Variational Equation was referred to. As the two-dimensional model at Eqn.2 has been formulated with C^0 continuity an isoparametric description can be referred to. The same shape functions are used for the different unknowns,

$$\{U_1^0, U_2^0, U_3^0, U_1^1, U_2^1, D_1, D_2\} = [N]\{Q\} \qquad (4)$$

$[N]$ is a diagonal matrix whose elements are the shape functions and $\{Q\}$ is the vector of the element unknows.

3. Governing FEM equations

Let us consider a multilayered plate of volume V subjected to external mechanical loads in the dynamic case. The principle of virtual displacement states,

$$\delta L_m = \delta L_e + \delta L_{in} \qquad (5)$$

$\delta L_m = \int_V \delta\{\epsilon\}_k^T \{\sigma\}_k dV = \delta\{Q\}^T[\mathcal{K}_S]\{Q\}$ is the contribution to the virtual work coming from internal deformations; $\delta L_e = \delta\{Q\}^T\{P\}$ is the variation of the work done by the applied external loads; where $\{P\}$ is the load vector equivalent in the finite element sense to the applied loads. The work done by inertia forces is: $\delta L_{in}= - \int_V \rho_k \{\delta u\}_k^T \{\ddot{u}\}_k dV = -\delta\{Q\}^T[\mathcal{M}]\{\ddot{Q}\}$; where ρ_k is the mass-density of the k-layer, double dot denotes accelerations and $[\mathcal{M}]$ is the mass matrix. Explicit version of matrices $[\mathcal{K}]$ and $[\mathcal{M}]$ are given in in Carrera and Krause (1997) and Carrera and Kröplin (1997). Finally, the approximate form of equilibrium reads:

$$[\mathcal{K}_S]\{Q\} + [\mathcal{M}]\{\ddot{Q}\} \quad = \{P\} \tag{6}$$

TABLE 1. Convergence characteristics of the used model.
MAT I, $a/h = 10$, $R = 5$

Theory	mesh	$\sigma_{xx} \times m_4$	$\tau_{xz} \times m_4$	$u_3 \times m_5$
HOST2	2×2	61.03	3.259	257.78
	5×5	60.52	3.953	257.37
S.R.	-	60.35	4.364	258.97
CLT	-	61.14	4.589	216.94
FSDT	2×2	61.27	2.905	236.60
	5×5	60.79	3.566	236.30
RMZC	2×2	62.40	2.898	253.90
	5×5	61.42	3.541	251.10

By referring to standard assembly finite element procedure (Zienkiewicz, 1986), such equation can be formally interpreted as the dynamic equilibrium condition at global-structure level. In the static case, the governing equations are obtained by neglecting the inertia terms in the previous equation: $[\mathcal{K}_S]\{Q\} = \{P\}$, which consist in a nonlinear system of algebraic equations. This system is here solved approximately by Newton-Raphson linearization (Carrera, 1994, Carrera and Kröplin, 1997). The applied loadings are supposed to vary proportionally by means of a load parameter λ from a certain reference configuration $\{P_{rif}\}$: $\{P\} =\lambda\{P_{rif}\}$. The free vibrational behavior coming from the linearized dynamical equation is governed by the following system of equations: $[\mathcal{K}_l]\{Q\} + [\mathcal{M}]\{\ddot{Q}\} = 0$. By assuming an exponential solution vibrating with circular frequency ω in the time domain, the following eigenvalue problem is obtained:

$$|[\mathcal{K}_l] + \omega^2[\mathcal{M}]| \quad = 0 \tag{7}$$

where $|...|$ denotes determinant and $[\mathcal{K}_l]$ is the linear part of $[\mathcal{K}_S]$.
The classical linearized buckling equation of Euler-type (Zienkiewicz, 1986)

can be obtained by writing an eigenvalue problem to the tangent stiffness matrix and neglecting their displacement dependent parts,

$$||[\mathcal{K}_l] + [\mathcal{K}_{\sigma_l}]|| \quad = 0 \tag{8}$$

$[\mathcal{K}_{\sigma_l}]$ is the linear stress matrix. Such equation takes sense if and only if the prebuckling path does not include any transverse deflections.

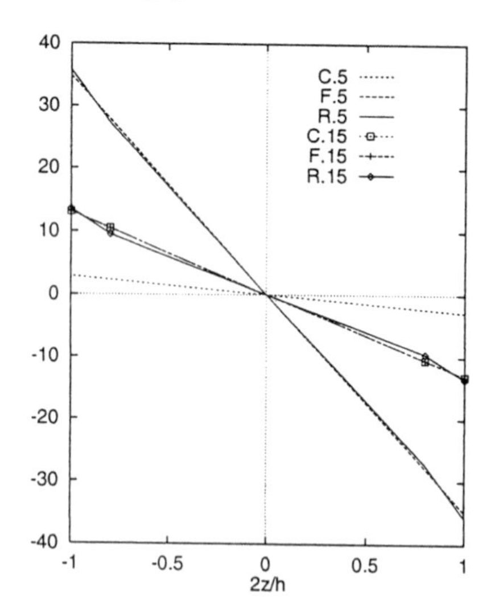

Figure 2. In-plane diplacements u_1 vs $2z/h$. MAT I, $\frac{a}{h}$=10 R=5, 15.

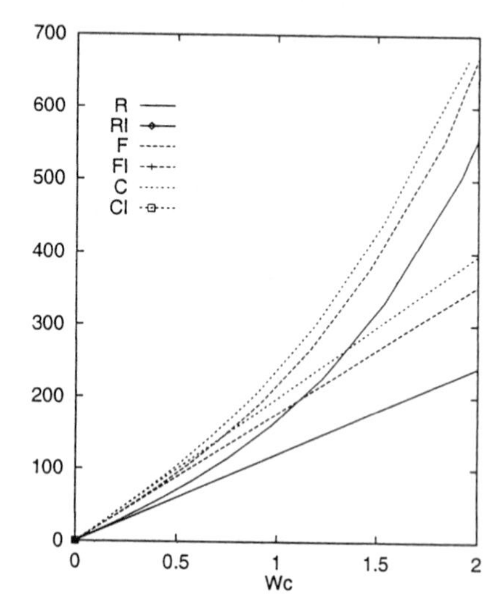

Figure 3. Large deflection behavior. Load parameter $\lambda \times a^4$ vs transverse deflections u_3.

4. Numerical Results and Discussion

The introduced coefficients, geometrical and mechanical data and boundary conditions and related acronyms are the same as described in Pandya and Kant (1988). Two material are considered: MAT I ($E_l = 1$, $E_t = 0.52$, $\nu = 0.44$, $G_{lt} = 0.26$, $G_{tt} = 0.26$); MAT II(faces: $E_l = 18$,$E_l = 2$,$\nu = 0.24$ $G_{lt} = 0.19$,$G_{tt}0.72$, core:$E_l = 1.1$,$E_t = 1.1$,$\nu = 0.21$ $G_{lt} = 0.19$,$G_{tt} = 0.75$) MAT I will be mainly used in conjunction with several values of the stiffness ratio R=E_{l_f}/E_{l_c} between the longitudinal Young moduli of facings and core. Consistent units will be applied in all the performed calculations. a and b will denote the plate length in x and y directions, respectively. FSDT results always refer to the value χ=5/6 of the shear correction factor. CLT results have been obtained by implementation of a penalty technique on χ (Carrera, 1996). Obviously, RMZC model does not require any shear correction factor. Nine node isoparametric plate elements have been used everywhere with 4×4 meshes. Selective integration technique (Zienkiewicz, 1986) has been implemented to contrast shear locking phenomenon. Whether not differently written it is intended that the transverse shear stresses are calculated via integration of the 3D indefinite equilibrium equations. Exception made for the postbuckling results where a point transverse load is applied with correspondence to the plate center, the sandwich plates are always bended by uniform distribution of transverse pressure q. Unit values of the reference in-plane loading $\{P_{rif}\}$ are considered. The parameter used in the ordinates and abscissa of the several diagrams will be directly quoted in their captions.

TABLE 2. Frequency parameter and normalized buckling loadings
MAT I, $\bar{\omega}$: a/b=1,b/h=10 ;K_x: a/b=5, b/h=10

R	$\bar{\omega} \times 10^2$				K_x			
	Present Analysis			Exact	Present Analysis			Exact
	CLT	FSDT	RMZC		CLT	FSDT	RMZC	
5	8.806	8.150	7.818	7.7148	5.349	4.636	4.214	4.046
	(14.14)	(5.654)	(1.345)		(32.20)	(14.58)	(4.15)	
10	11.90	10.62	9.947	9.8104	6.281	5.331	4.212	4.200
	(21.31)	(8.333)	(1.399)		(49.54)	(26.94)	(0.308)	
15	14.34	13.08	11.40	11.203	6.7230	5.651	4.019	4.037
	(28.03)	(16.77)	(1.767)		(66.53)	(39.99)	(-0.441)	

Converge characteristics of a bended thick plate by uniformly distributed pressure loads are presented in Tab.1 ($m_4 = \frac{1}{q}$,$m_5 = \frac{1}{hq}$). Conducted analyses are compared to the exact solution by Srinivas and Rao (1970) and to the cubic higher order models by Pandya and Kant (1988). These last models even though account for a better description of plate deformability with respect to FSDT analysis, they do not include interlamina continuity neither zig-zag. Several meshes are considered. Maximum transverse displacement and both in-plane stress and transverse shear stress components are compared. As in Srinivas and Rao (1970), the location of the written values coincides to the the plate points where they assume their maximum values. The good performance of the RMZC element with respect to CLT and FSDT is registered. Better description and convergence rate are obtained for the in-plane characteristics with respect to transverse shear evaluation. Two values of the relative stiffness ratio R are considered in Fig.2 for the distribution in the thickness direction z. CLT, FSDT and RMZC values are compared. Where available, 3D solutions have been quoted too. The in-plane displacement component u_1 has been traced. It becomes evident in this diagram the capability of the RMZC model to include the zig-zag effect. As expected, such effect is more evident for plates corresponding to higher R-value. Few results on buckling and vibration of sandwich panels are given in Tab.2 that compares CLT,FSDT,RMZC results to 3D exact so-

lutions. The fundamental frequency parameter $\bar{\omega} = \omega\sqrt{\rho_c h^2/E_{lc}}$ (subscript c denotes values referred to the core) and the values of the axial buckling load in the x-direction $K_x = \frac{12}{\pi^2}\frac{b^2}{h^2}\frac{P_{cr}}{E_l}$ (P_{cr} is the buckling load) are quoted for three value of the thickness parameter. The error with respect to 3D solution is quoted in brackets. The reliability of the RMZC model to analyze linearized stability behavior of sandwich plate is then confirmed.
Large deflection behavior (in von Kármán sense) has been investigated for the analysis in Figs.3-4. The load parameter $\lambda \times a^4$ vs maximum plate deflection is traced in Fig.3 (MAT I, R=25, $\frac{a}{h}$=10 values are treated) and linear (Rl,Fl,Cl) and nonlinear (R,F,C) results are compared. Deflection increasing the importance of nonlinear analysis becomes evident. The distribution of the in-plane displacements u_1 in the thickness direction are traced in Fig.4 (MAT I, $\frac{a}{h}$=10 and $\lambda \times a^4$=100 values are used). Linear analysis is also traced in this figure. Very different results for the three theories are obtained. The importance of the zig-zag effect is evident.

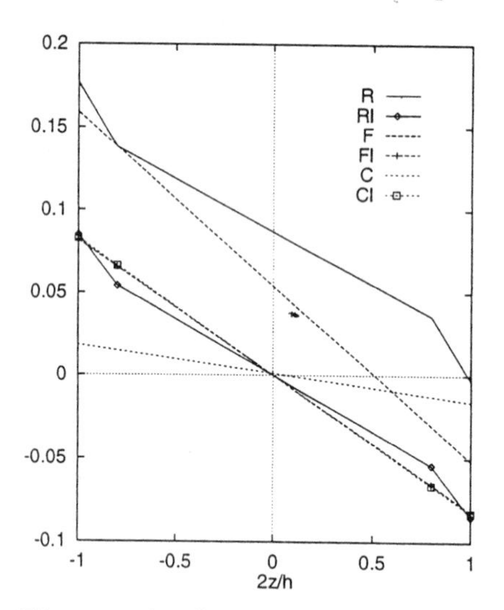

Figure 4. Large deflection behaviour. In-plane displacements u_1 vs $2z/h$.

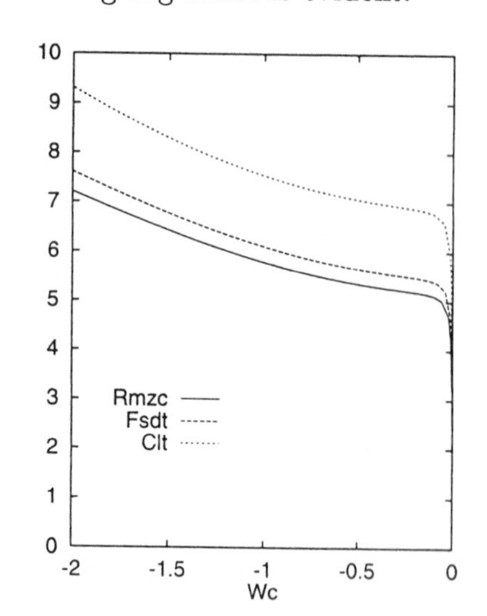

Figure 5. Postbuckling behaviour. Load parameter $\lambda \times a$ vs center deflection u_3.

Results related to the conducted investigation on the postbuckling behavior have been traced in Fig.5 (SC boundary condition and $\hat{P} = 0$ and $P_d = 5 \times 10^-3$ loadings are reffered). In-plane axial loadings have been applied. Buckled paths have been calculated through application of a transverse disturb load P_d at the plate center. Simply supported/clamped plates subjected to biaxial loading are considered to notice the effectiveness of the developed nonlinear model to trace the postbuckling behavior of in-plane compressed sandwich plates. It is confirmed that RMZC model leads to more deformable plates with respect to FSDT and CLT results.

The few presented investigations have shown that RMZC analysis, in conjunction to finite element procedure, consist in a very efficient and reliable tool to analize linear and non linear response of sandwich panels.

References

Hallen, H.G. (1969) *Analysis and design of structural sandwich panels*, Pergamon Press LTD.

Rao,R.S. (1986) Large deflection and non linear vibration of multilayered sandwich plates, *AIAA Journal* **15**, 876-895.

Srinivas, S., and Rao, A.K. (1970) Bending,vibration and buckling of simply supported thick orthotropic rectangular plates and laminates, *International Journal Solid Structures* **6**, 1463-1481.

Pandya, N.B., and Kant, T. (1988) High order shear deformable theories for flexure of sandwich plates-finite element evaluation, *International Journal Solid Structures* **24**, 1267-1286.

Burton, W.S., and Noor, A.K. (1994) Three-dimensional solutions for thermomechanical stresses in sandwich panels and shells, *Journal Engineering Mechanics ASCE* **120**, 2044-2071.

Noor, A.K., Burton, W.S., and Bert, C.W. (1996) Computational models for sandwich panels and shells, *Applied Mechanics Review* **9**, 155–199.

Zienkiewicz, O.C. (1986) *The Finite Element Method*, McGraw Hill Book Company, 1986.

Carrera, E. (1996) C^o Reissner-Mindlin multilayered plate elements including zig-zag and interlaminar stresses continuity, *International Journal Numerical Methods Engineering* **39**, 1797–1820.

Carrera, E., and Kröplin, B. (1997) Zig-Zag and interlaminar equilibria effects in large deflection and postbuckling analysis of multilayered plates, *Mechanics of Composite Materials and Structures* **4**, 69–94

Carrera, E., and Krause, H. (1997) An investigation on nonlinear dynamics of multilayered plates accounting for C_z^0 requirements, *CEAS Forum, Rome*, **III**, 183–192

Carrera E. (1995) A class of two-dimensional theories for anisotropic multilayered plates analysis, *Atti Accademia delle Scienze Torino*, **19–20**, 49-87

Carrera, E. (1994) A study on arc-length methods and their operation failures illustrated by a simple model, *Computers & Structures* **50**, 217-230.

A REFINED SHEAR-DEFORMATION SANDWICH FINITE ELEMENT

O. POLIT*-** & M.TOURATIER*
*LM^{2}S - URA CNRS 1776 - ENSAM - 151 Bd de l'Hopital - 75013 Paris
**Université Paris X - IUT - Dép. GMP - 1 Chemin Desvallières - 92410 Ville d'Avray

Abstract. This paper presents a new C^1 81-degrees-of-freedom triangular finite element for geometrically and materially linear elastic multilayered composite, moderately thick plates. The element has six nodes and is of triangular shape. It is based on a new kind of kinematics and built from Argyris interpolation for bending, and Ganev interpolation for membrane displacements and transverse shear rotations. This kinematics allows to exactly ensure, the continuity conditions for displacements and transverse shear stresses at the interfaces between layers of a laminated structure and the boundary conditions at the upper and lower surfaces of the plates. The representation of the transverse shear strains by cosine functions allows to avoid shear correction factors. The element performances are evaluated on some standard tests and also in comparison with an exact three-dimensional solution for a sandwich plate in statics and dynamics.

1. Introduction

The aim of this work is to analyse the mechanical behavior of multilayered structures by plate finite elements including transverse shear effects and continuity requirements between layers in order to predict displacements and stresses of such structures for design applications. Many of the existing analysis methods for multilayered anisotropic plates are direct extensions of those developed earlier for homogeneous isotropic and orthotropic plates, see Noor (NOO89). In fact, many approaches utilize a displacement field which does not account for the contact requirement at the interfaces of multilayered structure. A synthesis is given in Reference (RED89). A finite

A. Vautrin (ed.), Mechanics of Sandwich Structures, 71–78.

element of triangular shape has been proposed by Di Sciuva (DIS95) based on a third-order refined shear deformation theory satisfying interlaminar continuity. In this last work, piecewise linear functions are introduced using the Heaviside operator.

We present a new C^1 plate finite element based on a refined kinematic model, see Touratier (TOU91), incorporating :

- the transverse shear strains with cosine distributions,
- the continuity conditions between layers of the laminate for both displacements and transverse shear stresses,
- the satisfaction of the boundary conditions at the top and bottom surfaces of the plate, without shear correction factors and using only five independent generalized displacements (three translations and two rotations).

The element has a triangular shape and the generalized displacements are approximated by higher-order polynomials based on :

1. Argyris (ARG68) interpolation for the transverse normal displacement,
2. Ganev (GAN80) interpolation for the membrane displacements and for the transverse shear rotations.

2. The displacement field for laminated plates

Let $(x_1, x_2, x_3 = z)$ denote the cartesian co-ordinates such that x_1 and x_2 are in the midplane $(z = 0)$ of the plate, while z is the transverse normal co-ordinate. We denote by ${u_i}^{(k)}(x_1, x_2, z)$, $i \in \{1, 2, 3\}$ the cartesian components of the displacement field for the kth layer of a multilayered plate, and we suppose that the transverse normal strain denoted by ϵ_{33} is negligible, according to the moderately thick plate hypothesis. Since the material behavior is admitted linearly elastic and this work is limited to small disturbances (small displacements), strains and stresses are classically denoted by ${\epsilon_{ij}}^{(k)}$ and ${\sigma_{ij}}^{(k)}$.

From (BEA93), we assume the following distribution of transverse stresses in the kth layer :

$$\begin{aligned}
{\sigma_{13}}^{(k)}(x_1,x_2,z) = & \left({\bar{C}_{55}}^{(k)}\left(f'(z) - \frac{h}{\pi}b_{55}f''(z)\right) + {a_{55}}^{(k)}\right){\gamma_1}^0(x_1,x_2) + \\
& \left({\bar{C}_{45}}^{(k)}\left(f'(z) - \frac{h}{\pi}b_{44}f''(z)\right) + {a_{54}}^{(k)}\right){\gamma_2}^0(x_1,x_2) \\
{\sigma_{23}}^{(k)}(x_1,x_2,z) = & \left({\bar{C}_{45}}^{(k)}\left(f'(z) - \frac{h}{\pi}b_{55}f''(z)\right) + {a_{45}}^{(k)}\right){\gamma_1}^0(x_1,x_2) + \\
& \left({\bar{C}_{44}}^{(k)}\left(f'(z) - \frac{h}{\pi}b_{44}f''(z)\right) + {a_{44}}^{(k)}\right){\gamma_2}^0(x_1,x_2)
\end{aligned} \tag{1}$$

In these equations $f(z) = \frac{h}{\pi} \sin \frac{\pi z}{h}$ and $f'(z) = df(z)/dz$; h is the thickness of the plate ; ${\gamma_1}^0$ and ${\gamma_2}^0$ are the transverse shear strains at $z = 0$; ${\bar{C}_{ij}}^{(k)}$ are the modulii of the material for the kth layer taking into account of the zero transverse normal stress hypothesis. The constitutive law is expressed as :

$$[\sigma^{(k)}] = [\bar{C}^{(k)}][\epsilon^{(k)}]$$

$$\text{with} \left\{ \begin{array}{lll} {\bar{C}_{ij}}^{(k)} & = & {C_{ij}}^{(k)} - {C_{i3}}^{(k)}{C_{j3}}^{(k)}/{C_{33}}^{(k)} \ \text{ for } i,j = 1,2,6 \\ {\bar{C}_{ij}}^{(k)} & = & {C_{ij}}^{(k)} \ \text{ for } i,j = 4,5 \end{array} \right. \tag{2}$$

In equation (2), ${C_{ij}}^{(k)}$ are three-dimensional moduli of the material for the kth layer, and the equations (1) and (2) account for layers having orthotropic axes oriented at various angles with respect to the plate axes.

Otherwise, $\epsilon_{33} = 0$ allows to write ${u_3}^{(k)}(x_1, x_2, z) = u_3(x_1, x_2, z) = v_3(x_1, x_2)$.

We have on the one hand transverse shear strain definitions and on the other hand the constitutive law relating transverse shear strains to transverse shear stresses. Equating those two expressions for strains, introducing the above assumed transverse shear stresses and integrating with respect to the z co-ordinate gives the shear bending part of the displacement field. Finally, adding membrane displacements, it follows :

$$\left\{ \begin{array}{lll} {u_1}^{(k)} & = & v_1 - z v_{3,1} + \left(f_1 + {g_1}^{(k)}\right)(v_{3,1} + \theta_1) + {g_2}^{(k)}(v_{3,2} + \theta_2) \\ {u_2}^{(k)} & = & v_2 - z v_{3,2} + {g_3}^{(k)}(v_{3,1} + \theta_1) + \left(f_2 + {g_4}^{(k)}\right)(v_{3,2} + \theta_2) \\ u_3 & = & v_3 \end{array} \right. \tag{3}$$

Functions $f_1, f_2, {g_1}^{(k)}, \ldots, {g_4}^{(k)}$ are immediately deduced from the equations (1) and the above integration performed with respect to the z coordinate. They depend on coefficients $(b_{ij}, {a_{ij}}^{(k)})$ and on functions $f(z)$ and $f'(z)$. Moreover, coefficients $(b_{ij}, {a_{ij}}^{(k)})$ are determined from the boundary conditions on the top and bottom surfaces of the plate, and from the continuity requirements at the layer interfaces for displacements and stresses, see Béakou (BEA93). Hereafter, the superscript (k) for ${u_\alpha}^{(k)}$ components is deleted in order to simplify the finite element description of the model.

3. The triangular six node finite element for semi-thick laminates

The discrete formulation of the boundary value problem in linear elasticity is classically deduced from the following functional :

$$a(\vec{u}^h, \vec{u}^{*h})_{\cup\Omega_e} = f(\vec{u}^{*h})_{\cup\Omega_e} + F(\vec{u}^{*h})_{\cup\mathcal{C}_e} , \quad \forall \vec{u}^{*h} \tag{4}$$

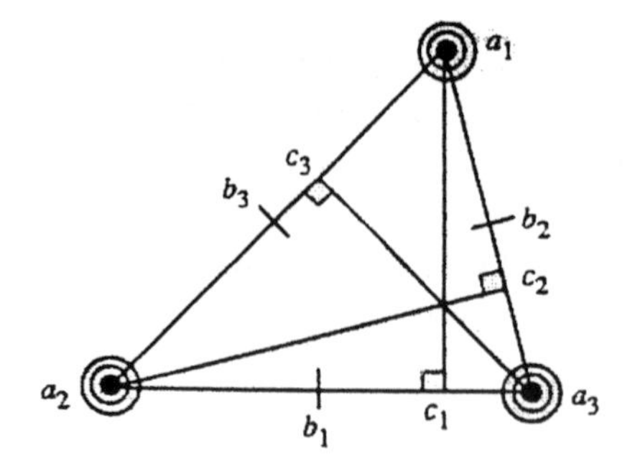

$$P_K = P_5(K)\ ;\ dim P_K = 21\ ;$$

$$\Sigma_K = \{p(a_i),\ Dp(a_i)(a_{i-1} - a_i),\ Dp(a_i)(a_{i+1} - a_i),\ 1 \leq i \leq 3\ ;$$

$$D^2p(a_i)(a_{j+1} - a_{j-1})^2,\ 1 \leq i,j \leq 3\ ; Dp(b_i)(a_i - c_i),\ 1 \leq i \leq 3\}$$

Figure 1. set Σ_K of the local degrees of freedom of a function p for Argyris triangle

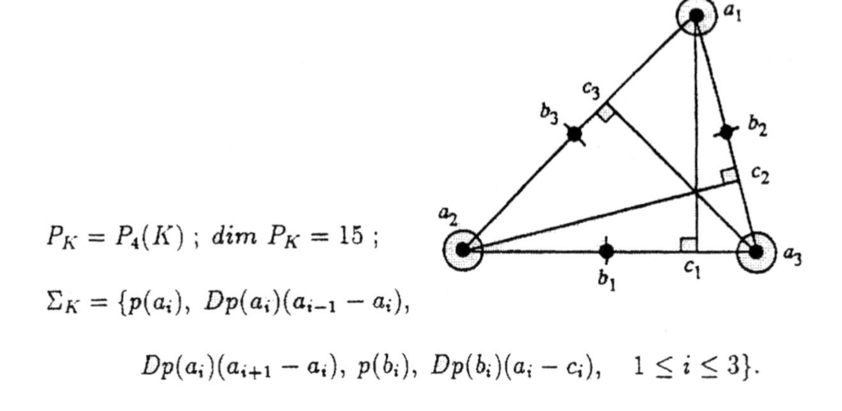

$$P_K = P_4(K)\ ;\ dim\ P_K = 15\ ;$$

$$\Sigma_K = \{p(a_i),\ Dp(a_i)(a_{i-1} - a_i),$$

$$Dp(a_i)(a_{i+1} - a_i),\ p(b_i),\ Dp(b_i)(a_i - c_i),\quad 1 \leq i \leq 3\}.$$

Figure 2. set Σ_K of the local degrees of freedom of a function p for Ganev triangle

where $\cup\Omega_e$ is the triangulation of the multilayered structure and $\cup\mathcal{C}_e$ is the edge of the meshed structure. In addition, $\vec{u}^h$ is the finite element approximation of the displacement field $\vec{u}$ given above by equation (3) and $\vec{u}^{*h}$ is the finite element approximation of the corresponding virtual velocity field $\vec{u}^*$. Linear functions f and F in equation (4) represent the body (including inertia terms) and surface external loads, actually surface and line loads respectively due to the integration performed throughout the thickness of the plate. The superscript h introduced in the equation (4) indicates the finite element approximation. Then, it is also used for finite element approximation of the generalized displacements in (3), denoted by ${v_i}^h$ and ${\theta_\alpha}^h$

where $i = 1, 2, 3$ and $\alpha = 1, 2$. Thus, the finite element approximation of functions v_i and θ_α are briefly indicated hereafter. It is recalled that the element is of triangular shape, and from (3) continuity C^1 is required for v_3.

The elementary stiffness matrix is obtained by computing at the element level the bilinear form given in (4) as :

$$a(\vec{u}^h, \vec{u}^{*h})_{\Omega_e} = \int_{\Omega_e} \left[E_e^{*h}\right]^T [A_e] \left[E_e{}^h\right] d\Omega_e \tag{5}$$

where $[A_e]$ is the material behavior matrix for a multilayered finite element resulting of the integration with respect to the thickness co-ordinate, and $\left[E_e{}^h\right]$ is the following vector :

$$\left[E_e{}^h\right]^T = \Big[\ v_1{}^h{}_{,1} \quad v_1{}^h{}_{,2} \ \vdots\ v_2{}^h{}_{,1} \quad v_2{}^h{}_{,2} \ \vdots\ v_3{}^h{}_{,1} \quad v_3{}^h{}_{,2} \quad v_3{}^h{}_{,11} \quad v_3{}^h{}_{,22} \quad v_3{}^h{}_{,12} \ \vdots\ \theta_1{}^h \quad \theta_1{}^h{}_{,1} \quad \theta_1{}^h{}_{,2} \ \vdots\ \theta_2{}^h \quad \theta_2{}^h{}_{,1} \quad \theta_2{}^h{}_{,2}\ \Big] \tag{6}$$

The matrix $\left[E_e^{*h}\right]$ is defined by an analogous expression introducing the superscript *. The form given by (6) for the vector $\left[E_e{}^h\right]$ may be seen as the vector of generalized strains for the proposed model. So, we must now explicit the interpolation for each approximated generalized displacement and its derivatives appearing in (6). In Figure 1 (from (BER94)) local degrees of freedom for the interpolation of the $v_3{}^h$ function are given, and Argyris interpolation for $v_3{}^h$ obtained from this set of degrees of freedom is P_5. In the same way, Figure 2 (from (BER94)) gives the set of local degrees of freedom for the other generalized displacements $v_\alpha{}^h$ and $\beta_\alpha{}^h$. Ganev interpolation constructs from this set is P_4. In these two Figures, the degrees of freedom in term of derivative are local because they are expressed with respect to directions associated with the edges of the triangle.

So, the discrete form of the vector $\left[E_e{}^h\right]$ can be written as :

$$\left[E_e{}^h\right] = [\Lambda]\,[T_e]\,[D_e]\,[Q_e] \tag{7}$$

where $[Q_e]$ is the vector of degrees of freedom in global co-ordinates, $[D_e]$ is a transformation matrix between local and global co-ordinates taking into account of the local derivative degrees of freedom. Finally, in the equation (7) the matrix product $[\Lambda]\,[T_e]$ gives interpolations for each component of the vector $\left[E_e{}^h\right]$ in terms of all barycentric monomial terms for the matrix $[\Lambda]$ and constant coefficients for the matrix $[T_e]$.

Then, it is evident from (5) and (7) that the stiffness matrix is obtained as :

$$[K_e] = \int_{\Omega_e} [D_e]^T [T_e]^T [\Lambda]^T [A_e][\Lambda][T_e][D_e] d\Omega_e \tag{8}$$

The elementary mass matrix $[M_e]$ is given after an integration with respect to the thickness as follow :

$$\int_{\Omega_e} \left[U_e^{*h}\right]^T [I_e] \left[\ddot{U}_e^{\,h}\right] d\Omega_e = [Q_e^*]^T [M_e] \left[\ddot{Q}_e\right] \tag{9}$$

where $[I_e]$ is the inertia matrix, $[\,\ddot{}\,] = \partial^2[\ \ ]/\partial t^2$, t is the time ; $\left[U_e^{\,h}\right]$ is the vector of generalized displacements deduced from the equation (3) and given by :

$$\left[U_e^{\,h}\right]^T = \left[\begin{array}{ccccccc} v_1^{\,h} & v_2^{\,h} & v_3^{\,h} & v_3^{\,h}{}_{,1} & v_3^{\,h}{}_{,2} & \theta_1^{\,h} & \theta_2^{\,h} \end{array}\right] \tag{10}$$

Finally, $\left[U_e^{*h}\right]$ is the vector of generalized virtual velocity consistently associated to (10).

Construction of the load vector does not need to be presented here as it is classic. The elementary matrices are integrated using a 16 point integration rule which integrates exactly eight order polynomials, see (DUN85).

4. Numerical evaluation of the element

This new finite element exhibits a very good behavior : six zero eigenvalues, no shear locking when the plate becomes thin and very thin, no sensitivity to mesh orientation with few elements so triangulation orientation doesn't need to be precised, very good convergence properties.

Sandwich plate in statics : the Srinivas' problem, see (SRI70), is considered to evaluate the performances of the element to compute deflection and stresses for a thick three-layered plate. The plate is simply supported and submitted to a transverse normal uniform load ($f_3 = -1.$ SI (International metric System)). The material properties of the three-layered (sandwich) square plate are given in (SRI70) and the skin by core ratio is $\beta = 15$ while geometric characteristics of the plate are :

- length of the side $a = 10$ SI,
- total thickness $e = 1.$ SI,
- thickness of the skin (symmetric plate) $e_s = 0.1$ SI,
- thickness of the core $e_c = 0.8$ SI.

Table 1 shows results deduced from the proposed triangular finite element. It is clear for deflection and stresses that the proposed model gives

TABLE 1. Deflections and stresses for an orthotropic sandwich plate under an uniform load

value for $N = 4$	ref. value (SRI70)	Present
$v_3(a/2, a/2, 0)E_{x_1}/p_0$	121.72	121.88
$\sigma_{11}(a/2, a/2, z)/p_0$		
top skin at top surface	66.787	66.742
top skin at interface	48.299	48.215
core at upper interface	3.2379	3.2143
core at lower interface	−3.2009	−3.2143
bottom skin at interface	−48.028	−48.215
bottom skin at bottom surface	−66.513	−66.742
$\sigma_{22}(a/2, a/2, z)/p_0$		
top skin at top surface	46.424	46.581
top skin at interface	34.955	35.109
core at upper interface	2.4941	2.3406
core at lower interface	−2.3476	−2.3406
bottom skin at interface	−35.353	−35.109
bottom skin at bottom surface	−46.821	−46.581
$\sigma_{13}(0, a/2, z)/p_0$		
top skin at top surface	0.0000	0.0000
top skin at interface	3.9559	3.5542
core at upper interface	3.9559	3.5542
at mid surface	3.9638	4.0841
core at lower interface	3.5768	3.5542
bottom skin at interface	3.5768	3.5542
bottom skin at bottom surface	0.0000	0.0000

very accurate results in comparison with the exact three-dimensional elasticity solution. Computations in Table 1 have been achieved using a $N = 4$ mesh (32 triangles) in a quarter of the plate. Convergence properties for deflection is reached only with two triangular elements in a quarter of the plate.

Sandwich plate in dynamics : still the above sandwich square plate is studied in dynamics for free vibrations in case of all simply supported edges. The non-dimensional fundamental frequency obtained is 0.11202, while the exact three-dimensional value from (SRI70) is 0.11203. The finite element result is obtained with only two elements in a quarter of the sandwich plate, and refined meshes give the same value.

5. Final remarks

In this paper a new conform six node multilayered triangular finite element has been presented to analyze the behaviour of composite multilayered structures. The novelty of this work is in using a refined shear deformation theory including interlaminar continuity, for displacements and transverse shear stresses, and exactly satisfying the boundary conditions at the top and bottom surfaces of a multilayered plate structure. In addition, the way to interpolate the generalized displacements using Argyris and Ganev interpolations assures the field compatibility for transverse shear strains. It also may be notice that all stresses are continuous at the corner nodes of two adjacent elements.

The element has good properties in the field of finite elements, and gives very good results compared to the exact three-dimensional elasticity solution for a sandwich plate.

References

A.K. Noor and W.S. Burton. Stress and free vibration analysis of multilayered composite plates. *Comp. and Struc.*, 11 (3):183–204, 1989.

J.N. Reddy. On refined computational models of composite laminates. *Int. Jour. Num. Meth. Eng.*, 27:361–382, 1989.

M. Di Sciuva. A third order triangular multilayered plate finite element with continuous interlaminar stresses. *Int. Jour. Num. Meth. Eng.*, 38:1–26, 1995.

M. Touratier. An efficient standard plate theory. *Int. J. Eng. Sci.*, 29:901–916, 1991.

J.H. Argyris, I. Fried, and D.W. Scharpf. The tuba family of plate elements for the matrix displacement method. *Aero. J. Royal Aeronaut. Soc.*, 72:701–709, 1968.

H.G. Ganev and Tch.T. Dimitrov. Calculation of arch dams as a shell using ibm-370 computer and curved finite elements. In *Theory of shells*, pages 691–696. North-Holland, Amsterdam, 1980.

A. Béakou and M. Touratier. A rectangular finite element for analysing composite multilayered shallow shells in statics, vibration and buckling. *Int. Jour. Num. Meth. Eng.*, 36:627–653, 1993.

M. Bernadou. *Méthodes d'Eléments Finis pour les Problèmes de Coques Minces.* Collection R.M.A. Masson, Paris, 1994.

D.A. Dunavant. Hight degree efficient symmetrical gaussian quadrature rules for the triangle. *Int. Jour. Num. Meth. Eng.*, 21:1129–1148, 1985.

S. Srinivas and A.K. Rao. Bending, vibration and buckling of simply supported thick orthotropic rectangular plates and laminates. *Int. J. Solids Struc.*, 6:1463–1481, 1970.

COMPARISON OF THREE SHEAR-DEFORMATION THEORIES IN THE NON-LINEAR ANALYSIS OF SANDWICH SHELL ELEMENTS

António J.M. Ferreira, A.Torres Marques, J. C. de Sá
Departamento de Engenharia Mecânica e Gestão Industrial,
Faculdade de Engenharia da Universidade do Porto, Rua dos Bragas,
4099 Porto Codex, Portugal

1. Introduction

Sandwich shell structures are tipically found in may structural applications. The correct modelling of its structural behaviour is of relevant interest. Shear-deformation effects are always present in sandwich shells due to the difference between the core and skin characteristics. In this work it is formulated and compared a 1st order, a 3rd order and a layerwise shear deformation theories in the geometric and material nonlinear range. In the first two theories both translational and rotational degrees of freedom are laminate dependent, while in the layerwise theory the rotational degrees of freedom are layer dependent. This last theory produces constant shear deformations in each layer, but different from one layer to another. In the 1st order theory the shear deformations are constant throughout the laminate. In the 3rd order theory, parabolic deformations are directly achieved. The finite element discretisation is made through the degenerated shell element, known as the Ahmad-Irons-Zienkiewicz element [1]. This element has proved to be very good in the analysis of not only arbitrary isotropic shells [2-6], but also composite layered shells [7-14]. Some modifications in the element basic matrices were made in order to follow the new kinematics according to each theory. Some examples are presented in order to discuss the performance of such theories in the analysis of sandwich shells.

2. Shear deformation theories

The displacement field for the first order shear deformation theory is based on the work of Mindlin [15] and is expressed for a plate as

$$\begin{aligned} u(x,y,z) &= u_o(x,y) + z\theta_x(x,y) \\ v(x,y,z) &= v_o(x,y) + z\theta_y(x,y) \\ w(x,y,z) &= w_o(x,y) \end{aligned} \tag{1}$$

where u_o and v_o are the middle surface displacements and θ_x , θ_y are the rotations of a normal. The strain-displacement relations are obtained as

$$\begin{bmatrix} \varepsilon_x \\ \varepsilon_y \\ \gamma_{xy} \end{bmatrix} = \begin{bmatrix} \varepsilon_x^o \\ \varepsilon_y^o \\ \gamma_{xy}^o \end{bmatrix} + z \begin{bmatrix} \kappa_x \\ \kappa_y \\ \kappa_{xy} \end{bmatrix} = \underline{\varepsilon}^o + z\underline{\kappa} \tag{2}$$

$$\begin{bmatrix} \gamma_{xz} \\ \gamma_{yz} \end{bmatrix} = \begin{bmatrix} w_{,x} + \theta_x \\ w_{,y} + \theta_y \end{bmatrix} \tag{3}$$

where

$$\begin{aligned} &\varepsilon_x^o = u_{o,x} \ , \ \varepsilon_y^o = v_{o,y} \ , \ \gamma_{xy}^o = v_{o,x} + u_{o,y} \\ &\kappa_x = \theta_{x,x} \ , \ \kappa_y = \theta_{y,y} \ , \ \kappa_{xy} = \theta_{x,y} + \theta_{y,x} \end{aligned} \tag{4}$$

A. Vautrin (ed.), Mechanics of Sandwich Structures, 79–88.

are the membrane strains and the plane curvatures. In each kth layer, the stress-strain relations are expressed as

$$\begin{bmatrix}\sigma_{11}\\ \sigma_{22}\\ \tau_{12}\end{bmatrix}_k = \begin{bmatrix}c_{11} & c_{12} & c_{13}\\ c_{12} & c_{22} & c_{23}\\ c_{13} & c_{23} & c_{33}\end{bmatrix}_k \begin{bmatrix}\varepsilon_{11}\\ \varepsilon_{22}\\ \gamma_{12}\end{bmatrix}_k$$

$$\begin{bmatrix}\tau_{13}\\ \tau_{23}\end{bmatrix}_k = \begin{bmatrix}c_{44} & c_{45}\\ c_{45} & c_{55}\end{bmatrix}_k \begin{bmatrix}\gamma_{13}\\ \gamma_{23}\end{bmatrix}_k \tag{5}$$

where c_{ij} are material coeficients to be defined later. In figure 1 the first order shear deformation theory is illustrated.

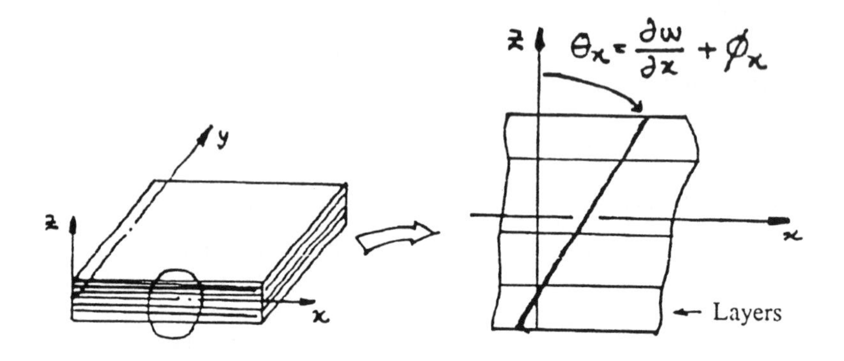

Figure 1 - First order shear deformation theory, one dimensional case.

Higher order shear deformation theories avoid the use of shear correction terms, through the use of higher order terms in the displacement field, particularly in the thickness direction. Several theories where published [16-24]. In this work it is used the Kant [23,24] approach for plates, with a implementation in the degenerated shell element. The third order theory that it is used in this work is illustrated in figure 2.
Several layerwise theories were proposed before [25,26] for plate analysis. The layerwise theory that was adopted in this work uses a rotation field for each layer [12]. Due to large number of unknowns that a general laminate would imply, this theory is restricted to a three or four layer sandwich laminated. The first order displacement field is imposed in each layer, with displacement continuity at the interfaces. In figure 3 the theory is illustrated.
The degenerated shell element that was used in this work is illustrated in figure 4. It is a general shell element, that can acomplish thickness variation, material nonlinearities and laminate stacking [1,7-9,12].

3. Displacement field for the first order shear deformation theory

In the shell element the displacements are obtained, for the first order shear deformation theory, as a function of three translational degrees of freedom and two local rotations about a normal to the middle surface, as can be seen in figure 4.

$$\underline{u} = \sum_{k=1}^{n} \underline{N}_k \underline{u}_k^{med} + \sum_{k=1}^{n} \underline{N}_k \zeta \frac{h_k}{2} [V_{1k}, -V_{2k}] \begin{Bmatrix} \beta_{1k} \\ \beta_{2k} \end{Bmatrix} \quad (6)$$

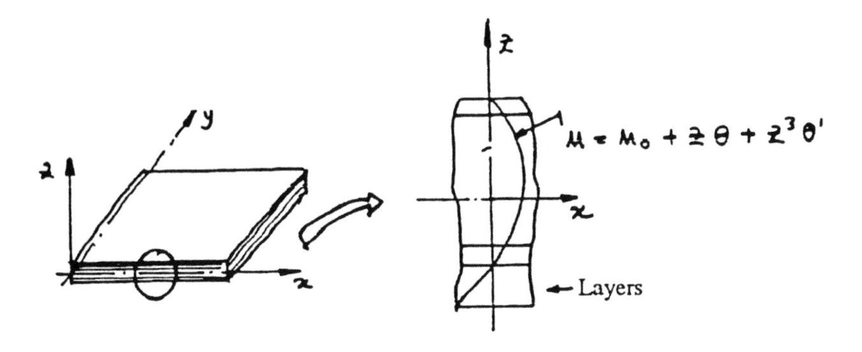

Figure 2 - Higher (third) order shear deformation theory, one dimensional case.

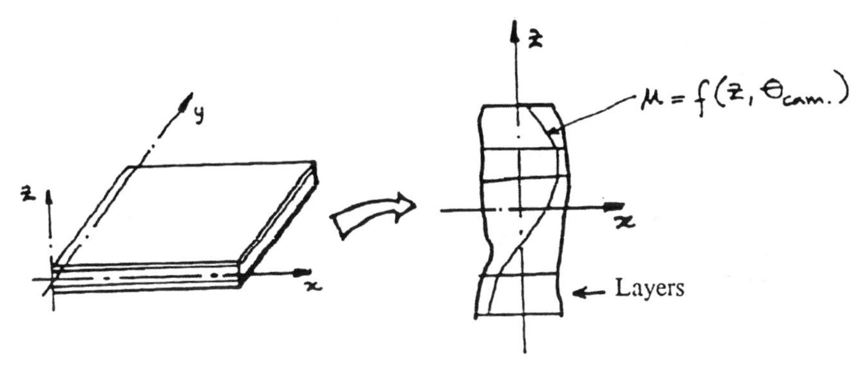

Figure 3 - Layerwise theory for one dimensional case.

4. Displacement field for the third order shear deformation theory

The third order theory, when applied to the shell element needs seven degrees of freedom, three translations and four rotations, two of the first degree and two of the third degree, as

$$\underline{u} = \sum_{k=1}^{n} \underline{N}_k \underline{u}_k^{med} + \sum_{k=1}^{n} \underline{N}_k \zeta \frac{h_k}{2} [V_{1k}, -V_{2k}] \begin{Bmatrix} \beta_{1k} \\ \beta_{2k} \end{Bmatrix} + \sum_{k=1}^{n} \underline{N}_k \left(\zeta \frac{h_k}{2} \right)^3 [V_{1k}, -V_{2k}] \begin{Bmatrix} \beta_{1k}^* \\ \beta_{2k}^* \end{Bmatrix} \quad (7)$$

This displacement field is equivalent to the one proposed by Kant [23,24] for plates.

5. Displacement field for the layerwise shear deformation theory

In the layerwise theory, the displacement field is obtained through the translations and through the two rotations of the normal in each layer middle surface, as

$$\underline{u}^{(1)} = \begin{Bmatrix} u_1 \\ u_2 \\ u_3 \end{Bmatrix}^{(1)} = \sum_{k=1}^{n} N_k \begin{Bmatrix} u_1 \\ u_2 \\ u_3 \end{Bmatrix}^{méd} + \sum_{k=1}^{n} N_k \zeta \frac{h_k^{(1)}}{2} \left[\underline{V}_{1k}, -\underline{V}_{2k} \right] \begin{Bmatrix} \beta_{1k}^{(1)} \\ \beta_{2k}^{(1)} \end{Bmatrix} \tag{8}$$

for the first layer, while for a general nth layer, .

$$\underline{u}^{(n)} = \begin{Bmatrix} u_1 \\ u_2 \\ u_3 \end{Bmatrix}^{(n)} = \sum_{k=1}^{n} N_k \begin{Bmatrix} u_1 \\ u_2 \\ u_3 \end{Bmatrix}^{med(n)} + \sum_{k=1}^{n} N_k \zeta \frac{h_k^{(n)}}{2} \left[\underline{V}_{1k}, -\underline{V}_{2k} \right] \begin{Bmatrix} \beta_{1k}^{(n)} \\ \beta_{2k}^{(n)} \end{Bmatrix} \tag{9}$$

where

$$\underline{u}^{med(n)} = \begin{Bmatrix} u_1 \\ u_2 \\ u_3 \end{Bmatrix}^{med(n)} = \sum_{k=1}^{n} N_k \begin{Bmatrix} u_1 \\ u_2 \\ u_3 \end{Bmatrix}^{med(n-1)} + \sum_{k=1}^{n} N_k \frac{h_k^{(n-1)}}{2} \left[\underline{V}_{1k}, -\underline{V}_{2k} \right] \begin{Bmatrix} \beta_{1k}^{(n-1)} \\ \beta_{2k}^{(n-1)} \end{Bmatrix} + \tag{10}$$

$$+ \sum_{k=1}^{n} N_k \frac{h_k^{(n)}}{2} \left[\underline{V}_{1k}, -\underline{V}_{2k} \right] \begin{Bmatrix} \beta_{1k}^{(n)} \\ \beta_{2k}^{(n)} \end{Bmatrix}$$

6 Strains

Strains are defined in the local system (figure 4), by assuming the usual shell constraint $\sigma_{z'}=0$. Locally, this enables the elimination in the local system of $\varepsilon_{z'}$ in the constitutive equation. So, the local strain vector is expressed as

$$\underline{\varepsilon}' = \begin{Bmatrix} \varepsilon_{x'} \\ \varepsilon_{y'} \\ \gamma_{x'y'} \\ \gamma_{x'z'} \\ \gamma_{y'z'} \end{Bmatrix} = \begin{Bmatrix} \frac{\partial u'}{\partial x'} \\ \frac{\partial v'}{\partial y'} \\ \frac{\partial u'}{\partial y'} + \frac{\partial v'}{\partial x'} \\ \frac{\partial u'}{\partial z'} + \frac{\partial w'}{\partial x'} \\ \frac{\partial w'}{\partial y'} + \frac{\partial v'}{\partial z'} \end{Bmatrix} \tag{11}$$

where u',v' and w' are the local displacements. These local strains can related to the global strains by proper transformations.

7 Elastic constitutive relations

Assuming $\sigma_{z'}=0$, the local stresses are obtained as

$$\underline{\sigma}' = \begin{Bmatrix} \sigma_{x'} \\ \sigma_{y'} \\ \tau_{x'y'} \\ \tau_{x'z'} \\ \tau_{y'z'} \end{Bmatrix} = \begin{Bmatrix} \underline{\sigma}_{p'} \\ \underline{\sigma}_{c'} \end{Bmatrix} \tag{12}$$

where $\sigma_{p'}$ and $\sigma_{c'}$ are the plane and shear components. Stresses and strains can be related through the constitutive law as,

$$\underline{\sigma}' = \underline{D}' \, \underline{\varepsilon}' \tag{13}$$

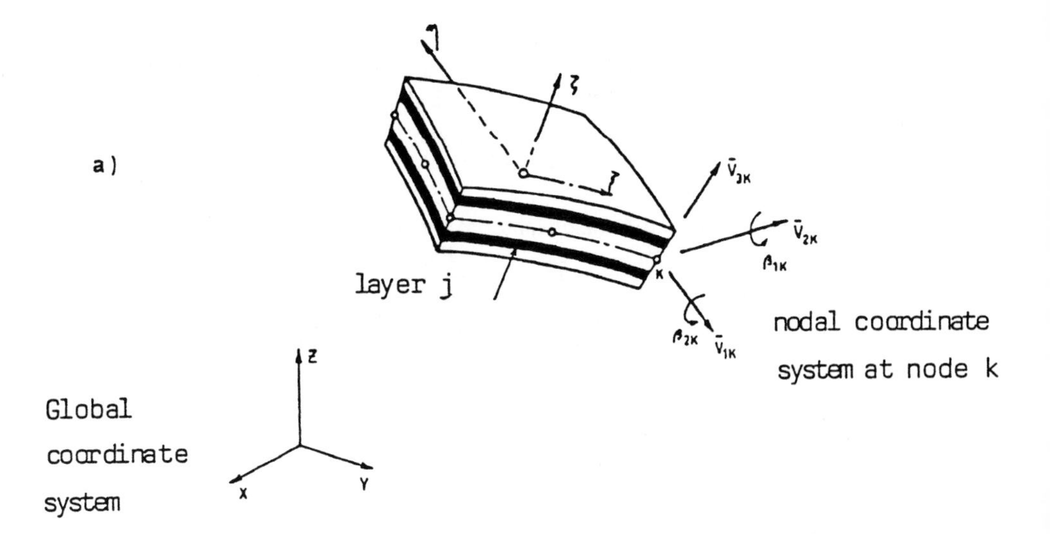

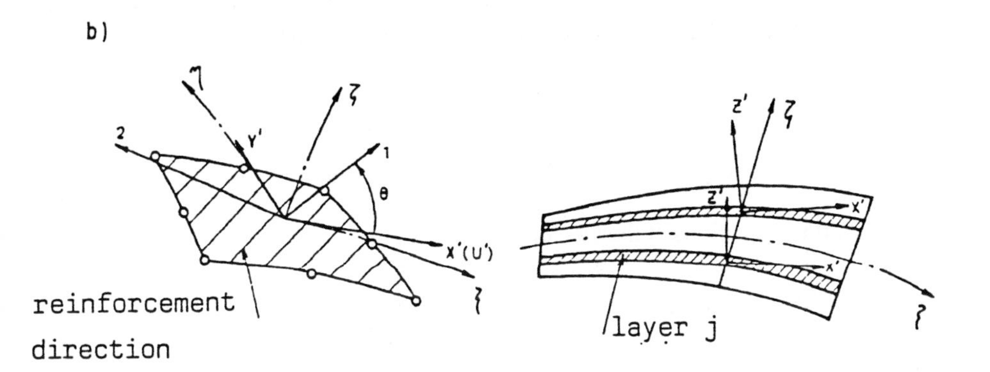

Figure 4 - The degenerated shell element

The local constitutive matrix can be obtained, supposing an orthotropic material behaviour. If the local system is parallel to the material system, we can write for a kth layer

$$\{\sigma_{1,2,3}\}_k = \underline{\overline{D}}_k \{\varepsilon_{1,2,3}\}_k \tag{14}$$

where

$$\{\sigma_{1,2,3}\}_k = \{\sigma_1, \sigma_2, \tau_{12}, \tau_{13}, \tau_{23}\}_k^T \tag{15}$$

$$\{\varepsilon_{1,2,3}\}_k = \{\varepsilon_1, \varepsilon_2, \gamma_{12}, \gamma_{13}, \gamma_{23}\}_k^T \tag{16}$$

$$\overline{\underline{D}}_k = \begin{bmatrix} \overline{D}_1 & \overline{D}_{12} & 0 & 0 & 0 \\ \overline{D}_{12} & \overline{D}_2 & 0 & 0 & 0 \\ 0 & 0 & \overline{D}_3 & 0 & 0 \\ 0 & 0 & 0 & \overline{D}_4 & 0 \\ 0 & 0 & 0 & 0 & \overline{D}_5 \end{bmatrix}_k \tag{17}$$

In the first order theory, the elastic material constants are obtained as

$$\overline{D}_{1_k} = \frac{E_{1_k}}{1-\nu_{12_k}\nu_{21_k}}, \overline{D}_{2_k} = \frac{E_{2_k}}{1-\nu_{12_k}\nu_{21_k}}, \overline{D}_{12_k} = \frac{\nu_{21_k}E_{1_k}}{1-\nu_{12_k}\nu_{21_k}} \tag{18}$$
$$\overline{D}_{3_k} = G_{12_k}, \overline{D}_{4_k} = K_1 G_{13_k}, \overline{D}_{5_k} = K_2 G_{23_k}$$

where E_{1k}, E_{2k}, ν_{12k}, G_{12k}, G_{13k} and G_{23k}, are the kth layer longitudinal and transversal modulii, the Poisson coefficient, the shear modulus and the transversal shear modulii. The shear correction terms K_1 and K_2 are associated to the first order theory and correct the computed formulation to the exact one [7,9,12]. In higher order theories, these terms are not introduced.

8 Stiffness matrix and equivalent nodal force vector

The Principle of Virtual Work can be written for a shell element as [9]:

$$\int_{V^{(e)}} \delta\underline{\varepsilon}'^T \underline{\sigma}' \, dv = \int_{V^{(e)}} \delta\underline{u}^T \underline{b} \, dv + \int_{S^{(e)}} \delta\underline{u}^T \underline{t} \, ds + \sum_{i=1}^{n} \left[\delta\underline{a}_i^{(e)}\right]^T \underline{q}^{(e)} \tag{19}$$

where the first equation member represenbts the internal virtual work, and where

$$\underline{b} = \left[b_x, b_y, b_z\right]^T \quad , \quad \underline{t} = \left[t_x, t_y, t_z\right]^T \quad , \quad \underline{q}_i^{(e)} = \left[X_i, Y_i, Z_i, M_{1i}, M_{2i}\right]^T \tag{20}$$

are the body force, surface and point load vectors. The element volume and surface are V_e e S_e.

The equilibrium equations at the element level result in

$$\underline{K}^{(e)}\underline{a}^{(e)} - \underline{f}^{(e)} = \underline{q}^{(e)} \tag{21}$$

where the stiffness matrix $\underline{K}_e$, and the equivalent force vector $\underline{f}_e$ are expressed by

$$\underline{K}_{ij}^{(e)} = \int_{V^{(e)}} \underline{B}_i'^T \underline{D}' \underline{B}_j' \, dv \tag{22}$$

$$\underline{f}_i^{(e)} = \int_{V^{(e)}} \underline{N}_i^T \underline{b} \, dv + \int_{S^{(e)}} \underline{N}_i^T \underline{t} \, ds \tag{23}$$

The solution of the integrals at the element level is made by the Gauss-Legendre quadrature

$$\underline{K}_{ij}^{(e)} = \sum_{p=1}^{n_\xi} \sum_{q=1}^{n_\eta} \sum_{r=1}^{n_\zeta} (\underline{B}_i'^T \underline{D}' \underline{B}_j' \left|\underline{J}^{(e)}\right|)_{p,q,r} W_p W_q W_r \tag{24}$$

$$\underline{f}_i^{(e)} = \sum_{p=1}^{n_\xi} \sum_{q=1}^{n_\eta} \sum_{r=1}^{n_\zeta} (\underline{N}_i^T \underline{b}' \left|\underline{J}^{(e)}\right|)_{p,q,r} W_p W_q W_r \tag{25}$$

where n_ξ and n_η are the number of gauss points, at the element surface and n_ζ is the number of points in the thickness direction ζ. The mid-ordinate rule is used for the thickness integration.

9 Numerical example - Square clamped plate under uniform pressure

A square clamped plate is subjected to uniform pressure. The geometry and mechanical properties are shown in figure 5. One quarter of the plate is analised, with 9 Serendipity elements. The uniform reduced integration is used. Non-linear convergence tolerance is of 0.25% in residual forces. In figures 6 to 9 the geometrical non-linear behaviour is compared, in terms of load-displacement curves and in terms of normal and shear stresses.

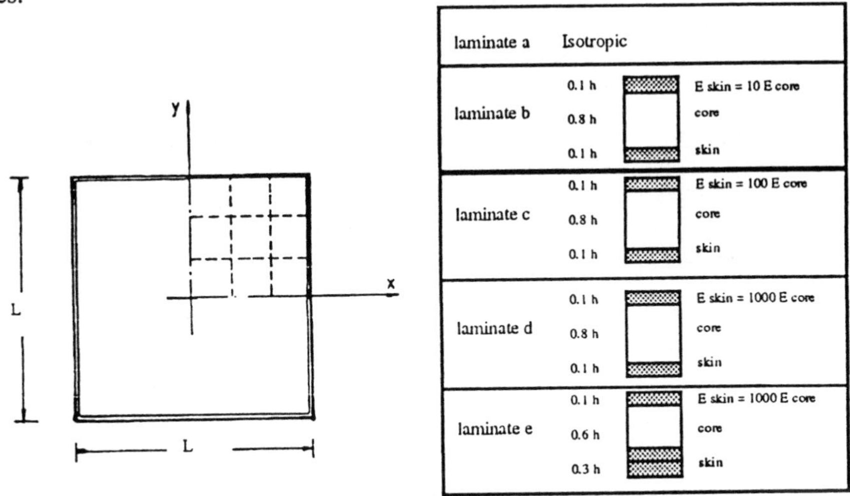

Figure 5 - Clamped square plate, geometry, mesh and materials

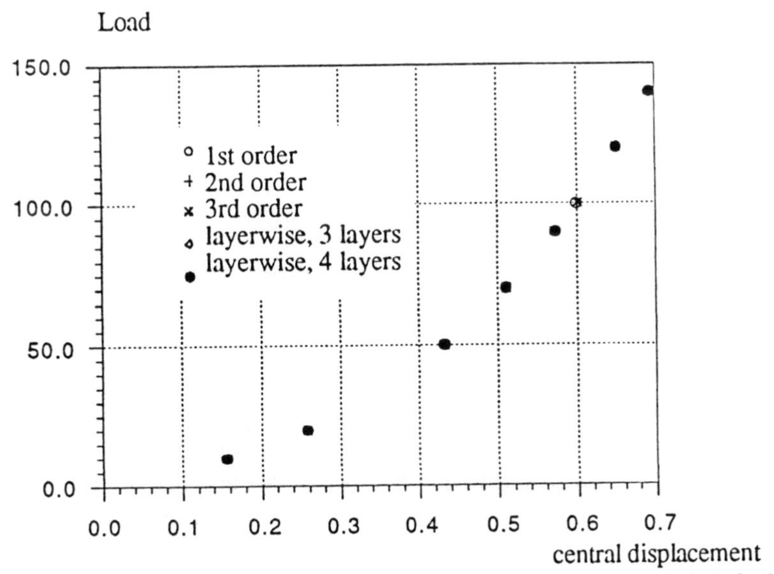

Figure 6 - Load-displacement curve for isotropic laminate (a), ratio L/h=20

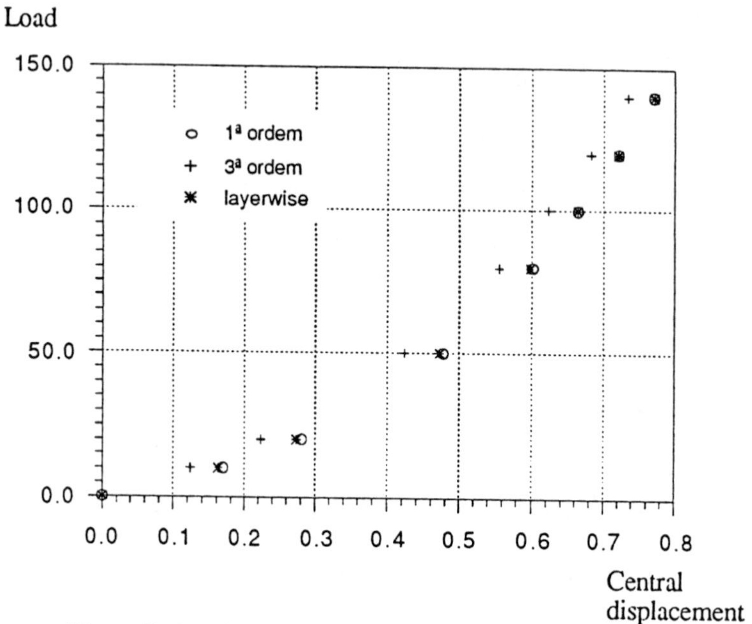

Figure 7 - Laminate (b), L/h=6, load-displacement curve

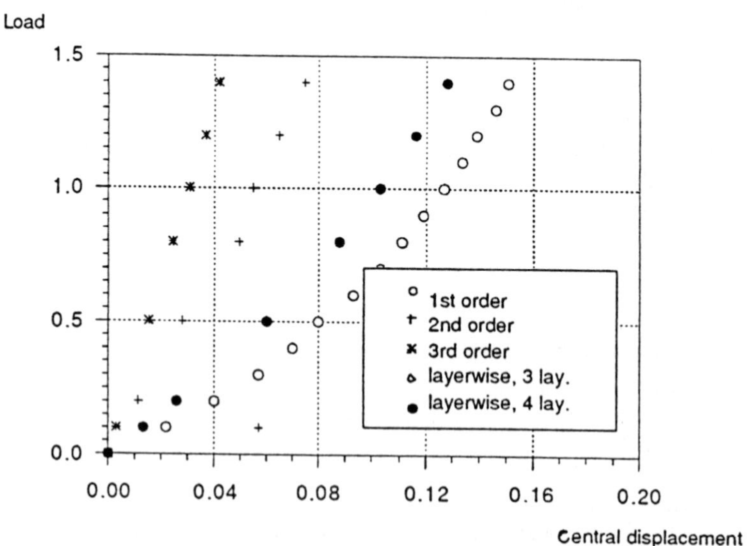

Figure 8 - Laminate (d), L/h=6, load-displacement

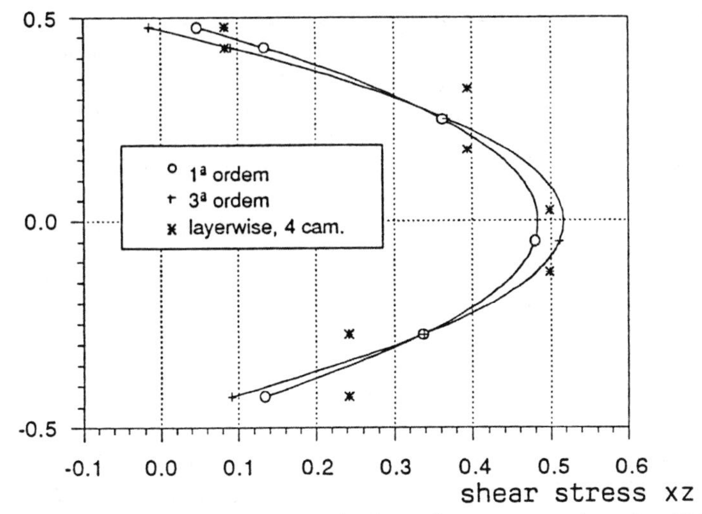

Figure 9 - Non-simetric laminate(e) , L/h=6, xz shear stress at location (L/2,0)

In this example can be seen a good approximation betweeen the three theories for the isotropic case. As the modulii ratio (skin to core) grows variations in response starts to increase, either in stiffness as in stresses. This is due to the different interpretation in the membrane response. It is believed that the layerwise formulation represents better the normal evolution through the thickness in sandwich laminates.

10 Conclusions

In this paper the geometric nonlinear analysis of laminated sandwich shells was made. Three deformation theories were presented an compared in the non-linear range. The degenerated shell element was used throughout. Variations in the response was noticed, either in stiffness, as in stresses, for the three theories.

11 References

1. S.Ahmad, B.M.Irons, O.C.Zienkiewicz, Analysis of thick and thin shell structures by curved finite elements, Int. J. of Num. Meth Eng., Vol.2, 419-451, 1970
2. O.C. Zienkiewicz, R.L.Taylor, J.M.Too, Reduced integration techniques in general analysis of plates and shells, Int. J. of Num. Meth Eng., Vol.3, 275-290, 1971
3. T.J.R. Hughes, R.L.Taylor, W. Kanoknukulchai, A simple and eficient element for plate bending, Int. J. of Num. Meth Eng., Vol.11, 1529-1543, 1977
4. T.J.R.Hughes, W.K.Liu, Nonlinear finite element analysis of shells, Part I. Three-dimensional shells, Comp. Meth.Appl. Mech. Eng., Vol.26, 331-362, 1981
5. E.N.Dvorkin, K.J.Bathe, A continuum mechanics based four-node shell element for general nonlinear analysis, Eng. Comp., Vol.1, 77-88, 1984

6. H.C.Huang, E.Hinton, A new nine node degenerated shell element with enhanced membrane and shear interpolation, Int. J. of Num. Meth Eng., Vol.22, 73-92, 1986
7.J. A. Figueiras, Ultimate load analysis of Anisotropic and Reinforced Concrete plates and shells, Ph. D. Thesis, c/ph/72/83, University College of Swansea, 1983.
8. D.R.J.Owen e J.A.Figueiras, Anisotropic elastoplastic finite element analysis of thick and thin plates, Int. J.Num.Meth.Eng., 19, 1983
9. J. M. A. Cesar de Sa, Numerical Modelling Of Incompressible Problems In Glass Forming And Rubber Technology, Ph.D. Thesis, University College Swansea, C/Ph/91/86, 1986.
10. E.Onate, Calculo de estructuras por el metodo de elementos finitos, CIMNE, 1992
11. A.J.M.Ferreira, Análise por elementos finitos de estruturas tipo casca em materiais compósitos, Tese de Mestrado, FEUP, 1990
12. A.J.M.Ferreira, A.T.Marques, J.C.de Sá, A degenerated shell element for the static linear analysis of sandwich structures, International Conference on Composite Materials, 1993
13. G. Stanley, Continuum-based shell elements, Ph.D.Thesis, Stanford University, 1985
14. D.R.J.Owen, Z.H.Li, A refined analysis of laminated plates by finite element displacement methods-I. Fundamentals and static analysis; II. Vibration and stability, Comp. Structures, Vol. 26, pag. 907-923, 1987
15. R. D. Mindlin, Influence of rotary inertia and shear on flexural motions of isotropic, elastic plates, J.Appl. Mech, vol.18(1), Trans. ASME, 31-38, 1951.
16. K.H.Lo, R.M.Christensen, E.M.Wu, A higher-order theory of plate deformation, J.Appl.Mech. Trans. ASME, Vol. 44, Part 1
17. K.H.Lo, R.M.Christensen, E.M.Wu, A higher-order theory of plate deformation, J.Appl.Mech. Trans. ASME, Vol. 44, Part 2
18. J.M.Whitney, A higher-order theory for extensional motion of laminated composites, J.Sound Vibration, Vol.30, 85-97, 1973
19. J.N.Reddy, A simple higher-order theory for laminated composite plates, J.of Applied Mech., Vol.51, 745-751, 1984
20. N.S.Putcha, A mixed shear flexible finite element for geometrically nonlinear analysis of laminated plates, Ph.D. Dissertation, V.P.I., 1984
21. N.S.Putcha, J.N.Reddy, A refined mixed shear flexible finite element for the nonlinear analysis of laminated plates, Comp. Struct., Vol.22, 529-538, 1986
22. N.S.Putcha, J.N.Reddy, Stability and vibration analysis of laminated plates by using a mixed element based on a refined plate theory, J.Sound Vibration, Vol.104, 285-300,1986
23. T.Kant e D.R.J.Owen, A refined higher-order C° plate bending element, Comp.Struct., Vol.15, 177-183, 1982
24. Mallikarjuna e T. Kant, On Transient response of laminated composite plates based on higher-order theory, Proc. 3rd Int Conf. on Recent advances in structural dynamics, 18-22, July 1988, Southampton, U.K.
25. A. S. Mawenya , J. D. Davies, Finite element analysis of multilayer plates, Int. J. Num. Meth. Eng., vol.8, pag. 215-225, 1974.
26. H. H. Al-Qarra, H. G. Allen, Finite deflections of sandwich beams and plates by the finite element method, Department of Civil Engineering, University of Southampton,UK, 1980

ON SHEAR AND BENDING-MEMBRANE COUPLING IN SANDWICH SHELLS WITH ELASTIC OR VISCOELASTIC CORE

A. BENJEDDOU

Structural Mechanics & Systems Laboratory, Chair of Mechanics, CNAM, 2 rue Conté, F-75003 Paris, France.

1. Introduction

A considerable body of literature exists on the modelling analysis and design of sandwich panels and shells [1,2]. However, there is still a need for new formulations and implementations of effective computational strategies and numerical techniques for the efficient generation of their response. The main aspects added by the sandwich construction are the membrane-bending coupling and transverse shear effects. The former is sometimes used to enhance the damping of composite structures [3], and the latter is the basis of shear damping mechanism in constrained layer treatments.

Several techniques were proposed to represent the shear effect. The simplest way is to use a first-order shear deformation theory with the proper shear-correction factors. These are generally determined in the framework of cylindrical bending and no-rotatory inertia assumptions [4], but can be overcome through the predictor-corrector approaches [1,2]. More sophisticated techniques could better represent the shear effect such as the refined theories [5] and the mixed stress-displacement formulations [6].

Extensive numerical studies [1,2] indicate that global response quantities can be predicted accurately by discrete three-layer models. Here a discrete model using Mindlin-Reissner first-order shear deformation theory for the core and Kirchhoff-Love for the face sheets is proposed. The transverse shear is represented explicitly by the relative tangential displacements of the skin middle surfaces without shear-correction factors. The membrane-bending coupling was found to be localised in the core, together with translation-rotatory inertial coupling.

A B-spline finite element approach is proposed to discretize the theoretical formulation. It enhances a recently developed one [7]. The main added originality is the use of special elements to avoid the use of non physical derivative degrees of freedom (dof) introduced by the B-spline interpolation. This was achieved through the development of new B-spline shape functions [8] which is a second originality. A third feature of the obtained formulation is the reduction to two the number of rotational dof. All other features of the previous B-spline finite elements [7,9] are preserved here.

2. Theoretical Aspects

The theoretical formulation is based on Kirchhoff-Love and Mindlin-Reissner first-order

A. Vautrin (ed.), Mechanics of Sandwich Structures, 89–96.

theories in the face sheets and the core respectively.

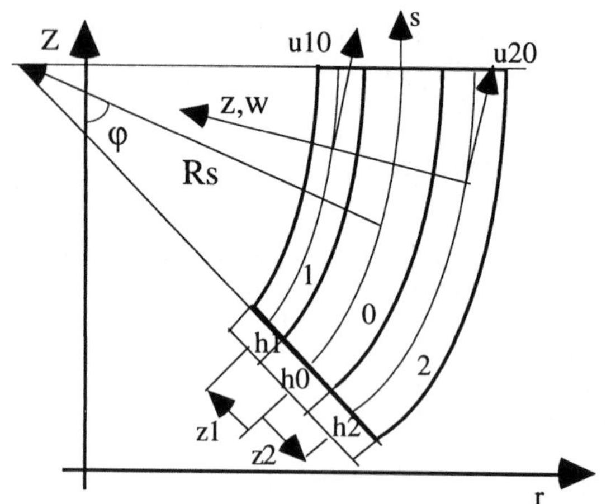

Figure 1. A sandwich element : geometry and notations.

The displacement continuity conditions together with the new variables (Figure 1.),

$$u_m = \frac{u_{10}+u_{20}}{2}, \quad \Delta u = u_{10} - u_{20}, \quad v_m = \frac{v_{10}+v_{20}}{2}, \quad \Delta v = v_{10} - v_{20} \tag{1}$$

lead to the in-plane displacement fields in the inner (i=1, +) and outer (i=2, -) face layers,

$$u_i = (u_m \pm \frac{\Delta u}{2}) + (z - z_i)(\beta_{sm} \pm \frac{\Delta\beta_s}{2}), \quad v_i = (v_m \pm \frac{\Delta v}{2}) + (z - z_i)(\beta_{\theta m} \pm \frac{\Delta\beta_\theta}{2}) \tag{2}$$

and, in the core layer,

$$\begin{aligned} u_0 &= (u_m - z_m\beta_{sm} - \frac{d-h_0}{4}\Delta\beta_s) + z(\frac{\Delta u}{h_0} - \frac{z_m}{h_0}\Delta\beta_s - (\frac{d}{h_0} - 1)\beta_{sm}) \\ v_0 &= (v_m - z_m\beta_{\theta m} - \frac{d-h_0}{4}\Delta\beta_\theta) + z(\frac{\Delta v}{h_0} - \frac{z_m}{h_0}\Delta\beta_\theta - (\frac{d}{h_0} - 1)\beta_{\theta m}) \end{aligned} \tag{3}$$

where, $z_m = \frac{z_1 + z_2}{2}$, $d = z_1 - z_2$

The deflection w is supposed constant in the shell thickness.

Due to Kirchhoff-Love hypotheses in the facings, the mean and relative rotations are dependent on the mean and relative displacements (1) and the deflection, w,

$$\beta_{sm} = -(\frac{\partial w}{\partial s} + \frac{u_m}{R_s}), \quad \beta_{\theta m} = -(\frac{1}{r}\frac{\partial w}{\partial \theta} + \frac{v_m}{R_\theta}), \quad \Delta\beta_s = -\frac{\Delta u}{R_s}, \quad \Delta\beta_\theta = -\frac{\Delta v}{R_\theta} \tag{4}$$

2.1. SHEAR EFFECT

Since only the core was supposed to resist shear stresses, the shear strains of the sandwich shell are found to be confined to the core and can be decomposed into mean and relative components,

$$\gamma_{sz0} = \gamma_{szm} + \gamma_{sz\Delta} \ , \ \gamma_{\theta z0} = \gamma_{\theta zm} + \gamma_{\theta z\Delta} \tag{5}$$

where,

$$\gamma_{szm} = -(\frac{d}{h_0} + \frac{z_m}{R_s})\beta_{sm} \ , \ \gamma_{\theta zm} = -(\frac{d}{h_0} + \frac{z_m}{R_\theta})\beta_{\theta m} \tag{6}$$

and

$$\gamma_{sz\Delta} = \frac{\Delta u}{h_0} - (\frac{z_m}{h_0} + \frac{d - h_0}{4R_s})\Delta\beta_s \ , \ \gamma_{\theta z\Delta} = \frac{\Delta v}{h_0} - (\frac{z_m}{h_0} + \frac{d - h_0}{4R_\theta})\Delta\beta_\theta \tag{7}$$

These equations show that the mean transverse shear strains (6) are defined by two tangential and meridian mean rotations. In contrast, the relative transverse shear strains (7) depend on four rotations : two tangential and meridian relative rotations and two rotations defined by the ratio of the tangential relative displacements to the core thickness. The latter two rotations represent the amount of transverse shear due to the sliding of the face sheets against the core.

2.2. ELASTIC COUPLINGS

The analytic through thickness integration of the potential energy of the sandwich shell leads to several elastic coupling phenomena. They could be classified to geometrical (different face thicknesses) and material (different elastic material properties) couplings. Furthermore, the present theory shows two types of elastic couplings, the classical membrane-bending coupling and unusual coupling between mean and relative strains.

The membrane and bending mean-relative couplings are represented by the following composite elastic matrices,

$$\left[\mathbf{D}^{e}_{m\Delta}\right] = \frac{1}{2}\left(h_1[\mathbf{D}_1] - h_2[\mathbf{D}_2]\right) \tag{8}$$

$$\left[\mathbf{D}^{\chi}_{m\Delta}\right] = \frac{1}{2}\left(\frac{h_1^3}{12}[\mathbf{D}_1] - \frac{h_2^3}{12}[\mathbf{D}_2]\right) + z_m \frac{(d - h_0)}{3} h_0[\mathbf{D}_0] \tag{9}$$

where, $z_m = \dfrac{\Delta h}{4}$, $d = h_0 + h_m$; $\Delta h = h_1 - h_2$, $h_m = \dfrac{h_1 + h_2}{2}$

and $[\mathbf{D}_i]$; i=0,1,2 are the elasticity matrices of the core, inner and outer face sheets. Eqns. (8,9) indicate that membrane and bending mean-relative couplings vanish for full symmetrical sandwich construction (identical face thicknesses and elastic material properties). However, the shear mean-relative coupling, present through, $[\mathbf{D}^{\gamma}_{m\Delta}] = h_0\ [\mathbf{D}_c]$ does not vanish for any symmetry. $[\mathbf{D}_c]$ is the core transverse shear elasticity matrix.

The membrane-bending coupling is composed of the mean and relative membrane-bending couplings which vanish for geometrical symmetry of the sandwich shell,

$$\left[\mathbf{D}^{e\chi}_{m}\right] = -z_m h_0[\mathbf{D}_0] \ , \ \left[\mathbf{D}^{e\chi}_{\Delta}\right] = -z_m \frac{h_0}{12}[\mathbf{D}_0] \tag{10}$$

and the geometrical and material mean-relative and relative-mean membrane-bending couplings which are always present and do not vanish even with full symmetric shell.

$$\left[\mathbf{D}^{e\chi}_{m\Delta}\right] = -\left(\frac{d - h_0}{4}\right)h_0[\mathbf{D}_0] \ , \ \left[\mathbf{D}^{e\chi}_{\Delta m}\right] = -(d - h_0)\frac{h_0}{12}[\mathbf{D}_0] \tag{11}$$

As can be seen from (10,11), all membrane-bending couplings are localised in the core, since they depend on the core elastic properties only.

2.3. INERTIAL COUPLINGS

The through thickness analytic integration of the kinetic energy of the sandwich shell gives arise to several inertial coupling phenomena. As for the elastic couplings, inertial couplings could be grouped into geometrical (different thicknesses) and material (different inertial properties) couplings. Beside, the present formulation provides two types of inertial couplings, the classical translation-rotatory inertial coupling and unusual inertial coupling phenomena between mean and relative translations and rotations.

The translation and rotatory mean-relative couplings can be seen from the following mass densities per unit area,

$$\rho^{u}_{m\Delta} = \frac{1}{2}\left(h_1\rho_1 - h_2\rho_2\right)\ ,\ \rho^{\beta}_{m\Delta} = \frac{1}{2}\left(\frac{h_1^3}{12}\rho_1 - \frac{h_2^3}{12}\rho_2\right) + z_m\frac{\left(d-h_0\right)}{3}h_0\rho_0 \tag{12}$$

where ρ_i ; i=0,1,2 are the mass densities of the core, inner and outer layers. These couplings vanish for full symmetric sandwich construction (identical face thicknesses and inertial material properties).

The translation-rotatory inertial coupling is composed of geometrical mean and relative translation-rotatory couplings which vanish simply for geometrical symmetry,

$$\rho^{u\beta}_{m} = -z_m h_0\rho_0\ ,\ \rho^{u\beta}_{\Delta} = -z_m\frac{h_0}{12}\rho_0 \tag{13}$$

and geometrical and inertial material mean-relative and relative-mean translation-rotatory inertial couplings that never vanish even for full symmetric construction,

$$\rho^{u\beta}_{m\Delta} = -\left(\frac{d-h_0}{4}\right)h_0\rho_0\ ;\ \rho^{u\beta}_{\Delta m} = -\left(d-h_0\right)\frac{h_0}{12}\rho_0 \tag{14}$$

Equations (13,14) indicate that all translation-rotatory inertial couplings are confined to the core layer since they depend on the mass density of the middle layer only.

It is also worthwhile to notice that there is a full analogy between elastic and inertial couplings in the present theory. The elasticity matrices should replace the mass densities in the present section to get the results of the previous one and *vice versa*.

3. Numerical Aspects

The present numerical treatment focuses on new features of the recently developed B-spline sandwich finite element [7]. New shape functions are proposed to suppress some derivative dof. Special end elements are then obtained and their formulation is briefly discussed. This technique was successfully tested on joined shells of revolution [8,10].

3.1. MODIFIED B-SPLINE SHAPE FUNCTIONS

The classical cubic B-spline interpolation needs four control dof per element due to the two fictive nodes that it induces. These could be avoided by adding two derivative dof at the end nodes. Nevertheless, they remain non physical and are cumbersome when boundary conditions are to be specified. Here, the number of control dof is reduced from four to three by making vanish the third parametric derivative. New modified B-spline functions are then obtained for the first element of a super element,

$$N_1^1(\xi)=F_2(\xi)+3F_1(\xi),\ N_2^1(\xi)=F_3(\xi)-3F_1(\xi)\ ,\ N_3^1(\xi)=F_4(\xi)+F_1(\xi) \quad (15)$$

and for the Nth element of a super element,

$$N_{4-i}^N(\xi)=N_i^1(-\xi)\ ;\ i=\ 1,2,3 \quad (16)$$

These new shape functions are of order three instead of four for the original basis functions, F_i ; i=1,...,4, used in [7]. Hence, they are to be used for C°-continuity interpolation only. Here they are used for tangential mean and relative displacements.

3.2. FINITE ELEMENT DISCRETIZATION

The meridian of the sandwich shell middle surface is divided into super elements (group of elements). For each element, the mean and relative displacements are represented by the product of Fourier series in the circumferential direction and uniform B-spline functions in the meridian direction. Usual cubic B-spline interpolations are used for the super element internal elements and the above modified ones for its end elements. Hence, the discretization of the symmetric n[th] order Fourier components of the mean and relative displacements inside the i-th (i=1,N) end element are written as,

$$\{\mathbf{d}_{mn}^i\}=[\mathbf{N}_m^i(\xi)]\{\mathbf{D}_{mn}^i\}\ ,\ \{\mathbf{d}_{\Delta n}^i\}=[\mathbf{N}_\Delta^i(\xi)]\{\mathbf{D}_{\Delta n}^i\} \quad (17)$$

where, $\langle\mathbf{d}_{mn}^i\rangle=\langle u_{mn}^i\ v_{mn}^i\ w_n^i\rangle$, $\langle\mathbf{d}_{\Delta n}^i\rangle=\langle\Delta u_n^i\ \Delta v_n^i\rangle$

$[\mathbf{N}_m^i(\xi)]$, $[\mathbf{N}_\Delta^i(\xi)]$ are interpolation matrices, given in the appendix, and $\{\mathbf{D}_{mn}^i\}$, $\{\mathbf{D}_{\Delta n}^i\}$ are non nodal mean and relative spline control dof vectors, respectively,

$$\langle\mathbf{D}_{mn}^1\rangle=\langle W_{n0}\,U_{mn1}\,V_{mn1}\,W_{n1}\,U_{mn2}\,V_{mn2}\,W_{n2}\,U_{mn3}\,V_{mn3}\,W_{n3}\rangle \quad (18)$$

$$\langle\mathbf{D}_{mn}^N\rangle=\langle U_{mnN-2}\,V_{mnN-2}\,W_{nN-2}\,U_{mnN-1}\,V_{mnN-1}\,W_{nN-1}\,U_{mnN}\,V_{mnN}\,W_{nN}\,W_{nN+1}\rangle \quad (19)$$

and

$$\langle\mathbf{D}_{\Delta n}^1\rangle=\langle\Delta U_{n1}\,\Delta V_{n1}\,\Delta U_{n2}\,\Delta V_{n2}\,\Delta U_{n3}\,\Delta V_{n3}\rangle \quad (20)$$

$$\langle\mathbf{D}_{\Delta n}^N\rangle=\langle\Delta U_{nN-2}\,\Delta V_{nN-2}\,\Delta U_{nN-1}\,\Delta V_{nN-1}\,\Delta U_{nN}\,\Delta V_{nN}\rangle \quad (21)$$

When these are put right in single vectors and transformed to nodal displacement dof, as indicated in [7], they become for each end element,

$$\langle\mathbf{d}_n^1\rangle=\left\langle u_{mn1}\,v_{mn1}\,w_{n1}\,\Delta u_{n1}\ \Delta v_{n1}(\frac{dw}{ds})_{n1}\ldots u_{mn3}\,v_{mn3}\,w_{n3}\,\Delta u_{n3}\,\Delta v_{n3}\right\rangle \quad (22)$$

$$\langle\mathbf{d}_n^N\rangle=\left\langle u_{mnN-2}\,v_{mnN-2}\,w_{nN-2}\,\Delta u_{nN-2}\ \Delta v_{nN-2}\ldots u_{mnN}\,v_{mnN}\,w_{nN}\,\Delta u_{nN}\,\Delta v_{nN}(\frac{dw}{ds})_{nN}\right\rangle \quad (23)$$

Hence, there is a gain of 8 dof for a super element compared to [7].

The rest of the numerical formulation follows closely that detailed in [7]. After the element assembly each internal node has five dof only, which are the mean and relative tangential displacements and the deflection. These are augmented by the bending rotation for the super element end nodes, leading to minimum rotational dof for the whole finite element model. The complex elastic modules associated to the Modal Strain Energy (MSE) method, using the associated conservative system, are retained to deal with the complex eigenvalues problem obtained for sandwich shells with viscoelastic core [11,12]. The eigenvalues problem is then solved with the sub-space iterative method.

4. Vibration and Damping Analyses

In order to illustrate the shear and membrane-bending coupling effects on free-vibrations and damping of sandwich shells of revolution, a ten B-spline finie element model was used for the modal analysis of a simply supported (S-S) symmetric sandwich short cylinder having the following properties:

$$R/L=2\ ;\ h/L=0.055\ ;\ h/R=0.0275\ ;\ h_0/h_1=20,$$
$$\rho_1/\rho_0=20,\ E_1/E_0=384.6\ ;\ G_1/G_0=385\ ;\ \nu_1=\nu_0=0.3\ ;\ \beta=0.1$$

Where, β is the loss factor of the viscoelastic material.

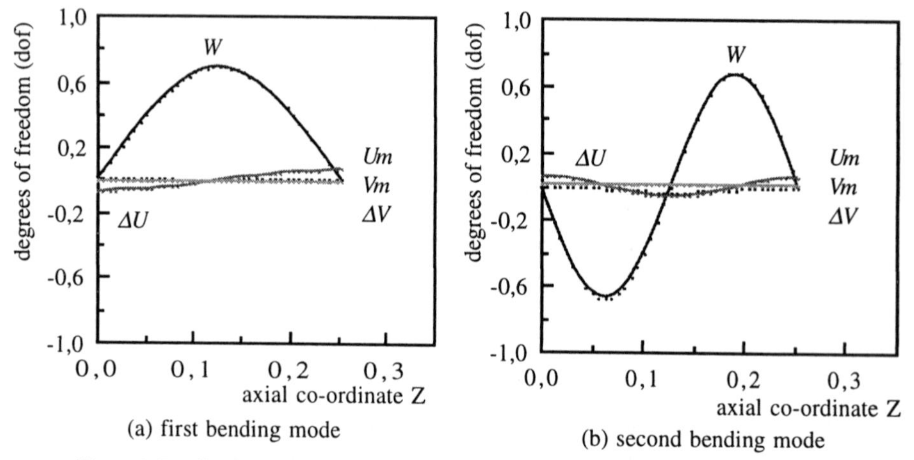

Figure 2. Bending-longitudinal shear coupling in the first modes of a S-S sandwich cylinder.

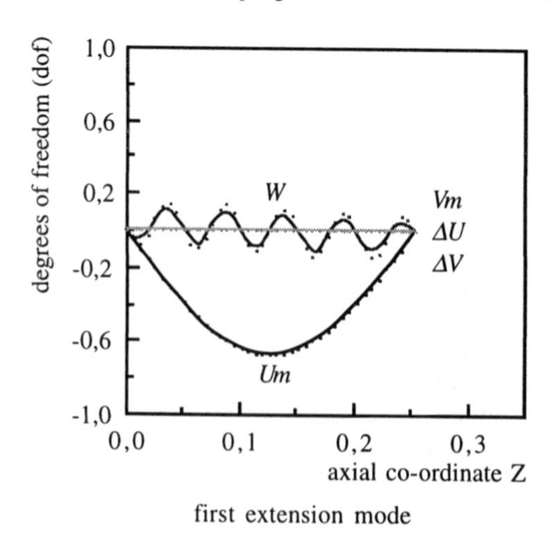

Figure 3. Membrane-bending coupling in the first extension mode of a S-S sandwich cylinder.

It was found that the first two bending modes were slightly coupled with longitudinal shear modes (Figure 2.) and the first extension mode was slightly coupled with higher bending modes (Figure 3.). Furthermore the most effective modal loss factors were obtained for the first two longitudinal and circumferential shear modes with very close frequencies (Figure 4.). They represent longitudinal and circumferential sliding of the face sheets against the core layer, respectively.

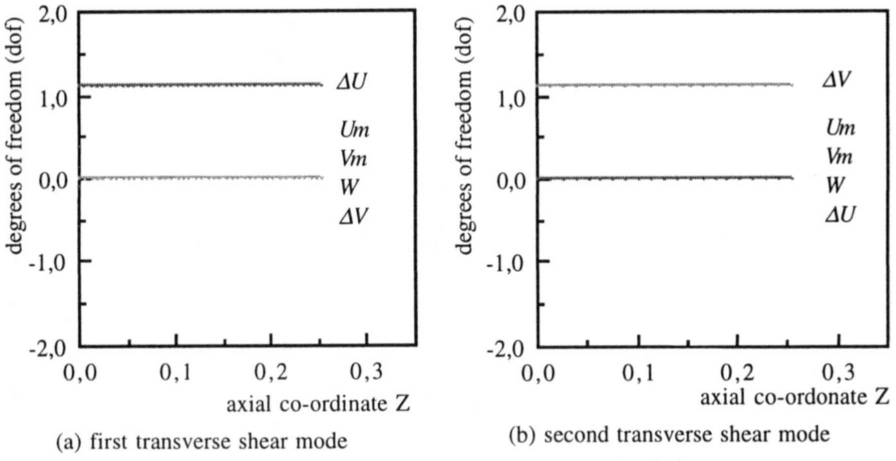

(a) first transverse shear mode (b) second transverse shear mode

Figure 4. Most damped modes of a S-S sandwich cylinder.

For this example, the use of the shear correction factor (5/6) enhanced slightly (2,3% of maximum) the natural frequencies compared to non corrected results. But these remain very satisfactory. Nevertheless, the influence of the shear correction factor is guessed to be greater for shear stresses. Contrary to [13,14], the present discrete-layer formulation is free from shear locking even when degenerated to single layer model. Beside, the influence of the rotational inertia was found to be weak except for the higher modes (0,4% maximum enhancement for the first fifteen modes).

5. Conclusion

Shear effect, elastic and inertial couplings were analysed by a three-layer discrete model discretized by a B-spline finite element method. The theoretical formulation assumes Kirchhoff-Love theory in the face sheets and Mindlin-Reissner theory for the core. The B-spline finite element used has minimum rotational degrees of freedom and represents explicitly the shear effect through the relative degrees of freedom. It was found that mean-relative and relative mean membrane-bending and translation -rotatory couplings do not vanish even for full (geometrical and material) symmetry of the sandwich shell.

The present methodology is being extended to smart structures [15], in particular to sandwich adaptive structures with viscoelastic and piezoelectric materials.

6. References

1. Noor A. K., Burton W.S. and Bert C. W. (1996) Computational models for sandwich panels and shells. *Appl. Mech. Rev.*, **49**, 155-199.
2. Burton W.S., Noor A. K. (1995) Assessment of computational models for sandwich panels and shells. *Comput. Methods Appl. Mech. Engrg.*, **124**, 125-151.
3. Barrett D. J.(1993) On the use of stress coupling in damped plates. *J. Sound Vibr.* **160**, 187-191.
4. Laitinen M. and Lahtinen G. (1995) Transverse shear corrections factors for laminates in cylindrical bending. *Comm., Num. Methods Engrg.*, **11**, 41-47.
5. Touratier M. and Faye J. P. (1995) On a refined model in structural mechanics : finite element approximation and edge effect analysis for axisymmetric shells. *Comput. Struct.*, **54**, 897-920.
6. Fages A. and Verchery G. (1986) Influence de la prise en compte du cisaillement transversal sur le calcul des fréquences propres de poutres sandwiches. *J. Méch. Théorique Appliquée*, **5**, 73-93.
7. Benjeddou A. and Hamdi M. A. (1996) A B-spline finite element for the dynamic analysis of sandwich shells of revolution. *Engrg. Computations*, **13**, 240-262.
8. Benjeddou A. (in press) Finite element modal analysis of joined shells of revolution. *Comput. Struct.*
9. Benjeddou A. and Hamdi M. A. (1994) Un nouvel élément fini du type B-spline pour l'analyse dynamique des coques de révolution. *Revue Europ. Éléments Finis*, **3**, 101-126.
10. Benjeddou A. (1997) Local finite element approach for the analysis of shell combinations, in L. A. Godoy, M. Rysz, L. E. Suarez (eds.), *Applied Mechanics in the Americas*, Univ. Iowa, IA, **4**, 147-150.
11. Benjeddou A. and Hamdi M. A. (1995) Viscoelastic damping of sandwich shells of revolution : an approximate analysis, in L. A. Godoy, S. R. Idelsohn, P. A. A. Laura, D. J. Mook (eds.), *Applied Mechanics in the Americas,* AAM and AMCA, Santa Fee, **1**, 331-336.
12. Zambrano A., Inaudi J. A. and Kelly J. M. (1994) Accuracy of the modal strain energy method, in C. D. Johnson (ed.) *Passive damping*, SPIE, **2193**, 284-295.
13. Gautham B. P. and Ganesan N. (1994) Vibration and damping characteristics of spherical shells with a viscoelastic core. *J. Sound Vibration*, **170**, 289-301.
14. Sivadas K. R. and Ganesan N. (1994) Free vibration and material damping analysis of moderately thick circular cylindrical shells. *J. Sound Vibration*, **172**, 47-61.
15. Rahmoune M., Osmont D., Benjeddou A. and Ohayon R. (1997) Finite element modelling of a smart structure plate system, in P. Santini, C. A. Rogers, Y. Morotsu (eds.), Seventh Int. Conf. Adaptive Struct. Tech., Technomic Pub. Co. Inc., pp. 463-473.

7. Appendix

For the first element of a super element, the modified mean and relative interpolation matrices are, respectrively,

$$\left[\mathbf{N}_m^I(\xi)\right]=\begin{bmatrix} 0 & N_1^I(\xi) & 0 & 0 & N_2^I(\xi) & 0 & 0 & N_3^I(\xi) & 0 & 0 \\ 0 & 0 & N_1^I(\xi) & 0 & 0 & N_2^I(\xi) & 0 & 0 & N_3^I(\xi) & 0 \\ F_1(\xi) & 0 & 0 & F_2(\xi) & 0 & 0 & F_3(\xi) & 0 & 0 & F_4(\xi) \end{bmatrix} \tag{A1}$$

and

$$\left[\mathbf{N}_\Delta^I(\xi)\right]=\begin{bmatrix} N_1^I(\xi) & 0 & N_2^I(\xi) & 0 & N_3^I(\xi) & 0 \\ 0 & N_1^I(\xi) & 0 & N_2^I(\xi) & 0 & N_3^I(\xi) \end{bmatrix} \tag{A2}$$

Similar matrices are obtained for the Nth element of a super element,

$$\left[\mathbf{N}_m^N(\xi)\right]=\begin{bmatrix} N_1^N(\xi) & 0 & 0 & N_2^N(\xi) & 0 & 0 & N_3^N(\xi) & 0 & 0 & 0 \\ 0 & N_1^N(\xi) & 0 & 0 & N_2^N(\xi) & 0 & 0 & N_3^N(\xi) & 0 & 0 \\ 0 & 0 & F_1(\xi) & 0 & 0 & F_2(\xi) & 0 & 0 & F_3(\xi) & F_4(\xi) \end{bmatrix} \tag{A3}$$

and

$$\left[\mathbf{N}_\Delta^N(\xi)\right]=\begin{bmatrix} N_1^N(\xi) & 0 & N_2^N(\xi) & 0 & N_3^N(\xi) & 0 \\ 0 & N_1^N(\xi) & 0 & N_2^N(\xi) & 0 & N_3^N(\xi) \end{bmatrix} \tag{A4}$$

BENDING, BUCKLING AND FREE VIBRATION OF SANDWICH COMPOSITE BEAMS WITH A TRANSVERSE SHEAR STRESS CONTINUITY MODEL

M. KARAMA, B. ABOU HARB, S. MISTOU AND S. CAPERAA
Laboratoire Génie de production, CMAO
ENIT BP 1629, 65016 Tarbes Cedex, France

Abstract - This work presents a new composite beam model based on discrete layer theory. It allows to verify automatically the continuity of transverse shear stresses by taking into account the Heaviside step function. Besides, the transverse shear is represented by a sine function (Touratier, 1991). Moreover, this model introduces membrane refinement (Ossadzow and al., 1995).

1. Introduction

Research on composite beams have resulted in diverse models the main problem of which is how to take into account the transverse shear effects. Most of the models do not satisfy the continuity of the stresses at layer interfaces whose mechanical properties are different. However, when designing the sandwich composite structures, it is necessary to verify the load transfer process to calculate the dimensions of the structure. That is why the development of refined models to describe the behaviour of beams is a fast expanding research axis.

Di Sciuva (1987,1993) and then Touratier (1991,1992) proposed simplified discrete layer models with only five variational unknowns (two membrane displacements, a transverse displacement and two rotations), allowing to represent the section warping in the deformed configuration for Touratier. Nevertheless, in these two cases, the compatibility conditions both at layer interfaces and on the frontiers cannot be satisfied. From Touratier's work, Idlbi (1995) proposed a plate model which satisfy both the stress continuity at interfaces and the zero at the frontiers. Finally, He (1994) introduced the Heaviside step function which allows automatic verification of the displacement continuity at interfaces between different layers.

The new discrete layer model presented comes from the work of Di Sciuva (1993), He (1994) and, Ossadzow and al. (1995), the displacement field is :

$$\begin{aligned} U_1(x_1, x_3, t) &= u_1^0(x_1, t) - x_3 w_{,1}(x_1, t) + h_1(x_3)\varphi_1(x_1, t) \\ U_2 &= 0 \\ U_3(x_1, t) &= w(x_1, t) \end{aligned} \tag{1}$$

A. Vautrin (ed.), Mechanics of Sandwich Structures, 97–104.

with transverse shear function :

$$h_1(x_3) = g(x_3) + \sum_{m=1}^{N-1} \lambda_1^{(m)} [-\frac{x_3}{2} + \frac{f(x_3)}{2} + (x_3 - x_{3_{(m)}}) H(x_3 - x_{3_{(m)}})] \tag{2}$$

and, $f(x_3) = \frac{h}{\pi} \sin \frac{\pi x_3}{h}$ shear refinement sine function, $g(x_3) = \frac{h}{\pi} \cos \frac{\pi x_3}{h}$ membrane refinement cosine function, $H(x_3 - x_{3_{(m)}})$ Heaviside step function.

The coefficients $\lambda_1^{(m)}$ are determined with the boundary conditions on top and bottom faces, and with the continuity conditions at layer interfaces :

$$\sigma_{13}(x_3 = 0) = \sigma_{13}(x_3 = h) = 0 \qquad \sigma_{13}^{(m)}(x_3 = x_{3_{(m)}}) = \sigma_{13}^{(m+1)}(x_3 = x_{3_{(m)}}) \tag{3}$$

2. Governing Equations

The virtual power principle permits to obtain the motion equations and the natural boundary conditions (Germain and al., 1995). The calculations are made in small perturbations (elastic domain) : $\varepsilon_{ij} = \frac{1}{2}(U_{i,j} + U_{j,i})$.

The principle is :

$$\underbrace{\int_\Omega \rho U^{*T} \ddot{U} d\Omega}_{P^*_{(a)}} = \underbrace{-\int_\Omega \overline{\overline{D}}^{*T} : \overline{\overline{\sigma}} d\Omega}_{P^*_{(i)}} + \underbrace{\int_\Omega U^{*T} f d\Omega + \int_\Gamma U^{*T} \hat{F} d\Gamma}_{P^*_{(e)}} \tag{4}$$

From the divergence theorem (or integration by parts) and the kinematic (1), the different components of the principle (4) can be obtained.

The virtual power of acceleration quantities is :

$$P^*_{(a)} = \int_0^L (\Gamma^{(u)} u_1^{0*} + \Gamma^{(w)} w^* + \Gamma^{(\varphi)} \varphi_1^*) dx_1 + \overline{\Gamma}^{(w)} w^* \tag{5}$$

$$\begin{aligned}
\Gamma^{(u)} &= I_w \ddot{u}_1^0 + I_{uw'} \ddot{w}_{,1} + I_{u\omega} \ddot{\varphi}_1 & I_w &= \int_0^h \rho dx_3 & I_{uw'} &= -\int_0^h \rho x_3 dx_3 \\
\Gamma^{(w)} &= -I_{uw'} \ddot{u}_{1,1}^0 + I_w \ddot{w} - I_{w'} \ddot{w}_{,11} - I_{\omega w'} \ddot{\varphi}_{1,1} & I_{w'} &= \int_0^h \rho x_3^2 dx_3 & I_{u\omega} &= \int_0^h \rho h_1(x_3) dx_3 \\
\Gamma^{(\varphi)} &= I_{u\omega} \ddot{u}_1^0 + I_{\omega w'} \ddot{w}_{,1} + I_\omega \ddot{\varphi}_1 & & & & \\
\overline{\Gamma}^{(w)} &= I_{uw'} \ddot{u}_1^0 + I_{w'} \ddot{w}_{,1} + I_{\omega w'} \ddot{\varphi}_1 & I_\omega &= \int_0^h \rho h_1^2(x_3) dx_3 & I_{\omega w'} &= -\int_0^h \rho x_3 h_1(x_3) dx_3
\end{aligned} \tag{6}$$

The virtual power of internal effort is :

$$P^*_{(i)} = \int_0^L (N_{11,1} u_1^{0*} + M_{11,11} w^* + (P_{11,1} - P_{13}) \varphi_1^*) dx_1 - N_{11} u_1^{0*} - M_{11,1} w^* - P_{11} \varphi_1^* + M_{11} w_{,1}^* \tag{7}$$

with,

$$\{N_{11}, M_{11}, P_{11}, P_{13}\} = \int_0^h \{\sigma_{11}, x_3\sigma_{11}, h_1(x_3)\sigma_{11}, h_{1,3}(x_3)\sigma_{13}\} dx_3 \tag{8}$$

The virtual power of external effort is :

$$P^*_{(e)} = \int_0^L (\overline{n}_1 u_1^{0*} + (\overline{n}_3 + \overline{m}_{1,1}) w^* + \overline{p}_1 \varphi_1^*) dx_1 + \overline{N}_1 u_1^{0*} + (\overline{N}_3 - \overline{m}_1) w^* + \overline{P}_1 \varphi_1^* - \overline{M}_1 w_{,1}^* \tag{9}$$

with,

$$\begin{aligned}
\overline{n}_1 &= \int_0^h f_1 dx_3 & \overline{n}_3 &= \int_0^h f_3 dx_3 & \overline{N}_1 &= \int_0^h F_1 dx_3 & \overline{N}_3 &= \int_0^h F_3 dx_3 \\
\overline{m}_1 &= \int_0^h x_3 f_1 dx_3 & \overline{p}_1 &= \int_0^h h_1 f_1 dx_3 & \overline{M}_1 &= \int_0^h x_3 F_1 dx_3 & \overline{P}_1 &= \int_0^h h_1 F_1 dx_3
\end{aligned} \tag{10}$$

The dimension x_2 is supposed unitary, and the effects of σ_{33} are neglected, the generalized constitutive law is :

$$\begin{bmatrix} N_{11} \\ M_{11} \\ P_{11} \\ P_{13} \end{bmatrix} = \begin{bmatrix} A_{11} & B_{11} & \tilde{K} & 0 \\ B_{11} & D_{11} & \tilde{T} & 0 \\ \tilde{K} & \tilde{T} & \tilde{S} & 0 \\ 0 & 0 & 0 & \tilde{Y} \end{bmatrix} \begin{bmatrix} u^0_{1,1} \\ -w_{,11} \\ \varphi_{1,1} \\ \varphi_1 \end{bmatrix} \tag{11}$$

with, $\{A_{11}, B_{11}, D_{11}, \tilde{K}, \tilde{T}, \tilde{S}\} = \int_0^h C'_{11}.\{1, x_3, x_3^2, h_1, h_1 x_3, h_1^2\} dx_3, \quad \tilde{Y} = \int_0^h C_{55} h_{1,3}^2 dx_3$ (12)

Gathering the different components of the principle (5)-(10) and (11), the motion equations obtained are, $\forall u_1^{0*}, \forall w^*, \forall \varphi_1^*$, :

$$\begin{aligned} I_w \ddot{u}_1^0 + I_{uw'} \ddot{w}_{,1} + I_{u\omega} \ddot{\varphi}_1 &= A_{11} u^0_{1,11} - B_{11} w_{,111} + \tilde{K}\varphi_{1,11} + \overline{n}_1 \\ -I_{uw'} \ddot{u}^0_{1,1} + I_w \ddot{w} - I_{w'} \ddot{w}_{,11} - I_{\omega w'} \ddot{\varphi}_{1,1} &= B_{11} u^0_{1,111} - D_{11} w_{,1111} + \tilde{T}\varphi_{1,111} + \overline{n}_3 + \overline{m}_{1,1} \\ I_{u\omega} \ddot{u}_1^0 + I_{\omega w'} \ddot{w}_{,1} + I_\omega \ddot{\varphi}_1 &= \tilde{K} u^0_{1,11} - \tilde{T} w_{,111} + \tilde{S}\varphi_{1,11} - \tilde{Y}\varphi_1 + \overline{p}_1 \end{aligned} \tag{13}$$

And the natural boundary conditions are, $\forall u_1^{0*}, \forall w^*, \forall \varphi_1^*, \forall w_{,1}^*$, :

$$\begin{aligned} 0 &= -A_{11} u^0_{1,1} + B_{11} w_{,11} - \tilde{K}\varphi_{1,1} + \overline{N}_1 \\ I_{uw'} \ddot{u}_1^0 + I_{w'} \ddot{w}_{,1} + I_{\omega w'} \ddot{\varphi}_1 &= -B_{11} u^0_{1,11} + D_{11} w_{,111} - \tilde{T}\varphi_{1,11} + \overline{N}_3 - \overline{m}_1 \\ 0 &= -\tilde{K} u^0_{1,1} + \tilde{T} w_{,11} - \tilde{S}\varphi_{1,1} + \overline{P}_1 \\ 0 &= B_{11} u^0_{1,1} - D_{11} w_{,11} + \tilde{T}\varphi_{1,1} - \overline{M}_1 \end{aligned} \tag{14}$$

The studied beam in the analysis below is a three-ply sandwich β=15, which mechanical characteristics of the mid ply are listed below (Srinivas and al., 1970) :

C_{11} = 10 GPa, C_{22} = 5,43103 GPa, C_{33} = 5,30172 GPa,
C_{12} = 2,3319 GPa, C_{13} = 0,10776 GPa, C_{23} = 0,98276 GPa,
C_{44} = 2,62931 GPa, C_{55} = 1,59914 GPa, C_{66} = 2,6681 GPa.

The density is 2950 Kg/m^3. The Abaqus reference model is composed by eight nodes plane stress elements (CPS8). The meshed model represents 4000 elements with 40 elements through the thickness.

3. Bending Analysis

In bending, the study is static, then the virtual power of acceleration quantities is cancelled, the first member of the motion equations (13) disappear. Three studies are going to be developped, each corresponding to a type of boundary problem. For the simply supported conditions, unknowns are deduced directly from the equilibrium equations. For clamped conditions, the kinematical boundary conditions are used, and finally in a free edge case, the natural boundary conditions (Idlbi, 1995). Integrating the equilibrium equation $\sigma_{13,1} + \sigma_{33,3} = 0$ permits to obtain an analytical value of σ_{33}.

3.1. SIMPLY SUPPORTED BEAM UNDER DISTRIBUTED SINUSOIDAL LOAD

The surface and volume forces components are cancelled except :

$$\overline{n}_3 = \int_0^h f_3 dx_3 = q = q_0 \sin\frac{\pi x_1}{L} \quad (15)$$

In the case of a boundary problem with simply supported conditions, the type of solution used is determined by the Levy technique (Timoshenko and al., 1989 and Karama and al., 1993). u_1^0, w and φ_1 are supposed by the following forms :

$$u_1^0 = u_0 \cos\frac{\pi x_1}{L} \qquad w = w_0 \sin\frac{\pi x_1}{L} \qquad \varphi_1 = \varphi_0 \cos\frac{\pi x_1}{L} \quad (16)$$

The equation system to solve is deduced from (13) and (15) with $P^*_{(a)} = 0$. The resolution permits to calculate the values of u_0, w_0 and φ_0.

TABLE 1. Bending of a simply supported beam under sinusoidal load (h/L=0,1)

	σ_{11} (L/2,h/10) (MPa)	σ_{13} (L,h/10) (MPa)	σ_{33} (L/2,h) (MPa)	U_1 (L,h) (mm)	U_3 (L/2) (mm)
Present	88,876	3,247	-0,998	2,5734	-22,096
Abaqus	89,273	3,318	-1,000	2,8518	-23,847
Error %	0,44	2,14	0,28	9,76	7,34

3.2. CLAMPED FREE BEAM UNDER DISTRIBUTED UNIFORM LOAD

In this case, the value of $\overline{n}_3$ is : $\quad \overline{n}_3 = \int_0^h f_3 dx_3 = q \quad (17)$

The equation system to solve is deduced from (13) and (17) :

$$\begin{aligned} 0 &= A_{11}u^0_{1,11} - B_{11}w_{,111} + \widetilde{K}\varphi_{1,11} \\ 0 &= B_{11}u^0_{1,111} - D_{11}w_{,1111} + \widetilde{T}\varphi_{1,111} + q \\ 0 &= \widetilde{K}u^0_{1,11} - \widetilde{T}w_{,111} + \widetilde{S}\varphi_{1,11} - \widetilde{Y}\varphi_1 \end{aligned} \quad (18)$$

Integrating the differential system (18) permits to obtain the solutions of u_1^0, w and φ_1. Eight constants are to be determined by the four natural boundary conditions at the free edge deduced from (14) with $P^*_{(a)} = 0$:

$$\begin{aligned} 0 &= A_{11}u^0_{1,1} - B_{11}w_{,11} + \widetilde{K}\varphi_{1,1} \\ 0 &= B_{11}u^0_{1,11} - D_{11}w_{,111} + \widetilde{T}\varphi_{1,11} \\ 0 &= \widetilde{K}u^0_{1,1} - \widetilde{T}w_{,11} + \widetilde{S}\varphi_{1,1} \\ 0 &= B_{11}u^0_{1,1} - D_{11}w_{,11} + \widetilde{T}\varphi_{1,1} \end{aligned} \quad (19)$$

and by the following kinematic boundary conditions at the clamped edge :

$$w(0)=0,\ u_1^0(0)=0,\ w_{,1}(0)=0,\ \varphi_1(0)=0 \quad (20)$$

TABLE 2. Bending of a clamped free beam under distributed uniform load (h/L=0,1)

	σ_{11} (L/10,h/10) (Pa)	σ_{13} (L/10,h/10) (Pa)	σ_{33} (L/2,h) (Pa)	U_1 (L,h) (mm)	U_3 (L) (mm)
Present	-376576	-9157	-995	0,0128	-0,224
Abaqus	-376420	-9441	-1001	0,0142	-0,244
Error %	0,04	3,00	0,57	10,18	8,06

3.3. CLAMPED FREE BEAM UNDER CONCENTRATED LOAD

In this case, the value of $\overline{N}_3$ is : $\overline{N}_3 = \int_0^h F_3 dx_3 = q$ (21)

The equation system to solve is deduced from (13) :

$$\begin{aligned} 0 &= A_{11}u^0_{1,11} - B_{11}w_{,111} + \widetilde{K}\varphi_{1,11} \\ 0 &= B_{11}u^0_{1,111} - D_{11}w_{,1111} + \widetilde{T}\varphi_{1,111} \\ 0 &= \widetilde{K}u^0_{1,11} - \widetilde{T}w_{,111} + \widetilde{S}\varphi_{1,11} - \widetilde{Y}\varphi_1 \end{aligned} \tag{22}$$

Integrating the differential system (22) permits to obtain the solutions of u^0_1, w and φ_1. Eight constants are to be also determined by the four natural boundary conditions at the free edge deduced from (14) and (21) :

$$\begin{aligned} 0 &= A_{11}u^0_{1,1} - B_{11}w_{,11} + \widetilde{K}\varphi_{1,1} \\ 0 &= B_{11}u^0_{1,11} - D_{11}w_{,111} + \widetilde{T}\varphi_{1,11} - q \\ 0 &= \widetilde{K}u^0_{1,1} - \widetilde{T}w_{,11} + \widetilde{S}\varphi_{1,1} \\ 0 &= B_{11}u^0_{1,1} - D_{11}w_{,11} + \widetilde{T}\varphi_{1,1} \end{aligned} \tag{23}$$

and by the following kinematic boundary conditions at the clamped edge :

$$w(0)=0,\ u^0_1(0)=0,\ w_{,1}(0)=0,\ \varphi_1(0)=0 \tag{24}$$

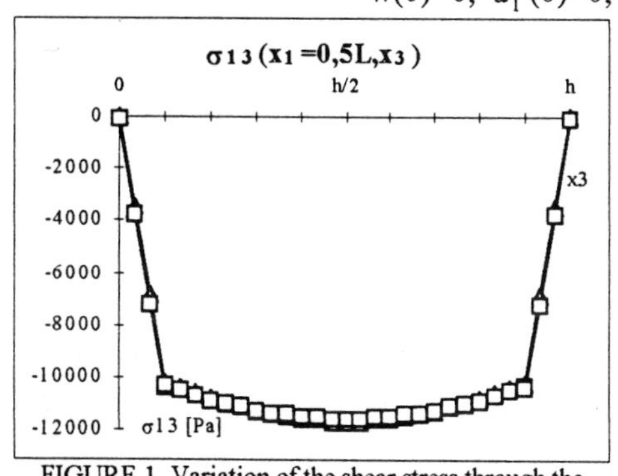

FIGURE 1. Variation of the shear stress through the thickness for x_1=L/2. (□ Abaqus, Δ Present)

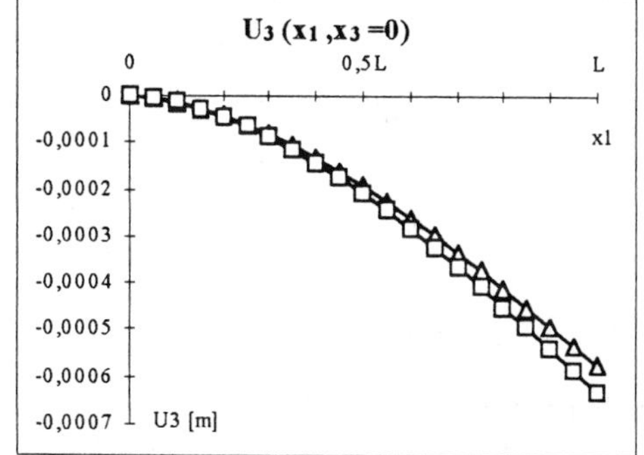

FIGURE 2. Variation of the deflection U_3 along the direction x_1 for x_3=0. (□ Abaqus, Δ Present)

The error obtained on displacements U_1 and U_3 in tables 1-2 and on figure 2 can be explained by the fact that the beam used for this analysis is thick (L/h=10).

4. Buckling Analysis of a Simply Supported Thin Beam

The analysis of the buckling behaviour underlines a succession of stable equilibrium configuration in a plane stress state. This fundamental equilibrium trajectory is followed by the structure from the loading. Then a critical load point where the equilibrium loses its stability. It is the junction threshold or the buckling critical threshold (Gachon, 1980 and,

Palardy and al., 1990). The buckling is a non linear static problem with large displacements. The ε_{11} strain is composed by a linear part and a non linear part (Preusser, 1984) : $\varepsilon_{11} = U_{1,1} + \frac{1}{2}(w_{,1})^2$.

The non linear term modifies the value of the virtual power of internal effort as follows :

$$P^{*}_{(i)} = P^{*L}_{(i)} + P^{*NL}_{(i)} \quad \text{with,} \quad P^{*NL}_{(i)} = -\int_{\Omega} \sigma_{11} w_{,1} w^{*}_{,1} d\Omega = \int_0^L N^0_{11} w_{,11} w^{*} dx_1 - N^0_{11} w_{,1} w^{*} \quad (25)$$

with, $N^0_{11} = \int_0^h \sigma^0_{11} dx_3$ = negative compression load.

The surface and volume forces components are cancelled. The equilibrium equations (13) are modified by the non linear term :

$$\begin{aligned} 0 &= A_{11} u^0_{1,11} - B_{11} w_{,111} + \widetilde{K}\varphi_{1,11} \\ 0 &= B_{11} u^0_{1,111} - D_{11} w_{,1111} + \widetilde{T}\varphi_{1,111} + N^0_{11} w_{,11} \\ 0 &= \widetilde{K} u^0_{1,11} - \widetilde{T} w_{,111} + \widetilde{S}\varphi_{1,11} - \widetilde{Y}\varphi_1 \end{aligned} \quad (26)$$

The Levy type solution supposes u^0_1, w and φ_1 as follows :

$$u^0_1 = U_n \cos(\lambda_n x_1) \quad w = W_n \sin(\lambda_n x_1) \quad \varphi_1 = \Phi_n \cos(\lambda_n x_1) \quad (27)$$

with, $\lambda_n = n\pi / L$, n is an integer.

To determine the buckling critical load, the value of N^0_{11} searched has to cancel the determinant of the system (26).

TABLE 3. Buckling of a simply supported beam (h/L=0,05)

Mode	Buckling Critical Loads (N)		
	Present	Abaqus	Error %
1	146266704	135615000	7,8
2	458550255	438447000	4,6
3	757364592	755132000	0,3
4	979065057	1015780000	3,6

For the first mode, the error maybe comes from the difficulty to introduce simply supported condition in Abaqus and because higher order terms are neglected in ε_{11}.

5. Free Vibration Analysis

The dynamic study is realised in free vibration case. The whole terms of the motion equations (13) are taking into account. Two studies in free vibration (without and with an initial load) are developped. In each problem it concerns simply supported thick and thin beams. The surface and volume forces components are cancelled in the two studies.

5.1. WITHOUT AN INITIAL LOAD

The equation system to solve is deduced from the motion equations (13) :

$$\begin{aligned} I_w \ddot{u}^0_1 + I_{uw'} \ddot{w}_{,1} + I_{u\omega} \ddot{\varphi}_1 &= A_{11} u^0_{1,11} - B_{11} w_{,111} + \widetilde{K}\varphi_{1,11} \\ -I_{uw'} \ddot{u}^0_{1,1} + I_w \ddot{w} - I_{w'} \ddot{w}_{,11} - I_{\omega w'} \ddot{\varphi}_{1,1} &= B_{11} u^0_{1,111} - D_{11} w_{,1111} + \widetilde{T}\varphi_{1,111} \\ I_{u\omega} \ddot{u}^0_1 + I_{\omega w'} \ddot{w}_{,1} + I_\omega \ddot{\varphi}_1 &= \widetilde{K} u^0_{1,11} - \widetilde{T} w_{,111} + \widetilde{S}\varphi_{1,11} - \widetilde{Y}\varphi_1 \end{aligned} \quad (28)$$

The Levy type solution supposes u_1^0, w and φ_1 as follows :

$$u_1^0 = U_n \cos(\lambda_n x_1).\exp(i\omega t), w = W_n \sin(\lambda_n x_1).\exp(i\omega t), \varphi_1 = \Phi_n \cos(\lambda_n x_1).\exp(i\omega t) \quad (29)$$

To determine the vibration frequency, the value of ω searched has to cancel the determinant of the system (28). When n changes, different equations in ω are obtained, in which the smallest positive root gives the vibration frequency of the associated mode n.

TABLE 4. Free vibration without an initial load

h/L	0,05			0,1		
Mode	Vibration Frequency (Hz)					
	Present	Abaqus	Error %	Present	Abaqus	Error %
1	5,562	5,295	5	19,674	18,931	4
2	19,674	18,931	4	57,498	56,551	1,7
3	37,916	36,903	2,8	96,693	96,903	0,2
4	57,498	56,551	1,7	134,34	137,21	2
5	77,232	76,691	0,7	170,09	177,49	4,2
6	96,693	96,903	0,2	203,84	218,06	6,5

5.2. WITH AN INITIAL LOAD

An initial axial load is added to study the influence of this initial constraint on the free vibration of the previous problem. This load has the same direction as the buckling study, but its value is lower than the first critical load. In this analytical resolution, the non linear term (25) defined in the buckling study is taking into account.

The equation system to solve is deduced from the motion equations (13) :

$$\begin{aligned} & I_w \ddot{u}_1^0 + I_{uw'} \ddot{w}_{,1} + I_{u\omega} \ddot{\varphi}_1 = A_{11} u^0_{1,11} - B_{11} w_{,111} + \widetilde{K} \varphi_{1,11} \\ & - I_{uw'} \ddot{u}^0_{1,1} + I_w \ddot{w} - I_{w'} \ddot{w}_{,11} - I_{\omega w'} \ddot{\varphi}_{1,1} = B_{11} u^0_{1,111} - D_{11} w_{,1111} + \widetilde{T} \varphi_{1,111} + N^0_{11} w_{,11} \\ & I_{u\omega} \ddot{u}_1^0 + I_{\omega w'} \ddot{w}_{,1} + I_\omega \ddot{\varphi}_1 = \widetilde{K} u^0_{1,11} - \widetilde{T} w_{,111} + \widetilde{S} \varphi_{1,11} - \widetilde{Y} \varphi_1 \end{aligned} \quad (30)$$

The Levy type solution supposes u_1^0, w and φ_1 the same as in the previous study (29). The resolution is the same as the previous problem.

TABLE 5. Free vibration with an initial load

h/L	0,05			0,1		
Mode	Vibration Frequency (Hz)					
	Present	Abaqus	Error %	Present	Abaqus	Error %
1	3,128	2,785	12,3	17,397	16,771	3,7
2	17,397	16,771	3,7	54,483	54,029	0,8
3	35,324	34,607	2	92,674	93,796	1,2
4	54,483	54,029	0,8	129,19	133,42	3,2
5	73,734	73,893	0,2	163,75	173,01	5,3
6	92,674	93,796	1,2	196,22	212,87	7,8

By the same reasons as in the buckling analysis the error for the first mode can be explained. In table 5, the error 12,3% is maybe the result of a small value of the frequency.

6. Conclusion

The proposed new model satisfies exactly and automatically the continuity conditions of displacements and stresses at the interfaces, as well as the frontier conditions, for sandwich composite beams in small elastic perturbations. In fact, the transverse shear stress continuity is automatically obtained thanks to the Heaviside step function.
Besides, the development of this model is simplified by the sine type transverse shear function. The sine function permits simple mathematical manipulations (differenciation, integration) while staying accurate because of the x_3 odds power terms which appears in its development. This model is also simple in so far as any correction factor is used in opposition to the higher order models.
The innovation in relation to the models proposed in the bibliography is also the introduction of the membrane refinement cosine function $g(x_3)$ which represent in fact the counterpart of the transverse shear sine function $f(x_3)$. Actually, the function $g(x_3)$ is the same as $f(x_3)$, but with a cosine instead of a sine, which permits to take into account the x_3 even power terms. This membrane refinement improves the warping of the straight section in bending deformations.

References

Di Sciuva, M. (1987). An Improved Shear-Deformation Theory for Moderately Thick Multilayered Anisotropic Shells and Plates, *ASME J. Appl. Mech.* **54**, 589-596.

Di Sciuva, M. (1993). A General Quadrilateral Multilayered Plate Element with Continuous Interlaminar Stresses, *Computers and Structures* **47**(1), 91-105.

Gachon, H. (1980). *Sur le flambage des plaques : modèle de calcul, modèles expérimentaux*, Construction Métallique.

Germain, P. and Muller, P. (1995). *Introduction à la mécanique des milieux continus*, Masson.

He, L.H. (1994). A Linear Theory of Laminated Shells Accounting for Continuity of Displacements and Transverse Shear Stresses at Layer Interfaces, *Int. J. Solids structures* **31**(5), 613-627.

Idlbi, A. (1995). *Comparaison de théories de plaque et estimation de la qualité des solutions dans la zone bord*, Thesis, ENSAM, Paris.

Karama, M. , Touratier, M. and Idlbi, A. (1993). An Evaluation of the Edge Solution for a Higher-order Laminated Plate Theory, *Composite Structures* **25**, 495-502.

Ossadzow, C. , Muller, P. and Touratier, M. (1995). *Une théorie générale des coques composites multicouches*, Deuxième colloque national en calcul des structures, Tome 1, Hermes.

Palardy, R.F. and Palazotto, A.N. (1990). Buckling and Vibration of Composite Plates using the Levy method, *Composite Structures* **14**, 61-86.

Preusser, G. (1984). *Ing. Arch.* **54**, 51-61.

Srinivas, S. and Rao, A.K. (1970). Bending, Vibration and Buckling of Simply Supported Thick Orthotropic Rectangular Plates and Laminates, *Int. J. Solids Structures* 6, 1463-1481.

Timoshenko, S. and Woinowsky-Krieger (1989). *Theory of Plates and Shells*, Mc Graw-Hill.

Touratier, M. (1991). An Efficient Standard Plate Theory, *Int. J. Eng. Sci.* **29**(8), 901-916.

Touratier, M. (1992). A Generalization of Shear Deformation Theories for Axisymmetric Multilayered Shells, *Int. J. Solids Structures* **29**(11), 1379-1399.

Touratier, M. (1992). A Refined Theory of Laminated Shallow Shells, *Int. J. Solids Structures* **29**(11), 1401-1415.

STRESS DISTRIBUTION IN A 4-POINT OR 3-POINT BENDING PROBE

H.-R. MEYER-PIENING
Prof. Dr.-Ing., ETH Zürich
Institute for Lightweight Structures and Ropeways
Leonhardstr. 25, CH-8092 Zürich

Abstract

An analytical method is presented, which allows the calculation of stresses and displacements in a multilayered sandwich beam satisfying stress and displacement continuities throughout the structural element, allowing for parametric studies without limitations regarding geometric dimensions or material properties. The model accounts for transverse compressibility as well as Poissons ratio effect, overhangs outside the supports and finite width at the loading zones. The method is evaluated for symmetrical 3pt. and 4pt. bending load cases but can easily be modified to cover more general cases. The results were found to agree well with fine-mesh FE results and are supported by test results.

Introduction

The application of soft core sandwich construction to structures like vehicles, buildings, ships, containers, a.s.o., require a fair knowledge of the stiffness properties and in some cases the local failure conditions.

Laboratory tests are frequently conducted on short four-point or three-point bending probes. The evaluation of the obtained test results relies on the knowledge of the actual stress distribution within the tested specimen. Several models exist in which attempts are made to describe the deflections throughout the sandwich beam in order to approximate the real stress and displacement distribution. So far the most promising results are related to multilinear displacement assumptions. However, the incompatibilities related to shear stress and shear strain distributions, known from Kirchhoff beams, are maintained.

The aim of this paper is to present an analytical tool to analyze a sandwich beam satisfying all specified stress and continuity conditions within each layer and at the layer interfaces in order to eliminate all possible geometric constraint for the analytical investigations. Specifically, there should be no restrictions related to the thickness of any layer, such that the faces may vary in thickness from zero up to a value which eliminates the core. And it should allow for unequal faces as a good portion of structural applications incorporate that feature.

This leads to the "exact" solution of the load equilibrium within each layer and at the outer and inner interfaces, as well as to the requirement to satisfy out-of-plane and inplane displacement continuities.

Nomenclature

a = length of overhanging ends
b = distance between outer edge of the beam and the center of the applied load
B = width of the beam
$E_x^{(i)}$ = Young's modulus in axial direction, $\bar{E}_x^{(i)} = E_x^{(i)}/(1-\nu_{xz}^{(i)}\nu_{zx}^{(i)})$
$E_z^{(i)}$ = Young's modulus in transverse direction, $\bar{E}_z^{(i)} = E_z^{(i)}/(1-\nu_{xz}^{(i)}\nu_{zx}^{(i)})$
$G_{xz}^{(i)}$ = transverse shear modulus
h_i = station of the layer interface (accumulative, starting from $z = 0$), $H = h_3$

A. Vautrin (ed.), Mechanics of Sandwich Structures, 105–112.

L = total length of the sandwich beam
m = halfwave number (in axial direction)
u, w = inplane and lateral deflection
$u^{(i)}_{jm}$, $w^{(i)}_{jm}$ = coefficients for layer (i) and series term m; $1 \le j \le 4$.
x_0 = width of the zone of load application
x_R = width of the contact zone at the supports
$\lambda^{(i)}_{jm}$ = exponent (in transverse direction), eigenvalue of Eqs. (11)
$v^{(i)}_{xz}$ = Poisson's ratio
σ = stress,
(i) = layer number, i = 1, 2 or 3 for a three-layered sandwich

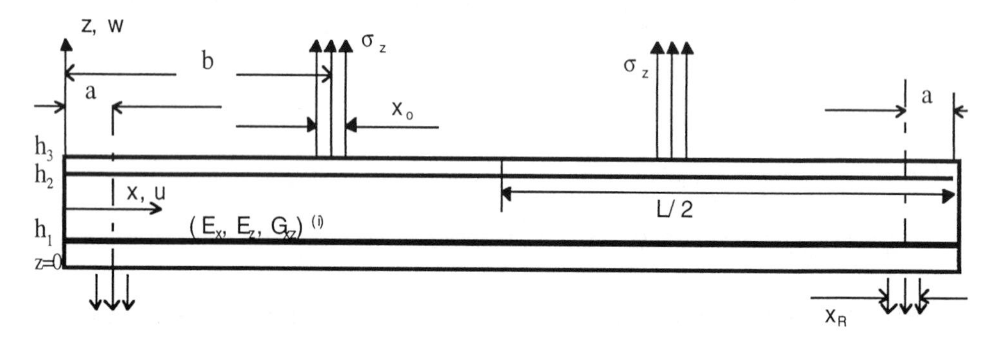

Fig. 1: Coordinates, dimensions

Governing equations

Each layer (i) of the sandwich beam is characterized by its axial and transversal Young's modulus, its shear modulus and the associated Poisson's ratios. The out-of-plane stress (y-direct.) is assumed to be negligible and, hence, the out-of-plane moduli do not appear in the equations.

Local internal equilibrium within layer (i) requires that

$$\frac{\partial \sigma^{(i)}_x}{\partial x} + \frac{\partial \tau^{(i)}_{xz}}{\partial z} = 0 \tag{1}$$

$$\frac{\partial \tau^{(i)}_{xz}}{\partial x} + \frac{\partial \sigma^{(i)}_z}{\partial z} = 0. \tag{2}$$

In view of the support conditions the resulting displacements shall be represented by

$$w^{(i)} = w^{(i)}_0 + w^{(i)}_1 * z + \sum_{(m)}\sum_{j=1}^{4} w^{(i)}_{jm} * e^{\lambda^{(i)}_{jm} z} \sin\frac{m\pi x}{L}; \quad \text{with } w\big|_{(x=a,L-a;z=0)} = 0 \tag{3}$$

$$\text{or } w^{(1)}_0 = w^{(i)}_0 = -\sum_{(m)}\sum_{j=1}^{4} w^{(1)}_{jm} * \sin\frac{m\pi a}{L}$$

$$u^{(i)} = \sum_{(m)}\sum_{j=1}^{4} u^{(i)}_{jm} * e^{\lambda^{(i)}_{jm} z} \cos\frac{m\pi x}{L}; \text{ with } \left.\frac{\partial u^{(i)}}{\partial x}\right|_{(x=0,L;z)} = 0 \tag{4}$$

Assuming linear strain-displacement and stress-strain relations, the corresponding stresses are found to be:

$$\sigma^{(i)}_x = \sum_{(m)} \sigma^{(i)}_{xm} = \overline{E}^{(i)}_x \left(\frac{\partial u^{(i)}}{\partial x} + v^{(i)}_{xz} \frac{\partial w^{(i)}}{\partial z} \right) \tag{5}$$

$$\sigma_z^{(i)} = \sum_{(m)} \sigma_{zm}^{(i)} = \overline{E}_z^{(i)}\left(\frac{\partial w^{(i)}}{\partial z} + \nu_{zx}^{(i)}\frac{\partial u^{(i)}}{\partial x}\right) \tag{6}$$

$$\tau_{xz}^{(i)} = \sum_{(m)} \tau_{xzm}^{(i)} = G_{xz}^{(i)}\left(\frac{\partial w^{(i)}}{\partial x} + \frac{\partial u^{(i)}}{\partial z}\right) \tag{7}$$

or, finally, for each harmonic component of the stresses

$$\sigma_{xm}^{(i)} = \overline{E}_x^{(i)} \sum_{j=1}^{4}\left[-\frac{m\pi}{L}u_{jm}^{(i)} + \nu_{xz}^{(i)} * \lambda_{jm}^{(i)} * w_{jm}^{(i)}\right]e^{\lambda_{jm}^{(i)}z} * \sin\frac{m\pi x}{L} \tag{8}$$

$$\sigma_{zm}^{(i)} = \overline{E}_z^{(i)} \sum_{j=1}^{4}\left[-\nu_{zx}^{(i)}\frac{m\pi}{L}u_{jm}^{(i)} + \lambda_{jm}^{(i)} * w_{jm}^{(i)}\right]e^{\lambda_{jm}^{(i)}z} * \sin\frac{m\pi x}{L} \tag{9}$$

$$\tau_{xzm}^{(i)} = G_{xz}^{(i)} \sum_{j=1}^{4}\left[\lambda_{jm}^{(i)} * u_{jm}^{(i)} + \frac{m\pi}{L} * w_{jm}^{(i)}\right]e^{\lambda_{jm}^{(i)}z} * \cos\frac{m\pi x}{L} \tag{10}$$

Substitution into the equilibrium equations Eq. (1) and (2) yields:

$$\left[-\overline{E}_x^{(i)}\left(\frac{m\pi}{L}\right)^2 + G_{xz}^{(i)}\lambda_{jm}^{(i)2}\right] * u_{jm}^{(i)} + \left[\overline{E}_x^{(i)}\nu_{xz}^{(i)} + G_{xz}^{(i)}\right] * \lambda_{jm}^{(i)}\frac{m\pi}{L} * w_{jm}^{(i)} = 0 \tag{11a}$$

$$\left[\overline{E}_x^{(i)}\nu_{xz}^{(i)} + G_{xz}^{(i)}\right] * \lambda_{jm}^{(i)}\frac{m\pi}{L} * u_{jm}^{(i)} + \left[-\overline{E}_z^{(i)}\lambda_{jm}^{(i)2} + G_{xz}^{(i)} * \left(\frac{m\pi}{L}\right)^2\right] * w_{jm}^{(i)} = 0 \tag{11b}$$

The set of linear equations, Eqs. (11a,b), can be solved for the condition of a vanishing determinant:

$$\lambda_{jm}^{(i)} = \pm\frac{m\pi}{L}\sqrt{\left(\frac{E_x^{(i)}}{2G_{xz}^{(i)}} - \nu_{zx}^{(i)}\right) \pm \sqrt{\left(\frac{E_x^{(i)}}{2G_{xz}^{(i)}} - \nu_{zx}^{(i)}\right)^2 - \frac{E_x^{(i)}}{E_z^{(i)}}}} \tag{12}$$

imposing some constraints on the allowable combination of the stiffness terms, see also [1[, [2], [3]. Subsequently, the unknowns $u_{jm}^{(i)}$ are eliminated in view of Eq. (11a):

$$u_{jm}^{(i)} = \frac{\left(\overline{E}_x^{(i)} * \nu_{xz}^{(i)} + G_{xz}^{(i)}\right)\frac{m\pi}{L}\lambda_{jm}^{(i)}}{\overline{E}_x^{(i)}\left(\frac{m\pi}{L}\right)^2 - G_{xz}^{(i)} * \lambda_{jm}^{(i)2}} * w_{jm}^{(i)} \tag{13}$$

Boundary and interface conditions

1. Loading at x = b, L-b, $z = h_3$ over a width of x_0:

$$\sigma_z^{(3)}\Big|_{z=h_3} = \overline{\sigma}\sum_{(m)}\frac{2}{m\pi}\left\{\begin{matrix} -\cos\left(\frac{m\pi(L-b+x_0/2)}{L}\right) - \cos\left(\frac{m\pi(b+x_0/2)}{L}\right) + \\ +\cos\left(\frac{m\pi(L-b-x_0/2)}{L}\right) + \cos\left(\frac{m\pi(b-x_0/2)}{L}\right)\end{matrix}\right\}\sin\left(\frac{m\pi x}{L}\right)$$

$$= \overline{E}_z^{(3)}\left\{w_1^{(3)} + \sum_{m=1,3,}\sum_{j=1}^{4}\left[-\upsilon_{zx}^{(3)}\frac{m\pi x}{L}u_{jm}^{(3)} + \lambda_{jm}^{(3)}w_{jm}^{(3)}\right]e^{\lambda_{jm}^{(3)}h_3}\sin\left(\frac{m\pi x}{L}\right)\right\} \tag{14}$$

or with $w_1^{(3)} = 0$ (as there is no constant transverse stress or strain considered)

$$\sigma_{zm}^{(3)}\Big|_{z=h_3} = \overline{E}_z^{(3)}\sum_{j=1}^{4}\left\{\frac{-\left(\overline{E}_x^{(3)}\nu_{xz}^{(3)} + G_{xz}^{(3)}\right) * \nu_{zx}^{(3)}\left(\frac{m\pi}{L}\right)^2}{\overline{E}_x^{(3)}\left(\frac{m\pi}{L}\right)^2 - G_{xz}^{(3)} * \lambda_{jm}^{(3)2}} + 1\right\}e^{\lambda_{jm}^{(3)}h_3} * \lambda_{jm}^{(3)} * w_{jm}^{(3)} * \sin\left(\frac{m\pi x}{L}\right)$$

2. Reaction load at x = a, L-a, z = 0 over a width of x_R:

$$\sigma_z^{(1)}\Big|_{z=0} = \bar{\sigma}\frac{x_0}{x_R}\sum_{(m)}\frac{2}{m\pi}\left\{\begin{matrix}-\cos\left(\frac{m\pi(L-a+x_R/2)}{L}\right)-\cos\left(\frac{m\pi(a+x_R/2)}{L}\right)+\\ +\cos\left(\frac{m\pi(L-a-x_R/2)}{L}\right)+\cos\left(\frac{m\pi(a-x_R/2)}{L}\right)\end{matrix}\right\}\sin\left(\frac{m\pi x}{L}\right)$$

$$= \bar{E}_z^{(1)}\left\{w_1^{(1)} + \sum_{m=1,3,}\sum_{j=1}^{4}\left[-\upsilon_{zx}^{(1)}\frac{m\pi}{L}u_{jm}^{(1)} + \lambda_{jm}^{(1)}w_{jm}^{(1)}\right]\sin\left(\frac{m\pi x}{L}\right)\right\} \tag{15}$$

or with $w_1^{(1)} = 0$ (as no constant transverse stress is applied):

$$\sigma_{zm}^{(1)}\Big|_{z=0} = \bar{E}_z^{(1)}\sum_{j=1}^{4}\left\{\frac{-\left(\bar{E}_x^{(1)}v_{xz}^{(1)}+G_{xz}^{(1)}\right)*v_{zx}^{(1)}\left(\frac{m\pi}{L}\right)^2}{\bar{E}_x^{(1)}\left(\frac{m\pi}{L}\right)^2 - G_{xz}^{(1)}*\lambda_{jm}^{(1)2}}+1\right\}*\lambda_{jm}^{(1)}*w_{jm}^{(1)}*\sin\left(\frac{m\pi x}{L}\right)$$

3., 4. Equal transverse load at $z = h_i$: (i = 1, 2)

$$\bar{E}_z^{(i)}\sum_{j=1}^{4}\left\{\frac{\left(\bar{E}_x^{(i)}v_{xz}^{(i)}+G_{xz}^{(i)}\right)v_{zx}^{(i)}\left(\frac{m\pi}{L}\right)^2}{\bar{E}_x^{(i)}\left(\frac{m\pi}{L}\right)^2 - G_{xz}^{(i)}*\lambda_{jm}^{(i)2}}-1\right\}\lambda_{jm}^{(i)}e^{\lambda_{jm}^{(i)}h_i}w_{jm}^{(i)} -$$

$$- \bar{E}_z^{(i+1)}\sum_{j=1}^{4}\left\{\frac{\left(\bar{E}_x^{(i+1)}v_{xz}^{(i+1)}+G_{xz}^{(i+1)}\right)v_{zx}^{(i+1)}\left(\frac{m\pi}{L}\right)^2}{\bar{E}_x^{(i+1)}\left(\frac{m\pi}{L}\right)^2 - G_{xz}^{(i+1)}*\lambda_{jm}^{(i+1)2}}-1\right\}\lambda_{jm}^{(i+1)}e^{\lambda_{jm}^{(i+1)}h_i}w_{jm}^{(i+1)} = 0; \tag{16, 17}$$

and

$$\sigma_{z0}^{(i)}(z=h_i) = \bar{E}_z^{(i)}*w_i^{(i)} = \sigma_{z0}^{(i+1)}(z=h_i) = \bar{E}_z^{(i+1)}*w_i^{(i+1)} \quad , \quad w_0^{(i)} = w_0^{(i+1)} = 0$$

5., 6. Vanishing shear load at z = 0 and z = h_3: (i = 1, 3; h_1 = 0 (in this case))

$$\tau_{xzm}^{(i)} \approx \sum_{j=1}^{4}\left\{\frac{\left(\bar{E}_x^{(i)}v_{xz}^{(i)}+G_{xz}^{(i)}\right)\lambda_{jm}^{(i)2}}{\bar{E}_x^{(i)}\left(\frac{m\pi}{L}\right)^2 - G_{xz}^{(i)}\lambda_{jm}^{(i)2}}+1\right\}e^{\lambda_{jm}^{(i)}h_i}*w_{jm}^{(i)} = 0; \tag{18, 19}$$

7., 8. Equal shear load at $z = h_i$: (i = 1, 2)

$$G_{xz}^{(i)}\sum_{j=1}^{4}\left\{\frac{\left(\bar{E}_x^{(i)}v_{xz}^{(i)}+G_{xz}^{(i)}\right)\lambda_{jm}^{(i)2}}{\bar{E}_x^{(i)}\left(\frac{m\pi}{L}\right)^2 - G_{xz}^{(i)}\lambda_{jm}^{(i)2}}+1\right\}e^{\lambda_{jm}^{(i)}h_i}*w_{jm}^{(i)} -$$

$$- G_{xz}^{(i+1)}\sum_{j=1}^{4}\left\{\frac{\left(\bar{E}_x^{(i+1)}v_{xz}^{(i+1)}+G_{xz}^{(i+1)}\right)\lambda_{jm}^{(i+1)2}}{\bar{E}_x^{(i+1)}\left(\frac{m\pi}{L}\right)^2 - G_{xz}^{(i+1)}\lambda_{jm}^{(i+1)2}}+1\right\}e^{\lambda_{jm}^{(i+1)}h_i}*w_{jm}^{(i+1)} = 0; \tag{20, 21}$$

9., 10. Equal inplane displacements: (i = 1, 2)

$$\sum_{j=1}^{4}\left\{\frac{\left(\overline{E}_x^{(i)}\nu_{xz}^{(i)}+G_{xz}^{(i)}\right)\lambda_{jm}^{(i)}}{\overline{E}_x^{(i)}\left(\frac{m\pi}{L}\right)^2-G_{xz}^{(i)}\lambda_{jm}^{(i)2}}\right\}e^{\lambda_{jm}^{(i)}h_i} * w_{jm}^{(i)} - \sum_{j=1}^{4}\left\{\frac{\left(\overline{E}_x^{(i+1)}\nu_{xz}^{(i+1)}+G_{xz}^{(i+1)}\right)\lambda_{jm}^{(i+1)}}{\overline{E}_x^{(i+1)}\left(\frac{m\pi}{L}\right)^2-G_{xz}^{(i+1)}\lambda_{jm}^{(i+1)2}}\right\}e^{\lambda_{jm}^{(i+1)}h_i} * w_{jm}^{(i+1)} = 0; \quad (22, 23)$$

11., 12. Equal out-of-plane displacements: (i = 1, 2)

$$\sum_{j=1}^{4} e^{\lambda_{jm}^{(i)}h_i} * w_{jm}^{(i)} - \sum_{j=1}^{4} e^{\lambda_{jm}^{(i+1)}h_i} * w_{jm}^{(i+1)} = 0; \quad (24, 25)$$

These twelve conditions result in a set of twelve linear equations for the twelve unknowns $w_{jm}^{(i)}$ for i = 1, 2, 3 and j = 1, 2, 3, 4, for a sandwich which consists of 3 layers:

$$\left[a_{jq}^{(i)}\right]_{(m)}\left\{w_{jm}^{(i)}\right\} = \left\{RS(\overline{\sigma}_z)\right\}_{(m)}, \quad (26)$$

yielding $\left\{w_{jm}^{(i)}\right\}$ and in view of Eqs. (11) $\left\{u_{jm}^{(i)}\right\}$.

For symmetric loading cases m = 1,3, ... m_e; m_e typically ranges from 51 to 101, see also Fig. 2.

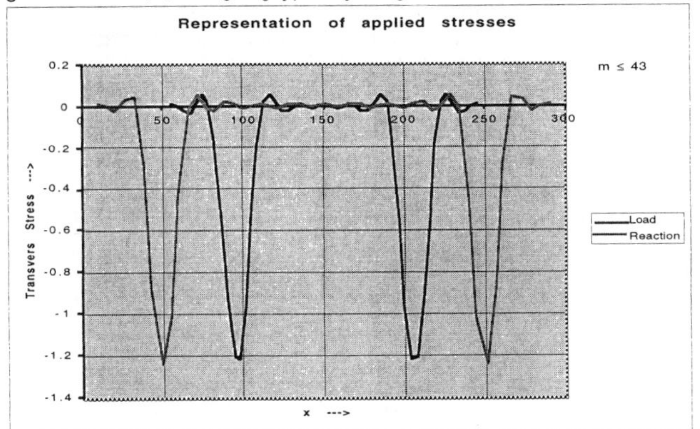

Fig. 2: Example for a 4-point loading representation (m ≤ 43), L = 300, a = 50, b =95, x_0=x_R=20

Numerical examples

1. Equal faces (H = 25, t_f = 2.5), 3pt. bending

Ex (1) =	12500	Ez (1) =	10300	Gxz (1) =	3850	nxz (1) =	0.28	h1 =	2.5
Ex (2) =	60	Ez (2) =	60	Gxz (2) =	25	nxz (2) =	0.18	h2 =	22.5
Ex (3) =	12500	Ez (3) =	10300	Gxz (3) =	3850	nxz (3) =	0.28	h3 =	25
L =	230	a =	15	b =	115	x0 =	20	xR =	20
w(0)=0	0		Lanczo's F:	m ≤	51	x (ref) =	60	Sigma z =	-1.06

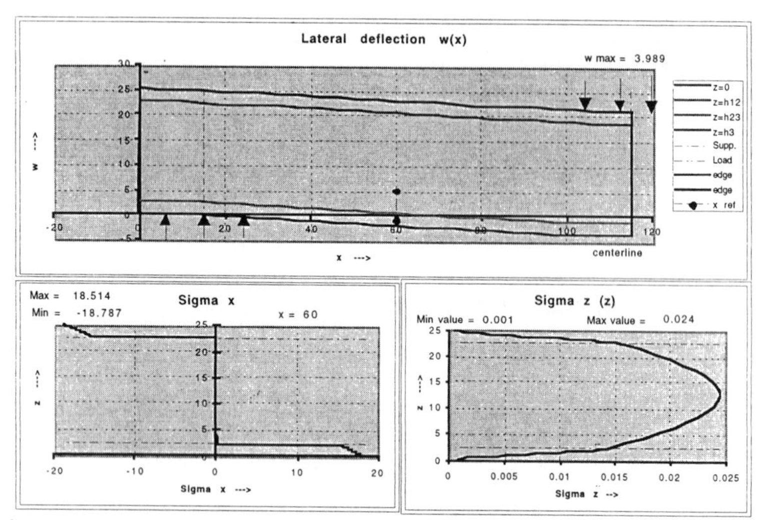

Fig. 3: Deflection and stress distribution (σ_x, σ_z) at station x = 60

In Fig. 3, the deflection of a halfspan of the symmetric beam is shown, while the area of load application is indicated by the arrows. The station at which the distribution functions are evaluated is indicated by a marker at x = 60. The distribution of the axial stress is rather linear throughout the individual layers, the transverse stress at this location close to zero.

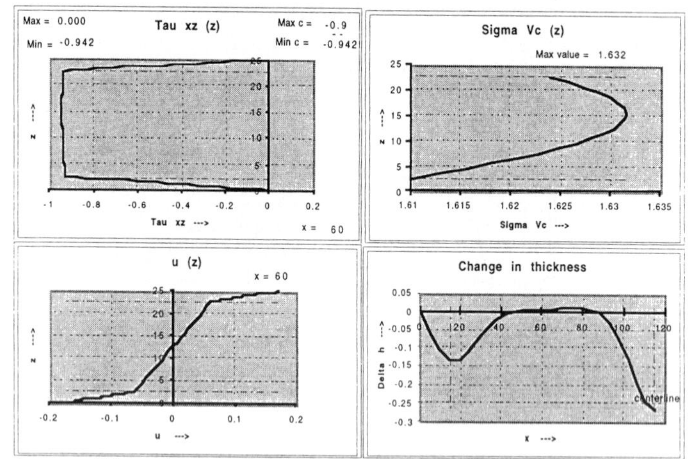

Fig. 4: Shear stress, v. Mises stress, axial displacement and thickness change

Fig. 4 shows the shear stress distribution and the calculated v. Mises stress in the core. The maximum of this stress was found at station x = 77 (somewhat beyond the station indicated in Fig. 4) with a maximum value of 1.66 at the lower interface ($z = h_1$). In the same figure the axial displacement is shown to be layerwise rather linear and the compressibility of the core results in local thickness changes, which could be interpreted as a deflection as result of a stiffness measurement if not being accounted for.

2. EQUAL FACES (H = 25, t_f = 2.5), 3-point bending, extended overhanging ends (a = 40)

The beam ist extended at its overhanging ends in order to maintain the span as well as all other properties and load conditions. From Fig. 5 it can be seen that the the maximum deflection is reduced by a minor amount from 3.989 (a = 15) to 3.978 (a = 40). In this case, the overhanging ends, although slightly distorted, contribute to the equivalent bending stiffness by only a minor amount.

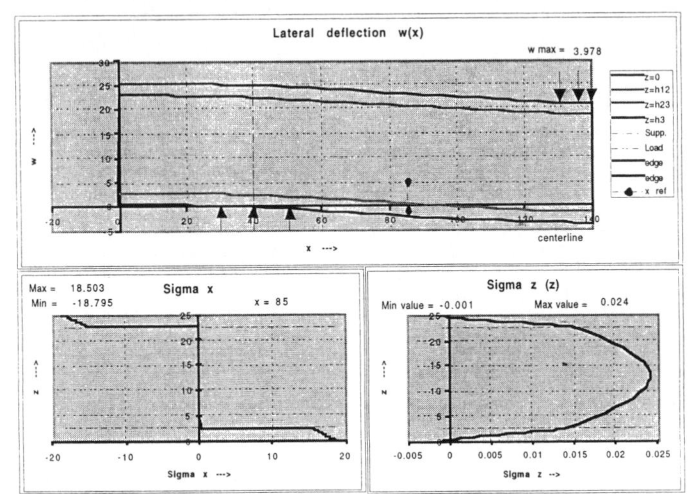

Fig. 5: Deflections and normal stresses for a probe with extended overhanging ends

3. UNEQUAL FACES (H = 25, t_f = 2.5 and 0.25)

The proposed analytical method permits to analyze (linearly) beams with very thin faces, i.e. cases for which the FE method usually is confronted with problems associated with element aspect ratios or mesh dimensions. Figs. 6 and 7 present results of an analysis for the same data as given for case 1, but with the upper face being reduced to 0.25 while the overall thickness of the beam remained 25 (the core height is increased accordingly).

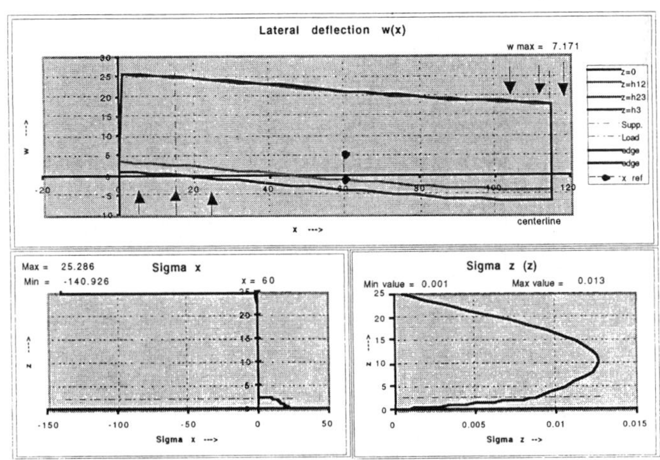

Fig. 6: Deflections and normal stresses for a beam with one very thin face

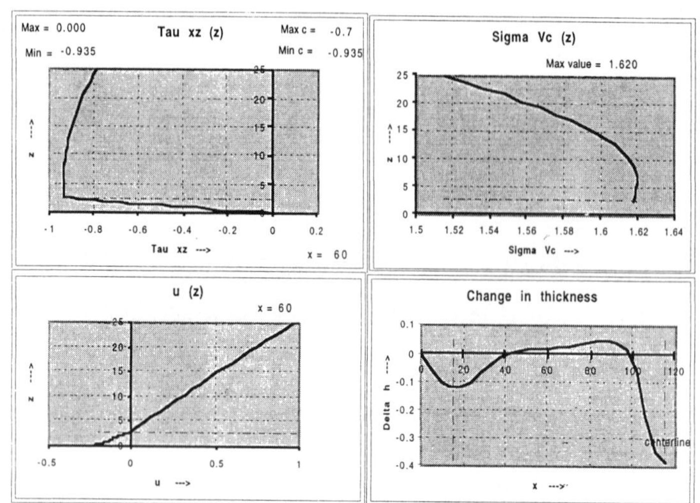

Fig. 7: Shear stress, v. Mises stress and axial displacement at x = 60 and change of thickness

Conclusion

It appears that a layerwise linear displacement assumption is a reasonable approximation for a wide range of geometries if the outlined rather elaborate method is to be avoided. In the considered example the length of the overhanging ends did not significantly contribute to the bending stiffness of the beam. The thickness change can influence significantly the measured deflection of the beam if the lateral displacement is measured as the displacement of the loading piston relative to the supports and not recorded from the beam at opposite faces.

Furthermore, the outlined method can be extended to cover overall and local buckling (crippling) of a sandwich column as will be presented in November at the occasion of an ASME conference in Dallas.

References

[1] Pagano, N.J., Exact solution for composite laminates in cylindrical bending. *J. Comp. Materials* **3** (1969), 398-411

[2] Bhimaraddi, A., Chandrashekhara, K., Comparison of elasticity and sandwich theories for a rectangular sandwich plate. *The Aeron. J. of the Aeron. Soc.,* (J/J 1984), 229-237

[3] Chattopadhyay, A., Gu, H., Exact elasticity solution for buckling of composite laminates. *Composite Structures* ***34*** (1996), 291-299

SAINT-VENANT'S PRINCIPLE FOR SANDWICH STRUCTURES

C. O. HORGAN AND S. C. BAXTER
Institute of Applied Mathematics and Mechanics
School of Engineering and Applied Science
University of Virginia
Charlottesville, VA 22903, USA

Abstract. A review is provided of results on Saint-Venant decay lengths for self-equilibrated edge loads in symmetric sandwich structures. In linear elasticity, Saint-Venant's principle is used to show that self-equilibrated loads generate local stress effects that decay away from the loaded end of a structure. For homogeneous *isotropic* linear elastic materials this is well-documented. Self-equilibrated loads are a class of load distributions that are statically equivalent to zero, i.e., have zero resultant force and moment. When Saint-Venant's principle is valid, pointwise boundary conditions can be replaced by more tractable resultant conditions. Saint-Venant's principle is also the fundamental basis for static mechanical tests of material properties. It is shown in the present paper that material inhomogeneity and anisotropy significantly affect the practical application of Saint-Venant's principle to sandwich structures.

1. Introduction

Designers of composite structures are constantly faced with the task of assessing stress decay effects associated with loading conditions, cut-outs, and other local discontinuities. Thus, a thorough understanding of Saint-Venant's principle as it applies to composite materials is of fundamental importance in expanding the development of composite structures technology. Previous work has shown that, even for *homogeneous* anisotropic materials, anisotropy can significantly affect the decay of Saint-Venant end effects. In particular, for the strongly anisotropic materials used in fiber-reinforced structural laminates, it has been shown that the *decay length* can be much longer than that for isotropic materials under the same loading conditions. The implications for the mechanics of composites have been widely discussed, (see e.g. the references cited at the end of the present paper).

A. Vautrin (ed.), Mechanics of Sandwich Structures, 113–122.

The effects of material *inhomogeneity* have also been investigated, though not as extensively. In [8] plane deformations of sandwich strips, with isotropic phases, were examined. In particular, for a sandwich strip with a relatively compliant core, the characteristic stress decay length is shown to be *much greater than* that for the homogeneous isotropic strip. An asymptotic estimate for the decay rate is also presented in [8] which agrees well with exact numerical results (see e.g., [12]). See also [21, 22, 23, 24, 25] for more recent results on the plane problem. Recently, an extensive study of analogous issues for anti-plane shear (APS) deformations has been carried out in [3, 4].

The relative analytic simplicity of the APS model allows for a more explicit analysis, and the results are relevant to Mode III fracture situations. (See [18] for a comprehensive review of APS.) Furthermore, it is shown in these references that the decay length for APS deformations can be longer than that for plane deformations in many cases.

The purpose of the present paper is to concisely review some of these results. In Section 2 the anti-plane shear problem is formulated for a linearly elastic sandwich structure with anisotropic phases, subject to self-equilibrated end loading only. Solutions which decay in the longitudinal direction are described and transcendental equations obtained for an eigenvalue λ on which the characteristic decay length predominantly depends. In Section 3, the special case of a sandwich structure with *isotropic* phases is examined. For this case the decay length is proportional to the half-width of the sandwich divided by λ. Asymptotic estimates for the decay length are obtained for the case of a relatively compliant core. Only one dimensionless material parameter δ, the ratio of the shear moduli of the two materials, appears in the analysis. This estimate is compared to the exact values obtained by numerical methods. Section 4 considers the effect of incorporating conditions of imperfect bonding at the layer interfaces into the model. The fully anisotropic problem is considered in Section 5, with discussion of a numerical example for sandwich structures with orthotropic phases.

2. Formulation of the Anti-Plane Shear Problem

Consider a three-layered symmetric sandwich structure with cross-section as shown in Figure 1. The outer layers are constructed of the same material and the inner core of a second material. The material in each layer is assumed to be homogeneous, anisotropic and *linearly elastic*. The cross-section of the structure is taken to be semi-infinite. It is convenient to establish a coordinate system for each layer, with common x and z axes and separate y-axes for each layer, which are then denoted by y_1, y_2, y_3.

The layers are numbered from top to bottom, 1,2, and 3.

To study Saint-Venant end-effects in anti-plane shear, a prescribed traction of the form $\mathbf{t}^* = (0, 0, t_3^*)$, shown schematically in Figure 1, is applied on the portion of the

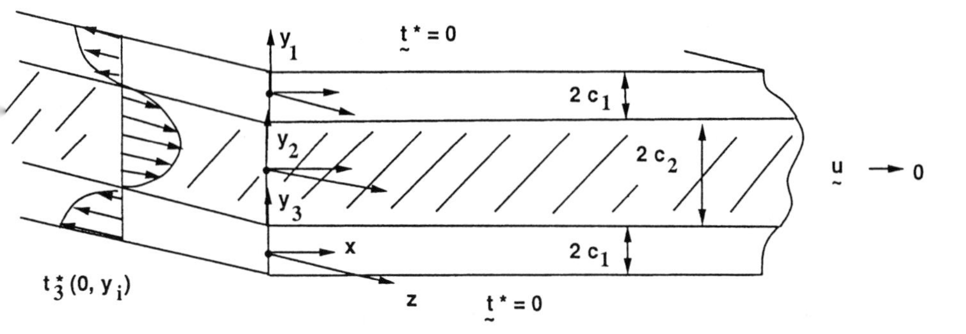

Figure 1. Typical cross-section of symmetric sandwich structure; self-equilibrated loads applied at $x = 0$.

boundary where $x = 0$. This shear is in the direction parallel to the z-axis and independent of the out-of-plane coordinate z, i.e., $t_3^* = t_3^*(0, y_i)$. The top and bottom surfaces of the sandwich are taken to be traction-free, and the applied tractions at $x = 0$ are assumed to be *self-equilibrated.* As $x \to \infty$, it is assumed that $\mathbf{u}= (u_1, u_2, u_3) \to 0$. Under such conditions it is shown in [17] that the deformation is an *anti-plane shear* (APS) deformation with

$$u_1 = 0, \quad u_2 = 0, \quad u_3 = u_3\,(x, y_i). \tag{1}$$

Since the out-of-plane displacement is the only non-zero component of displacement, the subscript 3 will be dropped where convenient, and u^i will signify the displacement in the ith layer.

For a wide class of anisotropic materials it is shown in [3, 4] that the governing differential equation becomes

$$A_{\alpha\beta}\, u_{,\alpha\beta} = 0 \qquad \text{on } \mathcal{D}, \tag{2}$$

where $\mathcal{D}$ denotes the cross-section of the composite and $A_{\alpha\beta} = A_{\beta\alpha}$ are the elastic constants which differ from layer to layer. The non-zero in-plane stresses are

$$T_{31} = A_{11}u_{,1} + A_{12}u_{,2}\ , \quad T_{32} = A_{21}u_{,1} + A_{22}u_{,2}. \tag{3}$$

The three-dimensional strain energy density is assumed positive definite in each layer, which requires that $A_{11} > 0, \qquad A_{11}A_{22} - A_{12}^2 \equiv A^2 > 0$, in each layer.

It is assumed that the materials are perfectly bonded at the layer interfaces so that the displacements and tractions are continuous there, so that

$$u^1(x,-c_1) = u^2(x,c_2)\ ,\ \ u^2(x,-c_2) = u^3(x,c_1) \tag{4}$$

and

$$a_{12}u^1_{,1}(x,-c_1) + a_{22}u^1_{,2}(x,-c_1) = b_{12}u^2_{,1}(x,c_2) + b_{22}u^2_{,2}(x,c_2) \tag{5}$$

$$b_{12}u^2_{,1}(x,-c_2) + b_{22}u^2_{,2}(x,-c_2) = a_{12}u^3_{,1}(x,c_1) + a_{22}u^3_{,2}(x,c_1), \tag{6}$$

respectively. In (5) and (6) the $A_{\alpha\beta}$ of equation (2) are replaced by $a_{\alpha\beta}$, representing the material constants in layers 1 and 3, and $b_{\alpha\beta}$, representing those of layer 2.

The traction-free boundary conditions on the top and bottom read

$$a_{12}u^1_{,1}(x,c_1) + a_{22}u^1_{,2}(x,c_1) = 0, \tag{7}$$

$$a_{12}u^3_{,1}(x,-c_1) + a_{22}u^3_{,2}(x,-c_1) = 0. \tag{8}$$

Since they are not used explicitly, the boundary conditions at $x = 0$ are not written down.

It is shown in [3, 4] that seeking exponentially decaying solutions of (2) leads to a transcendental equation governing eigenvalues, which in turn, characterizes the exponential *decay rates* in each layer. Dimensionless material parameters B_i are defined by

$$B_1 = B_3 = \frac{a_{11}a_{22} - a_{12}^2}{a_{22}^2}, \qquad B_2 = \frac{b_{11}b_{22} - b_{12}^2}{b_{22}^2}. \tag{9}$$

Introducing the notation

$$\frac{\lambda_1}{c_1\sqrt{B_1}} = \frac{\lambda_2}{c_2\sqrt{B_2}} \equiv \frac{\lambda}{2c_1\sqrt{B_1} + c_2\sqrt{B_2}} \equiv k, \tag{10}$$

the exponential decay rates, $\lambda_\alpha / c_\alpha\sqrt{B_\alpha}$, in each layer can be compared with that for a homogeneous strip of "weighted" total half-width $(2c_1\sqrt{B_1} + c_2\sqrt{B_2})$. If a non-dimensional weighted volume fraction is defined as

$$\hat{f} = \frac{2c_1\sqrt{B_1}}{2c_1\sqrt{B_1} + c_2\sqrt{B_2}}, \tag{11}$$

it is possible to express both λ_1 and λ_2 in terms of λ and $\hat{f}$:

$$\lambda_1 = \frac{\hat{f}\lambda}{2}, \quad \lambda_2 = \lambda(1 - \hat{f}). \tag{12}$$

The transcendental equation for the eigenvalues λ is given by (see [3, 4]):

$$\begin{aligned} &- \hat{\delta}^2 \sin^2 (\hat{f}\lambda) \sin [2\lambda(1-\hat{f})] \\ &+ \cos^2 (\hat{f}\lambda) \sin [2\lambda(1-\hat{f})] \\ &+ 2\,\hat{\delta} \cos (\hat{f}\lambda) \cos [2\lambda(1-\hat{f})] \sin (\hat{f}\lambda) = 0 \end{aligned} \tag{13}$$

where

$$\hat{\delta} \equiv \frac{a_{22}\sqrt{B_1}}{b_{22}\sqrt{B_2}} \equiv \sqrt{\frac{a_{11}a_{22} - a_{12}^2}{b_{11}b_{22} - b_{12}^2}} > 0. \tag{14}$$

It can be shown that the roots λ of (13) are all real. A complete solution to (2), subject to prescribed boundary conditions at $x = 0$, would involve an infinite series of eigenfunctions. As explained in [13, 14] for example, the *decay rate* k in each layer is given by (10), where λ is the smallest positive root of (13).

3. Specialization to Isotropy

If each of the layers is assumed to be *isotropic*, considerable simplification occurs. In this case the material constants in the outer layer are

$$a_{11} = a_{22} \equiv \mu_1, \qquad a_{12} \equiv 0, \qquad \text{so that } B_1 = 1, \tag{15}$$

while in the center core layer

$$b_{11} = b_{22} \equiv \mu_2, \qquad b_{12} \equiv 0, \qquad \text{and so } B_2 = 1. \tag{16}$$

Here μ_1 and μ_2 are the shear moduli of the material in the outer layers and the core, respectively. These shear moduli are the only elastic constants that appear in the analysis of the isotropic problem. The exponential decay factor e^{-kx} for each layer has a decay rate k given by

$$k = \frac{\lambda}{2c_1 + c_2}. \tag{17}$$

The *characteristic decay length* d, (i.e., the distance over which end effects decay to 1% of their value at $x = 0$) is defined by

$$d \equiv \frac{\ln(100)}{k}. \tag{18}$$

The weighted volume fraction, $\hat{f}$, given by (11), reduces to a simple volume fraction

$$f \equiv \frac{2c_1}{2c_1 + c_2}, \tag{19}$$

representing the relative thickness of an outer layer to the half-width of the entire structure. In the numerical results to follow we take $.1 \leq f \leq .9$. The ends of this interval correspond to the physical ideas of a *thin* and *thick* outer layer respectively. Similarly, the limits $f \to 0$ and $f \to 1$ correspond to a *homogeneous* strip composed of the core or face material only. For the isotropic sandwich, the material parameter $\hat{\delta}$, defined by (14), simplifies to the ratio of the two shear moduli

$$\delta \equiv \frac{\mu_1}{\mu_2}. \tag{20}$$

The parameters without the "carat" symbol will denote the isotropic case. The form of the transcendental equation (13) remains the same, with $\hat{f}, \hat{\delta}$ replaced by f, δ for the isotropic case. The parameter δ is referred to as the *core ratio*, with the understanding that a large core ratio means that the core is more compliant in shear than the outer layers and a small core ratio means that the core is stiffer in shear than the outer layers.

When $\mu_1 = \mu_2$ so that the strip is *homogeneous*, (20) yields $\delta = 1$. In this case, equation (13) reduces to $\sin(2\lambda) = 0$, whose smallest positive root yields the exact decay rate for a *homogeneous* isotropic strip, i. e.,

$$\lambda = \frac{\pi}{2}, \tag{21}$$

so that u decays as

$$e^{-kx}, \quad k = \pi/h, \tag{22}$$

where h is the strip width. The decay length d is thus approximately one and a half times the strip width. The foregoing results for *harmonic* functions on semi-infinite strips are well-known (see e.g. [13, 14]).

Analogous work on the plane problem in [8] shows that for a relatively compliant inner core ($\mu_1 >> \mu_2$), i.e., $\delta \to \infty$, the decay length tends to infinity. Numerical solutions verify that this is the case here also. An asymptotic analysis shows that

$$\bar{d} \sim \ln(100)\,[\delta\, f\,(1-f)]^{\frac{1}{2}} \quad \text{as } \delta \to \infty \quad (0 < f < 1), \tag{23}$$

where the *scaled decay length* is $\bar{d} \equiv d/(2c_1 + c_2)$. The exact $\bar{d}$ are shown by the solid curves in Figure 2, while the asterisk denotes values calculated from the asymptotic formula (23). Note that the asymptotic estimate is invariant with respect to interchanges of f and $(1-f)$. Figure 2 indicates that for values of $\delta \geq 100$, $.1 \leq f \leq .9$, the result (23) provides a very accurate estimate for the decay length.

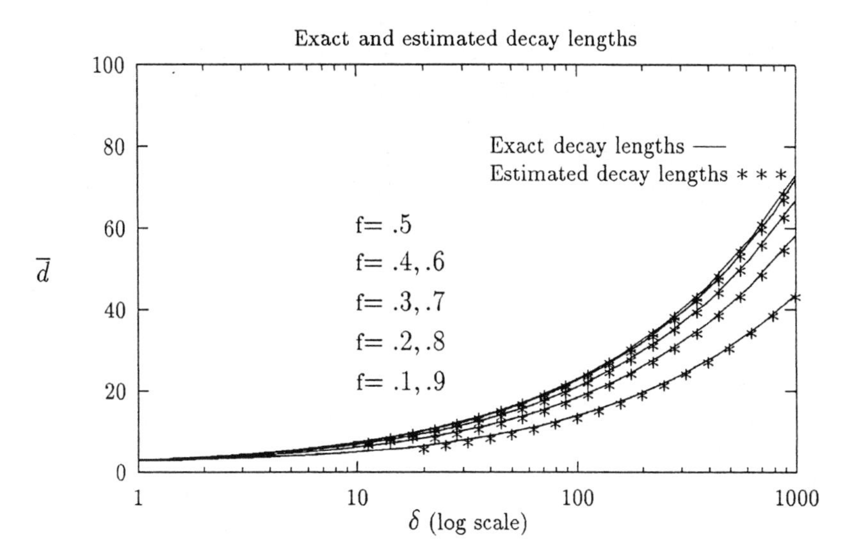

Figure 2. Scaled decay length vs. δ, for various f (from [4]).

4. Imperfect Bonding

The condition of perfect bonding used in Section 2 assumes that both displacements and tractions are continuous across the layer interfaces. The consequences of relaxing this condition are also examined in [3, 4]. Consider the case of *isotropic phases.* The condition of *imperfect bonding* (see e.g. [5, 1]) is modeled in the anti-plane shear context by modifying the interfacial conditions to allow jumps in the displacements proportional to the tractions at the interface. It is assumed that (5) and (6), the conditions for continuous traction, with $a_{12} = 0$, $b_{12} = 0$, $a_{11} = a_{22} = \mu_1$, $b_{11} = b_{22} = \mu_2$, still hold. Equations (4) are replaced by

$$u^1(x, -c_1) - u^2(x, c_2) = \mathcal{R}T_{32}, \qquad u^2(x, -c_2) - u^3(x, c_1) = \mathcal{R}T_{32}, \quad (24)$$

respectively, where $\mathcal{R}$ is a constant of proportionality and T_{32} serves to represent the components of traction at the interfaces. As $\mathcal{R} \to 0$, perfect bonding is recovered, while as $\mathcal{R} \to \infty$, perfectly lubricated contact ([1]) at the interfaces is obtained. It is convenient to define a new dimensionless parameter

$$\alpha \equiv \frac{(\mu_1 + \mu_2)\mathcal{R}}{2c_1 + c_2} \equiv \frac{(1+\delta)\mu_2\mathcal{R}}{2c_1 + c_2}, \quad (25)$$

where the definition of δ in (20) has been used to obtain the second expression for the *slip constant.* When $\mathcal{R} = 0$, so that the interfaces are perfectly bonded, then it follows that $\alpha = 0$. For typical slipping interfaces, α is a small number. For example, if the core material is glass and the face material an epoxy resin, the data given in [10] on pages 361 and 366 suggest that $\alpha\,(2c_1 + c_2) \doteq 3.138310^{-4}$ meters.

A detailed analysis of the dependence of the scaled decay length $\bar{d}$ on the parameters f , δ and α is given in [3, 4]. It is shown that the larger the amount of slip, the slower the attenuation of end effects. It is also shown that

$$\bar{d} \sim \ln(100)\sqrt{\frac{\alpha\,\delta\,f}{(1+\delta)}} \quad \text{as } \alpha \to \infty. \tag{26}$$

This asymptotic result is shown to be an accurate estimate for large values of slip.

5. The General Anisotropic Case

Under the most general anisotropy consistent with anti-plane shear, the decay rate k is given by $(10)_3$.

A special case of interest is that when $a_{\alpha\beta} \equiv b_{\alpha\beta}$ so that the strip is *homogeneous*. In this case one finds that the decay length is

$$d = \frac{\ln(100)}{\pi}\sqrt{B}\,h = \frac{\ln(100)}{\pi}\left(\frac{\sqrt{a_{11}a_{22} - a_{12}^2}}{a_{22}}\right)h, \tag{27}$$

where h is the strip width. The result (27) was obtained previously in [17].

An illustrative example of a sandwich structure with *orthotropic* phases is considered in [4]. The data used was supplied by W.B. Avery, Boeing Commercial Airplane Group, Seattle WA, and is used to indicate the order of magnitude which might be expected for the decay length in practical aircraft structures. The sandwich is composed of thin face layers of Hercules' AS4/8552 and Hexcel's HRP Honeycomb core. The elastic constants in the face ($a_{\alpha\beta}$) and core layers ($b_{\alpha\beta}$) are then

$$a_{11} = 153.2, \qquad b_{11} = 6.3920, \tag{28}$$

$$a_{22} = 104.0, \qquad b_{22} = 3.0456, \tag{29}$$

where all units are 10^4 psi. The geometry is defined by

$$2c_1 = .0876\,'', \qquad 2c_2 = .75\,''. \tag{30}$$

The decay length is calculated to be

$$d \doteq 6.102\,''. \tag{31}$$

The corresponding result for a *homogeneous isotropic* strip is given by (17), (18), (21) as

$$d = \frac{2\ln(100)}{\pi}(2c_1 + c_2) \doteq 1.356\ ''. \tag{32}$$

Thus the characteristic decay length (31) for the orthotropic sandwich structure is seen to be approximately *four and a half* times longer than that for the homogeneous isotropic strip.

6. Concluding Remarks

Previous research [8] on problems involving *plane deformations* of symmetric sandwich structures predicted a much longer decay length for Saint-Venant end effects in isotropic phase structures with relatively compliant cores compared with that for a homogeneous isotropic strip. The present work shows that similar conclusions can be reached regarding *anti-plane shear* deformations of isotropic phase structures. Additionally, it is found that an analogous result holds for isotropic phase structures where the conditions of perfect bonding are relaxed. The anti-plane shear deformation, with its simpler kinematics, allows for a more complete treatment of the Saint-Venant decay lengths, in particular with regard to asymptotic results for compliant and stiff cores respectively. Moreover, results for perfectly bonded sandwich strips with *anisotropic* phases have also been obtained.

The plane problem for generally laminated orthotropic strips was recently investigated in [21] while the effects of imperfect bonding in this context were examined in [22]. The results are qualitatively similar to those outlined here.

Acknowledgements

This research was supported by the U.S. Air Force Office of Scientific Research under Grant AFOSR-F49620-95-1-0308, by the U.S. Army Research Office under Grant DAAH04-94-G-0189 and by NASA under Grant NAG-1-1854. The research of S.C.B. was also supported by a grant from the Virginia Space Grant Consortium and an AASERT grant sponsored by the U.S. Army Research Office. We are grateful to Dr.M.P.Nemeth, NASA Langley Research Center for several helpful discussions concerning this work.

References

1. Aboudi, J. (1987). Damage in composites-modeling of imperfect bonding. *Composites Science and Technology* **28**, 103-128
2. Arimitsu, Y. Nishioka, K., and Senda, T. (1995). A study of Saint-Venant's principle for composite materials by means of internal stress fields. *Journal of Applied Mechanics* **62**, 53-58

3. Baxter, S.C. and Horgan, C.O. (1995). End effects for anti-plane shear deformations of sandwich structures. *J. of Elasticity* **40**, 123-164
4. Baxter, S.C. and Horgan, C.O. (1997). Anti-plane shear deformations of anisotropic sandwich structures: end effects. *Int. J. Solids Structures* **34**, 79-98
5. Benveniste, Y. (1984). On the effect of debonding on the overall behavior of composite materials. *Mechanics of Materials* **3**, 349-358
6. Carlsson, L.A. and Pipes, R.B. (1987). *Experimental Characterization of Advanced Composite Materials.* Prentice-Hall, New Jersey.
7. Choi, I. and Horgan, C.O. (1977). Saint-Venant's principle and end effects in anisotropic elasticity. *Journal of Applied Mechanics* **44**, 424-430
8. Choi, I. and Horgan, C.O. (1978). Saint-Venant end effects for plane deformation of sandwich strips. *Int. J. Solids Structures* **14**, 187-195
9. Crafter, E.C., Heise, R.M., Horgan, C.O., and Simmonds, J.G. (1993). The eigenvalues for a self-equilibrated semi-infinite, anisotropic elastic strip. *Journal of Applied Mechanics* **60**, 276-281
10. Devries, F. (1993). Bounds on elastic moduli of unidirectional composites with imperfect bonding. *Composites Engineering* **3**, 349-382
11. Gibson, R.F.(1994). *Principles of Composite Material Mechanics.* McGraw-Hill, New York.
12. Horgan, C.O. (1982). Saint-Venant end effects in composites. *Journal of Composite Materials* **16**, 411-422
13. Horgan, C.O. and Knowles, J.K. (1983). Recent developments concerning Saint-Venant's principle. *Advances in Applied Mechanics* (ed. J.W. Hutchinson and T.Y. Wu) **23**, Academic Press, New York, pp. 179-269
14. Horgan, C.O. (1989). Recent developments concerning Saint-Venant's principle: an update. *Applied Mechanics Reviews* **42**, 295-303
15. Horgan, C.O. (1996). Recent developments concerning Saint-Venant's principle: a second update. *Applied Mechanics Reviews* **48**, S101-S111.
16. Horgan, C.O. and Simmonds, J.G. (1994). Saint-Venant end effects in composite structures. *Composites Engineering* **3**, 279-286
17. Horgan, C.O. and Miller, K.L. (1994). Anti-plane shear deformations for homogeneous and inhomogeneous anisotropic linearly elastic solids. *Journal of Applied Mechanics* **61**, 23-29
18. Horgan, C.O. (1995). Anti-plane shear deformations in linear and nonlinear solid mechanics. *SIAM Review* **37**, 53-81
19. Miller, K.L. and Horgan, C.O. (1995a). End effects for plane deformations of an elastic anisotropic semi-infinite strip. *Journal of Elasticity* **38**, 261-316
20. Miller, K.L. and Horgan, C.O. (1995b). Saint-Venant end effects for plane deformations of elastic composites. *Mechanics of Composite Materials and Structures* **2**, 203-214
21. Tullini, N. and Savoia, M. (1997). Decay rate of Saint-Venant end effects for multilayered orthotropic strips. *Int. J. Solids Structures* (in press)
22. Tullini, N. Savoia, M. and Horgan C.O. (1997). End effects in multilayered orthotropic strips with imperfect bonding. *Mechanics of Materials* (in press)
23. Wijeyewickrema, A.C. and Keer, L.M. (1994). Axial decay of stresses in a layered composite with slipping interfaces. *Composites Engineering* 4, 895-899
24. Wijeyewickrema, A.C. (1995). Decay of stresses induced by self-equilibrated end loads in a multilayered composite. *Int.J.Solids Structures* **32**, 515-523
25. Wijeyewickrema, A.C., Horgan, C.O. and Dundurs, J. (1996). Further analysis of end effects for plane deformations of sandwich strips. *Int. J. Solids Structures* **33**, 4327-4336

MEMBRANE ANALOGY FOR SANDWICH PANELS IN THERMALLY OR PIEZOELECTRICALLY INDUCED BENDING

K. Hagenauer, H. Irschik
Institute of Mechanics and Machine Design
Johannes Kepler University of Linz
Altenbergerstr.68
4040-Linz-Auhof (Austria)

1. Introduction and statement of problem

In literature, a large amount of work has been devoted to the analysis of sandwich structures. This is because sandwiches may be tailored such that particular structural requirements can be matched with little waste of material capability, see e.g. the monographs by Baltema [1], Stamm and Witte [2], Wiedemann [3]. Since, however, the corresponding mathematical boundary-value problems turn out to be rather complex, a wide range of problems concerning the mechanics of sandwich structures remains to be treated, even for the case of static conditions. One of these problems is the development of benchmark formulations for sandwich panels with a plan-view of a more complex shape than the rectangular one frequently treated in the literature.

The present contribution is devoted to the effect of temperature and/or piezoelectric actuation and in-plane forces upon the static flexural behaviour of slightly curved elastic sandwich panels with isotropic faces and a shear-deformable isotropic core. Utilising classical homogenisation techniques, the modelling of the sandwich structure is based on the kinematic assumptions of the Reissner-Mindlin theory for shear-deformable plates, [4], [5], cf. Refs [1]-[3]. As a starting point of the following derivations, the field equations for the geometrically linearized flexural deformation of sandwich plates with thick faces are utilized, cf. Eq.(6.20) of Stamm and Witte [2]:

$$-\frac{\partial}{\partial y}\Delta w+\left(\frac{\partial^2}{\partial y^2}+\frac{1-\nu}{2}\frac{\partial^2}{\partial x^2}-\frac{A}{B_S}\right)\gamma_{yz}+\frac{1+\nu}{2}\frac{\partial^2\gamma_{xz}}{\partial x\partial y}=\frac{1}{B_S}\frac{\partial}{\partial y}M^T, \tag{1.a}$$

$$-\frac{\partial}{\partial y}\Delta w+\left(\frac{\partial^2}{\partial y^2}+\frac{1-\nu}{2}\frac{\partial^2}{\partial x^2}-\frac{A}{B_S}\right)\gamma_{yz}+\frac{1+\nu}{2}\frac{\partial^2\gamma_{xz}}{\partial x\partial y}=\frac{1}{B_S}\frac{\partial}{\partial y}M^T, \tag{1.b}$$

$$-(B_u+B_o)\,\Delta\Delta w+A\left(\frac{\partial}{\partial x}\gamma_{xz}+\frac{\partial}{\partial y}\gamma_{yz}\right)+N\,(\Delta w+\Delta w_0)=0. \tag{1.c}$$

In Eq.(1), the deflection is denoted by w , *and* γ_{xz} , γ_{yz} are the (effective) shear-angles. Δ is the two-dimensional Laplace operator. The sandwich plate is considered to be built up from isotropic, linear elastic materials with a perfect bond between the faces and the core. The faces of the plate are assumed to be rigid in shear, while the influence of shear upon the structural behaviour is taken into account for the core. Hence, the above

A. Vautrin (ed.), Mechanics of Sandwich Structures, 123–130.

equations represent a special type of multi-layer, first-order shear-deformable theories for composite plates. Stiffness parameters present in Eq.(1) are given by

$$B_S = \frac{E_u t_u E_o t_o}{E_u t_u + E_o t_o} \frac{a^2}{1-\nu^2}, \qquad A = G_k \frac{a^2}{h}, \qquad B_{o,u} = \frac{E_{o,u} t^3_{o,u}}{1-\nu^2}, \qquad B_{o,u} = \frac{E_{o,u} t^3_{o,u}}{1-\nu^2}, \tag{2}$$

where the subscripts refer to the lower (u) and upper (o) faces of the sandwich plate, respectively. Young's modulus of a face is denoted by E, and ν is Poisson's ratio. The centroidal distance between the faces is denoted by a. The shear modulus of the core is denoted by G_k, and h is the thickness of the core. In Eq.(1.c), the influence of a hydrostatic state of in-plane forces upon the deformation is considered,

$$n_x = n_y = N, \quad n_{xy} = 0. \tag{3}$$

In order to take into account a global imperfection, the formulation of Ref. [2] is extended with respect to the initial deflection w_0, see Eq.(1.c). Furthermore, flexure of the plate is assumed to be produced by a difference in temperature between lower and upper face, characterised by the thermal moment

$$M^T = -B_s(1+\nu)\frac{\alpha_u T_u - \alpha_o T_o}{a}, \tag{4}$$

see Eqs.(1.a,b). Temperature is denoted by T, and α is the coefficient of thermal expansion. The thermal moment may vary throughout the plate, $M^T = M^T(x,y)$. Note that M^T may also be interpreted as a piezoelectrically induced moment, in case the faces are made from piezoelectric material and the inverse piezoelectric effect is utilized for actuation, see Rao and Sunar[13] for a review, and e.g. Irschik, Schlacher and Haas [14] for a recent application. The analogy between piezoelectrically induced strains and thermal strains may be viewed as key for the understanding of piezoelectric actuation. It has been stated in the literature that this analogy is extremely important because it enables the analyst and designer to utilize all available thermoelastic solutions to solve problems involving piezoelectric materials, Vinson[15].

Panels with a plan-view of polygonal but otherwise arbitrary shape and hinged boundaries are studied, see Fig.1. The hinged boundary conditions considered in the present paper are modelled in the form introduced by Mindlin, Schacknow and Deresziewicz[6]:

$$\Gamma: w = 0; \tag{5.a}$$

$$\Gamma: m_n = 0; \tag{5.b}$$

$$\Gamma: \gamma_{sz} = 0; \tag{5.c}$$

where (s,n) denotes a local Cartesian coordinate system at the boundary Γ, and m_n is the bending moment. In the case of thick faces, the three boundary conditions (5) have to be accompanied by a proper fourth one.

It is the scope of the present contribution to develop membrane analogies for the above stated boundary-value problem, in order to reduce the computational effort of the problem and to derive a class of benchmark solutions. The present contribution is especially devoted to plates with thin faces, where $B_u + B_o \approx 0$.

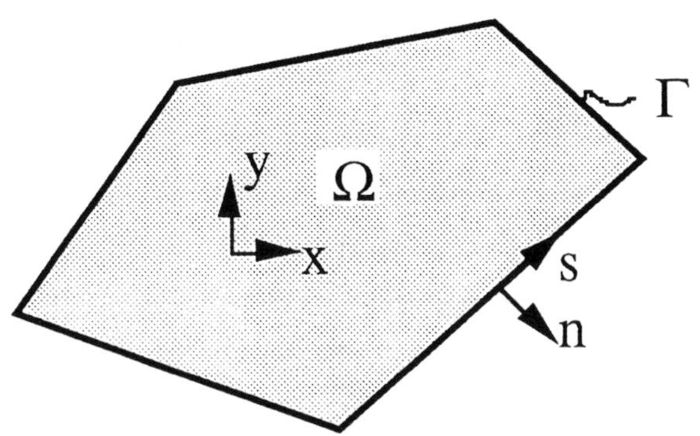

Fig. 1: Plan-view of polygonal panel

2.Differential equation for deflection

Eq.(1.a) is differentiated with respect to x, and Eq.(1.b) with respect to y. They are added,

$$-\Delta\Delta w + (\Delta - \frac{A}{B_S})(\frac{\partial}{\partial x}\gamma_{xz} + \frac{\partial}{\partial y}\gamma_{yz}) = \frac{1}{B_S}\Delta M^T . \tag{6}$$

From Eq.(1c), there is:

$$\frac{\partial}{\partial x}\gamma_{xz} + \frac{\partial}{\partial y}\gamma_{yz} = \frac{(B_u + B_o)}{A}\Delta\Delta w - \frac{N}{A}(\Delta w + \Delta w_0) . \tag{7}$$

Inserting Eq.(7) into (6) leads to

$$\Delta[\frac{(B_u + B_o)}{A}\Delta\Delta w - (1 + \frac{(B_u + B_o}{B_S} + \frac{N}{A})\Delta w + \frac{N}{B_S}(w + w_0) - \frac{N}{A}\Delta w_0 - \frac{1}{B_S}M^T] = 0 \tag{8}$$

The bracketted term in Eq.(8) is abbreviated by U:

$$U := \frac{(B_u + B_o)}{A}\Delta\Delta w - (1 + \frac{(B_u + B_o}{B_S} + \frac{N}{A})\Delta w + \frac{N}{B_S}(w + w_0) - \frac{N}{A}\Delta w_0 - \frac{1}{B_S}M^T , \tag{9}$$

such that Eq.(8) may be replaced by

$$\Delta U = 0 . \tag{10}$$

We seek for a boundary condition in U. From the boundary condition (5.a), it follows that

$$\Gamma: w = 0; \Rightarrow \Gamma: \frac{\partial^2}{\partial s^2}w = 0; \Rightarrow \Gamma: \Delta w = \frac{\partial^2}{\partial n^2}w; \tag{11}$$

while Eq.(5.c) gives

$$\Gamma: \gamma_{sz} = 0; \Rightarrow \Gamma: \frac{\partial}{\partial s}\gamma_{sz} = 0. \tag{12}$$

Eqs. (11) and (12) are inserted into the boundary condition of vanishing moment, Eq.(5.b):

$$\Gamma: m_n = -(B_S + B_u + B_o)\left(\frac{\partial^2 w}{\partial n^2} + \nu\frac{\partial^2 w}{\partial s^2}\right) + B_S\left(\frac{\partial}{\partial n}\gamma_{nz} + \nu\frac{\partial}{\partial s}\gamma_{sz}\right) - M^T =$$
$$= -(B_S + B_u + B_o)\,\Delta w + B_S\left(\frac{\partial}{\partial n}\gamma_{nz} + \frac{\partial}{\partial s}\gamma_{sz}\right) - M^T = 0 \tag{13}$$

Utilizing Eq.(7), this may be replaced by:

$$\Gamma: \frac{m_n}{B_S} = \frac{(B_u + B_o)}{A}\Delta\Delta w - \left(1 + \frac{B_u + B_o}{B_S} + \frac{N}{A}\right)\Delta w - \frac{N}{A}\Delta w_0 - \frac{M^T}{B_S} = 0 \tag{14}$$

Assuming that the initial deflection vanishes along the boundary,

$$\Gamma: w_0 = 0, \tag{15}$$

Eq.(14) may be replaced by a homogeneous Dirichlet boundary condition in U:

$$\Gamma: U = 0 \tag{16}$$

Together with the homogeneous Laplace-equation (10), it follows that U must vanish throughout the plate domain, $U = 0$. Hence, a fourth-order differential equation for the deflection is obtained, cf. Eq.(9):

$$\frac{(B_u + B_o)}{A}\Delta\Delta w - \left(1 + \frac{(B_u + B_o)}{B_S} + \frac{N}{A}\right)\Delta w + \frac{N}{B_S}(w + w_0) - \frac{N}{A}\Delta w_0 - \frac{1}{B_S}M^T = 0\,. \tag{17}$$

3.Membrane analogy:

In the following, sandwich plates with thin faces are considered, where

$$B_u + B_o \approx 0: \tag{18}$$

is used in Eq.(17). Together with the first of the boundary conditions, Eq. (5.a), a membrane-type boundary value problem of second order is immediately obtained from Eq.(17):

$$\Delta w - \frac{N}{B_S\left(1 + \frac{N}{A}\right)}w = -\frac{M^T}{B_S\left(1 + \frac{N}{A}\right)} + \frac{N}{B_S\left(1 + \frac{N}{A}\right)}\left(1 - \frac{B_S}{A}\Delta\right)w_0 \tag{19.a}$$

$$\Gamma: w = 0 \tag{19.b}$$

The deflection of the shear-deformable, initially curved sandwich plate thus may be obtained from the second-order boundary-value problem (19), instead of using the set of Eqs.(1), (5), (18). Recall that the type of boundary value problem stated in Eqs.(19) is well known from the classical theory of thin elastic single-layer membranes, see e.g.. Ziegler [7]. Similarities between shear-deformable theories of multi-layered structures and classical theories of single-layer structures date back to Ueng[8], Yan and Dowell[9]. A review on more recent results concerning such analogies has been presented by Irschik[10]. The new membrane analogy stated in Eqs.(19) is felt to be of a particular practical interest, since the inhomogeneous boundary value problem of originally sixth order is replaced by a single second-order one. It is thus possible, for the class of panels under consideration, to calibrate numerical routines developed for the full sixth-order theory by means of results for the more simple second-order problem, also in the case of

a complex shape of the panel domain. The following additional conclusions can be drawn from Eqs.(19):
Global buckling of the sandwich plate with complex polygonal planform can conveniently be tread utilizing Eqs.(19): Taking thermal and piezoelectric moment, as well as initial curvature to vanish,

$$M^T = 0; \; w_0 = 0, \tag{20}$$

such that Eq.(19.a) becomes homogeneous, Eqs.(19) are compared to the following Helmholtz-problem:

$$\Delta w_1 + \alpha_1 w_1 = 0, \tag{21.a}$$

$$\Gamma: w_1 = 0, \tag{21.b}$$

where α_1 denotes the smallest Helmholtz eigenvalue, and w_1 is the corresponding eigenfunction. It is thus seen that the buckling eigenform of the flat sandwich plate coincides with the Helmholtz-eigenfunction, and that the global critical buckling load N_{crit} of the plate becomes:

$$N_{crit} = -\frac{\alpha_1 B_S}{(1 + \alpha_1 \frac{B_S}{A})}. \tag{22}$$

In the case of a sandwich core rigid in shear, $A \to \infty$, one gets $N^*_{crit} = -\alpha_1 B_S$, where a star refers to the solution rigid in shear. Thus, Eq.(6) may be replaced by

$$N_{crit} = \frac{N^*_{crit}}{(1 - \frac{N^*_{crit}}{A})}. \tag{23}$$

For a more detailed study of membrane analogies for buckling eigenvalues and natural frequencies of single-layer polygonal Mindlin-plates, see Irschik [16], [17]. Equs. (22), (23) turn out to be special cases of the more general equations given in Ref.[16].
Static shape control of structures represents a topic of current interest, especially in aeronautics and in the field of smart structures. For static shape control of structures by an applied temperature distribution, see e.g. Haftka and Adelman[11], for piezoelectrically induced shape control see Refs.[18],[19]. In the present context, the problem of static shape problem may be stated as follows: What spatial distribution of thermal or piezoelectric moment should be imposed to the initially curved panel in order to nullify the deflection induced by inplane forces N ? For tensile in-plane loadings, or in the case of compressive loadings with $|N| < |N_{crit}|$, an exact solution is found from Eqs.(19) in the form:

$$M^T = N(1 - \frac{B_S}{A} \Delta) \, w_0 : \; w = 0. \tag{24}$$

Note, that in the case of vanishing in-plane force, $N=0$, there is no influence of shear upon the thermally induced panel deflection. This can be shown as follows. First consider the core to be rigid in shear, $A \to \infty$. Equs.(19) then become:

$$\Delta w^* - \frac{N}{B_S} w^* = -\frac{M^T}{B_S} + \frac{N}{B_S} w_0, \tag{25.a}$$

$$\Gamma: w^* = 0, \tag{25.b}$$

where a star refers to the solution rigid in shear, the so-called classical plate theory (CPT). Denote the difference between CPT and the present theory by w_d,

$$w = w^* + w_d. \tag{26}$$

The following boundary value problem is obtained for w_d:

$$(1 + \frac{N}{A})\, \Delta w_d - \frac{N}{B_S} w_d = -\frac{N}{A} \Delta (w^* + w_0), \tag{27.a}$$

$$\Gamma: w_d = 0. \tag{27.b}$$

Hence, if $N = 0$, then the difference to CPT vanishes, and there is no influence of shear upon the plate deflection:

$$N = 0: w_d = 0; \; w = w^* \tag{28}$$

For the case of flat single-layer plates, $w_0 = 0$, this fact has been already noted and confirmed by Finite Element calculations in Pachinger and Irschik [12].

4. Calculation of shear strains and moments from the analogy

In this section, the above analogy is extended to the calculation of shear strains and moments. We start from Eq.(1.c), taking into account Eq.(18):

$$A\, (\frac{\partial}{\partial x}\gamma_{xz} + \frac{\partial}{\partial y}\gamma_{yz}) + N\, (\Delta w + \Delta w_0) = 0 \tag{29}$$

This suggests to seek the shear-angles in the following form, satisfying Eq.(29):

$$\gamma_{yz} = -\frac{N}{A}\frac{\partial}{\partial y}(\, w + w_0\,) \tag{30.a}$$

$$\gamma_{xz} = -\frac{N}{A}\frac{\partial}{\partial x}(\, w + w_0\,) \tag{30.b}$$

In order to prove, whether Equs.(30) are the complete solution, they are inserted into Eqs.(1.a) and (1.b). After some calculations, taking into account Eq.(19.a), it is seen, that Eqs.(1.a,b) are also satisfied identically by Eqs.(30). What remains, is to study the boundary conditions. With $w + w_0 = 0$ all along the boundary, see Eqs.(5.a) and (15), it is seen that Eq.(5.c) is satisfied by Eq.(30). Furthermore, for the boundary condition of vanishing moment, Eq.(5.b), we note from Eqs.(30):

$$\Gamma: (\frac{\partial}{\partial n}\gamma_{nz} + \nu\frac{\partial}{\partial s}\gamma_{sz}) = -\frac{N}{A}(\frac{\partial^2}{\partial n^2} + \nu\frac{\partial^2}{\partial s^2})(w + w_0) = -\frac{N}{A}\Delta(w + w_0), \tag{31}$$

which holds since the boundary is straight. Eq.(31) is inserted into the expression for the boundary moment:

$$\Gamma: m_n = -B_S(\frac{\partial^2 w}{\partial n^2} + \nu\frac{\partial^2 w}{\partial s^2}) + B_S(\frac{\partial}{\partial n}\gamma_{nz} + \nu\frac{\partial}{\partial s}\gamma_{sz}) - M^T =$$

$$= -B_S\Delta w + B_S(\frac{\partial}{\partial n}\gamma_{nz} + \nu\frac{\partial}{\partial s}\gamma_{sz}) - M^T = 0, \tag{32}$$

which vanishes because of Eq.(19.a). It is thus concluded that Eqs.(30) represent the solution of the problem, since differential equations as well as boundary conditions remain satisfied. Hence, the shear-angles can be calculated directly from the deflection according to Eqs.(30), without the necessity of solving an additional boundary value problem. Bending and twisting moments then follow by an additional differentiation:

$$m_x = -B_S(\frac{\partial^2 w}{\partial x^2} + \nu\frac{\partial^2 w}{\partial y^2}) + B_S(\frac{\partial}{\partial x}\gamma_{xz} + \nu\frac{\partial}{\partial y}\gamma_{yz}) - M^T$$

$$= -B_S[(1 + \frac{N}{A})(\frac{\partial^2 w}{\partial x^2} + \nu\frac{\partial^2 w}{\partial y^2}) + \frac{N}{A}(\frac{\partial^2 w_0}{\partial x^2} + \nu\frac{\partial^2 w_0}{\partial y^2})] - M^T; \tag{33.a}$$

$$m_y = -B_S[(\frac{\partial^2 w}{\partial y^2} + \nu\frac{\partial^2 w}{\partial x^2}) + \frac{N}{A}(\frac{\partial^2(w + w_0)}{\partial y^2} + \nu\frac{\partial^2(w + w_0)}{\partial x^2})] - M^T; \tag{33.b}$$

$$m_{xy} = T_{xy}(-2\frac{\partial^2 w}{\partial x\,\partial y} + \frac{\partial}{\partial y}\gamma_{xz} + \frac{\partial}{\partial x}\gamma_{yz}) = -2T_{xy}\frac{\partial^2}{\partial x\,\partial y}[(1 + \frac{N}{A})w + \frac{N}{A}w_0]; \tag{33.c}$$

with

$$T_S = \frac{1-\nu}{2}B_S \tag{34}$$

Note that, in the case of N=0, there is no influence of shear on deflections, shear angles and moments of the thermally or piezoelectrically actuated sandwich plates under consideration, compare also Ref.[12] for single-layer flat plates:

$$N = 0: \gamma_{xz} = \overset{*}{\gamma}_{xz} = 0;\ \gamma_{yz} = \overset{*}{\gamma}_{yz} = 0; \tag{35}$$

$$N = 0: m_x = \overset{*}{m}_x, m_y = \overset{*}{m}_y, m_{xy} = \overset{*}{m}_{xy}. \tag{36}$$

5. Conclusions:

A membrane analogy has been presented for the deflection of thermally or piezoelectrically induced sandwich plates with slight initial curvature and a hydrostatic state of in-plane forces. It is thus possible, for the class of panels under consideration, to calibrate numerical routines developed for the full sixth-order theory by means of results for the more simple second-order problem presented in Eq.(19), also in the case of a complex shape of the panel domain. Global buckling and static shape control of the sandwich panel has been additionally considered utilizing this membrane analogy. It has been furthermore shown that shear angles and moments of the sandwich plate can be calculated directly from the deflection by an additional differentiation.

Acknowledgement: Support by the Austrian Scientific Research Promotion Found (FWF) under the contract P11993-TEC , Mechanics of Smart Structures, is gratefully acknowledged.

6. References:

[1] Plantema,F.J.: Sandwich Construction. John Wiley(1966)
[2] Stamm,K.,Witte,H.:Sandwichkonstruktionen. Springer-Verlag(1974)
[3] Wiedemann, J.: Leichtbau.Bd.1 Springer-Verlag(1986)
[4] Reissner,E.: Reflections on the Theory of Elastic Plates. Appl.Mech.Rev. 38(1985)
[5] Mindlin,R.D.: Influence of Rotatory Inertia and Shear on the Flexural Motions of Isotropic Elastic Plates. J.Appl.Mech.18(1951)
[6] Mindlin, R.D., Schacknow, A., Deresiewicz, H.: Flexural vibrations of rectangular plates. ASME J. Appl. Mech.(1956)
[7] Ziegler, F. Mechanics of Solids and Fluids. 2nd ed. Springer Verlag(1985)
[8] Ueng,C.E.S.: A Note on the Similarities between the Analyses of Homogenous and Sandwich Plates. J.Appl.Mech.(1966)
[9] Yan,M.J.,Dowell,E.H.: Elastic Sandwich Beam or Plate Equations Equivalent to Classical Theory.J.Appl.Mech.41(1974)
[10] Irschik, H. : On Vibrations of Layered Beams and Plates. ZAMM 73(1993)
[11] Haftka, R.T., Adelman, H.M., An Analytical Investigation of Static Shape Control of Large Space Structures by Applied Temperature, AIAA-Journal, 23 (1985)
[12] Irschik, H., Pachinger, F.: On Thermal Bending of Moderately Thick Polygonal Plates with Simply Supported Edges. Journal of Thermal Stresses 18 (1995)
[13]Rao, S.F. and Sunar, M., (1994), "Piezoelectricity and its use in disturbance sensing and control of flexible structures," Applied Mechanic Reviews, Vol. 47, pp.113-123.
[14]Irschik, H., Schlacher, K., Haas, W.: Output annihilation and optimal H_2 control of plate vibrations by piezoelectric actuation . In: Proc. IUTAM-Symp. on Interactions Between Dynamics and Control in Advanced Mechanical Systems (D. Van Campen, Ed.), p. 159-166.Dordrecht: Kluwer1997.
[15]Vinson, J. R., (1992), The Behavior of Shells Composed of Isotropic and Composite Materials, Kluwer.
[16]rschik, H.:Membrane-Type Eigenmotions of Mindlin Plates. IActa Mechanica 55 (1985), 1-20 .
[17]Irschik, H.: Stability of Skew Mindlin Plates Under Isotropic In-Plane Pressure. Disc. of ASCE Paper No. 3706. Journal of Engineering Mechanics 10, (1994), 2243-2245.
[18]Austin, F., Rossi, M.J., Van Nostrad, W., Knowles, G., and Jameson, A.: Static Shape Control of Adaptive Wings, *AIAA-Journal* 32 (1994), 1895-1901.
[19]Varadajan, S., Chandrashekara, K., and Agarwal, S.: Adaptive Shape Control of Laminated Composite Plates Using Piezoelectric Materials. Proc. AIAA/ASME/AHS Adaptive Structures Forum, Salt Lake C., UT, 1996 (Martinez, D., Chopra, I., eds.), 197-206, AIAA-Paper No. 96-1288.

HOMOGENIZATION OF PERIODIC SANDWICHES

Numerical and analytical approaches

S. BOURGEOIS*, **, P. CARTRAUD***, O. DÉBORDES*
**Laboratoire de Mécanique et d'Acoustique (CNRS, UPR 7051)*
& École Supérieure de Mécanique de Marseille
I. M. T., Technopôle de Chateau Gombert, 13451 Marseille Cedex 20, FRANCE.
*** PhD Thesis in cooperation with SOLLAC FOS,*
C.R.P.C., 13776 Fos-sur-Mer cedex, FRANCE
****Laboratoire de Mécanique et Matériaux*
École Centrale de Nantes, B.P. 92101, 44321 Nantes Cedex 3, FRANCE.

1. Introduction

Metallic sandwich plates (ribbed, honeycomb or corrugated plates) are widely used in aerospace and other industries due to their light weight and high stiffness. A three-dimensional accurate finite element model of such sandwich panels can become computationally expensive, and an equivalent plate formulation is therefore particularly useful.

When a panel with a regular periodic structure is considered, the effective plate properties can be obtained using the homogenization theory. In the literature, extensive work has been devoted to the mathematical aspects of the homogenization of periodic plates [1], [2], but the application of the method to practical situations appears to be limited, especially by means of a numerical approach. The purpose of this paper is to discuss different methods and to present our recent numerical developments. Numerical examples will illustrate the reliability of the homogenization process.

2. A literature survey

2.1. INTRODUCTION

For a periodic plate, one considers two small parameters e and ε, which respectively denote the ratio of the plate thickness and the period size, with respect to some characteristic length of the whole plate. The aim of the homogenization method is to replace the former thin heterogeneous structure by an equivalent homogeneous plate. This requires to make e and ε tend to zero, separately or together. Hence, three classes of methods may be developed with their own range of applicability [1], [3]:
First family of methods : e -> 0, then ε -> 0, when the thickness is much smaller than the size of the period.

A. Vautrin (ed.), Mechanics of Sandwich Structures, 131–138.

Second family of methods : e and ε -> 0 together, when the thickness and the size of the period have similar values.
Third family of methods : ε -> 0, then e -> 0, when the size of the period is much smaller than the thickness.

2.2. FIRST FAMILY OF METHODS

In this approach, a first step corresponding to the reduction of the transverse dimension (e -> 0) yields a periodic heterogeneous plate. The plate equations can be derived from the three-dimensional elasticity equations through a priori assumptions (e.g. Kirchhoff or Reissner-Hencky plate theories), or by means of the asymptotic expansion method, which results in Kirchhoff equations [4]. The effective stiffnesses of the plate are then obtained by homogenizing Kirchhoff equations (method noted m1 in the following) [1], [3], [5], or Reissner-Hencky equations (m1b) [3], [6], [7].

2.3. SECOND FAMILY OF METHODS

In these methods, the two small parameters tend to zero simultaneously. One obtains three-dimensional basic cell problems, the solution of which directly provides the effective stiffnesses of the plate (m2) [1], [2], [4], [8], [9].

In order to simplify this problem, approximate solutions are proposed in the literature. For the case in which the basic cell has the shape of a plate, two-dimensional basic cell problems are obtained by making assumptions about the displacement distribution across the basic cell thickness. If the Kirchhoff plate model is used, the method m1 is recovered [3]. With a Reissner-Hencky plate model, a method m2b is obtained, different from the method m1b [3], [10]. Approximate solutions can also be found in [8], [9] for basic cells consisting of several thin-walled parts made of homogeneous materials, but this approach will not be used in this paper.

2.4. THIRD FAMILY OF METHODS

At first, two-dimensional homogenization problems are solved on each constitutive slice of the basic cell (ε -> 0). The Kirchhoff plate model is then used to derive the effective stiffnesses of the plate, from the previous equivalent homogeneous slices (m3) [1], [4]. If the homogenization process in each slice is carried out using plane stress assumptions, the classical lamination theory is found (m3b) [7].

2.5. SUMMARY

The effective stiffness matrix depends on the method used. According to the method, the basic cell homogenization problems are two-dimensional (m1, m1b, m2b, m3, m3b) or three-dimensional (m2). The equivalent homogeneous plate is a Kirchhoff one (m1, m2, m2b, m3, m3b), or a Reissner-Hencky one (m1b). Several aforementioned works are restricted to symmetric plates in bending [3], [4], [5], [7] and in membrane behaviour [6]. In [6], [7], the method m1b is used, which provides the effective shear stiffnesses. Asymmetric plates are considered in [1], [2], [8], [9], [10], where the membrane, bending and reciprocal effective stiffnesses are obtained. The different methods used herein are summarized in table 1.

For unidirectional periodic structures, the above two-dimensional basic cell problems reduce to one-dimensional ones, and analytical solutions can be found [3], [10]. However, homogenization problems have to be usually solved numerically. To the authors knowledge, the numerical implementation of the computation of the effective plate stiffnesses is slightly discussed in the literature. Very few numerical results are available: a symmetric reinforced plate in bending is treated in [4], corrugated and ribbed plates are considered in [11], but these works are restricted to unidirectional periodic plates.

The structural elements mainly studied in this paper are sandwich plates, the period and the thickness of which are of the same order. In section 3, the method m2 is presented together with its finite element implementation.

TABLE 1. The different methods

First family of methods Homogenization of plate equations	Kirchhoff theory	m1
	Reissner-Hencky theory	m1b
Second family of methods 3D basic cell problems	3D solution	m2
	Reissner-Hencky approximation	m2b
Third family of methods 2D basic cell problems	Exact solution	m3
	Plane stress assumptions	m3b

3. The method m2. Numerical Implementation

3.1. THE BASIC CELL PROBLEMS

Starting from the three-dimensional elastic problem, and assuming that the two small parameters e and ε are equal, the asymptotic expansions method is used: up to the first order, a Kirchhoff-Love displacement field is found, which is the solution of a two-dimensional problem set on the plate midsurface, with homogeneous plate constitutive equations in the form of:

$$\begin{Bmatrix} \{N\} \\ \{M\} \end{Bmatrix} = \begin{bmatrix} [A^h] & [E^h] \\ [E^h]^T & [D^h] \end{bmatrix} \begin{Bmatrix} \{\varepsilon^M\} \\ \{\chi^M\} \end{Bmatrix} \tag{1}$$

where ε^M and χ^M are the membrane and bending (curvatures) generalized macroscopic strains: $\{\varepsilon^M\}^T=\{\varepsilon^M_{11}, \varepsilon^M_{22}, 2\varepsilon^M_{12}\}$ and $\{\chi^M\}^T=\{\chi^M_{11}, \chi^M_{22}, 2\chi^M_{12}\}$. The corresponding generalized macroscopic stresses are denoted N and M with $\{N\}^T=\{N_{11}, N_{22}, N_{12}\}$ and $\{M\}^T=\{M_{11}, M_{22}, M_{12}\}$. The membrane, bending and reciprocal effective stiffness matrices are respectively denoted $[A^h]$, $[D^h]$, and $[E^h]$.

These effective stiffnesses depend on some auxiliary fields solutions to problems set on the three-dimensional basic cell Y (Figure 1).

Denoting ∂Y^{up} and ∂Y^{low} the upper and lower faces of Y, T the holes in Y, Y^* the solid part of Y and a the elastic tensor, the basic cell localization problem is (for a given ε^M and χ^M constant on Y^*):

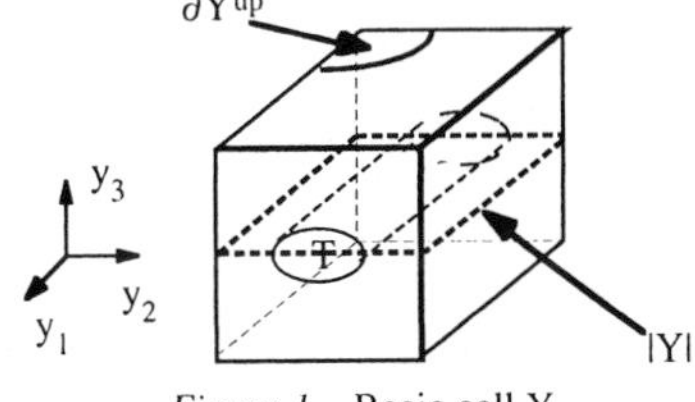

Figure 1. Basic cell Y

Find u^{per} such that:

$$\begin{cases} div\sigma = 0 \text{ on } \mathrm{Y}^* \\ \sigma = a : \varepsilon \text{ on } \mathrm{Y}^* \\ \varepsilon = \varepsilon^M - y_3 \chi^M + \varepsilon\left(u^{per}\right) \\ \sigma . n = 0 \text{ on } \partial \mathrm{Y}^{\mathrm{up}},\ \partial \mathrm{Y}^{\mathrm{low}} \text{ and } \partial \mathrm{T} \\ \sigma \text{ and } u^{per}\ \mathrm{y}_1,\ \mathrm{y}_2 \text{ periodic} \end{cases} \tag{2}$$

where the i3 (i=1, 2, 3) components of the generalized macroscopic strains ε^M and χ^M are zero. The generalized macroscopic stresses are ($|Y|$ is the middle surface of the basic cell, and the Greek indices run over 1 and 2):

$$N_{\alpha\beta} = \frac{1}{|Y|} \int_{\mathrm{Y}^*} \sigma_{\alpha\beta}\, dy \ ,\ M_{\alpha\beta} = \frac{1}{|Y|} \int_{\mathrm{Y}^*} -y_3\, \sigma_{\alpha\beta}\, dy \tag{3}$$

Six problems have to be solved to obtain the effective stiffnesses, corresponding to a unit generalized macroscopic strain.

3.2. NUMERICAL IMPLEMENTATION

In order to use the finite element method, the principle of virtual work is applied to the problem (2). Considering equation (3), one obtains the variational formulation of the basic cell problem:

Find u^{per} y_1,y_2 periodic such that $\forall$ δu^{per} y_1,y_2 periodic, $\delta\varepsilon^M$, $\delta\chi^M$,

$$\int_{Y^*} (\delta\varepsilon^M - y_3\,\delta\chi^M + \varepsilon(\delta u^{per})) : a : (\varepsilon^M - y_3\,\chi^M + \varepsilon(u^{per}))\, dy = |Y|\,(\delta\varepsilon^M : N + \delta\chi^M : M) \tag{4}$$

As it is shown in [12] for the classical periodic homogenization, equation (4) enables, in the finite element process, to treat the generalized macroscopic strains ε^M and χ^M as macroscopic degrees of freedom with the associated nodal forces $|Y|N$ and $|Y|M$. Thus, in each finite element of the basic cell mesh, the macroscopic degrees of freedom are considered as additional ones, which leads, using equation $(2)_3$, to modify the strain-displacements matrix. One then obtains the following linear system:

$$\begin{bmatrix} & & \\ & K & \\ & & \end{bmatrix} \begin{Bmatrix} \{u^{per}\} \\ \{\varepsilon^M\} \\ \{\chi^M\} \end{Bmatrix} = \begin{Bmatrix} \{\ 0\ \} \\ \{\ |Y|N\ \} \\ \{\ |Y|M\ \} \end{Bmatrix} \tag{5}$$

The periodic boundary conditions are satisfied by means of linear constraints. The method has been implemented in the finite element software SIC of UTC Compiegne and LMA Marseille. Three-dimensional quadratic elements are used because they provide a good description of the bending behaviour. In the following, all the results given for the method m2 have been obtained using this numerical model.

4. Numerical Examples

4.1. COMPARATIVE STUDY

At first, an example is studied in order to compare the results given by the different methods. Thus, a symmetric hollow plate, periodic in one direction and made of homogeneous material is chosen (Figure 2), for which closed-form formulae can be used.

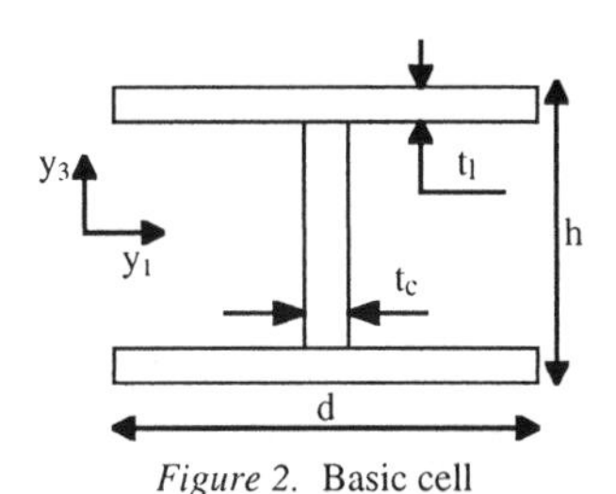

Figure 2. Basic cell

Due to the basic cell symmetry properties, the membrane-bending coupling vanishes, and the non-zero coefficients of the membrane and bending stiffness matrices reduce to the 1111, 2222, 1122 and 1212 components.

For such an unidirectional periodic structure, it is shown in [3] that the bending effective stiffnesses corresponding to the methods m1, m1b and m2b differ only in the twisting stiffness. Moreover, from [4], [7], one can prove that for this kind of basic cell, the methods m3 and m3b coincide. Analytical expressions for the membrane effective properties are available in [10] for the method m2b, and can be easily derived for the methods m1 or m1b and m3 or m3b. The results show that the methods m1, m1b and m2b provide the same values of the membrane stiffnesses. Due to the implementation of the method m2 for plates with properties varying in two directions, though the basic cell problems are in this case two-dimensional, the effective stiffnesses are computed from a three-dimensional mesh, with only one finite element in the y_2 direction.

Six different plates (given in table 2) are studied in order to investigate the influence of the ratio e/ε (which is equal to h/d) and the thickness reinforcement on the effective properties. The upper and lower layers thickness (t_l = 2 mm) and the period length (d = 50 mm) are fixed. The material properties are: Young's modulus E = 210000 MPa, Poisson's ratio $\nu = 0.3$.

TABLE 2. The basic cell dimensions

case	t_c (mm)	h (mm)	e/ε
1	1	5	0.1
2	1	50	1.
3	1	500	10.
4	20	5	0.1
5	20	50	1.
6	20	500	10.

Table 3 presents the plate effective stiffnesses.

For a thin reinforcement (cases 1, 2, 3), whatever the ratio e/ε may be, the results obtained with the different methods are very close, except for the twisting stiffness computed from the method m1 for $e/\varepsilon = 10$, and which is much overestimated.

For a thick reinforcement and $e/\varepsilon = 0.1$ (case 4), the differences between the methods are not significant, except for the membrane stiffnesses obtained with the method m3. These differences increase both for the membrane and bending effective stiffnesses when e/ε becomes larger (see cases 5 and 6), which corresponds to an increase in the reinforcement height. If case 5 is considered, the method m2 is fully justified ($e=\varepsilon$) and the corresponding values must be taken as a reference. It turns out that the other methods can not provide any satisfactory results because the three-dimensional effects which occur at the junction between the core and the face layers are not taken into account.

Comparing the results from the method m3 for t_c = 1 mm and t_c = 20 mm, as noted in [7], it can be seen that except the 2222 component, the effective stiffnesses are insensitive to the thickness reinforcement, which limits the validity of this method.

Moreover, it can be noticed that the strain energies corresponding to each family of methods are ordered as in [4]. Thus, the methods m1, m1b or m2b overestimate the effective stiffnesses obtained with the method m2, while the methods m3 or m3b underestimate them.

TABLE 3. Effective stiffnesses of the plates

	membrane stiffnesses $A^h_{\alpha\beta\gamma\delta}$			bending stiffnesses $D^h_{\alpha\beta\gamma\delta}$				
methods	m1, m1b and m2b	m2	m3 and m3b	m1	m1b	m2b	m2	m3 and m3b
case 1	(10^6 N.mm^{-1})			(10^6 N.mm)				
comp. 1111	0.927	0.924	0.923	2.385			2.385	2.385
comp. 2222	0.928	0.927	0.927	2.385			2.385	2.385
comp. 1122	0.278	0.277	0.277	0.716			0.715	0.715
comp. 1212	0.324	0.324	0.323	0.835	0.835	0.835	0.835	0.835
case 2	(10^6 N.mm^{-1})			(10^9 N.mm)				
comp. 1111	0.940	0.925	0.923	0.540			0.533	0.532
comp. 2222	1.118	1.116	1.116	0.567			0.566	0.566
comp. 1122	0.282	0.277	0.277	0.162			0.160	0.160
comp. 1212	0.329	0.324	0.323	0.199	0.189	0.189	0.187	0.186
case 3	(10^6 N.mm^{-1})			(10^{11} N.mm)				
comp. 1111	0.942	0.926	0.923	0.584			0.574	0.572
comp. 2222	3.008	3.007	3.006	1.000			1.000	0.999
comp. 1122	0.283	0.278	0.277	0.175			0.172	0.172
comp. 1212	0.330	0.324	0.323	0.365	0.204	0.204	0.201	0.200
case 4	(10^6 N.mm^{-1})			(10^6 N.mm)				
comp. 1111	1.003	0.999	0.923	2.392			2.392	2.385
comp. 2222	1.014	1.014	1.007	2.392			2.392	2.392
comp. 1122	0.301	0.300	0.277	0.718			0.718	0.715
comp. 1212	0.351	0.351	0.323	0.837	0.837	0.837	0.837	0.835
case 5	(10^6 N.mm^{-1})			(10^9 N.mm)				
comp. 1111	1.461	1.109	0.923	0.773			0.638	0.532
comp. 2222	4.835	4.804	4.787	1.235			1.223	1.213
comp. 1122	0.438	0.333	0.277	0.232			0.191	0.160
comp. 1212	0.511	0.450	0.323	0.448	0.270	0.290	0.261	0.186
case 6	(10^6 N.mm^{-1})			(10^{11} N.mm)				
comp. 1111	1.530	1.176	0.923	0.939			0.729	0.572
comp. 2222	42.64	42.61	42.59	9.147			9.128	9.114
comp. 1122	0.459	0.353	0.277	0.282			0.219	0.172
comp. 1212	0.536	0.458	0.323	3.486	0.329	0.333	0.285	0.200

4.2. VALIDATION OF THE HOMOGENIZATION PROCESS

A honeycomb plate is considered as a second example. The purpose of this section is to demonstrate the effectiveness and accuracy of the homogenization method. The thickness and the period size have similar values and then only the

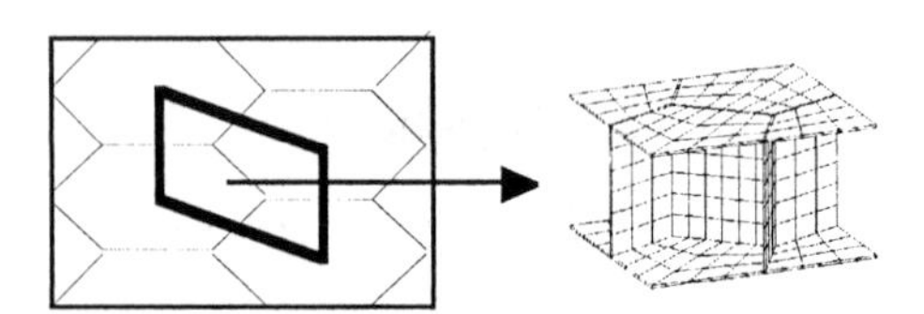

Figure 3. Basic cell with honeycomb core

method m2 is relevant. The basic cell mesh is shown in Figure 3.

To evaluate the accuracy of the effective stiffnesses, a plate made of 10x16 periods is used and the displacements under different loadings obtained with two kinds of computations are compared. The first computation (taken as a reference) consists in modeling the complete heterogeneous plate with DKQ-CSQ plate elements. In the second one, an homogeneous plate is used with the previously computed homogenized effective stiffnesses (method m2).

Each loading test consists in considering only one macroscopic generalized stress to be non equal to zero. On the homogeneous equivalent plate, the corresponding boundary conditions are easily derived and a uniform generalized force is applied to the edge. On the heterogeneous plate, there is no unique way of assigning a load to the edge that is equivalent to the load applied to the homogeneous plate. Uniform generalized forces to the edges of the upper and lower layers are applied. Figure 4 presents the deflected shapes of the heterogeneous and equivalent homogeneous plates for a pure bending moment M_{11} (about the x_2 axis). The deflections of the plates midsurfaces are compared, showing that the discrepancy is less than 2%. It must be noted that using the effective plate properties results in a substantial reduction in the number of degrees of freedom.

For some other load cases, the difference grows up near the boundary because of edge effects but the differences are of the same order.

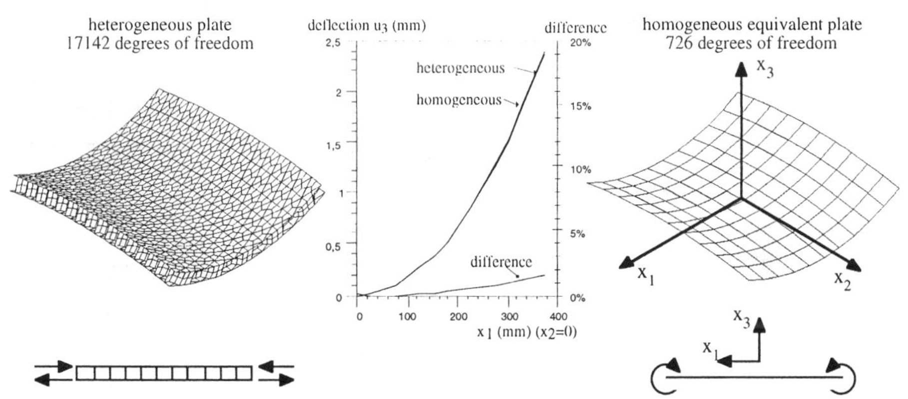

Figure 4. Compared deflections between the heterogeneous and homogeneous equivalent plates under a pure bending moment M_{11}

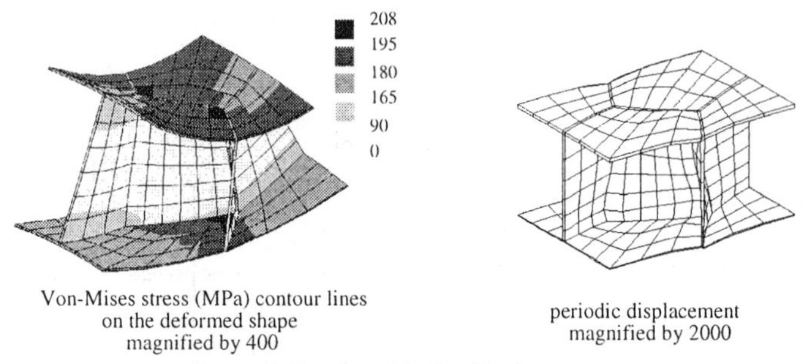

Von-Mises stress (MPa) contour lines on the deformed shape magnified by 400

periodic displacement magnified by 2000

Figure 5. Results of the localization process

Furthermore, this numerical method enables to compute local distributions of stresses and strains within the microscopic level. The microscopic stresses and strains are derived from the computation of the basic cell problems (5) with the given macroscopic generalized stresses or strains. Then, a criterion can be applied at the microscopic scale to ensure that the loads are not critical for the structure. Figure 5 shows the results of the localization process for the previously described example (pure bending moment M_{11}).

5. Conclusions

The results of the parametric study show that a homogenization method may yield a good approximation of the effective stiffnesses even beyond its theoretical range of applicability. Thus, for an unidirectional periodic structure made of thin-walled parts, the methods m1b, m3 or m3b which lead to very simple analytical expressions for the effective stiffnesses, can be widely used. These methods are therefore very attractive in the context of the optimization of such structures, rather than the method m2 which requires numerical solutions.

However, the method m2 (e and ε have the same order of magnitude), based upon very weak assumptions, can be considered as a reference method for many engineering applications. With this method, the basic cell problems are analytically untractable, and a numerical implementation using the finite element method has been proposed. The accuracy of this homogenization method has been proved through the homogenized honeycomb plate computations, which fits with a three-dimensional finite element analysis. Therefore, the effective plate stiffnesses can be used for coarse models of a complete structural entity such as a deck or an airframe.

Acknowledgements
The present research is supported by SOLLAC Co (1st and 3rd authors) and IRCN (French Shipbuilding Research Institute) (2nd author). The authors express their sincere appreciation to these supports.

References

1. Caillerie, D.: Thin elastic and periodic plates, *Math. Meth. Appl. Sci.* **6** (1984), 159-191.
2. Lewinsky, T.: Effective models of composite periodic plates, part I. Asymptotic solution, *Int. J. Solids Structures* **27** (1991), 1155-1172.
3. Lewinsky, T.: Effective models of composite periodic plates, part III. Two-dimensional approaches, *Int. J. Solids Structures* **27** (1991), 1185-1203.
4. Kohn, R.V. and Vogelius, M.: A new model for thin plates with rapidly varying thickness, *Int. J. Solids Structures* **20** (1984), 333-350.
5. Duvaut, G.: Analyse fonctionnelle des milieux continus. Application à l'étude des matériaux composites élastiques à structure périodique. Homogénéisation, *Theoretical and applied mechanics*, Koiter W.T. (ed.), North-Holland, Amsterdam, (1976), 119-132.
6. Tadlaoui, A. and Tapiero, R.: Calcul par homogénéisation des microcontraintes dans une plaque hétérogène dans son épaisseur, *J. Méc. Théor. Appl.* **7** (1988), 573-595.
7. Soto, C.A. and Diaz, A.R.: On the modelling of ribbed plates for shape optimization, *Struct. Optim.* **6** (1993), 175-188.
8. Kalamkarov, A.L.: *Composite and Reinforced Elements of Construction*, John Wiley and Sons, New-York, 1992.
9. Parton, V.Z. and Kudryavtsev, B.A.: *Engineering Mechanics of Composite Structures*, C.R.C. Press, 1993.
10. Lewinsky, T.: Effective stiffnesses of transversely non-homogeneous plates with unidirectional periodic structure, *Int. J. Solids Structures* **32** (1995), 3261-3287.
11. Mouftakir, L.: Homogénéisation des structures ondulées, *Thèse*, Université de Metz, 1996, 128p.
12. Débordes, O.: Homogenization computations in the elastic or plastic range; applications to unidirectional composites and perforated sheets, 4th Int. Symp. Innovative Num. Methods in Engng, *Computational Mechanics Publications*, Springer-Verlag (1986), Atlanta, 453-458.

CHARACTERIZATION AND SIMULATION OF THE MECHANICAL BEHAVIOUR OF MULTILAYERED COMPONENTS COMPOSING A FIBROUS CYLINDER HEAD GASKET

M.-D. DUPUITS
Renault D. R.
9-11 av du 18 Juin 1940, 92150 Rueil Malmaison, France

AND

M. BOUSSUGE AND S. FOREST
Centre des Matériaux Pierre-Marie FOURT
E.N.S.M.P. / CNRS, B.P. 87, 91003 Evry, France

This paper is devoted to the characterization and the simulation of the mechanical behaviour of multilayered components, that compose the fibrous cylinder head gasket of automotive engines. This structure is characterized for compressive loadings. In the present approach, the layered structure is considered as a transverse isotropic material, and a phenomenological description of the observed behaviour in the loading direction is developed. A non-unified elastoviscoplastic constitutive model is presented accounting for the observed responses for cyclic load/unload and step loading tests.

1. Introduction

Cylinder head gaskets are sandwich structures made of two layers of a fibre reinforced elastomeric composite material, pinned on a central steel sheet. During the engine assembly, this layered structure is clamped between the cylinder head and the cylinder block. While the engine is working, local compressive stresses must remain sufficient to ensure water, oil and gas tightnesses. In finite element simulations, which substitute more and more often for expansive engine tests, the mechanical behaviour of the gasket is generally simply considered as non-linear elastic. From the basis of compressive tests carried out at room temperature, this work proposes a more complete description of the actual behaviour of the gasket sandwich structure, taking irreversible deformations and time effects into account. In the first part of this paper, after a brief description of the test rig, the loading

A. Vautrin (ed.), Mechanics of Sandwich Structures, 139–146.

conditions of the cylinder head gasket in the engine are analysed. Subsequent test procedures are then defined, and the observed behaviour is described. The different components of the phenomenological, uniaxial model are presented in the second part, and discussed with respect to the different features of the experimentally observed behaviour. Finally, simulated results obtained through the identification of the constitutive equations are compared with experimental ones, and perspectives are proposed.

2. Experiment

2.1. DESCRIPTION OF THE TEST RIG

The experimental testing device has been designed in order to compress 19 mm diameter chips cut in gaskets. The sample is positionned between two plates, the upper one being ball-jointed. The displacement in the loading direction is recorded by averaging the informations of three inductive type transducers (LVDT), located around the sample (fig.1).

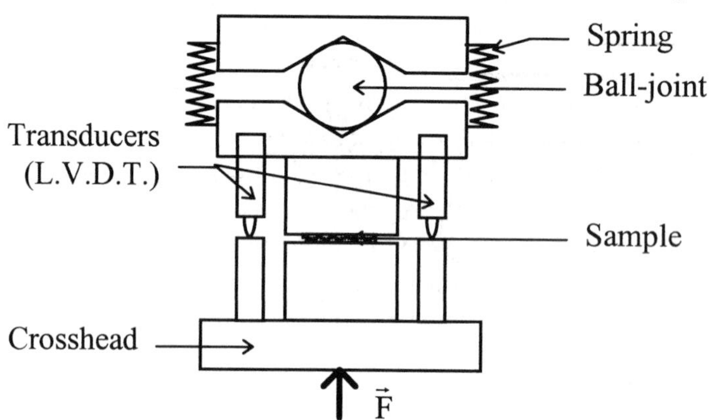

Figure 1. Compressive test rig used for testing gasket pieces.

2.2. LOADING CONDITIONS

Real loading conditions have been analysed during the assembly of the engine as well as during operation, to define realistic experimental procedures.

During the engine assembly, the tightening of the cylinder head is performed in such a way that a permanent compressive state of the gasket is obtained, after stress re-arrangement. Hence, before each mechanical test, a pre-stressing sequence of the specimens has been performed, which leads to the average compressive stress encountered on an actual gasket. As the stresses remain always compressive, contrary to the classical standard, we will later on consider them as positive. The preloading procedure consists in compressing the sample up to 10 MPa and then, maintaining the deformation constant. After a relaxation of four hours, the stress stabilizes at a value close to 3.5 MPa. To improve the reproducibility, the stress is then set to 3.5 MPa : that constitutes the starting state (origin of strain measurement) of further tests. The permanent state of compressive stresses induced, on the gasket, by the clamping forces may be affected by transient thermal regimes. When the engine is heated, the cylinder head, often made of aluminum alloy, dilates more than the other parts of the engine, inducing a stress increase on the gasket. This mechanical loading may be imposed slowly when the whole engine is gently heated up, or rapidly during sudden engine power changes (in that case, the temperature of the cylinder head changes simultaneously with those of exhaust gases, with a very low thermal inertia). From these data, triangle-type loading/unloading tests have been defined, with stressing rates ranging between 0.007 and 3.5 MPa/s. In order to obtain a more accurate identification of the time dependent part of the deformation, long-term step loading tests have been also performed.

2.3. OBSERVED BEHAVIOUR

During cycling, the recorded strain/stress loops (fig.2) are open, non-linear, and a residual permanent strain is observed at the end of each cycle. Even when a maximum stress lower than 1 MPa is applied during a cycle, a entirely reversible deformation is never observed. The structure becomes more and more stiff during cycling. For the cyclic tests, the total deformation increases with the number of cycles, up to reach a stable state characterized by a sligthly open loop. Besides these last observations, surprisingly, in the tested range, the stressing rate has no significant effect on the mechanical response of the material. During the unloading/loading transition, a rebound phenomenon, frequently observed in soils mechanics, arises : the loading slope is greater than the unloading one.

When submitted to a constant stress, the structure creeps, and partial unloadings may lead to reverse creep (fig.3). From the latter observation, we deduce that a viscoplastic component is obvious.

3. Constitutive Equations

Phenomenological approaches involving viscoplastic unified models have already been applied and validated in the particular case of laminated glass/epoxy composite materials (Lesné, Lesne & Maire 1994). A similar methodology is used in this work.

In this section, a simple, uniaxial formulation of the model is provided, in order to explain the role of the different strain components, parameters and internal variables in the description of the material behaviour.

The adjusted numerical parameters are indicated with a star.

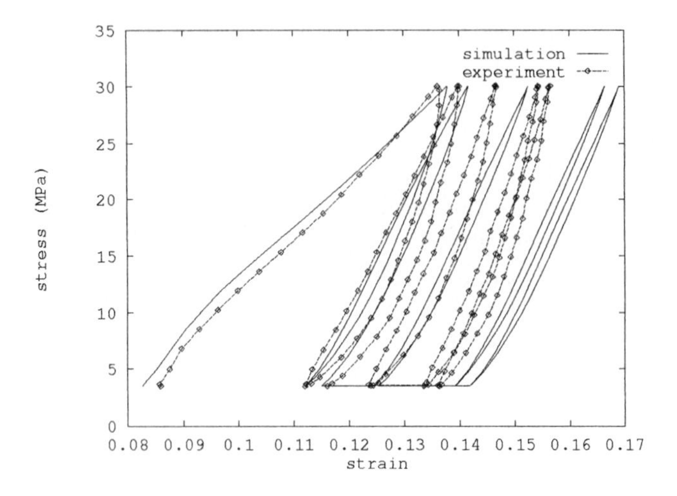

Figure 2. Stress/strain curves, compressive triangle-type loading at 3.5 MPa/s (cycles 1, 2, 10, 100 and 200 are represented) : comparison between simulation and experiment.

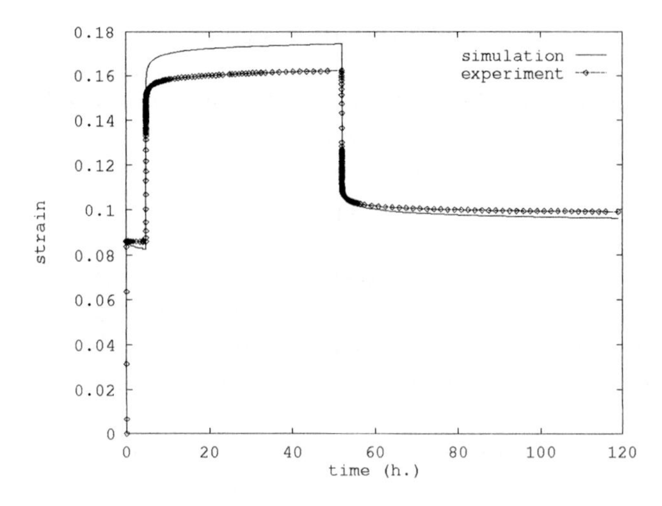

Figure 3. Response of the structure to a step- type (0-30-0 MPa) loading : comparison between simulation and experiment.

A multi-mechanism model involving three different strain components (elastic, plastic and viscoplastic) has been developed. This type of model, based on Mandel's works (1971), has been already developed for an application to stainless steel (Contesti 1988). It has been further generalized and extensively formulated (Cailletaud & Sai 1995). A plastic and a viscoplastic strains are added to the elastic strain :

$$\varepsilon_{tot} = \varepsilon_{el} + \varepsilon_p + \varepsilon_v \tag{1}$$

$$\varepsilon_{in} = \varepsilon_p + \varepsilon_v \tag{2}$$

3.1. ELASTICITY.

As mentioned above, the behaviour of our structure is never fully reversible and, hence, the yield stress is close to zero. The non-linear response is expressed by an hypoelastic law. As usual in soils mechanics, the elastic parameters are defined as a power function of the stress :

$$\dot{\sigma} = (E_0{}^\star(\beta^\star + \sigma)^{m_{el}{}^\star})\dot{\varepsilon}_{el} \tag{3}$$

$E_0{}^\star$, $\beta^\star$ and $m_{el}{}^\star$ are the elasticity parameters.

3.2. VISCOPLASTICITY

The viscoplastic component is defined by the following yield function :

$$f_v = |\sigma - X_v| - \sigma_0^{v\star} \tag{4}$$

Where X_v, the internal viscoplastic stress, has a classical non-linear kinematic expression :

$$\dot{X}_v = \frac{2}{3}C_v{}^\star(\dot{\varepsilon}_p - D_v{}^\star \alpha_v \dot{v}) \tag{5}$$

The viscoplastic component accounts for the creep behaviour of the structure. The viscoplastic strain rate (eq.6) decreases to zero when the internal stress reaches the applied stress. Strain recovery occurs once the applied stress becomes lower than the internal stress X_v.

$$\dot{\varepsilon}_v = < \frac{|\sigma - X_v| - \sigma_0{}^{p\star}}{K^\star} >^{n^\star} \text{sign}(\sigma - X_v) \tag{6}$$

v being the viscoplastic multiplier.

However, the present formulation is not sufficient to describe the cyclic behaviour. In particular, at this state of development, the model induces stress rate dependence of the shape of the loops, what is contrary to experimental observations.

3.3. PLASTIC MECHANISM

3.3.1. *Yield criterion and internal variables*

The introduction of a plastic deformation is necessary to describe the deformation loops of figure 2. A yield function, defined by a center and a radius

that is chosen very low, determines the plastic criteria, with a similar expression to the viscoplastic one (eq.4).
The back-stress, X_p, is the sum of two terms : $X_p = X_{1p} + X_{2p}$.

The plastic mechanism leads to residual strain after unloading but, in the particular case for which the elasticity limit is set to zero, the material retrieves its initial state if a reversible kinematic hardening is considered. Asaro (1975) already considered this configuration for which elasticity can not be distinguished from plasticity with reversible kinematic hardening. Therefore, the evolution laws of kinematic hardening is chosen with a dynamic recovery term for the first component to get open loops and a residual permanent strain at the end of the unloading :

$$X_{1p} = C_{1p}{}^{\star}\alpha_{1p} \quad \text{with} \quad \dot{\alpha}_{1p} = \dot{\varepsilon}_p - D_{1p}{}^{\star}\dot{p}\alpha_{1p} \tag{7}$$

$$X_{2p} = C_{2p}{}^{\star}\alpha_{2p} \quad \text{with} \quad \dot{\alpha}_{2p} = \dot{\varepsilon}_p \tag{8}$$

The plastic cumulative strain is defined by $\dot{p} = |\dot{\varepsilon}_p|$.

3.3.2. *Internal stress depending on inelastic strain*
In order to improve the concavity of the simulated loop, due to the elastic strain, the internal plastic stress X_{1p} is chosen as a power function of the inelastic strain :

$$X_{1p} = C_{1p}{}^{\star}\varepsilon_{in}^{m^{\star}}\alpha_{1p} \tag{9}$$

Figure 4a) shows a strain/stress curve given by the model with $D_{1p} = 0$ while simulating a triangle-type test. The observed opening of the loops is simulated when D_{1p} does not vanish as shown in figure 4b).

3.3.3. *Ratchetting effect*
In order to take the change in the shape of the hysteresis loop during cycling into account, the parameter D_{1p} is modified by a cumulative plastic strain dependent function. The latter has been introduced by Marquis (1979), in the case of metallic materials :

$$\begin{cases} D_{1p} = D_{1p}^{\star}\phi(p) \\ \phi(p) = \phi_s{}^{\star} - (1 - \phi_s{}^{\star})exp(-\omega^{\star}p) \end{cases} \tag{10}$$

During cycling, when the cumulative plastic strain is increasing, the dynamic recovery term reaches progressively the value of $D_{1p}{}^{\star}\phi_s{}^{\star}\dot{p}X_{1p}$. The opening and shape of loops stabilize. The comparison of figures 4b) and 4c) illustrates the influence of this function.

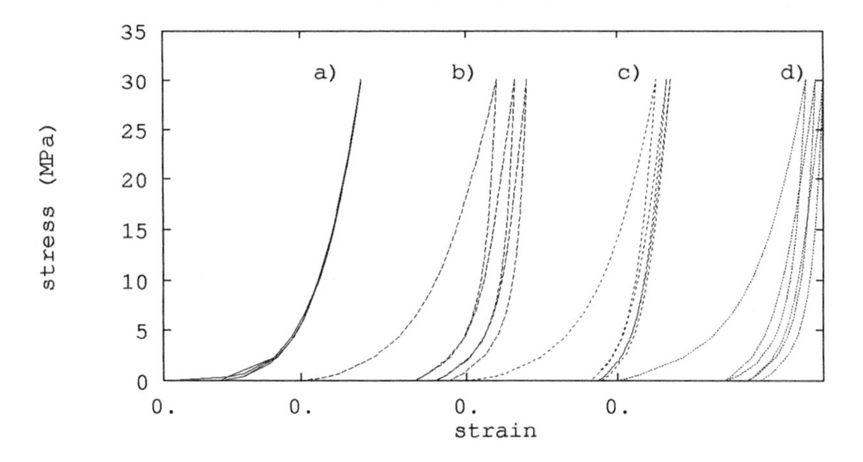

Figure 4. Simulation of three cycles : a) with a linear kinematic hardening depending on the inelastic strain : non-linear reversible behaviour, b) with a non-linear kinematic hardening : reversible behaviour with open loops, c) id b) with Marquis' function, d) id b) with rebound ($C_{2p} \neq 0$).

Progressive deformation is then limited. If $\phi_s{}^\star$ is set to zero, the ratchetting effect is stopped but the loops are closed. Thus, a linear component has to be introduced in the kinematic hardening, in order to describe the observed plastic shakedown, with a slightly open hysteresis loop.

Contrary to experiments (fig.2), after a first loading/unloading cycle, when a new loading branch is considered, the simulated response is entirely located below the one obtained in the previous unloading (fig.4b)), and no rebound is observed. The second internal variable can account for rebound (fig.4d)).

3.4. IDENTIFICATION OF MATERIAL PARAMETERS

The different numerical parameters involved in the present model have been identified with the code SiDoLo (Pilvin 88), that is a software of optimisation. It aims at minimizing an error function calculated as the difference between experimental results and simulation. Our experimental data base contains different types of test results : triangle-type loading/loading test at 3.5, 0.035 and 0.007 MPa/s and two creep/recovery tests at stress levels of 20 and 30 MPa.
Identification of all coefficients is carried out on the complete data base to get a single set of parameters and the comparison between simulation and experiment is given figure 2. Regarding creep/recovery test, the simu-

lations and the experiments are in a good agreement (fig.3). This feature of the behaviour is of prime importance when considering the permanent compressive stress applied on a gasket.

4. Conclusion and perspectives

In order to characterize the mechanical behaviour of the multi-layered structure that composes a cylinder head gasket, compressive loading experiments have been performed. These tests have been defined with respect to the loading conditions encountered in an engine, during assembly and operation. In the present work, a constitutive model has been proposed, that accounts for the numerous observed phenomena (non-linear open loops, stabilized ratchetting, rebound, creep and strain recovery).

In order to perform structural computations of industrial applications, a three dimensional formulation is necessary. This one is presently developed on the basis of the assumptions of a transverse isotropic material, and the absence of transverse plastic and viscoplastic yields. The shear components of the composite behaviour is planned to be identified using mechanical tests, in which shear and compressive loadings are simultaneously applied. A test rig has been designed in order to simulate on a simple structure the compressive stress gradients encountered in an actual gasket. The corresponding experimental results will be used to validate and discuss the overall three dimensional model.

Acknowledgments : The authors wish to thank Renault company (Direction de la Recherche, Pierre Lory) for supporting this work.

References

Asaro R. J. (1975) : Elastic-Plastic Memory and Kinematic-Type Hardening, *Acta Met.*, **23**, 1255-1265.

Cailletaud G. & Sai K. (1995) : Study of Plastic/Viscoplastic Models with Various Inelastic Mechanisms, *Int. Journal of Plasticity*, **11-8**, 991-1005.

Contesti E. & Cailletaud G. (1989) : Description of Creep Plasticity Interaction with non Unified Constitutive Equations : Application to an Austenitic, Stainless Steel *Nuclear Eng. and Design*, 116-265.

Lesné O., Lesne P.M. & Maire J.F. (1994) : A Viscoplastic Constitutive Model for an Organic Matrix Composite, *9èmes Journées Nationales sur les Composites proc.; ed. Favre J.P. and Vautrin A.*, AMAC, St Etienne, 485-494.

Mandel J. (1971) : Plasticité Classique et Viscoplasticité, *Courses and Lectures N^0 97*, Udine, Springer Verlag.

Marquis D. (1979) : Etude Théorique et Vérification Expérimentale d'un Modèle de Plasticité Cyclique, Thèse de Doctorat, Univ. de Paris VI.

Pilvin Ph. (1988) : Identification des Paramètres de Lois de Comportements, *Inelastic behaviour of solids models and utilization, Mécamat'88*, Besançon (France), 155-164.

PREDICTION OF ELASTIC PROPERTIES OF SANDWICH PANELS USING A HOMOGENIZATION COMPUTATIONAL MODEL.

G. VOUGIOUKA,
Department of Engineering Science
National Technical University of Athens
Athens, Greece

H. RODRIGUES, J. M. GUEDES
IDMEC/IST
Instituto Superior Técnico
Av. Rovisco Pais 1
1096 Lisboa Codex, Portugal

1. Introduction

One of the major difficulties in the analysis of composite materials is to characterize its average or equivalent elastic mechanical properties. There are many different approaches to address the problem, ranging from experimental to analytical and computational methods. A description of some general classical methods, and methods for laminate theory can be found, for example, in [1] and [2] and references therein.

This paper presents computational models to predict the equivalent material properties of sandwich panels. These models use the homogenization theory or the combination of this theory with classical energy based methods.

First the homogenization theory is introduced, then the energy based methods are described, and finally the computational models are presented and applied to some examples.

2. Homogenization Method

The homogenization technique is a mathematical method to characterize the material properties of composite materials. The fundamentals of this mathematical theory can be found, among others, in [3,4,5]. Here the characterization of the homogenized material properties is made using an asymptotic expansion method. A summary of this approach is presented. More details can be found in [6] and references therein.

Consider a loaded structure made of composite material, with microstructure formed by the repetition in space of a very small unit cell (representative volume), then the displacement field is assumed to be:

$$\mathbf{u}^{\varepsilon}(\mathbf{x}) = \mathbf{u}^{0}(\mathbf{x}) + \varepsilon\mathbf{u}^{1}(\mathbf{x},\mathbf{y}) + \varepsilon^{2}\mathbf{u}^{2}(\mathbf{x},\mathbf{y}) + \ldots, \mathbf{y} = \frac{\mathbf{x}}{\varepsilon} \tag{1}$$

A. Vautrin (ed.), Mechanics of Sandwich Structures, 147–154.

ε is a small parameter representing the ratio of the size of the unit cell and the size of the structure. The function $\mathbf{u}^0(\mathbf{x})$ represents the average displacement of a point. The functions $\mathbf{u}^i(\mathbf{x},\mathbf{y})$, Y-periodic (periodic with the unit cell) in the $\mathbf{y}$ variable, represent corrections to the average field $\mathbf{u}^0(\mathbf{x})$ in the scale of the unit cell.

The overall equilibrium of the structure subjected to body forces $\boldsymbol{f}$ in the domain Ω, traction $\boldsymbol{t}$ on part of the boundary Γ_t, and fixed on part Γ_u of the boundary can be expressed through the principle of virtual work as

$$\int_{\Omega^\varepsilon} E_{ijkm} \frac{\partial u_i^\varepsilon}{\partial x_j} \frac{\partial v_k^\varepsilon}{\partial x_m} d\Omega - \int_{\Omega^\varepsilon} f_i v_i^\varepsilon d\Omega - \int_{\Gamma_t} t_i v_i^\varepsilon d\Gamma = 0, \tag{2}$$

for all admissible test functions $\mathbf{v}$.

Introducing the asymptotic expansion (1) in the equilibrium equation (2), and assuming similar expansion for the test functions, separating the different powers of the parameter ε, it can be shown (e.g. [6]) that the average response of the solution is characterized by

$$\int_{\Omega} E_{ijkm}^H \frac{\partial u_i^0}{\partial x_j} \frac{\partial v_k^0}{\partial x_m} d\Omega - \int_{\Omega} f_i v_i^0 d\Omega - \int_{\Gamma_t} t_i v_i^0 d\Gamma = 0, \tag{3}$$

for all admissible test function $\mathbf{v}^0$.

From (3) it can be seen, that the average solution depends on the equivalent material properties (homogenized material properties) E_{ijkm}^H. These properties are given by

$$E_{ijkm}^H = \frac{1}{|Y|} \int_{Y_s} E_{ijkm} - E_{ijpq} \frac{\partial X_p^{km}}{\partial y_q} dY \tag{4}$$

where $|Y|$ is the total volume of the unit cell, Y_s is the corresponding solid part, and the X^{km} functions are solution of the following problems in the unit cell Y

$$\int_{Y_s} E_{ijpq} \frac{\partial X_p^{km}}{\partial y_q} \frac{\partial v_i^1}{\partial y_j} dY = \int_{Y_s} E_{ijkm} \frac{\partial v_i^1}{\partial y_j} dY \tag{5}$$

for all admissible test functions $\mathbf{v}^1$.

3. Energy Based Methods

The energy based methods used here are the strain energy and the complementary strain energy methods for composite materials. A description of these methods can be found in [1] and references therein.

Let Y be a representative volume (RVE) of the composite. Then the (complementary) strain energy methods states that the equivalent material properties can be found assuming that the (complementary) strain energy of the equivalent material is equal to the true (complementary) strain energy in the RVE, for homogeneous boundary conditions, i.e.:

$$U = \frac{1}{2} E^{eq}_{ijkm} \bar{\varepsilon}_{ij} \bar{\varepsilon}_{km} |Y| = \frac{1}{2} \int_Y E_{ijkm} \varepsilon_{ij} \varepsilon_{km} dY \tag{6}$$

and

$$U^* = \frac{1}{2} C^{eq}_{ijkm} \bar{\sigma}_{ij} \bar{\sigma}_{km} |Y| = \frac{1}{2} \int_Y C_{ijkm} \sigma_{ij} \sigma_{km} dY \tag{7}$$

where $\bar{\varepsilon}_{ij}$ and $\bar{\sigma}_{ij}$ are average fields, and $C_{ijkm} = E^{-1}_{ijkm}$ is the material compliance tensor. By a suitable choice of the average fields it is possible to identify the equivalent material properties. Note that, in general, the values obtained for the equivalent properties are not the same, i.e., $C^{eq}_{ijkm} \neq E^{eq^{-1}}_{ijkm}$.

4. Computational Models

A sandwich panel is usually a laminate whose exterior layers (skin) are made of fiber reinforced material, and a middle layer (the core) made usually by some sort of cellular material as shown in Figure 1. The skin layers are usually made of one or several layers of fiber reinforced material with different orientations.

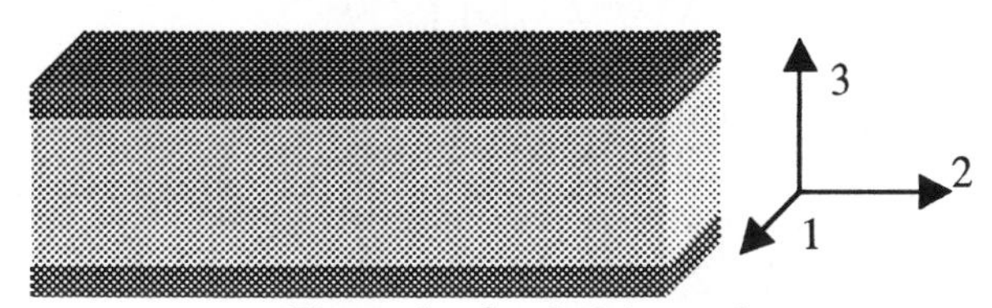

Figure 1 – Sandwich panel

In this study, three computational models are introduced. All the computational models use a two step procedure: Step 1 – computing the equivalent material properties of the skin layers and the core layers, Step 2 – computing the equivalent material properties of the sandwich panel.

Step 1 is equal to all models. In this step unit cells are established for the skin and core layers. Then the homogenization theory is used to compute the full set of homogenized (equivalent) material properties for each layer, by solving problems (5) and using equation (4).

Step 2 is different for the three models:

- Model 1 – Uses the strain energy method (6) to compute the equivalent material properties. For this model the average strain fields $\bar{\varepsilon}_{ij}$ are imposed in the RVE by specifying homogeneous displacement fields on its boundaries. By suitably choosing these fields the material properties E^{eq}_{ijkm} are easily computed. For example, to determine E^{eq}_{1111} it is enough to set $\bar{\varepsilon}_{11} = 1$ ($\Leftrightarrow \bar{u}_1 = x_1 + const$), and set all other components equal to zero. Then from (6), computing the total strain energy density in the RVE for this boundary condition, the value of E_{1111}^{eq} is easily determined.

- Model 2 – Uses the complementary strain energy method (7) to compute the equivalent material properties. In this case, specifying uniform stresses on the boundaries of the RVE imposes the average stress field $\overline{\sigma}_{ij}$. Choosing the applied stresses, the equivalent compliance material properties C_{ijkm}^{eq} are computed.

- Model 3 – Uses the homogenization theory again, considering the RVE as a unit cell. As before, the material properties are computed by solving (5) and using (4).

The implementation of these computational models is made using the finite element method. To solve the homogenization problems (5) and compute the homogenized material properties (4), the code PREMAT described in [6] is used. To compute the strain energy densities in the RVE for Step 2 in models 1 and 2 the code ABAQUS (a finite element code from Hibbit, Karlsson & Sorensen, Inc.) is used.

When applying these models to sandwich panels, since they are layered materials, each layer made of an "homogeneous" material (whose properties were computed in Step 1) the computations of Step 2 may be simplified.

In the case of Model 1, if the true strain field were equal to the average field, then, from expression (6), the equivalent properties would be the average of the component properties:

$$E_{ijkm}^{eq} = \left\langle E_{ijkm} \right\rangle = \frac{1}{|Y|}\int_Y E_{ijkm}\, dY \qquad (8)$$

In (8) the brackets $\langle\ \rangle$ are used to designate the average within the RVE. However this is not true in general, and the previous expression will only produce some bounds for the material properties ([1]).

In the case of Model 2, some simplification can be made if the layers are orthotropic with principle directions of orthotropy aligned with each other. In this case the shear stresses 13 and 23 are equal in all layers and from (7) one can conclude that the corresponding shear moduli properties are:

$$\frac{1}{G_{13}^{eq}} = \left\langle \frac{1}{G_{13}} \right\rangle \qquad \frac{1}{G_{23}^{eq}} = \left\langle \frac{1}{G_{23}} \right\rangle \qquad (9)$$

In the case of Model 3, if the layers are orthotropic with the principle directions of orthotropy aligned with each other, then it can easily be shown, for sandwich panels as in figure 1, that the homogenized material properties are given by:

$$E_{1111}^{H} = \left\langle E_{1111} \right\rangle - \left\langle \frac{E_{1133}^2}{E_{3333}} \right\rangle + \frac{\left\langle \frac{E_{1133}}{E_{3333}} \right\rangle^2}{\left\langle \frac{1}{E_{3333}} \right\rangle} \qquad E_{2222}^{H} = \left\langle E_{2222} \right\rangle - \left\langle \frac{E_{2233}^2}{E_{3333}} \right\rangle + \frac{\left\langle \frac{E_{2233}}{E_{3333}} \right\rangle^2}{\left\langle \frac{1}{E_{3333}} \right\rangle}$$

$$E^H_{1122} = \langle E_{1122} \rangle - \left\langle \frac{E_{1133}E_{2233}}{E_{3333}} \right\rangle + \frac{\left\langle \frac{E_{1133}}{E_{3333}} \right\rangle \left\langle \frac{E_{2233}}{E_{3333}} \right\rangle}{\left\langle \frac{1}{E_{3333}} \right\rangle} \qquad E^H_{3333} = \frac{1}{\left\langle \frac{1}{E_{3333}} \right\rangle}$$

$$E^H_{1133} = \frac{\left\langle \frac{E_{1133}}{E_{3333}} \right\rangle}{\left\langle \frac{1}{E_{3333}} \right\rangle} \qquad E^H_{2233} = \frac{\left\langle \frac{E_{2233}}{E_{3333}} \right\rangle}{\left\langle \frac{1}{E_{3333}} \right\rangle}$$

$$E^H_{1313} = \frac{1}{\left\langle \frac{1}{E_{1313}} \right\rangle} \qquad E^H_{2323} = \frac{1}{\left\langle \frac{1}{E_{2323}} \right\rangle} \qquad E^H_{1212} = \langle E_{1212} \rangle \qquad (10)$$

all other components being zero.

5. Examples

5.1 EXAMPLE 1

Consider a square sandwich plate as in figure 1. The dimensions are L=0.4m, thickness of the core t_{core}=0.018m, thickness of the skin layers, t_{skin}=0.001m. The skin is made of fiber reinforced material and the core by a honeycomb type cellular material. The respective unit cells are shown in figure 2.

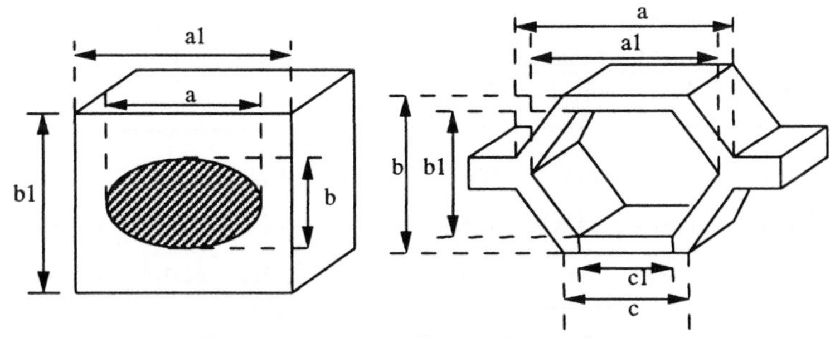

Figure 2 – Unit cells for skin and core

The geometry and material parameters for both unit cells are shown in table 1.

In Step 1 of the computational models, the homogenization method is used, equations (5) are solved and the homogenized material properties computed using (4) (code PREMAT). Note that the honeycomb structure was rotated in order to be vertical (aligned with 3 axis). The results (in MPa) are summarized in table 2.

Table 1 – Unit Cells Properties

	a/b	a/c	a_1/b_1	a_1/c_1	Matrix V.F.	E_f MPa	ν_f	E_m MPa	ν_{μ_2}
Skin	1		1		0.7	400	0.3	100	0.3
Core	1	1.5	1	1.5	0.7			400	0.3

Table 2 – Homogenized material properties

	E_1	E_2	E_3	G_{23}	G_{31}	G_{12}	ν_{21}	ν_{31}	ν_{32}
Skin	187.48	145.66	145.66	51.656	55.044	55.044	.23307	.23307	.3
Core	106.67	83.08	220.55	47.111	62.972	33.94	.35895	.3	.3

In Step 2 depending on the model, different computations were performed. For Model 1, the code ABAQUS was used to compute the strain energy densities in the RVE with a finite element mesh of 1600 20 node elements. Since there are symmetries in the problem (geometric and material) the resulting properties will be orthotropic and, consequently, only nine computations are necessary. The average strain fields are obtained by choosing suitable sets of displacement boundary conditions for each computation.

For Model 2, she shear terms are computed using (9) and the other properties by solving with ABAQUS six RVE problems with the same finite element mesh as before. The average stress fields for these problems are imposed through a uniform pressure distribution on the boundaries of the RVE.

For Model 3, expressions (10) were used.

The results (in MPa) are summarized in table 3.

Table 3 – Material properties of the Sandwich Panel

Model	E_{1111}	E_{2222}	E_{3333}	E_{1122}	E_{1133}	E_{2233}	E_{1313}	E_{2323}	E_{1212}
1	151.25	116.88	245	59.375	62.186	53.125	62.031	35.156	47.968
2	150.98	116.84	244.69	59.564	62.278	53.280	62.078	47.529	
3	151.19	116.89	244.7	59.613	62.346	53.320	62.078	47.529	36.050

5.2 EXAMPLE 2

This example considers a sandwich panel with the same dimensions as in example 1. The goal here is to try to predict the sandwich panel properties, by adjusting the material homogenized material properties obtained in Step 1 of our computational models, with some available experimental data.

The unit cell used in Step 1 for the skin is shown in figure 3, and for the core a cell of the same type as in previous example is used. The properties, after adjusting the geometry parameters, are summarized in table 4

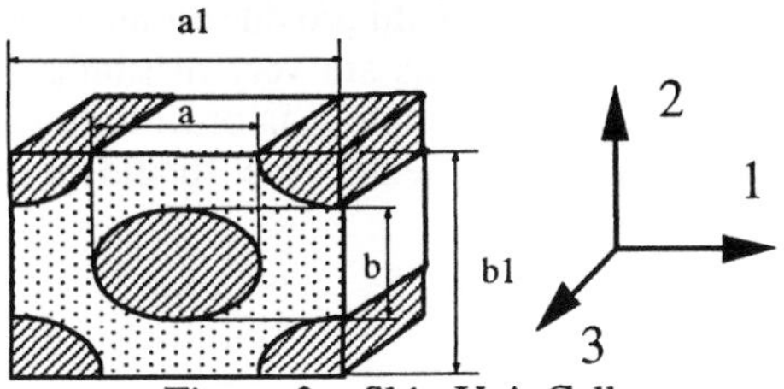

Figure 3 – Skin Unit Cell

Table 4 – Unit Cells Properties

	a/b	a/c	a_1/b_1	a_1/c_1	Matrix V.F.	E_f GPa	ν_f	E_m GPa	$\nu_{\mu 2}$
Skin	1		1		0.67	255	0.2	3.6	0.34
Core	1.1547	12	1.1547	2	0.0582			4	0.34

The homogenized properties obtained in Step 1, as well as the available experimental data are summarized in table 5 (in MPa for the Skin and KPa for the Core)

Table 5 – Homogenized and experimental material properties

	E_1	E_2	E_3	G_{23}	G_{31}	G_{12}	ν_{21}	ν_{31}	ν_{32}
Skin^H	9.5365	9.5137	136.33	4.3828	4.3817	5.5278	0.52486	0.25527	0.25557
Skin^E	9.8		133.7						0.286
Core^H	2.0428	2.0168	156.36	22.387	33.714	1.1967	.99088	.34000	.34000
Core^H				21	35				

Note that the fibers are assumed aligned with direction 3, while in the sandwich panel they are aligned with direction 1. Finally the predictions for the sandwich panel are shown in table 6 (in MPa).

Table 6 – Material properties of the Sandwich Panel

Model	E_{1111}	E_{2222}	E_{3333}	E_{1122}	E_{1133}	E_{2233}	E_{1313}	E_{2323}	E_{1212}
1	13.688	1.0043	0.2252	0.2259	0.0561	0.0543	0.0510	0.0384	0.4383
2	1.473	0.5818	0.1942	0.0623	0.0310	0.0345	0.0249	0.0374	
3	13.734	0.9982	0.1945	0.2834	0.0356	0.0382	0.0249	0.0374	0.4383

These two examples show how the computational models can be used to characterize the full complete set of the elastic equivalent properties of sandwich panels. It should be noted that in general model 1 and 3 should predict bounds on the value of the material properties (see e.g. [1]), and model 2 should provide an intermediate value. Also, when the material properties of the components are very dissimilar, these bounding values tend to be far apart, as can be seen in example 2. In order to produce better estimates, an adjustment of the models with experimental data should be done, and better finite element meshes can be used.

Acknowledgements

This work was partially supported by the Human Capital and Mobility program HCM-CHRX-CT93-0222 (GV), AGARD P119 (SMP/ASP 31) project, and JNICT project N° PBIC/C/TPR/2404/95 (HR, JMG)

References

1. Hashin, Z., "Analysis of Composite Materials – A Survey", Journal of Applied Mechanics, Vol. 50, 481-505, 1983
2. Reddy, J.N., "Mechanics of laminated Composite Plates", CRC Press, Boca Raton, 1997
3. Lions, J. L., "Some Methods in the Mathematical Analysis of Systems and their Control", Science Press, Beijing, and Gordon and Breach, New York, 1981
4. Sanchez-Palencia, E., "Non Homogeneous Media and Vibration Theory", Lecture Notes in Physics, No. 127, Springer, Berlin, 1980
5. Murat, F., Tartar, L., "Calcul des Variations et Homogénéisation", Les Méthodes de l'Homogénéisation: Theorie et Aplications en Physique, Eyrolles, 127, 319-369.
6. Guedes, J.M., Kikuchi, N., "Preprocessing and Postprocessing for Materials Based on the Homogenization Method with Adaptive Finite Element Methods", Computer Methods in Applied Mechanics and Engineering 83, 143-198, 1990

EFFECTS OF BOUNDARY CONDITIONS IN HIGH-ORDER BUCKLING OF SANDWICH PANELS WITH TRANSVERSELY FLEXIBLE CORE

V. SOKOLINSKY AND Y. FROSTIG
Fac. of Civ. Engng., Technion, I.I.T, Haifa, Israel 32000

Abstract. The effects of boundary conditions on the buckling behavior and bifurcation load level of sandwich panels/beams with a "soft" core due to in-plane loads are presented. The study is conducted using a closed-form high-order linearized buckling analysis, that includes the influence of the transverse flexibility of the core as well as of the localized effects on the overall sandwich panel/beam behavior, and allows the use of different boundary conditions for the upper and the lower skin at the same section. The panel/beam construction is general and consists of two skins (not necessary identical), metallic or composite laminated symmetric, and a "soft" core made of foam or a low strength honeycomb which is flexible in the vertical direction. The closed-form high-order analysis yields the general buckling behavior of the structure which means that the solutions obtained allow for interaction between the skins and the core. The solutions are general and are not based on separation of the buckling response on several types of uncoupled buckling phenomena, such as overall buckling, skins wrinkling, etc., as commonly used in the literature. The finite differences technique has been applied to approximate the governing equations of the closed-form high-order formulation and transform the set of the linearized governing differential equations to an eigenvalue problem that is solved using the deflated iterative Arnoldi procedure. The influence of various boundary conditions, including the different support types throughout the height of the same section and non-identical conditions at the upper and the lower skin, as well as of the core properties, on the buckling behavior of the sandwich panels/beams is considered. The discrepancy between the Timoshenko-Reissner model and the present closed-form high-order formulation is discussed. In particular, a partial fixity phenomenon due to the

A. Vautrin (ed.), Mechanics of Sandwich Structures, 155–162.

existence of the pinned boundary conditions, i.e. simply-supported conditions, at the upper and the lower skins at the edge is demonstrated. It is shown that the core properties affect the buckling loads and the corresponding modes of the panel/beam in such a way that the structures with identical boundary conditions but with different cores may undergo different types of buckling such as overall and local as well as interactive loss of stability. The effect of edge concentrated moment induced by a couple of forces exerted on the skins only is also studied.

1. Introduction

Nowadays every field of industry resorts to the use of sandwich construction. Combination of a high bending stiffness for a small weight penalty with a good thermal and acoustical insulation, as well as ease of mass production, make the sandwich construction very attractive to airspace, machine-building, chemical, ship-building and building industries. Introduction of new materials such as laminated composites for the face sheets and solid low-density materials (eg., plastic foams, nonmetallic honeycombs) for the core opens up new possibilities in the design of sandwich constructions while poses some problems that need to be dealt with.

Hunt *et al.* [4] point to the tendency of sandwich structures with transversely flexible ("soft") core for more complex and catastrophic forms of failure which is a result of the great efficiency of carrying load. They approximate a sandwich strut by the simple six degree-of-freedom pilot model which tends to have a triggering bifurcation into an overall mode of buckling rapidly followed by secondary bifurcation into an unstable combination of local modes.

Frostig *et al.* [2] and Frostig and Baruch [3] proposed a refined closed-form high-order theory for sandwich constructions. With the aid of this theory the behavior of a general sandwich construction with a transversely flexible core subjected to various types of external actions can be described in the general sense. The term "general" means that the analysis does not separate the complicated sandwich response into isolate problems but rather gives the general geometrically non-linear picture of that response in the overall and local sense. It should be particularly emphasized that the non-linear behavior of the core is the *result* of the theory rather than a presumed displacement fields. This is one of the main advantages of the theory under consideration in comparison with other refined ones where the displacement patterns through the section height are given *a priory*. It must be emphasized that the high-order theory under consideration enables to analyze the sandwich panel/beam as a *compound* structure rather than merely an ordinary panel/beam element, approximated by its central axis, that is the case in the open literature. While the simplified panel/beam classical approaches are relevant within certain limits when applied to the sandwich panels/beams with transversely stiff cores (eg., honeycomb), they lead to the absolutely inadequate and inconsistent picture of the structure response in the case of the "soft" core.

The present paper uses the sandwich constructions model described above to analyze the effect of different types of boundary conditions on buckling behavior of sandwich panels/beams with transversely flexible core. The distinctive feature of the paper as compared with the work [3] consists in analyzing the sandwich panels/beams with non-membrane regime in the prebuckling state which leads to a non-uniform internal resultants at this stage. The buckling of ordinary panels/beams, i.e. bifurcation, means the transition of the panel/beam from the unstable membrane state to the stable bending one at some critical value of the external compressive load called the buckling load. However, since the sandwich panel/beam is a compound structure, with in general non-membrane regime in the prebuckling state, its buckling behavior implies the transition of the structure from the unstable equilibrium path to the stable one at the critical value of the external longitudinal load pattern. In both cases bifurcation of the solutions takes place. The mathematical formulation of the problem appears in [8].

2. Numerical Study

The panel/beam example presented in [3] is considered first for the purpose of testing the numerical tools in use. [1] It is a longitudinally compressed sandwich panel/beam with geometric and physical characteristics given in Fig. 2, and the critical load corresponding to the symmetrical wrinkling buckling. Comparison between the analytical result of the cited paper and the numerical solution obtained were in close agreement.

The classical approach to the analysis of sandwich panels/beams in [1, 6] follows the usual way of replacing the real structure by its centerline, see Figs. 1 (a), (b); in so doing the possibility of analyzing effects of the *real* boundary conditions on the structure behavior is lost. This observation can be clearly demonstrated by Figs. 1 (c), (d) which represent the buckling behavior of stiff-core sandwich panels/beams. (Choosing the stiff core in these examples is due to the possibility of an immediate comparison of the numerical results with known solutions). At first glance it would seem that cases (c) and (d) are equivalent. However, a close look at the

[1]The numerical study was conducted with the aid of the program "FSAN" working in the "MATLAB" software environment, see [5]

boundary conditions of case (d) reveals that the horizontal reactions developing at the pinned supports of the right panel/beam edge give rise to a fixing couple. As a consequence, Fig. 1 (d) represents the well-known Euler's clamped-roller case for a shear deformable panel/beam (Timoshenko beam). The panel/beam of Fig. 1 (c), on the other hand, has a roller at its upper right support that prevents the arising of the fixing couple and the case turns to be equivalent to the Euler's shear deformable simply-supported panel/beam. Thus the representation of the sandwich structure by its centerline is more than misleading when compared to the actual case.

The next example considers a general case of boundary conditions, see Fig 2. The left edge of the panel/beam rests on the roller support at the lower skin, being prevented from the vertical displacements at the core edge face that in turn prevents the vertical displacement of the upper skin. The boundary conditions for the left edge thus read as $-N_{xxt} = -N_{xxb} = N/2; M_{xxt} = M_{xxb} = 0; w_t = w_b = 0; \tau_{,x} = 0$. The right panel/beam edge is clamped at the lower skin and free at the upper skin and the core. The resulting buckling response is a combination of the overall mode and the localized buckle pattern at the region of the right support.

The peculiarities of the buckling responses discussed above are caused by the "soft" core non-linear displacement pattern which increases the peeling and shear stresses in the core as well as in the skins-core interface layers. Therefore, it is of interest to see how the variations in the core elastic modulus, E_c, and its shear modulus, G_c, influence the sandwich panel/beam buckling response. The non-dimensional critical load, N_{cr}/N_e, where N_e is Euler's buckling load for a simply-supported Timoshenko panel/beam, versus E_c/E_t appears in Fig. 3 for various ratios of E_c/E_t. The range of E_c/E_t is between 0.001 to 0.01 which coincide with the properties of a so called "soft" core, see [2]. It can be clearly seen, see Fig. 3, that the sandwich panel/beam with a low shear resistance ($E_c/G_c = 25$) buckles into an overall mode with a small value of buckling load. The longitudinal shear deformations are dominant hence the vertical core resistance remains untapped (see the upper picture in the graph frame). Moreover, it should be particularly emphasized that for the small values of the ratio E_c/E_t the buckling load in the case being considered ($E_c/G_c = 25$) is the same as for the simply-supported—clamped Timoshenko beam while as E_c increases, the influence of the clamping at the lower skin decays and the sandwich panel/beam more and more approaches the buckling behavior of the simply-supported Timoshenko beam (see Fig.3).

When the shear modulus increases, the buckling behavior is changed from an overall pattern (see the upper picture in the graph frame) to an interactive one (see the lower picture in the graph frame). However, the localized buckle pattern at the right panel/beam end tends to disappear

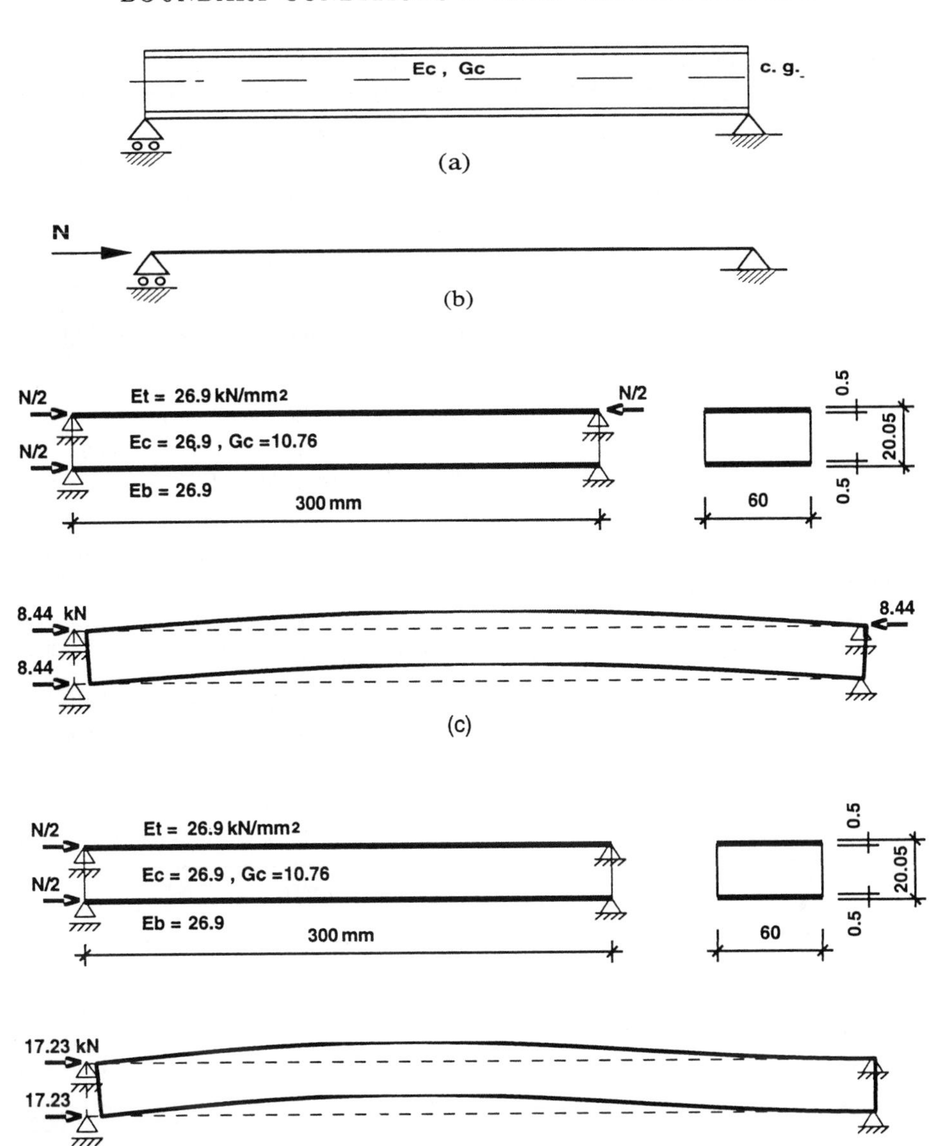

Figure 1. Buckling of a longitudinally compressed sandwich panel/beam with a stiff core and a symmetrical section

with increase of E_c. Any increasing of the core moduli ratio above $E_c/G_c = 1$ does not lead to an appreciable change in the critical load.

It must be emphasized that neglecting of the non-zero bending in the prebuckling state in the all cases described above leads to the negligible change in the results (within 1.5 – 2%). This is in conformation with the

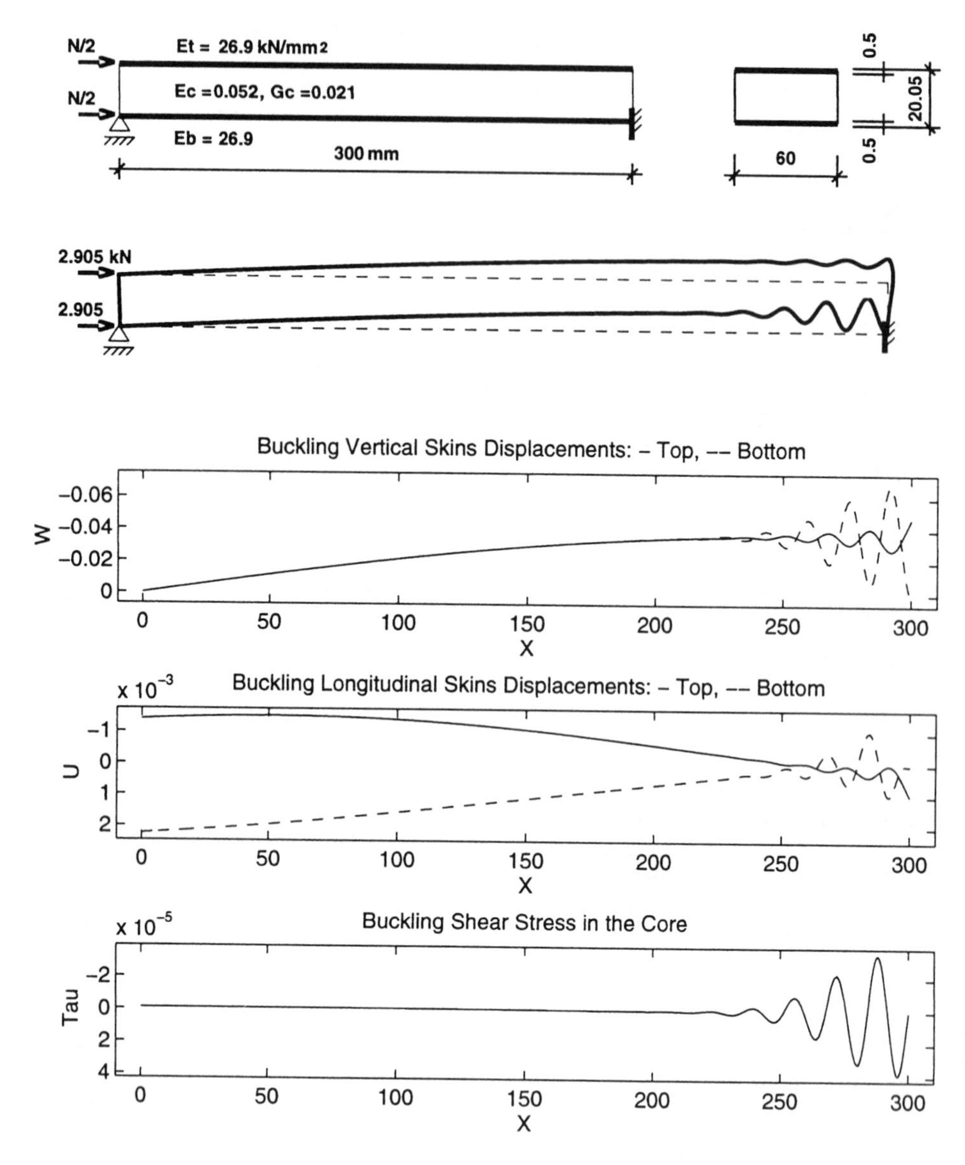

Figure 2. Buckling of a longitudinally compressed sandwich panel/beam with a symmetrical section and non-identical boundary conditions

assumption of the minor influence of the prebuckling non-linearity on the structure behavior which justifies the linear approach at the prebuckling state.

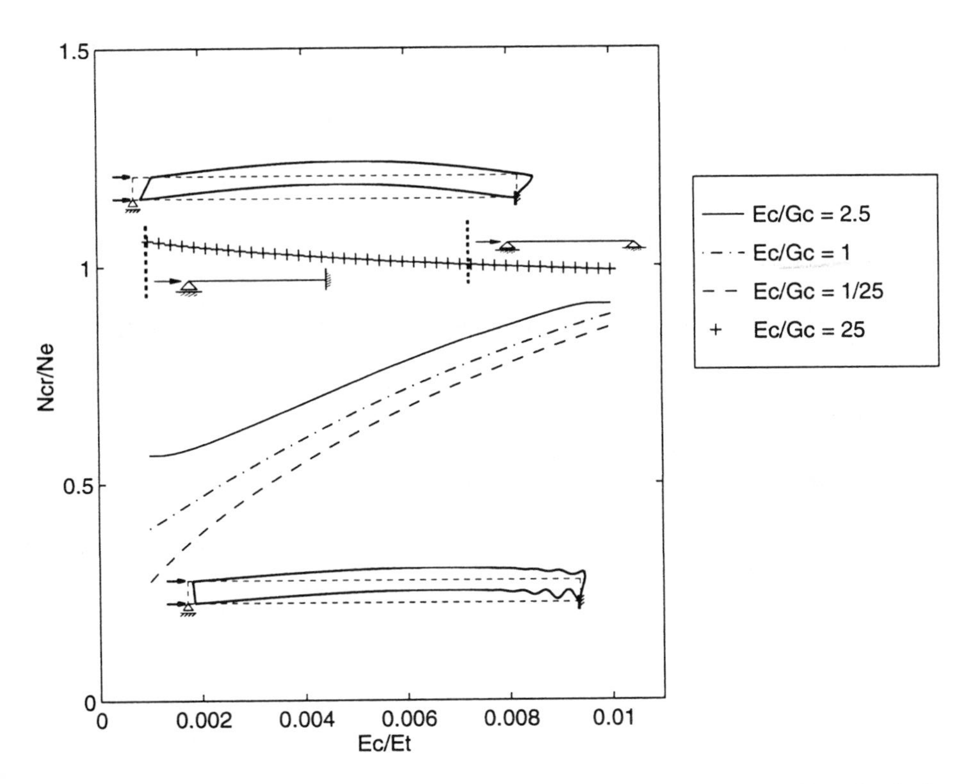

Figure 3. Effect of the core mechanical properties variation on the buckling load of a sandwich panel/beam (see Fig. 2)

3. Conclusions

The closed-form high-order sandwich theory developed in [2, 3] is used for studying the effects of boundary conditions on buckling behavior of sandwich panels/beams with transversely flexible core. The approach enables to analyze these constructions from the general point of view without separating the complicated buckling response into isolated problems and resorting to assumed buckling patterns.

It is shown that the account for the possibility of different support conditions at the same edge of the sandwich panel/beam (as it is the case in practice) has brought about a qualitative and quantitative change in buckling response of the sandwich panels/beams. The way of applying the edge longitudinal loads is also found to be an essential factor influencing the panel/beam buckling behavior.

The present paper shows the influence of the boundary conditions and edge loading patterns on buckling behavior of the sandwich panels/beams with transversely flexible core and gives the qualitative and quantitative *general* picture of this complicated phenomenon. The analysis presented enables practical engineers to predict the complicated response of the sandwich panels/beams with the "soft" core and *real* boundary conditions.

References

1. Allen, H. G.: *Analysis and Design of Structural Sandwich Panels,* Pergamon Press, London, U. K., 1969.
2. Frostig Y., Baruch, M., Vilnay, O., and Sheinman, I.: High-Order Theory for Sandwich-Beam Behavior with Transversely Flexible Core, *J. Engng. Mech. Div. Proc. ASCE,* **118**(1992), 1026–1043.
3. Frostig Y., and Baruch, M.: High-order Buckling Analysis of Sandwich Beams with Transversely Flexible Core, *J. Engng. Mech. Div. Proc. ASCE,* **119**(1993), 476–495.
4. Hunt, G. W., da Silva, L. S., and Manzocchi, G. M. E.: Interactive Buckling in Sandwich Structures, *Proceedings of the Royal Society of London,* **417A**(1988), 155–177.
5. *MATLAB, High-Performance Numerical Computation and Visualization Software, Reference Guide,* The Math Works Inc, 1993.
6. Reissner, E.: Finite Deflections of Sandwich Plates, *J. Aero. Sci.,* **15**(1948), 435–440.
7. Saad, Y.: *Numerical Methods for Large Eigenvalue Problems,* Manchester University Press, U. K., 1992.
8. Sokolinsky, V., Frostig Y.: Buckling of Sandwich Panels With "Soft" Core Using Closed-Form High Order Theory— Boundary Conditions Effects, (*Submitted*).

A Study on the Influence of Local Buckling on the Strength of Structural Core Sandwich Structures

Poorvi Patel*, Tomas Nordstrand* and Leif A. Carlsson**
*SCA Research AB, Box 3054, 850 03 Sundsvall, Sweden.
**Department of Mechanical Engineering, Florida Atlantic University, Boca Raton, FL 33431 USA.

ABSTRACT

The collapse mechanism of biaxially loaded corrugated board has been experimentally examined. Biaxial loading was accomplished by subjecting board cylinders to axial compression, torque, external pressure and combinations thereof. Failure analysis of the board was based on the stress state in the facings in conjunction with a combined stress failure criterion. The study demonstrates that local buckling is the dominant failure mechanism of the corrugated board when there are large inplane compressive stresses acting in a direction across the corrugations and in the facings. Reduction in strength of the corrugated board due to local buckling was also observed when large shear stresses are present in the facings. In this study we examined the implications of the strength degradation due to local buckling on the performance of various corrugated panels. It is demonstrated that modification of the Tsai-Wu failure criterion to include these failure modes significantly improves the strength prediction of the panels, especially if they are loaded in compression transverse to the corrugations.

KEYWORDS: Corrugated board, local buckling, Tsai-Wu failure criterion, failure envelope, biaxial loading.

1. INTRODUCTION

Corrugated board is a sandwich structure used in a wide variety of packaging applications [1]. The principal directions of elastic symmetry are defined as those of the paper constituents, i.e. the machine direction (MD), cross direction (CD) and thickness direction (z), **Fig**. 1. It has long been recognized that corrugated board panels loaded in compression display local buckling of the facings between the corrugations [2-4]. Local buckling and failure of such a panel has been found to initiate in the facing in regions where the facings are under high shear and compressive stresses [3,4]. A comprehensive experimental program has therefore been performed to examine under which stress conditions local buckling of the facings would initiate. The reduction in strength of the board due to local buckling was inferred by comparison of the actual strength data with ideal strength calculated without

A. Vautrin (ed.), Mechanics of Sandwich Structures, 163–173.

consideration of local buckling. A wide range of combined uniform stress states was achieved by subjecting cylinders to axial load, torque and external pressure. The failure mechanism of corrugated board was examined using visual inspection and video recordings. In this study, we will briefly review the previous experimental results. Based on the observed strength degradation, a modification of the Tsai-Wu failure criterion was incorporated, and the implications of local buckling on the performance of corrugated board panels was examined.

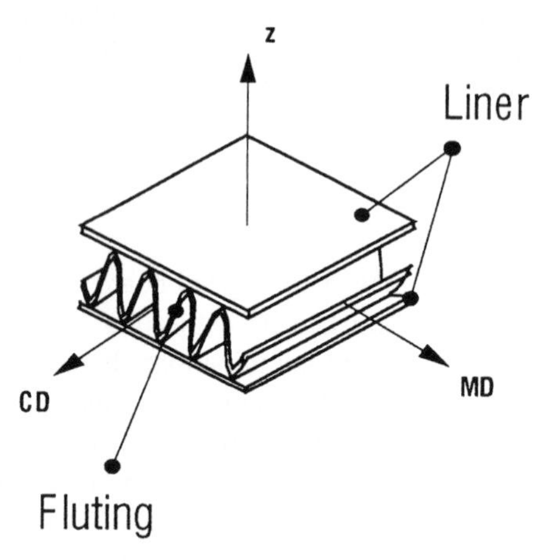

Figure 1. Single wall corrugated board and its principal directions.

2. MATERIALS

The board considered previously [5,6] was further examined herein. The geometry and material properties of the board are listed in **Tables** 1 and 2. The effective core properties listed in **Table** 2 are homogenized properties obtained as explained in Ref.[11].

TABLE 1. Geometry parameters of the corrugated board examined

Parameters	(mm)
t_o	0.266
t_i	0.231
t_C	0.252
t_f	3.6
λ	7.2

t_o, t_i and t_C are the thicknesses of the outer, inner facings and fluting, resp. t_f is the flute height and l is the flute wavelength.

TABLE 2 . Material properties of the board constituents (at 23°C and 50%RH).

	Outer liner	Fluting	Inner liner	Effect. core prop.*
E_x (MPa)	8510	5270	8220	0.5
E_y (MPa)	3810	2260	3010	227
G_{xy} (MPa)	2200	1340	1920	0.23
G_{xz} (MPa)	45	45	45	3.5
G_{yz} (MPa)	45	45	45	35
ν_{xy}	0.37	0.37	0.37	0.004
SCT_x (kN/m)	8.6	-	7.0	-
SCT_y (kN/m)	5.4	-	3.8	-

*) Effective core properties for use in FE-analysis (core is considered homogenous).

3. LOADING OF A CYLINDER AND STRESSES IN THE STRUCTURE

A global cylindrical coordinate system (r-ϕ-z) was assigned to the cylinder, and a local Cartesian system (x-y-z) to the cylinder wall, **Fig**. 2. The radial direction of the cylinder coincides with the thickness coordinate of the board. The machine direction (MD) of the board is along the hoop direction of the cylinder and the cross direction (CD) of the board is aligned with the cylinder axis. The axial force, F, corresponds to normal stress, σ_y, and the torque, T, to shear stress, τ_{xy}. Hoop stress, σ_x, is achieved by internally or externally pressurizing the cylinder. Normal stresses, σ_y and σ_x, and shear stress, τ_{xy}, in the facings are calculated as functions of torque, T, axial force, F, and pressure, P. The corrugated board is considered symmetric i.e. the facings are assumed to be identical. Furthermore, the board is assumed to be under uniform strain (equal in-plane strains in both facings and core).

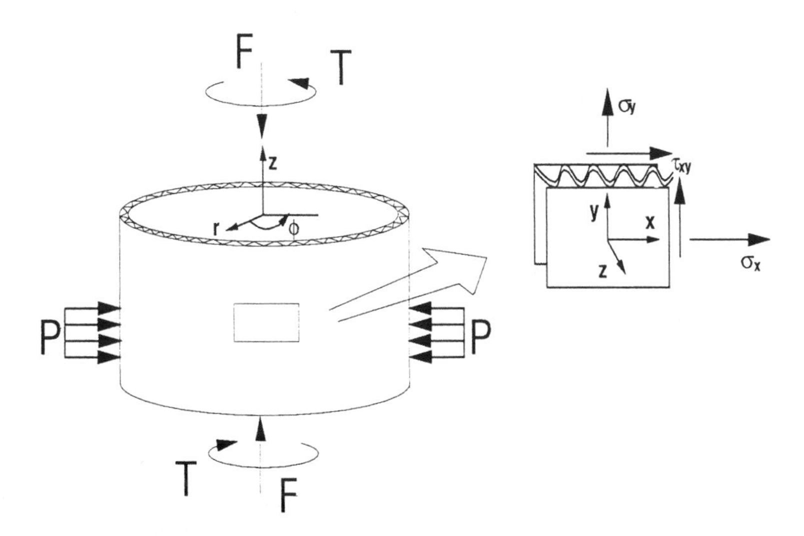

Figure 2. Coordinate systems and loading of the cylinder.

The axial force per unit width of an axially compressed cylinder may be expressed as

$$f = \frac{F}{2\pi R} \tag{1}$$

where F is the load applied, and R is the mean radius. If α is the take-up factor of the core, i.e. the ratio between arc length and wavelength of the corrugated sheet, equilibrium in the axial direction requires

$$f = 2t_L \sigma_{yL} + \alpha \sigma_{yC} t_C \tag{2}$$

where t_L and t_C are the facing (liner) and fluting thickness, and σ_{yL} and σ_{yC} are normal stresses in the facings (L) and core sheet (c)

$$\sigma_{yL} = E_{yL} \varepsilon_{yL} \tag{3a}$$

$$\sigma_{yC} = E_{yC} \varepsilon_{yC} \tag{3b}$$

where E_{yL} and E_{yC} are the Young´s moduli of the facing and core in the CD, and ε_{yL} and ε_{yC} are the strains in the facings and core. Assuming uniform strain, $\varepsilon_{yC} = \varepsilon_{yL}$ in eqs.(3) yields

$$\sigma_{yC} = \frac{E_{yC}}{E_{yL}} \sigma_{yL} \tag{4}$$

Substitution of eqs.(2) and (4) in (1) gives

$$\sigma_{yL} = \frac{F}{2\pi R\left(2t_L + \alpha \frac{E_{yC}}{E_{yL}} t_C\right)} \tag{5}$$

For a cylinder under externally pressure, P, the stresses σ_x and σ_y in the facings are

$$\sigma_x = \frac{PR}{2t_L} \tag{6}$$

where subscript L on the stress has been dropped from now on. Notice, that the corrugated core is assumed not to carry hoop stress, σ_x, in eq.(6).
For a cylinder in torsion, the shear stress in the facings is

$$\tau_{xy} = \frac{T}{4\pi R^2 t_L} \tag{7}$$

where T is the torque applied. The contribution of the corrugated core to the torque is small compared to that from the face sheets [9] and is therefore neglected.

4. TSAI-WU FAILURE CRITERION

Failure prediction of corrugated board structures such as panels under combined stresses is commonly performed based on the hypothesis that the board fails when the stresses in the facings reach their ultimate combination. For anisotropic materials such as paper, the Tsai-Wu failure criterion [12] is commonly applied [13,14]. For plane stress the Tsai-Wu criterion [12] is formulated as follows

$$F_1 \sigma_x + F_2 \sigma_y + F_{11} \sigma_x^2 + F_{22} \sigma_y^2 + F_{66} \tau_{xy}^2 + 2F_{12} \sigma_x \sigma_y = 1 \qquad (8)$$

The parameters F_i and F_{ij} in eq.(8) are related to the tension and compression strengths in the x and y directions, and the shear strength in xy plane. The parameters F_i and F_{ij} used in this study are listed in **Table** 3.

TABLE 3. Parameters used in the Tsai-Wu failure criterion, eq.(1)

F_i and F_{ij}	Value mm^4/N^2
F_{11}	0,00037
F_{22}	0,00153
F_{66}	0,0017
F_{12}	-0.00027
	mm^2/N
F_1	-0,0202
F_2	-0,0261

To examine the potential for failure and identify failure initiation site, the left hand side of eq.(8) was calculated at any applied loading situation. The left-hand side is denoted by the symbol "k", here called the "Tsai-Wu factor". Consequently, if $k<1$ material failure has not yet initiated, and if $k \geq 1$ material failure has occurred. This approach, however, neglects possible strength reduction caused by local buckling of the facings.

5. EXPERIMENTAL SET-UP

The corrugated board cylinders were fabricated as detailed in Ref.[5]. Eighteen cylinders were tested on a servohydralic MTS test frame, and three cylinders were tested in pure torsion in a specially designed frame [6]. External pressure loading of a cylinder was achieved by applying vacuum inside the cylinder [15]. To examine and record local buckling and failure of both facings of the torqued cylinders, three high shutter speed video cameras (25 pictures/sec) were placed on each side of the cylinder (front and back) and inside the cylinder. The cylinders, which were tested in torsion and external pressure, were conditioned at 50 ± 2 %RH and 24 ± 1 °C. The remaining cylinders were conditioned between 20 and 26 %RH and 23 ± 1 °C. The previous humidity range is not expected to change the strength of the board significantly [16].

6. EXPERIMENTAL RESULTS

Figure 3 shows a failure envelope σ_y vs. τ_{xy} (σ_x = 0) generated by the Tsai-Wu prediction and a Tsai-Wu curve fitted to the experimental collapse stresses. For cylinders subjected to axial compression combined with small torques, the experiments were in good agreement with predictions based on the Tsai-Wu failure criterion [5]. In pure shear, however, local buckling was observed just before the cylinder collapsed at 14.1 MPa [6]. This indicates that in shear local buckling is an operative failure mechanism. Cylinders subjected to external pressure failed at a hoop stress, σ_x between -7.5 and -8.5 MPa. The outer facing buckled and collapsed between two fluting tips and the cylinder wall buckled inwards severely limiting the strength of the cylinder. Figure 4 displays the experimental and Tsai-Wu failure envelopes for σ_y versus σ_x quadrant. Comparing the experimental results with the collapse stresses predicted by the Tsai-Wu criterion reveals that this criterion overpredicts the collapse stresses by almost a factor of 3.

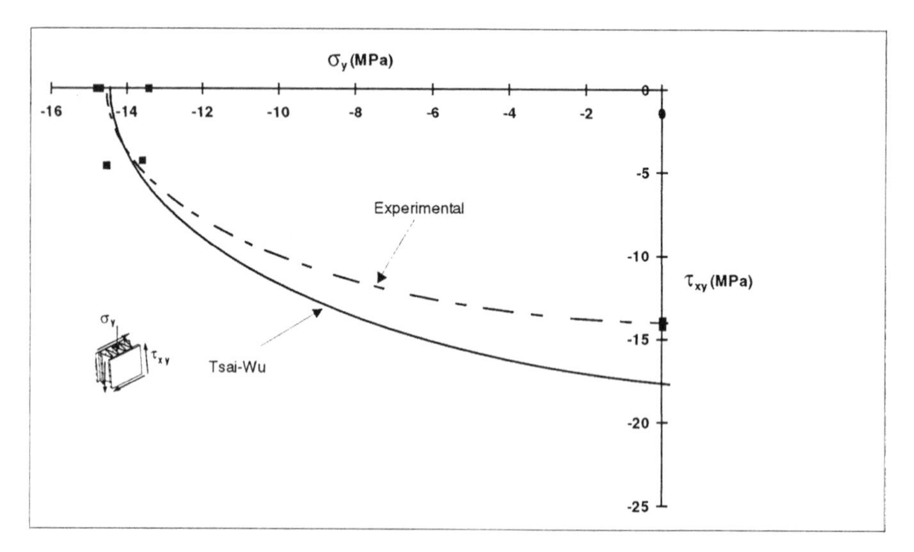

Figure 3. Experimental and analytical Tsai-Wu failure envelopes for the facings (liners) of corrugated board. σ_y and τ_{xy} stress quadrant

The strength loss is attributed to the strong influence of local buckling. Figure 5 shows experimental and Tsai-Wu failure envelopes in the σ_x- τ_{xy} quadrant. The Tsai-Wu failure criterion overpredicted the collapse stresses by a factor of about four in this case because the cylinders failed by global buckling.

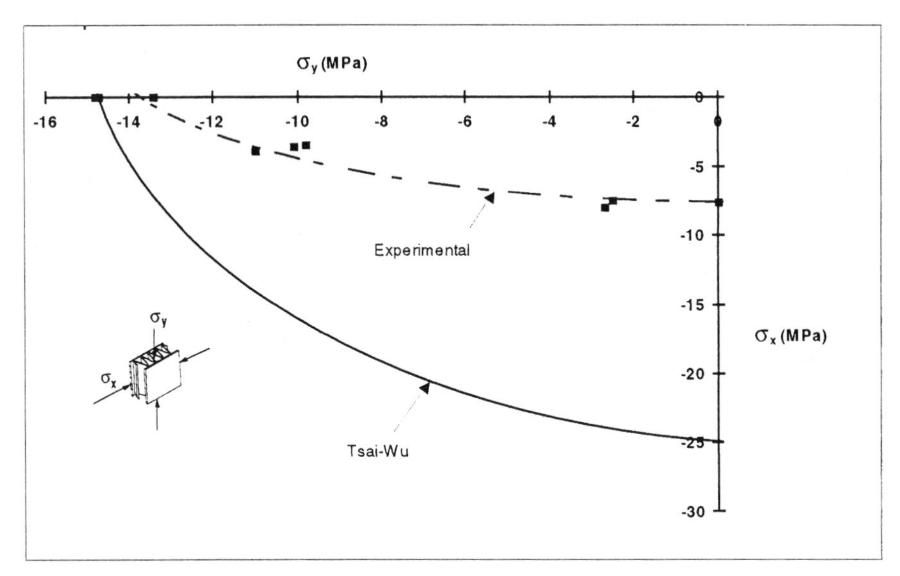

Figure 4. The σ_x and σ_y stress quadrant.

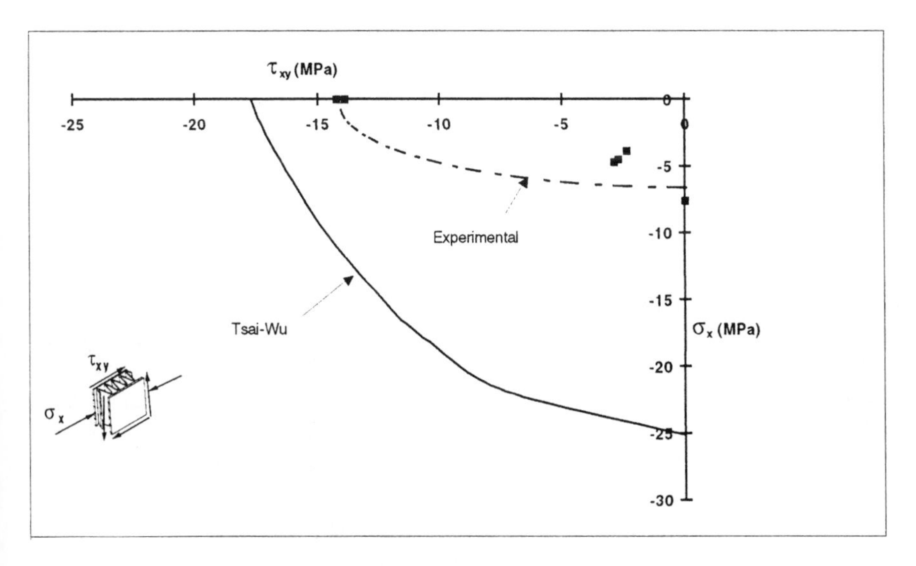

Figure 5. The σ_x and τ_{xy} stress quadrant.

7. THE INFLUENCE OF STRENGTH DEGRADATION DUE TO LOCAL BUCKLING ON PANEL COMPRESSION STRENGTH

The Tsai-Wu failure criterion was modified to accommodate strength reduction due to local buckling by using the experimental collapse stresses from tests of the cylinders. The parameters used in the modified Tsai-Wu failure criterion are listed in **Table** 4. Compression loaded board panels were analyzed using the finite element code ANSYS [17] to identify the failure regions and collapse loads. Analysis of panels compressed in both MD and CD were considered. The dimensions of the panel studied were 400 mm x 290 mm (width x height) and the thickness was 4.3 mm. The same panel board constituents as for the cylinder tests were used. An eight-noded layered shell element (STIF99) was employed for modeling the panel. It was also assumed that each layer of the multi-layered shell element is homogeneous and orthotropic.

TABLE 4. Modified parameters used in the Tsai-Wu failure criterion. eq.(1)

F_i and F_{ij}	Value mm^4/N^2
F_{11}	0,01309
F_{22}	0,00306
F_{12}	0,002
F_{66}	0,005
	mm^2/N
F_1	-0,0202
F_2	-0,0261

Consequently, the corrugated core was transformed to an equivalent homogenous core layer with effective material properties according to the procedure outlined in Ref. [11]. A quarter of the panel was modeled, since the panel is expected to buckle in the first mode. The model consists of 133 nodes (6 dof per node) and 36 elements. At the loaded edge and unloaded supported edge of the quarter panel the out-of plane displacement is prevented. Symmetrical boundary conditions were assigned to the nodes at the bottom and left symmetry cross-section of the quarter panel. Initially approximately 20% of the critical displacements on the nodes at the loaded edge is applied in the simulation. Failure is predicted by gradually increasing the axial displacement until the failure criterion is fulfilled. **Figure** 6 shows a contour plot of a panel loaded in the CD at a compression load of 1.15 kN. Notice that failure is expected to initiate where the Tsai-Wu factor is equal to one i.e. somewhere between 65-145 mm from the loaded end.

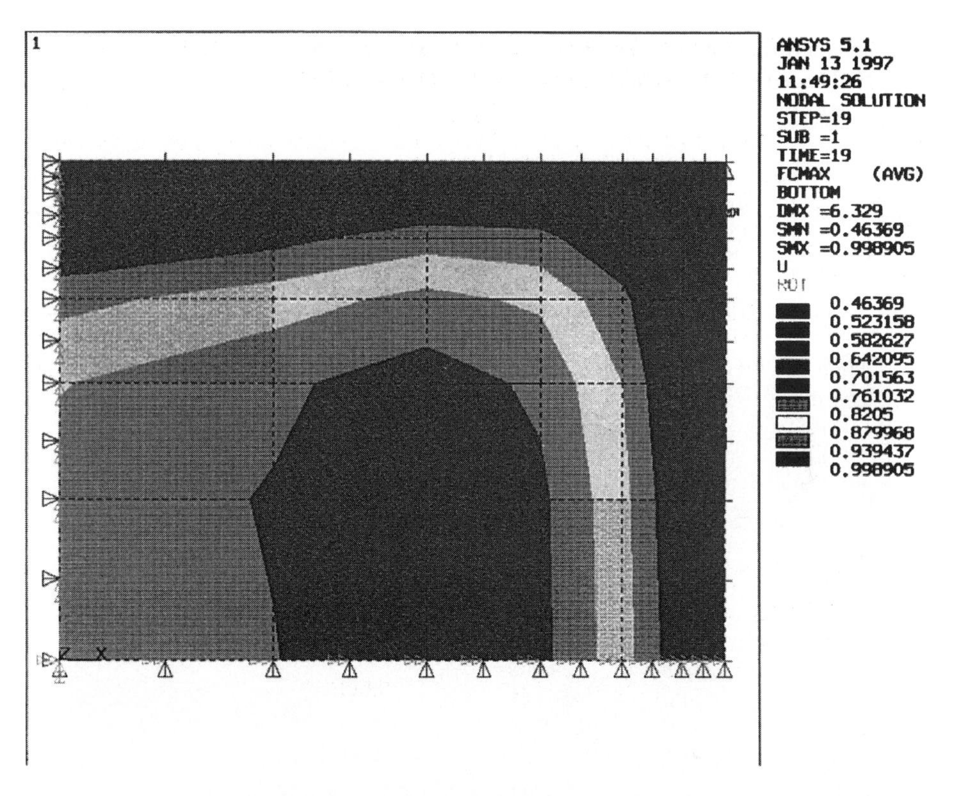

Figure 6. Contour plot of a panel loaded in CD at a compression load of 1.15 kN. Notice failure is initiated where the Tsai-Wu factor is equal to 1.

In this region the inside liner is subjected to stresses σ_y = -7.25 MPa, σ_x = -5.60 MPa and τ_{xy} = 0.5 MPa. For panels loaded in the MD the failure region is somewhere near the center of the panel. In this case the calculated collapse stresses were σ_y = -1.40 MPa, σ_x = -7.90 MPa and τ_{xy} = 0.05 MPa. The compression load at collapse was in this case 810 N. It was noticed that the collapse stresses obtained from FEA for panels loaded in the CD and MD are close to the Tsai-Wu failure envelope. In order to examine the validity of the modified Tsai-Wu failure criterion, panel compression tests were performed. The panel compression test fixture is described in Refs.[3,4]. Two panels were tested in MD and CD. **Figure** 7 shows a panel loaded in compression in the CD. It is observed that local buckling occurs at approximately 70-80 mm from the loaded edges of the panel, which is in close proximity of the failure region predicted by FEA, **Fig**. 6. Similar observation could be made for a panel loaded in compression in the MD where the panel buckled and collapsed near the center region. The experimental collapse loads of the panel and the collapse loads predicted by FEA are listed in **Table** 5. **Table** 5 reveals that the experimental failure loads in both directions are close to the predicted failure loads. Based on these preliminary experiments, proposed modification of the Tsai-Wu failure criterion seems to be viable for corrugated board structures that display local buckling.

TABLE 5. Compression load at collapse of panel loaded in the MD or CD.

Test number	Experimental test results		FEA+Modified Tsai-Wu		FEA+Unmodified Tsai-Wu	
#	F_{CD}	F_{MD}	F_{CD}	F_{MD}	F_{CD}	F_{MD}
	(kN)	(kN)	(kN)	(kN)	(kN)	(kN)
1	1,16	0,72	1,14	0,81	1,46	1,61
2	1,09	0,75	1,14	0,81	1,46	1,61

Figure 7. Panel loaded in compression in the CD. Local buckling can be observed 70-80 mm from the loaded edges of the panel.

8. CONCLUSIONS

An experimental study of failure of corrugated board under combined stress was presented. Biaxial loading of the board was obtained by subjecting corrugated board cylinders with the corrugations along the cylinder axis to combinations of axial compression, torsion and external pressure. For loading combinations where a compressive stress acts perpendicular to the corrugations (MD) and for shear dominated loading, the collapse of the board was found to be governed by local buckling of the facings. In order to study the influence of strength degradation due to local buckling on panel compression strength, the Tsai-Wu predictions were modified by using the experimental collapse stresses. Panels loaded in compression in the CD and the MD were analyzed using FEA. Two panels in each direction (CD and MD) were also tested in compression. It was found that the observed failure regions and the experimentally measured failure loads for the panel were in good agreement with predictions from FEA, which seems to justify the approach to accommodate strength degradation due to local buckling.

9. ACKNOWLEDGMENT

Financial support provided to the first two authors by NUTEK, SCA Research and Lund Institute of Technology through Prof. Gunilla Jönson is gratefully appreciated. Participation of the third author acknowledges support from the Department of Mechanical Engineering of Florid Atlantic University.

REFERENCES

[1] Jönsson, G. "Corrugated Board Packaging".Pira International, 1993.

[2] Johnson, M.W., and Urbanik, T.J. "Analysis of the Localized Buckling in Composite Plate Structures with Application to Determining the Strength of Corrugated Board", J. Compos. Tech. & Res., Vol. 11, No. 4, 1989, pp121-127.

[3] Hahn, E.K. and Carlsson, L.A. and Westerlind, B.S., "Edge Compression Fixture for Buckling Studies of Corrugated Board Panels", Exper. Mech., Vol. 32, 1992, pp 252-258.

[4] Hahn, E.K., de Ruvo, A. and Carlsson, L.A., "Compressive Strength of Edge-loaded Corrugated Board Panels", Exp. Mech., Vol. 32.1992, pp 259-265.

[5] Patel, P., Nordstrand T. and Carlsson L.A., "Instability and Failure of Corrugated Core Sandwich Cylinders Under Combined Stresses", in Multiaxial Fatigue and Deformation Testing Techniques, ASTM STP 1280 (S. Kalluri and P.J. Bonacuse, Eds.), 1997, pp. 264-289.

[6] Patel, P., Nordstrand T. and Carlsson L.A., "Local Buckling and Collapse of Corrugated Board under Biaxial Stresses", Submitted to Comp. Structures.

[7] Zenkert, D., "An Introduction to Sandwich Construction", Eng. Mater. Advis. Serv. (EMAS), London, 1995.

[8] Anderson, M.S., "Local Instability of the Elements of Truss-Core Sandwich Plate", NASA Technical Report TR R-30, 1959.

[9] Zahn, J.J. "Local Buckling of Orthotropic Truss-Core Sandwich", USDA Forest Service, Forest Products Laboratory, Research Paper FPL 220, 1973.

[10] Starlinger, A., PhD. Dissertation (Inst. of Lightweight Structures and Aerospace Eng., Vienna University of Technology) "Sandwich Shell Element", VDI -Verlag, Düsseldorf, 1991.

[11] Nordstrand, T., "Parametric Study of the Postbuckling Strength of Corrugated Core Sandwich Panels", Compos. Struct., Vol. 30, 1995, pp 441-451.

[12] Tsai, S.W. and Wu, E.M., "A General Theory of Strength for Anisotropic Materials", J. Compos. Mater.,Vol. 5, 1971, pp 58-80.

[13] de Ruvo, A., Carlsson, L. and Fellers, C.,"The Biaxial Strength of Paper", Tappi J, Vol. 63, No. 5, May 1980, p 133-136.

[14] Fellers, C., Westerlind, B.S. and de Ruvo, A.,"An Investigation of the Biaxial Failure of Paper: Experimental Study and Theoretical Analysis", Transaction of the Symposium held at Cambridge: September 1981, Vol. 1, 1983, pp 527-559.

[15] CRC Handbook of Chemistry and Physics, 65th ed., CRC Press, Boca Raton, 1984-1985, pp F19.

[16] Fellers, C. and Brange, A., "Impact of Water Sorption on Compression Strength of Paper", Papermak. Raw Mater., Mech. Eng. Publns. Ltd., London, 1985, Vol. 2, pp 529-539.

[17] ANSYS User´s Manual, Swanson Analysis System Inc., Houston, PA, 1989.

BOUNDARY EFFECTS AND LOCAL STABILITY OF SANDWICH PANELS

V.SKVORTSOV
Professor
Department of Strength of Materials,
State Marine Technical University of St.-Petersburg
101, Leninski Prospect, 198262, St.-Petersburg, Russia

Plane deformation of sandwich panels with symmetric faces in local areas near bounds or concentrated loads is described using combination of the simplest plate theory for faces and 3D elasticity for core. Solutions are obtained in the form of series of eigen functions. Analogous approach is used to assess influence of initial stresses on behaviour of the structure, including loosing a local stability.

1. Concepts of global and local description in asymptotic analysis.

Three main approaches to mathematical description of deformation of sandwich structures are usually referred to in surveying literature (e.g. [1]):
(a) using classical models of plates, shallow panels or shells for a whole stack of layers - a model of Kirchoff-Love type (of 8th order), a reduced model of Reissner-Mindlin type (of 8th order), a complete model of Reissner-Mindlin type (of 10th order), where an order is defined as a sum of orders of derivatives;
(b) using non-classical models for a whole stack of layers - models which take account of effect of core transverse shear deformation (of 12th order) or shear and normal deformation (of 16th order), models of higher order (18th, 24th, etc.);
(c) using 3D elasticity or combination of the simplest classical model for faces and 3D elasticity for a core.

Choice of a model depends on an objective of analysis (which variables under which load are to be determined) and also on magnitudes of dimensionless geometrical-physical parameters. For a structure characterized by initial parameters E_f, t_f, E_c, t_c, a (modulus and thickness of material for faces and core and in-plane size), the main dimensionless parameters are

A. Vautrin (ed.), Mechanics of Sandwich Structures, 175–182.

$$\varepsilon_1 = (t_f + t_c)^2 / a^2,\ \varepsilon_2 = t_f^2 / (t_f + t_c)^2,\ \varepsilon_3 = E_f t_f / E_c t_c$$

In asymptotic analysis of global deformation [2], $a = a_1$ is interpreted as a size of a structure or load variability, then ε_1 is a small parameter ($O(\varepsilon_1) < 1$). Within these assumptions, areas of validity of the models in respect to other small parameters are determined as follows. If $\varepsilon_1\varepsilon_3$ and $\varepsilon_1^2\varepsilon_2\varepsilon_3$ are finite or large (i.e. $O(\varepsilon_1\varepsilon_3, \varepsilon_1^2\varepsilon_2\varepsilon_3) > 1$) then the simplest model which can be asymptotically accurate is that of 16th order. Otherwise, if $O(\varepsilon_1\varepsilon_2\varepsilon_3) < 1$ and $O(\varepsilon_2) = 1$ then this is a model of 12th order; if $O(\varepsilon_1\varepsilon_2\varepsilon_3,\ \varepsilon_2) < 1$ then - of Reissner-Mindlin type; and if $O(\varepsilon_1\varepsilon_3) < 1$ then - of Kirchoff-Love type.

An analogous classification cannot be used when $a = a_2$ is a scale of deformation variability, i.e. when an objective of analysis is to obtain maximum stresses near bounds and concentrated forces (in areas of boundary effects) or to predict a possible local buckling (wrinkling) of the faces. The reason is that the size a_2 is governed only by thickness and combinations of parameters $\varepsilon_2, \varepsilon_3$.

2. Equations for a potential boundary layer.

An exact complete solution for a deformed plate consists of three items:
(a) partial solution brought on by non-self-balanced boundary loads distributed along z according to a certain law and also loads on faces;
(b) general solution with in-plane displacements u_x, u_y represented as gradients of a normal potential function $u_x = \partial\phi / \partial x,\ u_y = \partial\phi / \partial y$;
(c) general solution with in-plane displacements represented as vortices of one more potential function $u_x = \partial\psi / \partial y,\ u_y = -\partial\psi / \partial x$ and zero deflection u_z.
Both general solutions necessarily contain decaying functions of in-plane co-ordinates and therefore describe some boundary layers. First of them is usually called "potential layer" whilst second one - "vortexal layer".

Main features of potential boundary layer reveal themselves in plane strain state. With the above simplifications for faces, deformation of semi-infinite plate ($|z| \le t/2 \equiv t_c/2 + t_f,\ x \ge 0,\ y\forall$) of piece-wise homogeneous constitution

$$|z| < t_c/2:\ E_p = \frac{E_c}{1-\nu_c^2},\ E_n = \frac{E_c(1-\nu_c)}{(1+\nu_c)(1-2\nu_c)},\ G_n = \frac{0.5E_c}{1+\nu_c},\ \nu_n = \frac{\nu_c}{1-\nu_c};$$

$$t_c/2 < |z| \le t/2:\ E_p = E_f/(1-\nu_f^2),\ 1/E_n = 1/G_n = \nu_n = 0 \tag{1}$$

is governed by equations

$$\sigma_{x,x} + \tau_{xz,z} = \sigma_{z,z} + \tau_{xz,x} + \sigma_0 u_{z,xx} = 0, \tag{2}$$

$$\sigma_x = \nu_n \sigma_z + E_p u_{x,x}, \ \tau_{xz} = G_n(u_{z,x} + u_{x,z}), \ u_{z,z} = -\nu_n u_{x,x} + \sigma_z / E_n \tag{3}$$

where $\sigma_x, \tau_{xz}, \sigma_z$ are stresses to be obtained and σ_0 is the initial in-plane stress.

General solution of (2), (3) is given by a set of eigen complex functions expressed by complex transcendental functions of $\lambda_k z$ multiplied by decaying (oscillating or non-oscillating) functions of $\lambda_k x$:

$$(\sigma_x, \tau_{xz}, \sigma_z, u_x, u_z) = \sum_k c_k (p, \tau, \sigma, u, w)_k \exp(-\lambda_k x), \operatorname{Re}\lambda_k > 0 \tag{4}$$

Hereafter, let us consider only the most realistic case when parameter ε_2 is small and ε_3 is large whilst their product is arbitrary. If the initial stress $\sigma_0 = \sigma_{f0}$ is uniform in faces, equations for eigen values are the following:

$$\frac{\chi^3}{\alpha^3} - 3k_0 \frac{\chi}{\alpha} + f_{1(2)} = 0, \ \alpha \equiv \sqrt[3]{\frac{E_{nc} t_c^3}{E_{pf} t_f^3}}, \ k_0 \equiv \frac{\sigma_0}{\sqrt[3]{E_{pf} E_{nc}^2}}, \chi \equiv \frac{\lambda t_c}{2} \tag{5}$$

where, for symmetric and anti-symmetric deformations, respectively,

$$f_1 = \frac{3(1-2\nu_c)\cos^2\chi}{(3-4\nu_c)\cos\chi\sin\chi - \chi}, f_2 = \frac{3(1-2\nu_c)\sin^2\chi}{(3-4\nu_c)\cos\chi\sin\chi + \chi} \tag{6}$$

If forces in faces are far from critical values, only two pairs among the roots of (5), (6) may be asymptotically finite or small (so that $O(\operatorname{Re}(a_1\lambda)) \le 1$, i.e. boundary effect occupies a large part of a plate). According to limiting equations (necessarily for a small α, i.e. for a large product $\varepsilon_2\varepsilon_3$)

$$\chi^4 - 3k_0\alpha^2\chi^2 + 1.5\alpha^3 = 0, \tag{7}$$

$$\chi^2 - 3k_0\alpha^2 - f_{1\nu}\alpha^3 = 0, \ f_{1\nu} \equiv 3(1-2\nu_c)/4(1-\nu_c) \tag{8}$$

the roots are

$$\lambda_1 = \lambda_{11} \pm i\lambda_{12}, \ \lambda_{11} = \lambda_{10}\sqrt{1+k_f}, \ \lambda_{12} = \lambda_{10}\sqrt{1-k_f},$$

$$i \equiv \sqrt{-1}, \ \lambda_{10} = \frac{1}{t_c}(6\alpha^3)^{1/4}, \ k_f = \frac{\sigma_0}{\sigma_{cr}} = \frac{\sigma_0\sqrt{3}\, t_c}{\sqrt{2E_{pf}E_{nc}t_f}}, \tag{9}$$

$$\lambda_2 = \lambda_{20}\sqrt{1+k_f}, \ \lambda_{20} = \frac{2}{t_c}\sqrt{f_{1\nu}\alpha^3}, \ k_f = \frac{\sigma_0}{\sigma_{cr}} = \frac{2\sigma_0 t_f}{G_{nc} t_c} \tag{10}$$

They completely coincide with analogous values in the models of 12th and 16th orders. This means that, for $O(\alpha) < 1$, the models are adequate both for global solution and for local effects. In all other cases, local solution decays rapidly, so that a model of semi-infinite plate is valid for modeling the local effects.

In another limiting case (when α is large enough), there are specific roots (common for both types of deformations) which formally correspond to a plate on elastic half-space. The roots can be obtained from limiting equation

$$\chi^3 - 3k_0\alpha^2\chi - f_{2\nu}\alpha^3 i\,\mathrm{sgn}(\mathrm{Im}\,\chi) = 0, \quad f_{2\nu} \equiv 3(1-2\nu_c)/(3-4\nu_c) \tag{11}$$

and presented by formulae

$$\lambda_3 = \lambda_{31} \pm i\lambda_{32},\ \lambda_{31} = \lambda_{30}(s_1 + s_2)\sqrt{3},\ \lambda_{32} = \lambda_{30}(s_1 - s_2),$$

$$\lambda_{30} = \left(\frac{E_{nc}f_{2\nu}}{E_{pf}t_f^3}\right)^{1/3},\ s_{1,2} = \left(\frac{\sqrt{1+k_f^3} \pm 1}{2}\right)^{1/3},\ k_f = \frac{4^{1/3}\sigma_0}{(E_{pf}E_{nc}^2 f_{2\nu}^2)^{1/3}} \tag{12}$$

All other roots can be obtained by asymptotic or numerical analysis.

3. Solutions for a potential boundary layer.

Within the above assumptions for ε_{1-3}, partial solution brought on by non-self-balanced boundary loads ensures distributions of all the variables along z according to hypothetic distributions in a model of Reissner-Mindlin type. Partial solution generated by smooth facial load in areas between line-concentrated forces ensures the same distribution within accuracy of $\varepsilon_{1,2}$. Therefore, local disagreements may arise on bounds or conjugating lines. Residuals (disagreement functions) of stresses σ_x, τ_{xz} and displacements u_x, u_z are given by two functions from a set $\widetilde{p}, \widetilde{\tau}$ (self-balanced), $\widetilde{u}, \widetilde{w}$. Boundary layers have to compensate the residuals. For this purpose, for each generalized bound, coefficients c_k in solutions (4) are to be obtained.

Eigen functions are orthogonal to each other as follows:

$$k \neq m:\ \left\{\tau_m^* w_k - u_m p_k\right\} = 0,\ \{\ldots\} \equiv \int_{-t/2}^{t/2} \ldots dz,\ \tau_m^* \equiv \tau_m + \lambda_m \sigma_0 w_m \tag{13}$$

If pairs $\widetilde{u}, \widetilde{\tau}$ or $\widetilde{w}, \widetilde{p}$ are given, then c_k are determined by the simplest formulae, for example,

$$c_k = \left\{\widetilde{\tau}\, w_k - \widetilde{u} p_k\right\} / \left\{\tau_k^* w_k - u_k p_k\right\} \tag{14}$$

Otherwise, an infinite system of equations for c_k is to be solved.

There are several realistic types of boundary fixing and internal concentrated loading: one-face supporting; supporting through boundary insert; clamping; one-face concentrated loading; loading through insert, where contact between a core and insert (bound) can be with or without adhesion. Typical boundary problems for symmetric and anti-symmetric compensating solutions are illustrated by

Fig.1. Some of them are solved by a number of authors [3-7]: (a) - for large α using a model of Winkler's foundation, (c-e) - for small α using a model of 12th order. The above technique enables to solve the problems without any constraint for α, but a closed-form solution can be derived only if boundary functions $\tilde{u},\tilde{\tau}$ are given. Let us present final formulae for maximum bending face stress and interface stress in the problems (a,c) for small α (actual range of validity α<0.9-1.0) and large α (actual range of validity α>2.9-3.0):

$$\sigma_f^a = \frac{3P}{2\sqrt{1+k_f}}\sqrt[4]{E_{pf}t_c/6E_{nc}t_f^5}, \sigma_c^a = \frac{P}{2\sqrt{1+k_f}}\sqrt[4]{6E_{nc}/E_{pf}t_f^3 t_c},$$
$$\sigma_f^c = \gamma_0\sqrt{1+k_f}\sqrt{3E_{pf}G_{nc}(t_f+t_c)^2/t_f t_c}; \tag{15}$$

$$\sigma_f^a = \sqrt{3}P/2t_f^2\lambda_{30}(s_1+s_2),\ \sigma_c^a = 2P\lambda_{30}/\sqrt{3}(s_1^2-s_2^2),$$
$$\sigma_f^c = \sigma_f^a t_c/(t_f+t_c)+\gamma_0 E_{pf}t_f\lambda_{30}(s_1^2+s_2^2+4s_1s_2)/\sqrt{3}(s_1+s_2), \tag{16}$$
$$\sigma_c^c = \sigma_c^a t_c/(t_f+t_c)+2\gamma_0 E_{pf}t_f^3\lambda_{30}^3(s_1-s_2)/3\sqrt{3}(s_1+s_2)$$

Qualitative curves for maximum stresses versus α are shown in Fig.2.

Note that exact solutions for large α do not completely coincide with results of using the Winkler's foundation model. That is illustrated by the inequality $\lambda_{31} \neq \lambda_{32}$ for $k_f = 0$, (see (12)). The imposed equivalence of λ_{31} and λ_{32} in the model would give an error for σ_c of about 25%. An alternative simplified effective model of a higher accuracy would assume that an additional membrane is inserted between a plate and a foundation. A tensile membrane force and a foundation stiffness are determined by equalities

$$N_{eff} = E_{pf}t_f^3\lambda_{30}^2/3(s_1-s_2),\ C_{eff} = 4E_{pf}t_f^3\lambda_{30}^4/3(s_1-s_2)^2 \tag{17}$$

4. Influence of compressive forces on a boundary layer.

A decay rate of boundary effect decreases as initial stress σ_0 decreases and becomes negative. Critical values for wrinkling $\sigma_0 = -\sigma_{cr}$ $(k_f = -1)$ lead to a situation when $\mathrm{Re}\,\lambda_k = 0$ even for one k. For equations (5), (6), this approach gives expressions for σ_{cr} coinciding with results presented in literature for equal stresses in faces [3 ,5, 8] (see curves *1* and *5* in Fig.3).

One can see from (15), (16), (12) that, for static boundary conditions, a limit $k_f \to -1$ yields $\sigma_{bend} \to \infty$ whilst for kinematic conditions - $\sigma_{bend} \to 0$.

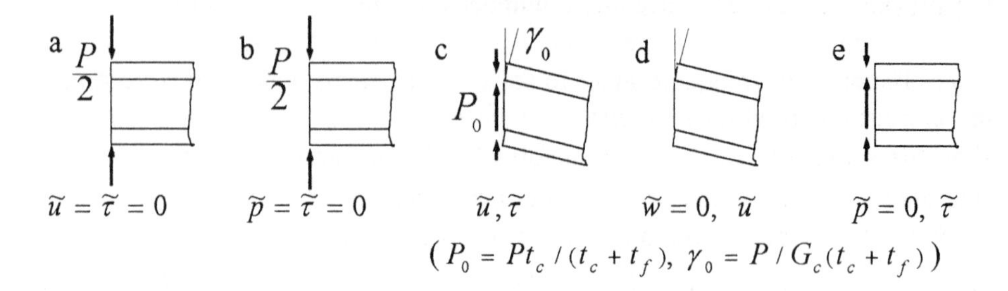

Figure 1. Schemes of loading on bounds or conjugating lines.

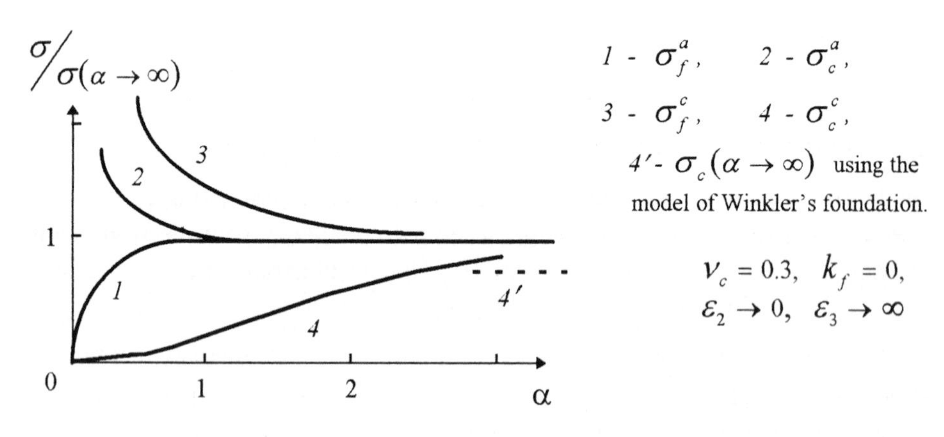

Figure 2. Dimensionless maximum stresses in boundary layer.

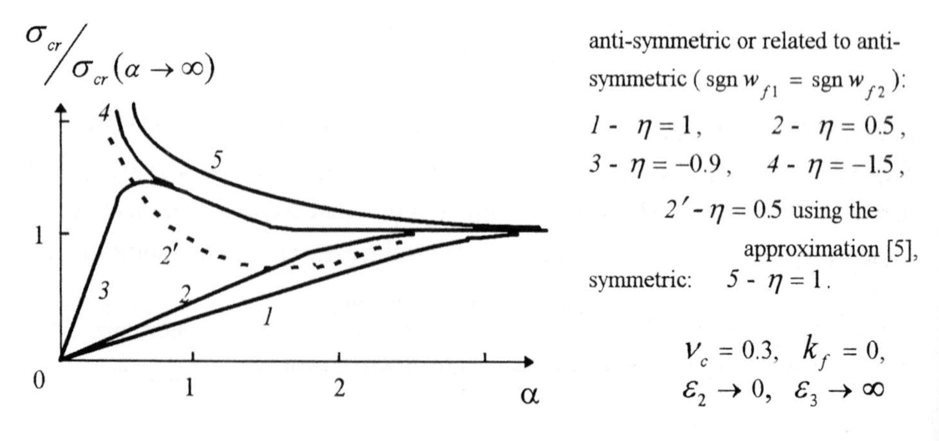

Figure 3. Dimensionless critical values of stresses.

5. Influence of non-uniform loading and curvature on critical values.

Wrinkling may occur not necessarily for a straight-shaped structure under uniform membrane stresses in faces 1, 2. Therefore, an analysis of local stability should be complemented by assumptions $\sigma_{10} \neq \sigma_{20} \neq \text{const}(x)$ and/or $\kappa \neq 0$, where κ is a sum of initial and bending curvatures.

First, let us assume that $\sigma_{20} = \eta\sigma_{10}$ $(\eta \leq 1)$. In wrinkling, face deflections and interface stresses vary as sinusoids $(w_{f1}, w_{f2}, \sigma_{c1}, \sigma_{c2})\sin\omega x$. For the known parameters of effective symmetric and anti-symmetric stiffness, b_s, b_c, obtained with 3D elasticity [3, 5], conditions of solvability of equations of a coupled bending of faces as plates of cylindrical stiffness D_f

$$\begin{pmatrix} D_f\omega^4 + \sigma_{10}t_f\omega^2 + 0.5(b_c + b_s), & 0.5(b_c - b_s) \\ 0.5(b_c - b_s), & D_f\omega^4 + \eta\sigma_{10}t_f\omega^2 + 0.5(b_c + b_s) \end{pmatrix}\begin{pmatrix} w_{f1} \\ w_{f2} \end{pmatrix} = 0 \quad (18)$$

give two functions for critical stress $\sigma_{cr}(\alpha, \eta, \nu_c) = \min\limits_{\omega}(-\sigma_{10})$. Similarly to equal-stress loading, the lowest among them corresponds to wrinkling with $\operatorname{sgn} w_{f1} = \operatorname{sgn} w_{f2}$ (anti-symmetric mode as a limit), and all the functions tend to Hoff's value [3, 4] when $\alpha \to \infty$. There is a considerable difference between the exact solution obtained using (18) and an approximation proposed in [5] and used in [8] $(b(\eta) \cong 0.5((1+\eta)b_1 + (1-\eta)b_{-1})$, where $b_{-1} = 0.5(b_s + b_c)$ for pure bending ($\eta = -1$) and $b_1 = b_c$ for pure compression ($\eta = 1$)). It is easy to see from Fig.3 that the approximation is adequate only for large α.

In a performed analysis of wrinkling under varied stress, a critical value of conventional dimensionless membrane compressive stress (a maximum of periodical function $k_f(\xi) = k_{\max} f(\xi)$) which ensures solvability of equation

$$d^4w / d\xi^4 + k_f(\xi)d^2w / d\xi^2 + w = 0 \quad (19)$$

is compared with a critical value for $k_f = \text{const}$. In the latest case, the value is obviously $k_{cr} = 1$ and the critical solution is $w = w_0 \sin\xi$ (where ξ is conventional dimensionless co-ordinate). The obtained partial periodical solutions define w and k_f as elliptical functions of ξ ($f(\xi)$ varies from 0 to 1). If $f(\xi) \geq \mu$ within interval of the length l, then critical value of $k_{\max}$ can be presented as $k_{cr}(\mu, l)$. For example, in the solution, if k_f varies in 4 times within quarter-period of wrinkling for $k_f = \text{const}$, then $k_{cr}(0.25, \pi / 2) = 2\sqrt{3}$.

Wrinkling in a case of nearly constant (within typical length of boundary layer) curvature κ is analysed using equations similar to (2), (3) in polar co-ordinates. The solution is presented as a series

$$\sigma_{cr} = \Sigma\sigma_k(\alpha, \eta, \nu_c)\varepsilon_0^k, \quad \varepsilon_0 \equiv \kappa t_c \tag{20}$$

where two terms are assessed. According to the assessments, the influence of curvature is considerable (more than 10-15%) only for $\varepsilon_0 > (0.25 - 0.35)$.

6. Other boundary layers.

Local solutions of boundary layer type show up also in other situations. They can be produced by strong curvature of shell, strong tension, etc. However, these layers have no specific features intrinsic exactly for sandwich structures, their substance is independent on distribution of modules along thickness, and technique of mathematical description is the same that for any other structure.

Acknowledgements

The author wishes to thank the Organizing Committee and RusNord Institute for financial support received to participate in the colloquium EUROMECH-360.

References

1. Noor, A.K., Burton, W.S. and Bert, C.V. (1996) Computational models for sandwich panels and shells, *J. of Applied Mechanics Review* **49**, No 3, 155-198.
2. Ustinov, Yu.A. (1976) On structure of boundary layer in laminated plates, *Doklady Akademii Nauk SSSR* (Russian) **229**, No 2, 325-328.
3. Allen, H.G. (1969) *Analysis and Design of Structural Sandwich Panels*, Pergamon Press, Oxford, UK.
4. Zenkert, D. (1995) *An Introduction to Sandwich Construction,* EMAS Ltd, UK.
5. Stamm, K. and Witte, H. (1974) *Sandwichkonstruktionen - Berechnung, Fertigung, Ausführung*, Springer-Verlag, Wien, Austria.
6. Thomsen, O.T. (1992) Analysis of local bending effects in sandwich panels subjected to concentrated loads, in K.-A.Ollson and D.Weismann-Berman (eds), *Sandwich Construction 2*, Proceedings of the 2nd International Conference on Sandwich Construction, University of Florida, Gainesville, Florida, USA, 417-440.
7. Meyer-Piening, H.-R. (1989) Remarks on higher-order stress and deflection analyses, in K.-A.Ollson and R.P.Reichard (eds), *Sandwich Construction 1*, Proceedings of the 1st International Conference on Sandwich Construction, Stockholm, Sweden, 107-127.
8. Starlinger, A. And Rammerstorfer, F.G. (1992), A finite element formulation for sandwich shells accounting for local failure phenomena, in K.-A.Ollson and D.Weismann-Berman (eds), *Sandwich Construction 2*, Proceedings of the 2nd International Conference on Sandwich Construction, University of Florida, Gainesville, Florida, USA, 161-185.

OVERALL BEHAVIOUR OF LATERALLY LOADED SHALLOW SANDWICH PANELS

V. SKVORTSOV
State Marine Technical University of St-Petersburg, Lotsmanskaya 3, 190008 St-Petersburg, Russia
E. BOZHEVOLNAYA
Institute of Mechanical Engineering, Aalborg University, Pontoppidanstraede 101, DK-9220 Aalborg East, Denmark

Sandwich plates and shells are being increasingly used in a variety of technical applications. Singly-curved sandwich panels appear to be inevitable parts of transport units and building constructions. In the course of industrial exploitation, sandwich structures are affected by different loads. Accurate knowledge of the deformation response to these loads and of the appropriate buckling limits is essential for obtaining a reliable and optimal structural design.

Here, we study the load-deformation behaviour and global stability of shallow singly-curved sandwich panels subjected to lateral loading on the basis of first order shear deformation theory. The appropriate mathematical technique is developed and compact formulae are obtained, that are suitable to describe nonlinear chracteristics of the panel in a wide range of initial geometrical and constitutive characteristics.

Explicit Formulae for Overall Behaviour

The configuration under consideration is shown in fig. 1, where the cross-section of a singly-curved sandwich panel and its geometrical parameters are indicated. Mechanical properties of the materials constituting the panel are as follows: E_f and E_c are Young's moduli of the faces and core; G_c is the shear modulus of the core. The panel is simply supported along its straight edges. One of the supports is fixed, whereas the other is fastened to a wall by a spring with stiffness C per unit length.

To describe the overall behaviour of the panel, a non-linear model

A. Vautrin (ed.), Mechanics of Sandwich Structures, 183–190.

on the basis of the Reissner plate theory [1] is used. The model is based on the following assumptions: 1) the total thickness of curved panel is small in comparison with its span: $(2t_f + t_c)^2 \ll a^2$; 2) the bending stiffness of faces is much smaller than the total bending stiffness of sandwich section: $t_f^2 \ll 3(t_f + t_c)^2$; 3) the relative core compliance is small: $E_f t_f^3 t_c \ll E_c (t_f + t_c)^2 a^2$; 4) the panel is shallow: $h^2 \ll a^2$. Note, however, that the ratios $E_f t_f^3 / E_c t_c^3$ and $E_f t_f t_c / E_c a^2$ can be arbitrary.

A complete system of nonlinear equations for large deflections

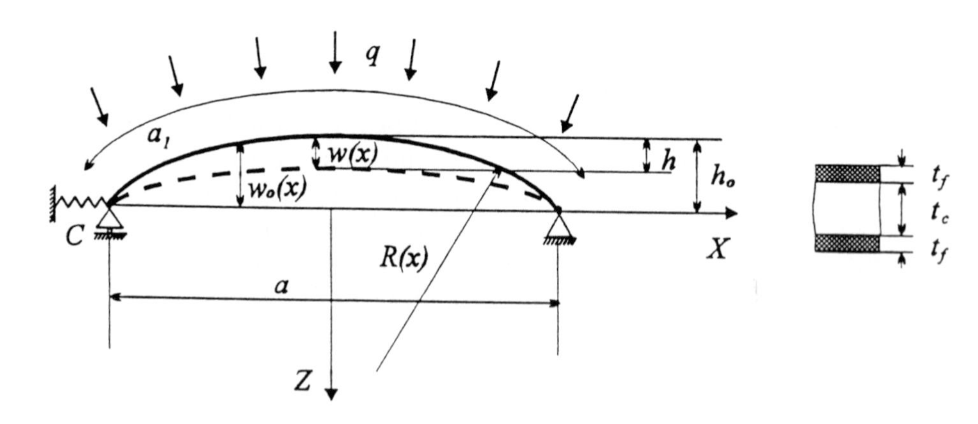

Figure 1: Geometry of the model and the panel lateral cross-section.

of the curved plate includes the equilibrium, kinematic, and one-dimensional elasticity relations:

$$\begin{cases} N_x' \equiv dN/dx = 0 \\ Q_{xz}' + N_x(w'' - w_0'') + q = 0 \\ M_x' - Q_{xz} = 0 \end{cases} \quad \begin{cases} \varepsilon_x = u' + \frac{1}{2}((w_0' - w')^2 - w_0'^2) \\ \gamma_{xz} = w' + \theta \\ \kappa_x = \theta' \end{cases}$$

$$N_x = B\varepsilon_x \ , \quad Q_{xz} = S\gamma_{xz} \ , \quad M_x = D\kappa_x \ , \quad (1)$$

where u, w and θ are the generalized displacements, i.e. the displacements along the x and z axes and the angle of rotation; N_x, Q_{xz} and M_x are the generalized forces per unit width of the panel: the membrane force, the tangential shear force and the bending moment, respectively; ε_x, γ_{xz} and κ_x are the generalized strains, i.e. the normal strain, the shear strain and the curvature, respectively; q is the pressure acting on the panel; $B = 2E_f t_f + E_c t_c$, $S = G_c(t_f + t_c)^2/t_c$

and $D = E_f t_f (t_f + t_c)^2/2 + E_f t_f^3/6 + E_c t_c^3/12$ are the membrane, shear and bending (cylindrical) stiffnesses. In the case of a weak core, all other terms except the first one in the expressions for B and D can be neglected.

The underlined terms in Eqs. (1) are responsible for the nonlinearity of the model whereby the membrane force is dependent not only on the external forces, but also on the strain in the panel. The term underlined twice in Eqs. (1) takes into account the shear compliance of the core, thus generalizing the conventional theory according to Reissner [1]. For $S \to \infty$, Eqs. (1) constitute the classical Kirchoff-Love model, which disregards deformation due to the transverse stress component.

The compressive force $P = -N_x$ can be expressed through the effective stiffness B_{eff} as follows

$$P = \frac{B_{eff}}{a} \int_0^a (w_0' w' - \frac{w'^2}{2}) dx \ , \qquad B_{eff} = \frac{BCa}{B + Ca} \ . \tag{2}$$

Equations (1) allow us to express all the variables in terms of w and to obtain the following governing equation

$$D[(1 - \frac{P}{S}) w'''' + \frac{P}{S} w_0''''] + P(w'' - w_0'') = q - \frac{D}{S} q'' \ . \tag{3}$$

Boundary conditions for a panel simply supported along its straight edges are:

$$x = 0,\ a: \qquad w = 0\ , \qquad M_x = 0\ . \tag{4}$$

The governing equation (3) is equivalent to the equation derived by Reissner except for the additional terms containing w_0 which are due to the initial curvature of the plate (cf. [2, 268-269]). The solution technique is described in details in [3], where the load-deflection relations are derived in the form of two implicit equations for arbitrary initial panel geometry and lateral load distribution. It has been shown also that for an initially-symmetric geometry of arch and a smooth pressure distribution a solution of Eqs. (2)-(4) is considerably simplified. Equations that determine the overall behaviour of the panel with constant radius of curvature under uniform pressure q_0 read

$$q_0 = 4\pi^2 \frac{E_f t_f (t_f + t_c)^2 h_0}{a^4} \bar{q} \ , \quad \bar{q} = \xi \left(\frac{1}{1 + \delta_1} + \frac{1}{\delta_2} (2 - \xi)(1 - \xi) \right) \ ,$$

$$P = \frac{\pi^2}{2} \frac{E_f t_f (t_f + t_c)^2}{a^2} \bar{p} \ , \quad \bar{p} = \frac{1}{\delta_2} (2 - \xi) \xi \ , \tag{5}$$

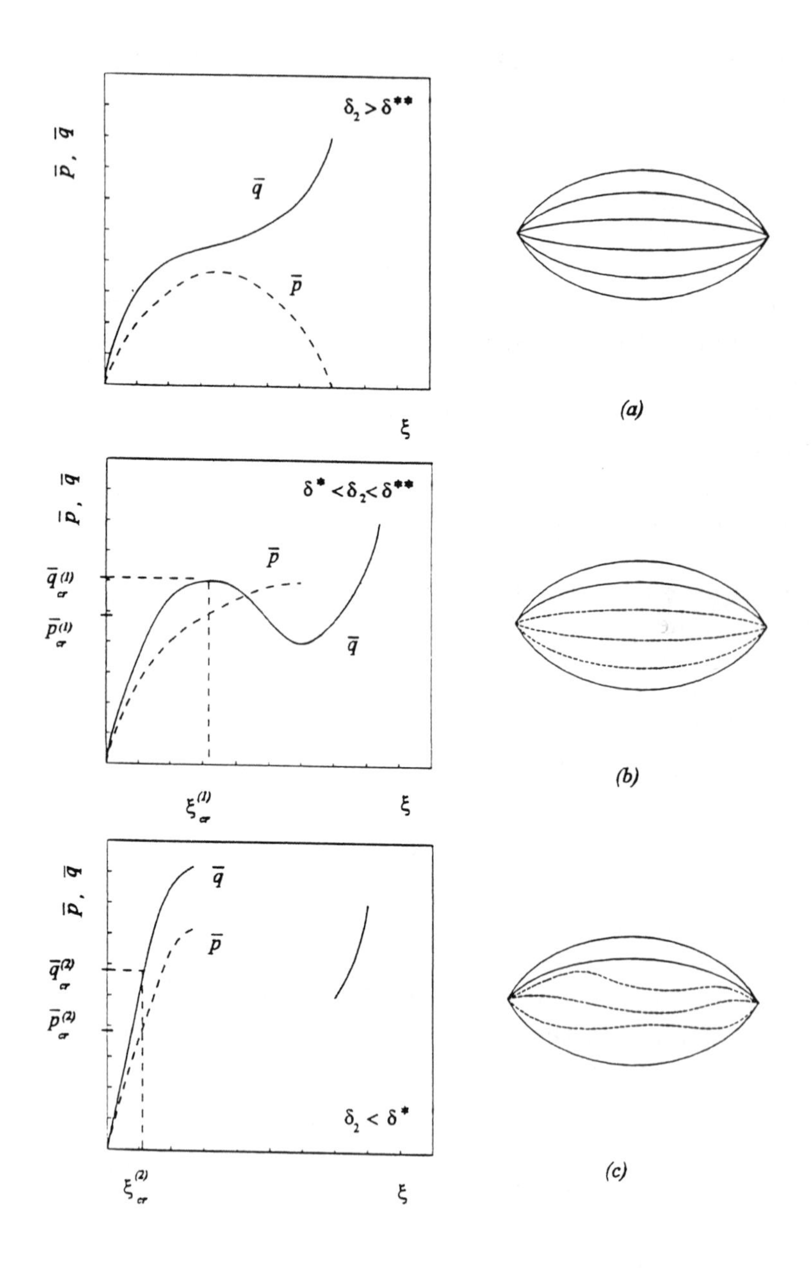

Figure 2: Instability patterns of curved panel under uniform pressure: no snap-through (a); symmetric (b) and antisymmetric (c) modes.

where the following dimensionless parameters have been introduced

$$\xi = \frac{\pi^3}{32}\frac{h}{h_0}, \quad \delta_1 = \frac{\pi^2}{2}\frac{E_f t_f t_c}{G_c a^2}, \quad \delta_2 = \left(\frac{\pi^3}{32}\right)^2 \frac{(t_f + t_c)^2}{h_0^2}\left(1 + \frac{2E_f t_f}{Ca}\right). \tag{6}$$

Analysis of Global Stability

The explicit form of Eqs. (5) enables us to analyze the stability of the curved panel under uniform pressure. It is obvious that $\bar{q}(\xi)$ has no extremum, if the parameter δ_2 satisfies the following inequality: $\delta_2 > \delta^{**} \equiv 1 + \delta_1$. Such a relation between δ_1 and δ_2 implies that the panel compliance is sufficiently small and/or a curved panel is sufficiently flat. As a consequence, the transition of the panel from its initial undeformed upper position into a (deformed) lower position occurs smoothly, without snap-through. The qualitative relations $\bar{q}(\xi)$ and $\bar{p}(\xi)$ are shown in fig. 2a. The midpoint deflection of panel ξ grows with an increase in the pressure $\bar{q}$. The compressive force $\bar{p}$ attains its maximum, when the panel is deformed into a straight line, but it does not achieve any critical value, so that the panel does not buckle. Successive steps in the gradual deformation of the panel are also shown in fig. 2a.

For δ_2 smaller than δ^{**}, the relation $\bar{q}(\xi)$ has an extremum $\bar{q}_{cr}^{(1)}$ at

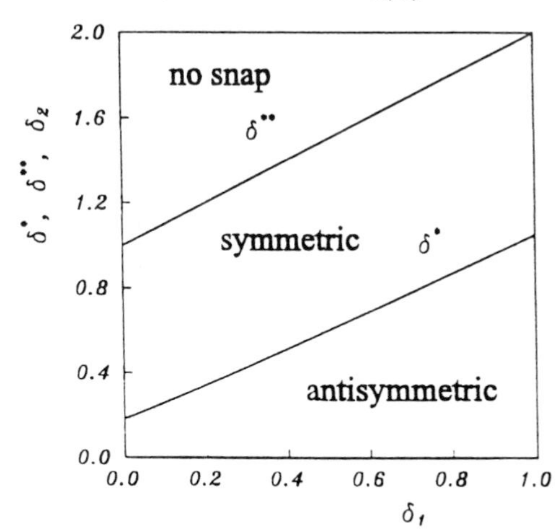

Figure 3: Critical buckling parameters and areas of different buckling modes.

the point $\xi_{cr}^{(1)}$ (see fig. 2b). In order to avoid loss of stability via the

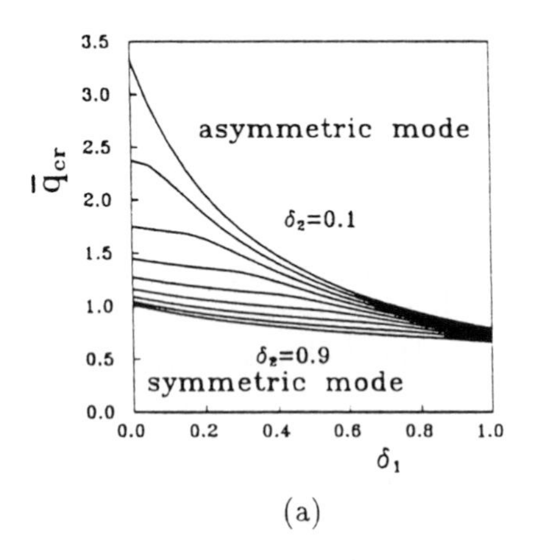

(a)

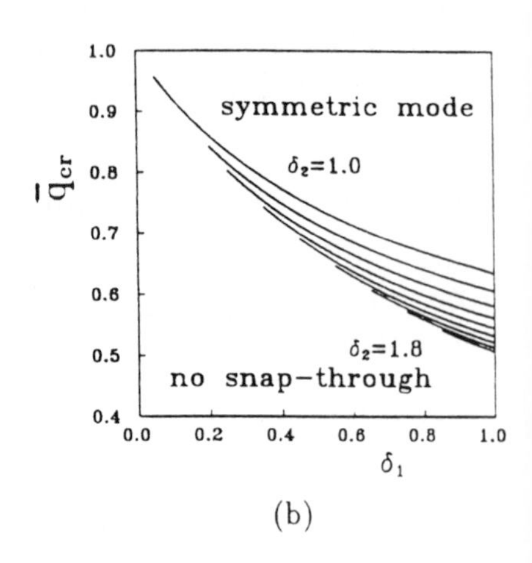

(b)

Figure 4: Critical buckling lateral load.

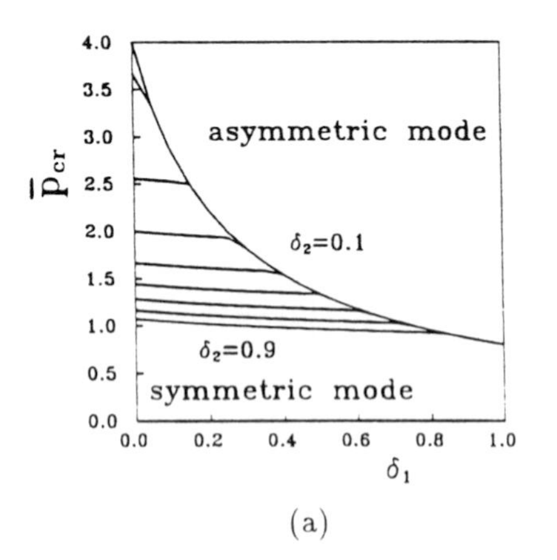

(a)

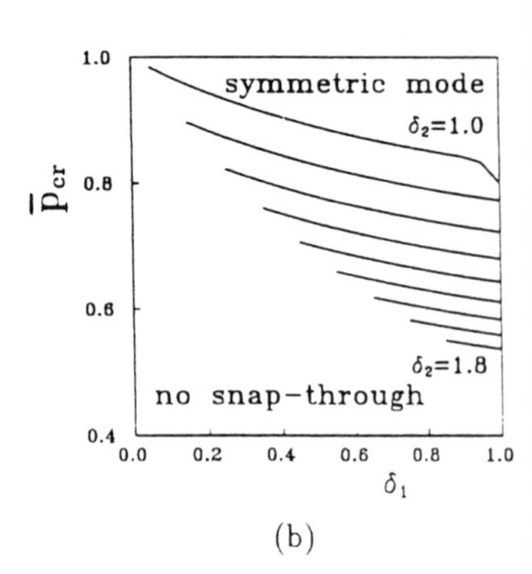

(b)

Figure 5: Critical buckling compressive force.

antisymmetrical mode, it is important that the accompanying force $\bar{p}(\xi)$ does not exceed the critical buckling force $\bar{p}_{cr}^{(2)}$ for $\xi \leq \xi_{cr}^{(1)}$. By equating $\bar{p}_{cr}^{(1)}$ (corresponding to $\bar{q}_{cr}^{(1)}$ at $\xi = \xi_{cr}^{(1)}$) to the critical compressive force $\bar{p}_{cr}^{(2)}$, one can find a relation between δ_1 and δ_2 for which the above holds

$$\delta^* \equiv \frac{2(1+\delta_1)(1+4\delta_1)}{11+8\delta_1} < \delta_2 < \delta^{**} .$$

If δ_2 satisfies the above inequality, the panel snaps through symmetrically in a half-sine wave when the pressure on the panel reaches its maximum $\bar{q}_{cr}^{(1)}$ (see fig. 2b). In this case, the critical values of the dimensionless pressure, deflection and force are

$$\bar{q}_{cr}^{(1)} = \frac{1}{1+\delta_1} + \frac{2}{\delta_2}\left(\frac{1}{3}\left(1 - \frac{\delta_2}{1+\delta_1}\right)\right)^{3/2} , \tag{7}$$

$$\bar{p}_{cr}^{(1)} = \frac{1}{3}\left(\frac{1}{1+\delta_1} + \frac{2}{\delta_2}\right) , \quad \xi_{cr}^{(1)} = 1 - \sqrt{\frac{1}{3}\left(1 - \frac{\delta_2}{1+\delta_1}\right)} ,$$

If the parameter δ_2 satisfies the inequality $\delta_2 < \delta^*$, then the compressive force $\bar{p}(\xi)$ corresponding to the extremum of lateral load $\bar{q}(\xi)$ is larger than $\bar{p}_{cr}^{(2)}$ (see fig. 2c). In this case, the panel loses its stability in an antisymmetric two-half wave mode before the pressure reaches its maximum. The critical values of the dimensionless pressure, deflection and force are

$$\bar{q}_{cr}^{(2)} = \frac{1 + 3/(1+4\delta_1)\sqrt{1 - 4\delta_2/(1+4\delta_1)}}{1+\delta_1} , \tag{8}$$

$$\bar{p}_{cr}^{(2)} = \frac{4}{1+4\delta_1} , \quad \xi_{cr}^{(2)} = 1 - \sqrt{1 - \frac{4\delta_2}{1+4\delta_1}} ,$$

It should borne in mind, that the structural parameters δ_1 and δ_2 are of great importance for the overall behaviour and stability of singly-curved panels. The parameter δ_1 is a material property; it determines the relative shear compliance of the sandwich panel. If the sandwich is very stiff, $S \to \infty$ and $\delta_1 \to 0$, thus the present model reduces to the Kirchhoff-Love description. The parameter δ_2 is rather of a geometrical nature and is related to the shallowness of the panel arch; it decreases when the panel becomes steeper.

Figure 3 gives a graphical representation of the relationships between δ^*, δ^{**} and δ_1, δ_2. The importance of the structural shear compliance parameter δ_1 is obvious in the analysis of the buckling pattern of panels with low arches: $\delta_2 > 1$. For these panels, the Kirchoff-Love model predicts a gradual transition from the upper to the lower position, while our refined model shows that both symmetrical and antisymmetrical buckling modes are possible for certain values of δ_1.

Appropriate critical values of pressure and compressive force can be determined with the help of figs.4 and 5. The critical buckling pressures are presented for two ranges of δ_2: $0.1 < \delta_2 < 0.9$ (fig.4a) and $1.0 < \delta_2 < 1.8$ (fig.4b). One can see that for any $\delta_2 < 1.0$ both symmetric and asymmetric modes can be realized depending on δ_1. For any $\delta_2 \geq 1.0$ either symmteric mode or gradual deformation without failure are possible. Fig.5 is a graphic representation of the appropriate critical compressive forces calculated for different δ_1 and δ_2. Note that the load-deflection and thrust-deflection dependencies might be calculated with the help of Eqs.(5)-(6) only up to a definite critical deflection, which is defined for the symmetric and asymmetric buckling modes by Eqs. (7)-(8).

Conclusions

- Equations which describe the deformation behaviour of shallow sandwich panels with constant curvature under a uniform pressure have been obtained in the explicit form.

- The global stability of such panels has been analyzed. Three buckling patterns depending on relation between structural parameters δ_1 and δ_2 have been shown to be possible. Critical lateral loading and accompanying compressive forces have been calculated.

References

[1] Reissner, E.(1948) Finite Deflections of Sandwich Plates. *J. of Aerospace Sci.*,**15**(7)435-440.

[2] Allen, H.G.(1969) *Analysis and Design of Structural Sandwich Panels*, Pergamon Press, Oxford,.

[3] Skvortsov, V. and Bozhevolnaya, E.(1997) Overall Behaviour of Shallow Singly-Curved Sandwich Panels, *Composite Structures* (in press).

GEOMETRIC NON-LINEAR ANALYSIS OF SANDWICH STRUCTURES

A Comparison Between the Ahmad and the Marguerre Shell Elements in the Elastic Buckling and Post-Buckling Range

J.T. BARBOSA, A.M. FERREIRA, J. CÉSAR SÁ, A.T. MARQUES
DEMEGI, Faculdade de Engenharia, Universidade do Porto
Rua dos Bragas, 4099 Porto Codex, Portugal

1. Introduction

In this paper it is analysed the elastic buckling and post-buckling behaviour of sandwich shells resorting to the Finite Element Method. If transversely loaded to its mid-surface and in special support conditions, those structures can exhibit a snapping behaviour.

A formulation of both the Ahmad and the Marguerre shell elements are compared in the geometric non-linear behaviour. Those elements are suitable for the analysis of thin or moderately thick shallow shells, although the formulation of the Ahmad element is far more complex. The consideration of laminated shells can be accomplished by a layered discretization of the shell in the thickness direction, through a mid-ordinate rule.

2. Marguerre and Ahmad Shell Elements

2.1. BASIC ASSUMPTIONS

Two basic assumptions are employed in both the Ahmad and Marguerre shell elements. Firstly it is assumed that 'normals' to the mid-surface remain practically straight after deformation. Secondly, the stress component normal to the shell mid-surface is constrained to be zero in the constitutive equations. When applied to plates this formulation is equivalent to that developed by Mindlin [1] in that it accounts for transverse shear deformation effects. The elements displacement field are expressed in terms of three displacements of the reference (mid) surface and two rotations of the 'normal', defined at each element node, and appropriate shape functions [2].

Two main co-ordinate systems are used in both elements [3-5]. The global Cartesian co-ordinate system (x,y,z), in relation to which the nodal co-ordinates and displacements are defined, and the curvilinear co-ordinate system (ξ,η,ζ), where the shape functions N_k are expressed. The mid-surface of the elements are defined by the ξ and η co-ordinates. In the Ahmad shell element ζ is a linear co-ordinate in the thickness direction and is only

A. Vautrin (ed.), Mechanics of Sandwich Structures, 191–198.

approximately normal to the shell mid-surface.

2.2. MARGUERRE SHELL ELEMENT

The Marguerre shell element, figure 1, results of the combination of Marguerre and Mindlin plate theories [3-4]. The displacement vector at a generic point of the plate, **u**, is defined through the reference surface translations $\hat{u}$, $\hat{v}$, $\hat{w}_1$, the rotations of the 'normal' to the planes x-z and y-z, θ_x and θ_y respectively, and the deflection $\hat{w}_0$, which defines the initial configuration of the plate before deformation [4], that is

$$\mathbf{u} = \begin{bmatrix} u \\ v \\ w \end{bmatrix} = \begin{bmatrix} \hat{u}(x,y) - z\,\theta_x(x,y) \\ \hat{v}(x,y) - z\,\theta_y(x,y) \\ \hat{w}_1(x,y) + \hat{w}_0(x,y) \end{bmatrix} \tag{1}$$

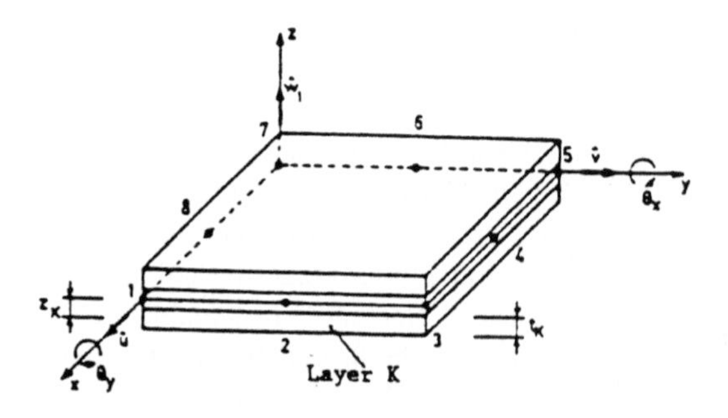

Figure 1. The Marguerre shell element: geometry and nodal variables.

The strain vector ε results of the application of Mindlin and von Kármán [6] hypotheses to the Green deformation tensor and is defined in the global co-ordinate system as [3,4]

$$\varepsilon = \left[\varepsilon_x, \varepsilon_y, \gamma_{xy}, \gamma_{xz}, \gamma_{yz}\right]^T = \begin{bmatrix} \varepsilon_p \\ \varepsilon_s \end{bmatrix} = \begin{bmatrix} \varepsilon_m + z\,\varepsilon_b + \varepsilon_I + \varepsilon_L \\ \varepsilon_s \end{bmatrix} \tag{2}$$

where ε_m, ε_b, ε_s are, respectively, the membrane, bending and shear strains, ε_L are the non-linear components of in-plane strains, ε_p, and, finally, ε_I are the linear contributions of the initial deformation to the in-plane strains.

The constitutive equations are derived for an orthotropic material with the material axes 1, 2 parallel to the plane x-y and rotated by some angle α in relation to the axes x and y, being the material axe 3 parallel to z. Taking into account that $\sigma_z = 0$, the stresses σ at a layer of the element are related to the deformations ε by [7]

$$\sigma = \left[\sigma_x, \sigma_y, \tau_{xy}, \tau_{xz}, \tau_{yz}\right]^T = \mathbf{D}\,\varepsilon = \mathbf{T'}^T\,\overline{\mathbf{D}}\,\mathbf{T'}\,\varepsilon \tag{3}$$

where

$$\mathbf{T}' = \begin{bmatrix} \mathbf{T}_1' & \mathbf{O} \\ \mathbf{O} & \mathbf{T}_2 \end{bmatrix} \qquad \mathbf{T}_1' = \begin{bmatrix} c^2 & s^2 & cs \\ s^2 & c^2 & -cs \\ -2cs & 2cs & c^2 - s^2 \end{bmatrix} \qquad \mathbf{T}_2 = \begin{bmatrix} c & s \\ -s & c \end{bmatrix} \tag{4}$$

is the strain transformation matrix, in which $c = \cos\alpha$ and $s = \sin\alpha$, and $\overline{\mathbf{D}}$ is the material elasticity matrix, defined for the material axes 1 , 2 , 3 , having non-zero terms

$$\overline{D}_{11} = \frac{E_1}{1 - \nu_{12}\nu_{21}} \qquad \overline{D}_{22} = \frac{E_2}{1 - \nu_{12}\nu_{21}} \qquad \overline{D}_{33} = G_{12} \tag{5a}$$

$$\overline{D}_{12} = \nu_{12}\,\overline{D}_{22} \qquad \overline{D}_{44} = k_{13}\,G_{13} \qquad \overline{D}_{55} = k_{23}\,G_{23} \tag{5b}$$

in which E_1 and E_2 are the Young's moduli in the 1 and 2 directions respectively, ν_{ij} is Poisson's ratio for transverse strain in the i-direction when stressed in the j-direction and G_{13} and G_{23} are the shear moduli in the 1-3 and 2-3 planes respectively; the terms k_{13} and k_{23} are shear correction factors in the 1-3 and 2-3 planes respectively.

2.3. AHMAD SHELL ELEMENT

Ahmad degenerated a three-dimensional brick element to a curved shell element, which has nodes at the mid-surface [5]. This element defines, in addition to the co-ordinate systems above referred, two more sets as can be seen in figure 2 .

A nodal Cartesian co-ordinate system (v_{1k}, v_{2k}, v_{3k}) is associated with each nodal point of the element and has its origin at the mid-surface [7]. The unit vector $\overline{\mathbf{v}}_{3k}$, constructed from the nodal co-ordinates of the top, $\mathbf{x}_k^{top}$, and bottom, $\mathbf{x}_k^{bot}$, surfaces at node k , determines the 'normal' direction v_{3k} , which is not necessarily perpendicular to the reference surface. The unit vectors $\overline{\mathbf{v}}_{1k}$ (direction v_{1k}) and $\overline{\mathbf{v}}_{2k}$ (direction v_{2k}) define the nodal rotations, β_{2k} and β_{1k} respectively, of the referred 'normal'.

The local Cartesian co-ordinate system (x',y',z') is used to define local stresses and strains at the sampling points wherein they must be calculated. The vector direction **z'** is taken to be orthogonal to the surface ζ = constant , the direction **x'** is defined similarly as the direction of $\overline{\mathbf{v}}_{1k}$ and, finally, **y'** is obtained by the cross product of **z'** and **x'** [7]. This co-ordinate system varies along the thickness of the element and it is useful to define the direction cosine matrix $\boldsymbol{\theta}$, which relates the transformations between the local, and global co-ordinate systems; this matrix is defined by

$$\boldsymbol{\theta} = \begin{bmatrix} \overline{\mathbf{x}'} & \overline{\mathbf{y}'} & \overline{\mathbf{z}'} \end{bmatrix} \tag{6}$$

where $\overline{\mathbf{x}'}$, $\overline{\mathbf{y}'}$ and $\overline{\mathbf{z}'}$ are unit vectors along the directions **x'** , **y'** and **z'** respectively.

The co-ordinates **x** of a point within the element are obtained as

$$\mathbf{x} = \left[x\,,\,y\,,\,z\right]^T = \sum_{k=1}^{n} N_k(\xi,\eta)\left[\mathbf{x}_k^{mid} + h_k\,\zeta\,/\,2\,\overline{\mathbf{v}}_{3k}\right] \tag{7}$$

and the displacements by

$$\mathbf{u} = \sum_{k=1}^{n} N_k(\xi,\eta)\left[\mathbf{u}_k^{mid} + h_k\,\zeta\,/2\left[\bar{\mathbf{v}}_{1k} \quad -\bar{\mathbf{v}}_{2k}\right]\begin{bmatrix}\beta_{1k}\\ \beta_{2k}\end{bmatrix}\right] \tag{8}$$

where n is the number of nodes per element, h_k is the shell thickness at node k , i.e. the respective 'normal' length, $\mathbf{u}_k^{mid}$ and $\mathbf{x}_k^{mid}$ are, respectively, the displacements and the global co-ordinates at the mid-surface, and $N_k(\xi,\eta)$ are the element shape functions.

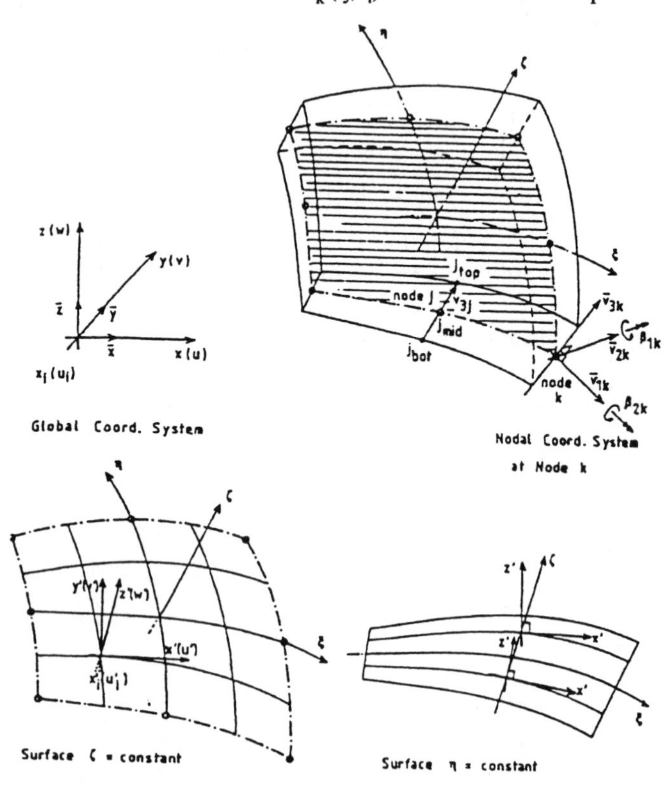

Figure 2. The Ahmad shell element.

The strain vector ε' is expressed in the local co-ordinate system, through the derivatives of the local displacements u' , v' and w' in order to x' , y' and z' ; these local derivatives are obtained from the curvilinear derivatives of u , v and w as [7]

$$\begin{bmatrix} u'_{,x'} & v'_{,x'} & w'_{,x'} \\ u'_{,y'} & v'_{,y'} & w'_{,y'} \\ u'_{,z'} & v'_{,z'} & w'_{,z'} \end{bmatrix} = \theta^T \begin{bmatrix} x_{,\xi} & y_{,\xi} & z_{,\xi} \\ x_{,\eta} & y_{,\eta} & z_{,\eta} \\ x_{,\zeta} & y_{,\zeta} & z_{,\zeta} \end{bmatrix}^{-1} \begin{bmatrix} u_{,\xi} & v_{,\xi} & w_{,\xi} \\ u_{,\eta} & v_{,\eta} & w_{,\eta} \\ u_{,\zeta} & v_{,\zeta} & w_{,\zeta} \end{bmatrix} \theta \tag{9}$$

Taking into consideration the assumption $\sigma_{z'} = 0$, the stress vector σ' in the local co-ordinate system is obtained from ε' through an expression similar to (3) , where α is now the angle that the material axes 1 , 2 are rotated in relation to the axes x' and y' .

3. Non-Linear Solution Algorithms

The non-linear equilibrium equations to be solved are set using a Total Lagrangean Formulation [8] of the Finite Element Method, where the nodal displacements **d** are refered to the initial structure configuration as

$$\Psi(\mathbf{d}) = \int_V \mathbf{B}^T \sigma \, dV - \mathbf{f} = \mathbf{r}(\mathbf{d}) - \mathbf{f} \neq \mathbf{O} \tag{10}$$

in which **r** (**d**) are the equivalent nodal forces and **f** are the external forces; both the stress vector σ and the deformation matrix **B** are dependent on **d** [2-4,7].

The solution of the equations (10) is searched using incremental and iterative techniques, where the loads are incrementally applied. For each load level an iterative process is performed in order to reduce the errors to be transfered to the next load level.

In the Newton-Raphson method [2] the equations (10) are approximated, at each load level, by a set of linear solutions. If $\mathbf{d}_i$ is the solution of the nodal displacements for the ith iteration of the nth load level, an improved solution for **d** can be obtained as

$$\mathbf{d}_{i+1} = \mathbf{d}_i + \delta \mathbf{d}_i \tag{11}$$

where

$$\delta \mathbf{d}_i = - \left. \frac{\partial \Psi(\mathbf{d})}{\partial \mathbf{d}} \right|_i \Psi(\mathbf{d}_i) = - \mathbf{K}_i^{-1} \Psi(\mathbf{d}_i) \tag{12}$$

is the variation of the nodal displacements and $\mathbf{K}_i$ is the tangential stiffness matrix of the structure [2-4,7]. The updated displacements $\mathbf{d}_{i+1}$ are used to evaluate the current stresses σ_{i+1} and hence the residual forces Ψ_{i+1} from (10). The iteration process will be stopped and the converged solution will be reached for the nth load level, when a suitable predefined convergence criteria will be attained and the residual forces are sufficiently close to zero.

In high geometric instability conditions, where ‘snap-through’ and ‘snap-back’ phenomena are possible [9], the Newton-Raphson method is not suitable to obtain the full structure equilibrium path. In this case it is adopted the spherical formulation of the arc-length methods presented by Crisfield [9-11], where the iterative process is forced to remain on a ‘sphere’ of radius Δl , centered on the converged final solution of the previous load level.

The variation of the nodal displacements $\delta \mathbf{d}_i$ can now be written as

$$\delta \mathbf{d}_i = - \mathbf{K}_i^{-1} \Psi(\lambda_i) + \delta \lambda_i \mathbf{K}_i^{-1} \mathbf{f} = \delta \bar{\mathbf{d}}_i + \delta \lambda_i \mathbf{d}_{Ti} \tag{13}$$

where λ_i is the load level factor for the ith iteration, $\delta \bar{\mathbf{d}}_i$ is the variation of the nodal displacements related to $\Psi(\lambda_i)$, as obtained in (12), and $\mathbf{d}_{Ti}$ are the nodal displacements due to the total reference load **f** . The iterative variation of the applied load factor, $\delta \lambda_i$, is reached by solving the quadratic equation

$$\mathbf{d}_{Ti}^T \mathbf{d}_{Ti} (\delta\lambda_i)^2 + 2\,\Delta\bar{\mathbf{d}}_i^T \mathbf{d}_{Ti}\, \delta\lambda_i + \Delta\bar{\mathbf{d}}_i^T \Delta\bar{\mathbf{d}}_i - (\Delta l)^2 = 0 \tag{14}$$

with

$$\Delta\bar{\mathbf{d}}_i = \Delta\mathbf{d}_i + \delta\bar{\mathbf{d}}_i \tag{15)a$$

which results from the substitution of the updated relations

$$\Delta\lambda_{i+1} = \Delta\lambda_i + \delta\lambda_i \tag{15)b$$

$$\Delta\mathbf{d}_{i+1} = \Delta\mathbf{d}_i + \delta\mathbf{d}_i = \Delta\bar{\mathbf{d}}_i + \delta\lambda_i\, \mathbf{d}_{Ti} \tag{16}$$

in the constrained equation of the spherical formulation

$$(\Delta\mathbf{d}_{i+1})^T \Delta\mathbf{d}_{i+1} + b\,(\Delta\lambda_{i+1})^2\, \mathbf{f}^T\mathbf{f} = (\Delta l)^2 \tag{17}$$

where b = 0 according to Crisfield [10].

To avoid 'doubling back' on the equilibrium path, will be kept the solution of $\delta\lambda_i$ in (14) that corresponds to a minimum positive value [9,10] of

$$\varphi = (\Delta\mathbf{d}_{i+1})^T \Delta\mathbf{d}_i \tag{18}$$

For all increments other than the first, the initial incremental loading parameter $\Delta\lambda_1$ would be obtained from

$$\Delta\lambda_1 = \pm\,\Delta l \,/\, \sqrt{\mathbf{d}_{T1}^T \mathbf{d}_{T1}} \tag{19}$$

The sign to be chosen in (19) must be the same of the previous increment unless the determinant of **K** has changed sign, in which case, a sign reversal is applied [9,10].

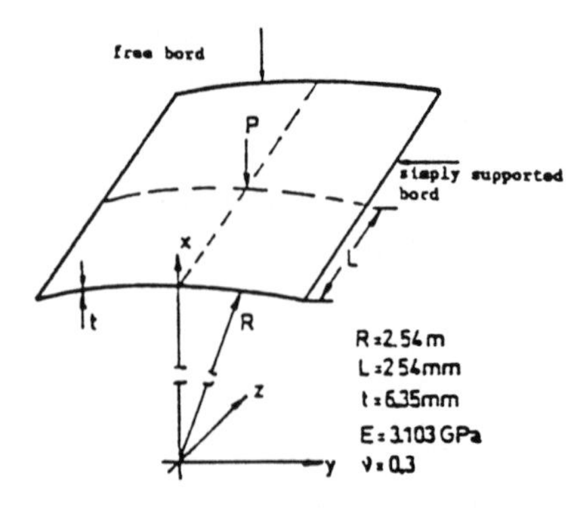

Figure 3. Elastic cylindrical shell subjected to central point load.

4. Numerical example

The performance of both elements are tested on a cylindrical laminated shell, subjected to a central point load, as can be seen in figure 3. The full load-displacement path is displayed for isotropic (figure 4) and sandwich (figure 5) shells.

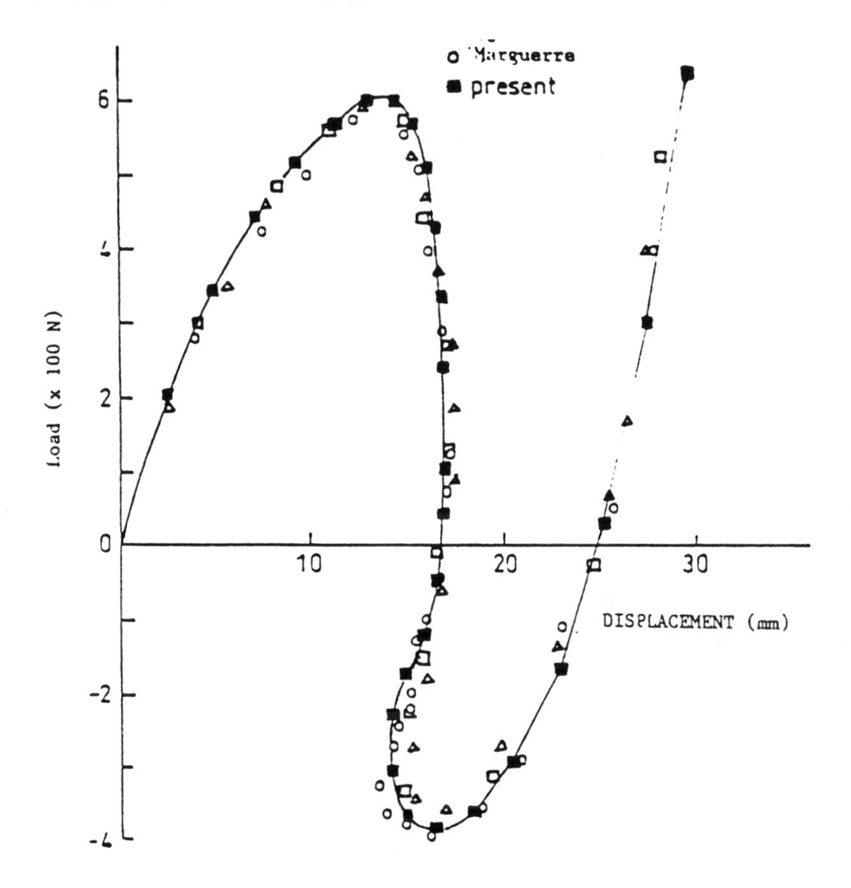

Figure 4. Load-displacement path for isotropic shell.

5. Conclusions

In this paper the Ahmad and Marguerre shell elements were compared in the analysis of laminated sandwich shells. The full non-linear load-displacement path was achieved, by the use of proper non-linear solution methods, such as the Ramm-Crisfield methods.

The performance of both elements was considered to be quite similar, although it was only tested shallow shells problems. In future papers it will be compared their performance on higher curvature shells.

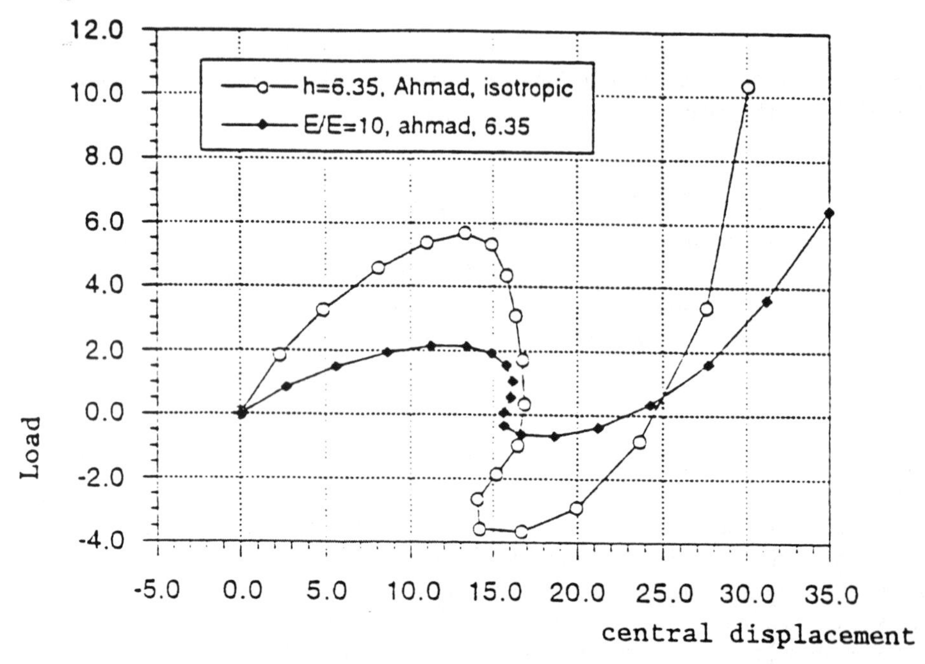

Figure 5. Load-displacement path for sandwich shell.

6. References

1. Mindlin, R.D.: Influence of rotatory inertia and shear in flexural motions of isotropic elastic plates, *J. Appl. Mech.* **18** (1951), 31-38.
2. Zienkiewicz, O.C.: *The Finite Element Method*, McGraw-Hill, N. Y., USA, 1977.
3. Pica, A., Wood, R.D. and Hinton, E.: Finite element analysis of geometrically nonlinear plate behaviour using a mindlin formulation, *Computers & Structures* **11** (1980), 203-215.
4. Pica, A. and Wood, R.D., Postbuckling behaviour of plates and shells using a mindlin shallow shell formulation, *Computers & Structures* **12** (1980), 759-768.
5. Ahmad, S., Irons, B.M. and Zienkiewicz, O.C.: Analysis of thick and thin shell structures by curved finite elements, *Int. J. Num. Meth. Eng.* **2** (1970), 419-451.
6. Fung, Y.C.: *Foundations of Solid Mechanics*, Prentice-Hall, N. J., USA, 1965.
7. Hinton, E. and Owen, D.R.J.: *Finite Element Software for Plates and Shells*, Pineridge Press, Swansea, UK, 1984.
8. Bathe, K.-J.: *Finite Element Procedures in Engineering Analysis*, Prentice-Hall, N. J., USA, 1982.
9. Crisfield, M.A.: A fast incremental/iterative solution procedure that handles 'snap-through', *Computers & Structures* **13** (1981), 55-62.
10. Crisfield, M.A.: An arc-length method including line searches and acceleration, *Int. J. Num. Meth. Engng.* **19** (1983), 1269-1289.
11. Crisfield, M.A.: Solution procedures for non-linear structural problems, in E. Hinton, D.R.J. Owen and C. Taylor (eds.), *Recent Advances in Non-Linear Computational Mechanics*, Pineridge Press Ltd., U.K., 1982, pp. 1-39.

PLASTIC INTERFACE STABILITY IN BIMETALLIC LAYERS SUBJECTED TO BIAXIAL LOADING

J.L. ALCARAZ[1], J. GIL-SEVILLANO[2]

[1] *Departamento de Ingeniería Mecánica- Escuela Técnica Superior de Ingenieros- Alameda Urquijo, s/n- 48013 Bilbao (SPAIN)*

[2] *Escuela Superior de Ingenieros Industriales (Universidad de Navarra)- Pº Manuel de Lardizábal, 15- 20009 San Sebastián (SPAIN)*

Abstract. The occurrence of thickness fluctuations in the forming of a bimaterial layer is considered in the present work. For an orthotropic, incrementally-linear solid, the existence of non-uniform solutions near the bimaterial interface is investigated by assuming an initial perturbation along the interface. The bifurcation equation for the problem is established and solved numerically to obtain firstly the critical strain of the process in terms of a characteristic wavenumber. Then the bifurcation equation is solved for strains above the critical strain to analyse the growth of the perturbation. This is accomplished by means of an instability parameter introduced in the analysis. A set of values for several selected parameters of the process and three constitutive equations are considered in the computations. Their influence in the stability of the process is discussed in detail.

1. Introduction

One problem encountered in metal forming of bimaterial layers is the appearance of thickness fluctuations and occasional decohesion along the interface. Once originated, it should be guaranteed that the growth of these fluctuations remains below a critical level. The analysis of this phenomenon, under the point of view of bifurcation and plastic stability, has motivated the present paper.

Several works on this subject are found. Steif [1] has studied the periodic necking instabilities of a solid composed of alternating material layers under uniaxial tension. More recently, Tomita and Kim [2] have considered the nonaxisymmetric bifurcation behaviour of bilayered tubes subjected to uniform shrinkage at the external surface in plane strain. Independently, Suo *et al.* [3] investigated the stability for two semi-infinite solids bonded along a planar interface. Apart from these studies, Triantafyllidis and Lehner [4] analysed the interfacial instabilities of density-stratified two-layer systems under initial stress.

In this paper, a dynamic analysis of stability is applied to a bimaterial layer, placed between rigid surfaces and subjected to biaxial loading. The problem is tackled as a

A. Vautrin (ed.), Mechanics of Sandwich Structures, 199–206.

bifurcation problem, by considering a dynamic term in the equilibrium equations. An instability parameter defined in the analysis contemplates the growth of the initial interface undulations.

Three different constitutive models are considered in the calculations (the so named Voce, Prager and Hollomon models). In the bibliography, special attention is given to the last one, the Hollomon model, which is not adequate to describe the actual material behaviour at high plastic strains, especially at high temperatures. The two other models seem, however, more appropriate under such conditions.

2. Formulation of the problem

A plane bilayer with finite thickness and infinite length is subjected to biaxial loading in plane strain between two fixed rigid surfaces, as shown in Fig. 1.

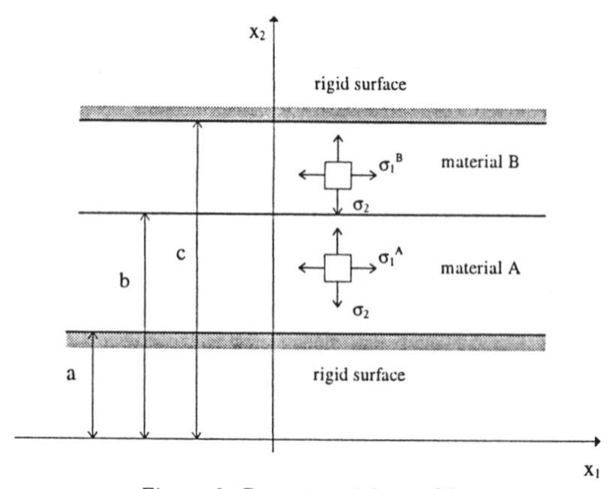

Figure 1. Geometry of the problem.

Incompressibility, plane strain conditions and time independent behaviour are considered in the model. Deformation up to a certain instant is assumed homogeneous, with principal axes x_1 and x_2, and both materials orthotropic along these axes.

Consider then the stability of the initially plane interface, subjected at t = 0 to a perturbation of small amplitude (ε) and arbitrary wavelength, around the equilibrium state. The linearized equations of motion, in the absence of body forces, yield:

$$n_{ij,i} = \rho \ddot{u}_j \tag{1}$$

where n_{ij} represents the nominal stress tensor, $(\ )_{,i}$ meaning partial differentiation with respect to x_i; $\ddot{u}_i$ is the acceleration vector, u_i the displacement vector, and ρ the density.

All the quantities in the perturbed system can be written as a sum of their corresponding fundamental values, plus a term which depends on the amplitude of the initial disturbance, ε. In particular:

$$n_{ij} = n_{ij}^{o} + \varepsilon\tilde{n}_{ij} + O(\varepsilon^2) \qquad u_i = u_i^o + \varepsilon\tilde{u}_i + O(\varepsilon^2) \tag{2}$$

where n_{ij}^o, u_i^o are the fundamental values, and $\varepsilon\tilde{n}_{ij}, \varepsilon\tilde{u}_i$ are the linear terms of the perturbed magnitudes.

By inserting (2) in (1), the first order approximation yields:

$$\tilde{n}_{ij,i} = \rho\ddot{\tilde{u}}_j \tag{3}$$

On the other hand, the continuity conditions at the bimaterial interface can be written in terms of the perturbed values:

$$\langle \tilde{n}_{ij} N_i \rangle = 0, \quad \langle \tilde{u}_i \rangle = 0 \tag{4}$$

where the sign $\langle\ \rangle$ indicates the jump in the enclosed magnitude across the interface, and N_i are the components of the outward normal. The boundary conditions imply that the velocity component $\dot{u}_2$ vanishes at $x_2 = a$ and $x_2 = c$.

Assuming that the hydrostatic pressure does not affect the relation between the deviators of stress rate and strain rate, the constitutive equation for incrementally-linear solids is expressed in the following form [5]:

$$\hat{\sigma}_{11} - \hat{\sigma}_{22} = 2\mu^*(\varepsilon_{11} - \varepsilon_{22}) \qquad \hat{\sigma}_{12} = 2\mu\varepsilon_{12} \qquad \hat{\sigma}_{21} = \hat{\sigma}_{12} \tag{5}$$

where $\hat{\sigma}_{ij}$ is the Jaumann derivative (referred to the rotating axes) of the true stress, ε_{ij} is the eulerian strain rate, $\varepsilon_{ij} = (\dot{u}_{i,j} + \dot{u}_{j,i})/2$, $\dot{u}_i$ being the velocity. μ, μ^* are two incremental shear moduli. The linearization applies similarly to (5) yielding:

$$\tilde{\hat{\sigma}}_{11} - \tilde{\hat{\sigma}}_{22} = 2\mu^*\left(\dot{\tilde{u}}_{1,1} - \dot{\tilde{u}}_{2,2}\right) \qquad \tilde{\hat{\sigma}}_{12} = \tilde{\hat{\sigma}}_{21} = \mu\left(\dot{\tilde{u}}_{1,2} + \dot{\tilde{u}}_{2,1}\right) \tag{6}$$

As the coefficients of the linearized equations do not depend on time, we can find solutions in the form:

$$\begin{aligned} \tilde{u}_i(x_1, x_2, x_3, t) &= e^{i\xi t}\bar{u}_i(x_1, x_2, x_3) \\ \tilde{n}_{ij}(x_1, x_2, x_3, t) &= e^{i\xi t}\bar{n}_{ij}(x_1, x_2, x_3) \end{aligned} \tag{7}$$

By inserting (7), (3) turns into

$$\bar{n}_{ij,i} + \rho\xi^2\bar{u}_j = 0 \tag{8}$$

and the continuity conditions at the interface (4) can be written as

$$\langle \bar{n}_{ij} N_i \rangle = 0, \quad \langle \bar{u}_i \rangle = 0 \tag{9}$$

Note that ξ^2 is the linear eigenvalue of (8). When the minimum of these eigenvalues, ξ_m, satisfies $\xi_m^2 > 0$, solution (7) will remain bounded with time and the interface will be stable. For $\xi_m^2 < 0$, (7) provides unbounded solutions. In that case, the system is said to be unstable. Finally, if $\xi_m^2 = 0$ for a certain perturbation, the problem reduces to a pure bifurcation case.

At the moment of the analysis, the stress components under the assumed biaxial loading are $\sigma_{11} = \sigma_1$ and $\sigma_{22} = \sigma_2$. Both components are supposed uniform in each material and σ_1 can be different, in general, in A and B.

By expressing (5) in terms of the nominal stresses rates and using the incompressibility condition determined by $\dot{u}_{1,1} + \dot{u}_{2,2} = 0$, linearization yields:

$$\begin{aligned} \bar{n}_{11} - \bar{n}_{22} &= \left[2\mu^* - \frac{1}{2}(\sigma_1 + \sigma_2)\right]\left(\frac{\partial \bar{u}_1}{\partial x_1} - \frac{\partial \bar{u}_2}{\partial x_2}\right) \\ \bar{n}_{12} &= \left[\mu + \frac{1}{2}(\sigma_1 - \sigma_2)\right]\frac{\partial \bar{u}_2}{\partial x_1} + \left[\mu - \frac{1}{2}(\sigma_1 + \sigma_2)\right]\frac{\partial \bar{u}_1}{\partial x_2} \\ \bar{n}_{21} &= \left[\mu - \frac{1}{2}(\sigma_1 + \sigma_2)\right]\frac{\partial \bar{u}_2}{\partial x_1} + \left[\mu - \frac{1}{2}(\sigma_1 - \sigma_2)\right]\frac{\partial \bar{u}_1}{\partial x_2} \end{aligned} \tag{10}$$

Let us now introduce a flow function $\psi(x_1,x_2)$ such that

$$\bar{u}_1=\frac{\partial\psi}{\partial x_2},\qquad \bar{u}_2=-\frac{\partial\psi}{\partial x_1} \tag{11}$$

Then, with (10) and (11) in (8) and (9), the following partial derivative equation can be derived:

$$(R+S)\frac{\partial^4\psi}{\partial x_1^4}+2(1-R)\frac{\partial^4\psi}{\partial x_1^2\partial x_2^2}+(R-S)\frac{\partial^4\psi}{\partial x_2^4}+e^2\left(\frac{\partial^2\psi}{\partial x_1^2}+\frac{\partial^2\psi}{\partial x_2^2}\right)=0 \tag{12}$$

where, for concision, the following notation has been used:

$$R=\frac{\mu}{2\mu^*},\qquad S=\frac{\sigma_1-\sigma_2}{4\mu^*},\qquad e^2=\frac{\rho\xi^2}{2\mu^*} \tag{13}$$

Equation (12) is said to be elliptic, hyperbolic or parabolic according to whether there are zero, four or two real roots, respectively, for its characteristic equation. Real roots provide characteristic planes of (12), allowing the possibility of strain discontinuity in the form of shear bands, and imaginary roots imply the existence of sinusoidal solutions around the interface. Here we are concerned with this last type of instability, the only mode encountered in the elliptic regime. The analysis that follows is centred on this regime, because it provides the most general solutions (complex conjugates) and, for this reason, could be considered to include the remaining regimes.

3. Eigenmodes. Deduction of the governing equation.

To determine the eigenmodes of the problem, let us seek for solutions of the flow function, ψ, in the form:

$$\psi=f(x_2)\sin\frac{2\pi x_1}{\lambda} \tag{14}$$

where λ is the wavelength of the deformation in the x_1 direction and $f(x_2)$ has to be selected so that $\dot{u}_1$ and $\dot{u}_2$ verify the continuity conditions.

By inserting (14) into (12), the following differential equation for function f is obtained:

$$(R-S)f^{IV}+\left[-2(1-R)\omega^2+e^2\right]f^{II}+\left[(R+S)\omega^4-e^2\omega^2\right]f=0 \tag{15}$$

where $\omega=2\pi/\lambda$. In the elliptic regime, the roots of the characteristic equation of (15) are complex. After lengthy manipulations, it is obtained:

$$(R_A - S_A)\left\{\frac{s_p^A}{p_A}\left[X_A(1-S'_A-b_A/2)-S''_A+b_A/2\right]+\frac{s_r^A}{r_A}\left[X_A(1-S'_A-b_A/2)+S''_A-b_A/2\right]\right\}\left[\frac{s_p^B}{p_B}-\frac{s_r^B}{r_B}\right]+$$

$$+\zeta^2(R_B - S_B)\left\{\frac{s_p^B}{p_B}\left[X_B(1-S'_B-b_B/2)-S''_B+b_B/2\right]+\frac{s_r^B}{r_B}\left[X_B(1-S'_B-b_B/2)+S''_B-b_B/2\right]\right\}\left[\frac{s_p^A}{p_A}-\frac{s_r^A}{r_A}\right]-$$

$$-\zeta\left\{2(R_A - S_A)(R_B - S_B)\left[X_A(c_p^A+c_r^A)(c_r^B-c_p^B)+X_B(c_r^A-c_p^A)(c_r^B+c_p^B)\right]+2(1-S'_A-b_A/2)(R_B - S_B)\left[\frac{s_p^A}{p_A}-\frac{s_r^A}{r_A}\right]\right.$$

$$(p_B s_p^B + r_B s_r^B)+(1-S'_A-b_A/2)(1-S'_B-b_B/2)\left[\frac{s_p^A}{p_A}-\frac{s_r^A}{r_A}\right]\left[\frac{s_p^B}{p_B}-\frac{s_r^B}{r_B}\right]+$$

$$\left.+4(R_A - S_A)(R_B - S_B)(p_A s_p^A + r_A s_r^A)(p_B s_p^B + r_B s_r^B)+2(1-S'_B-b_B/2)(R_A - S_A)\left[\frac{s_p^B}{p_B}-\frac{s_r^B}{r_B}\right](p_A s_p^A + r_A s_r^A)\right\}=0 \tag{16}$$

where

$$S'=\frac{\sigma_1+\sigma_2}{4\mu^*} \quad S''=\frac{RS'-(S^2+S'^2)/2}{R-S} \quad \zeta=\frac{\mu_B^*}{\mu_A^*} \quad X=\sqrt{\frac{R+S-b}{R-S}} \quad s_p=\sin 2pq$$

$$s_r=\sinh 2rq \quad \alpha=p_A+ir_A \quad c_p=\cos 2pq \quad c_r=\cosh 2rq \quad \beta=p_B+ir_B$$

$$q_A=\frac{2\pi}{\lambda}(b-a) \quad b_A=\frac{\rho_A\xi^2(b-a)^2}{2\mu^* q_A^2} \quad q_B=\frac{2\pi}{\lambda}(c-b) \quad b_B=\frac{\rho_B\xi^2(c-b)^2}{2\mu^* q_B^2}$$

In equation (16) the bifurcation strain is obtained by imposing $\xi=0$, i.e. $b_A=b_B=0$, and finding then the minimum of the strain eigenvalues as a function of the wavenumber q_A. This minimum denotes a critical equivalent strain below which no bifurcation can occur.

4. Results

In this section, results are presented for materials with a constitutive equation in the form $\bar{\sigma}=\bar{\sigma}(\bar{\varepsilon})$, where $\bar{\sigma}$ and $\bar{\varepsilon}$ are the equivalent stress and strain, respectively. Three constitutive models have been considered in the calculations:

(i) Voce model: $\bar{\sigma}=C(1-me^{-n\bar{\varepsilon}}$ (17)

(ii) Hollomon model: $\bar{\sigma}=k\bar{\varepsilon}^n$ (18)

(iii) Prager model: $\bar{\sigma}=C\tanh(n\bar{\varepsilon})$ (19)

where C, k, m, n are constants. Details of the choice of values for these constants are given in [6]. Table 1 gives the expressions for R, S, S' and ζ corresponding to each constitutive equation.

4.1. INFLUENCE OF THE CONSTITUTIVE MODELS

Figure 2 shows a comparison among the three assumed constitutive models for two typical cases: (1) a bilayer with an upper location of the harder material, a yield stress ratio $k_r = Y_B/Y_A = 2$ and a thickness ratio $r_e = (c-b)/(b-a) = 1/3$ (in Figs. 2a-b), and (2) a

TABLE 1. Expressions of the parameters in the bifurcation equation for the three constitutive models.

Voce $\bar{\sigma} = C\left(1 - me^{-n\bar{\varepsilon}}\right)$	$R = \frac{\sqrt{3}}{2n}\left(\frac{e^{n\bar{\varepsilon}}}{m} - 1\right)\coth\left(\sqrt{3}\bar{\varepsilon}\right)$ $S' = \frac{3}{2}\frac{\sigma_h}{Cmne^{-n\bar{\varepsilon}}}$	$S = \frac{\sqrt{3}}{2n}\left(\frac{e^{n\bar{\varepsilon}}}{m} - 1\right)$ $\zeta = \frac{C_B m_B n_B}{C_A m_A n_A} e^{-(n_B - n_A)\bar{\varepsilon}}$
Hollomon $\bar{\sigma} = k\bar{\varepsilon}^n$	$R = \frac{\sqrt{3}}{2}\frac{\bar{\varepsilon}}{n}\coth\left(\sqrt{3}\bar{\varepsilon}\right)$ $S' = \frac{3}{2}\frac{\sigma_h}{kn\bar{\varepsilon}^{n-1}}$	$S = \frac{\sqrt{3}}{2}\frac{\bar{\varepsilon}}{n}$ $\zeta = k_r \frac{n_B}{n_A}\bar{\varepsilon}^{n_B - n_A}$
Prager $\bar{\sigma} = C\tanh(n\bar{\varepsilon})$	$R = \frac{\sqrt{3}}{2}\frac{1}{2n}\sinh(2n\bar{\varepsilon})\coth\left(\sqrt{3}\bar{\varepsilon}\right)$ $S' = \frac{3}{2}\frac{\sigma_h}{Cn}\cosh^2(n\bar{\varepsilon})$	$S = \frac{\sqrt{3}}{2}\frac{1}{2n}\sinh(2n\bar{\varepsilon})$ $\zeta = \frac{C_B n_B}{C_A n_A}\frac{\cosh^2(n_A\bar{\varepsilon})}{\cosh^2(n_B\bar{\varepsilon})}$

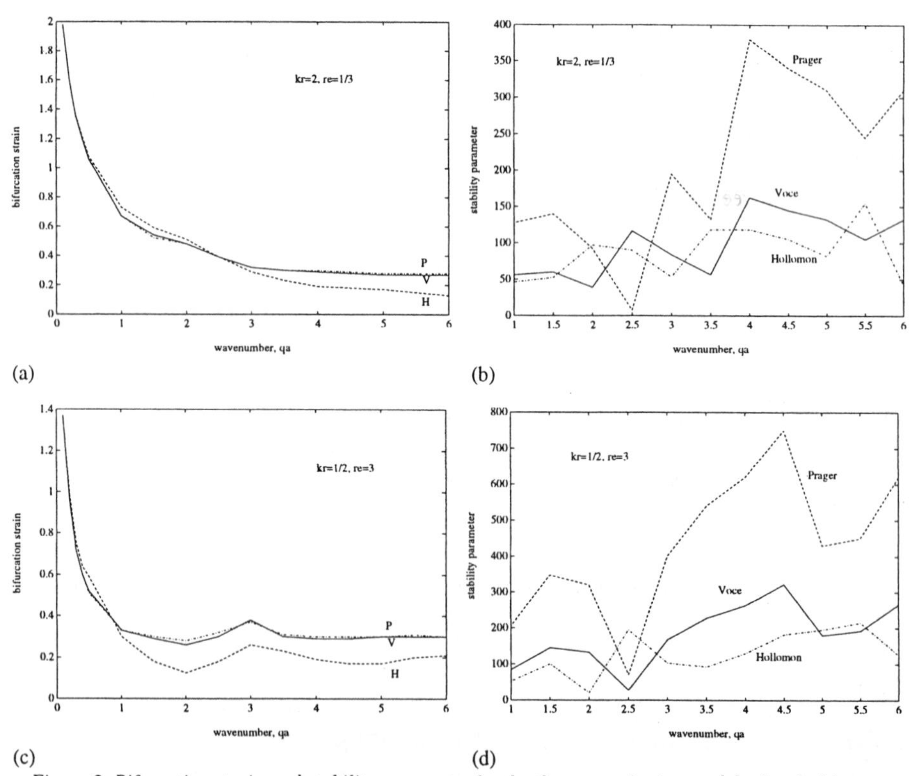

Figure 2. Bifurcation strain and stability parameter for the three constitutive models, for: (a-b) an upper location of the harder material with $kr=Y_2/Y_1=2$, $re=e_2/e_1=1/3$, (c-d) a lower location of the harder material with kr= 0.5, re= 3. ('P', 'V' and 'H' refer to Prager, Voce and Hollomon models, respectively).

By comparing Figs. 2a and 2c, it can be seen that there is a wider range of elliptic regions in the case of a lower harder material. From Figs. 2b and 2d, it is drawn that the Hollomon equation leads to a more stable behaviour. Note that Prager and Voce equations seem to be more adequate to represent the elastic-perfectly plastic behaviour in metal forming at high temperature, in the sight of their asympthotic trend at high strains. In the following, the Voce model will be selected.

4.2. INFLUENCE OF SEVERAL PARAMETERS OF THE PROBLEM

The influence of other parameters are now considered. In particular, the yield stress ratio, $k_r = Y_B/Y_A$, and the hardening parameter n_a are selected.

Assumed reference values of all the parameters for the two possible locations of the harder material are given in Table 2, where p_r stands for the hydrostatic pressure, d for the density and ε_{max} for the maximum strain in the process.

TABLE 2. Reference values for both locations of the harder material.

location	k_r	r_e	n_a	n_b	p_r^a	p_r^b	d_a/d_b	ε_{max}
upper	2	1/3	12	12	1	1	1	1
lower	0.5	3	12	12	1	1	1	1

The influence of the yield stress ratio, k_r, is shown in Fig. 3, for the standard case. It is noticed in Fig. 3a that above $q_A = 1$ the higher the yield stress ratio, the higher the critical strain. Figure 3b shows that the maximum stability parameter increases as the discrepancy between the yield stresses of the two materials increases.

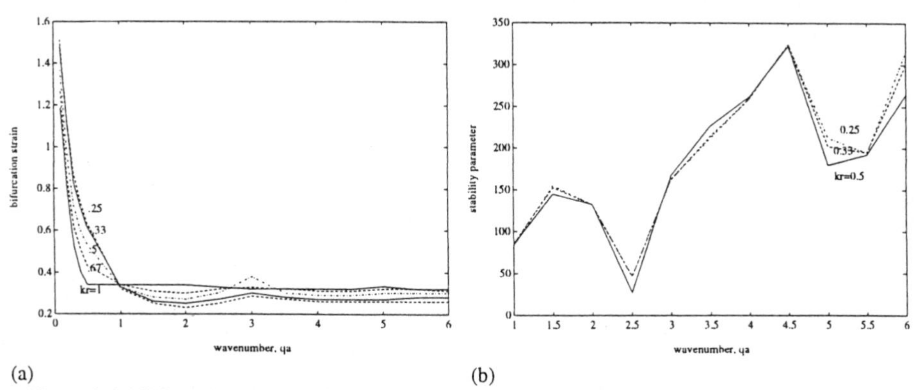

(a) (b)

Figure 3. (a) Bifurcation strain and (b) stability parameter for different yield stress ratios, kr, for a lower location of the harder material.

Figure 4 illustrates the influence on instability of the hardening parameter n_a. Figure 4a shows that the bifurcation strain decreases as the hardening parameter increases, this leading to the conclusion that instability is promoted with a more hardening material. On the other hand, Fig. 4b indicates that the stability parameter ξ considerably increases as the hardening parameter increases.

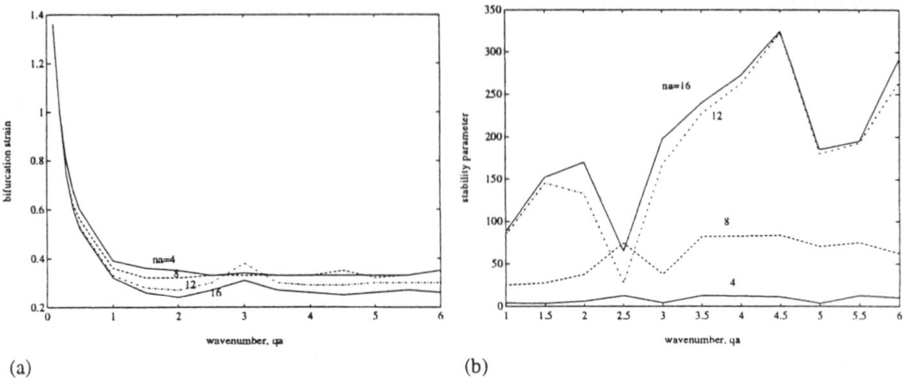

(a) (b)

Figure 4. (a) Bifurcation strain and (b) stability parameter for different hardening parameters na, in the case of a lower location of the harder material.

5. Conclusions

The analysis carried out in this paper involves both a bifurcation and a stability analysis of bimetallic layers between rigid surfaces under biaxial plastic loading. A stability parameter is established to determine the growth of an initial perturbation originated at the interface, in terms of the strain rate, the total plastic strain and other geometric and material parameters of the process. Detailed results are presented about the effect of two variables related to the problem (the yield stress ratio and the hardening parameter), as well as three different constitutive equations (Voce, Prager and Hollomon).

In summary, it can be concluded that instability is originated and promoted as the yield stress ratio between the harder and softer material increases. The same yet stronger tendency is found on increasing the hardening parameter of the constitutive law. It is also shown that instability is generally favoured with a lower location of the harder material. Finally, the influence of the selected constitutive model is illustrated. Among the three models considered (Voce, Prager and Hollomon), the Prager model appears to provide the worst conditions for stability.

6. References

1. P.S. Steif. Periodic necking instabilities in layered plastic composites. *Int. J. Solids Struct.* **22**, 1571-1578, 1986.
2. Y. Tomita, Y. Kim. Bifurcation behaviour of bilayered tubes subjected to uniform shrinkage under plane strain condition. *Int. J. Solids Struct.* **29**, 2723-2733, 1992.
3. Z. Suo, M. Ortiz, A. Needleman. Stability of solids with interfaces. *J. Mech. Phys. Solids* **40**, 613-640, 1992.
4. N. Triantafyllidis, F.K. Lehner. Interfacial instability of density-stratified two-layer systems under initial stress. *J. Mech. Phys. Solids* **41**, 117-142, 1993.
5. M.A. Biot. *Mechanics of incremental deformation*. Ed. Wiley, New-York, 1965.
6. J.L. Alcaraz. Doctoral Thesis, Faculty of Engineering, University of Navarre, Spain, 1993.

DYNAMIC BEHAVIOUR OF SANDWICH STRUCTURES USED IN AIR TRANSPORT

A. MORENO
Service Durcissement et Vulnerabilité
Département Structures Impacts
Centre d'Etudes de Gramat, 46500 Gramat France

1. Introduction

The "Centre d'Etudes de Gramat (C.E.G.)" is working on the modelling of the behaviour of double walled aeronautical structures submitted to impulsive loading. These structures are made of a composite lining panel, on which the pressure loads apply, and an aluminum alloy skin. The lining panel and the skin are both attached on frames which separate them (figure 1).

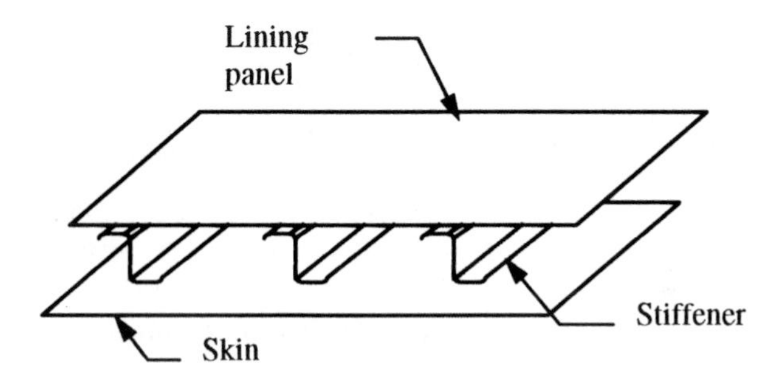

Figure 1. Double walled structure

The effort profile that applies to the skin depends, on the one hand, on the stiffener and skin responses and, on the other hand, on the liner deterioration.

The object of the study is to develop a lining panel simplified modelling element for predicting more particularly the skin damage. The analysis approach reported here uses the ABAQUS finite element code to model the structure response; it is supported by a set of static and dynamic experimental studies.

These works are part of a wider research project backed by the "Direction Générale de l'Aviation Civile" to assess the vulnerability of an airliner fuselage submitted to an internal explosion.

2. Technical data

The lining panels are manufactured by bonding two composite skins over a cardboard honeycomb core. The composite skin is a two - ply woven consisting of resin matrix

A. Vautrin (ed.), Mechanics of Sandwich Structures, 207–214.

reinforced by glass fibres whose directions are 0 et 90° : it is covered with a thin aluminum alloy sheet (figure 2).

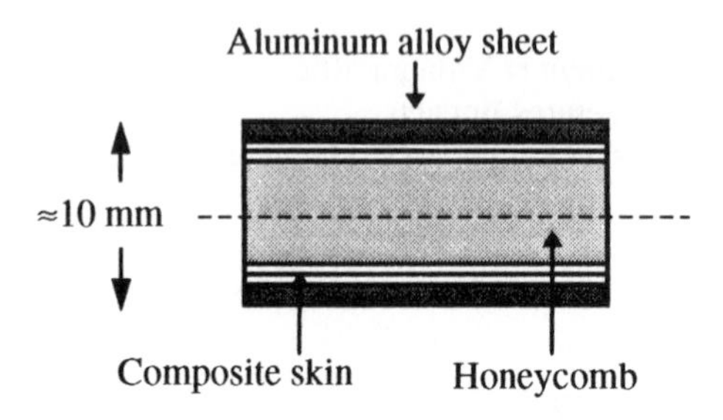

Figure 2. Lining panel section

In a first analysis, the behaviour law of the different components of the lining panel is defined without considering the effect of the strain rate.

The honeycomb, of nidanomex type, is defined as an orthotropic material [1].

The aluminum alloy is described by an elastoplastic law that takes into consideration the material softening beyond the failure stress; the plasticity criterium is a VON MISES' criterium with isotropic hardening.

The composite skin is considered as a linear orthotopic material [1, 2].

3. Process

The process used to model the lining panel (figure 3, page 3) consists of studying :

(1) The linear static behavior until failure of lining panel strips under tensile, shearing, bending and peeling loads to define elastic characteristics and rupture levels of lining panel components,

(2) The dynamic behavior of the lining panel, first single (local approach) then supported by stiffeners (global approach) to study the liner response sensitiveness to different parameters (boundary conditions, load profile,...).

3.1. STATIC STUDY

30 trials were performed on three kinds of lining panel [3, 4]:

- Tensile tests of composite skin and aluminum alloy sheet to define their Young modulus.
- Composite skin shearing tests to calculate skin shearing modulus (figure 4a, page 3).
- Double shearing test of sandwich to estimate the out of plane honeycomb characteristics (figure 4b).
- Peeling test on sandwich to characterize the skin - honeycomb disbonding (figure 4c).
- Bending tests on long and short sandwich specimens to determine the shearing modulus of honeycomb (figure 4d).

An optimization numerical tool was used to determine elastic and failure characteristics of each sandwich layer. Bending configurations allowed to size the effort to fail the panel by skin-core interface failure by suiting numerical deflections to computational ones [5].

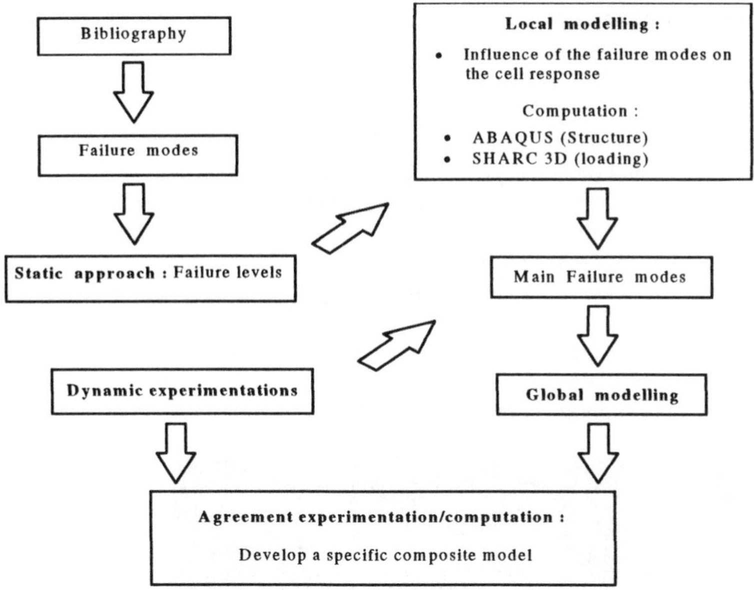

Figure 3. Process

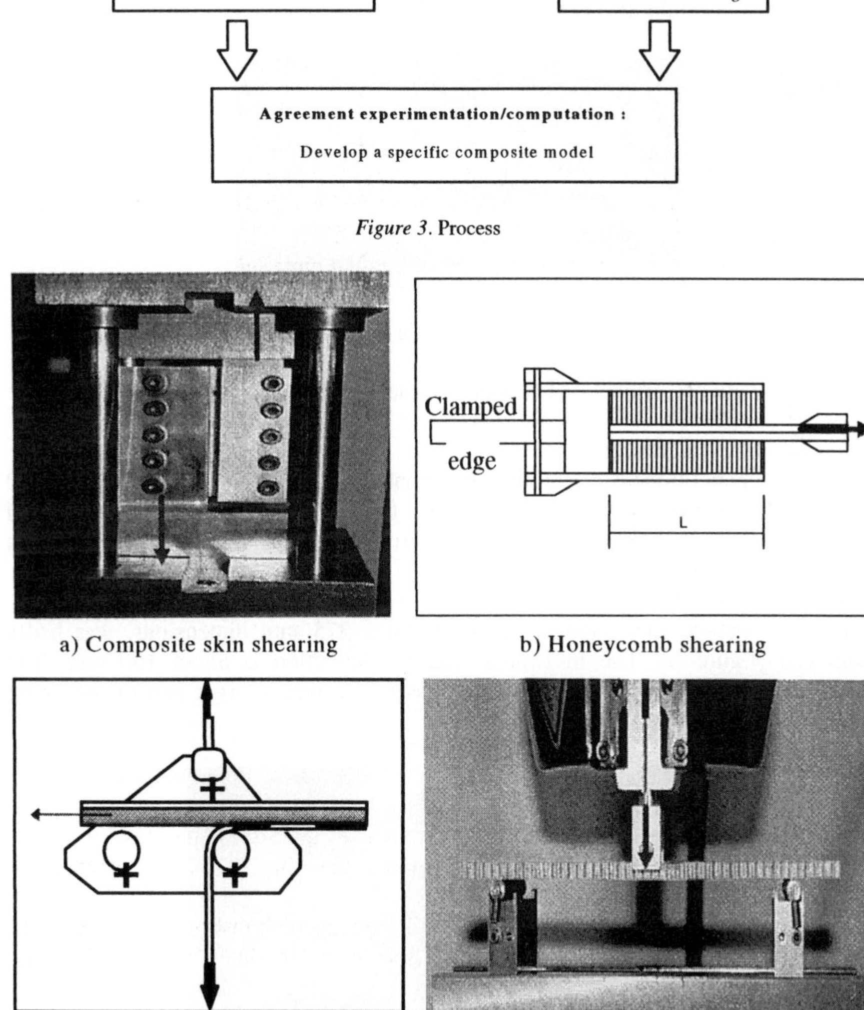

Figure 4. Static tests on sandwich specimens

3.2. DYNAMIC STUDY

3.2.1. *Experimental analysis (global approach)*

Dynamic trials are required to identify the failure modes of the sandwich under shock loads. The experimental device is presented in figure 5.
The liner is attached on 6063 T5 aluminum alloy I-beams in the same manner as in the aircraft in order to investigate the influence of the lining panel response on the effort profile transmitted by this panel on its support.

Two loading levels were selected to either preserve the integrity or generate severe damage of the liner.

Two plates, made respectively of foam and rubber, separate the support from the concrete paving; these plates are flexible enough to suppose that their response time is larger than that of the support. They were to get simply supported conditions for I-beams.

Figure 5. Experimental device

Three shots were performed. The results of these shots are defined as follows :

(a) Weak loading trial results
Two lining panel dimensions were experimented : 1585 mm x 1440 mm and 2080 mm x 375 mm. The test panels show, around their attachment points, small residual deformations of the metallic skin, cracks through the thickness of the liner and a delamination area.

(b) High loading trial results
The experimented lining panel, of 2080 mm x 375 mm dimensions, was badly damaged (figure 6). The maximum residual deflection is about 100 mm. The interface between the skins and the core has failed over a larger part of the lining panel.

Figure 6. Residual deformation of the experimented liner submitted to high loads

The high liner deflection seemed to show a possibility of contact between the liner and I-beam support. For a lined fuselage panel submitted to the same high loads, the liner-skin contact would increase the skin stress and then generate more severe damage to skin. Moreover, the experimental results shows the requirement to model :

- The out of plane shear response of the core because it contributes to reduce the global stiffeness of the liner and the effort levels transmitted to the stiffeners.
- The contacts between the lining panel and the beams supporting it that may modify the effort distribution at the supports.
- The failure of the interface composite skin-honeycomb.
- The sandwich panel damping that may affect the global structure response.

3.2.2. *Numerical analysis*

Failure and elastic characteristics previously estimated from static tests are herein introduced to make restitution of the liner mechanical behavior under impulsive loads.

Loading conditions. The pressure loads are calculated by means of an hydrodynamic code. The pressure characteristics vary in space and time (figure 7): the rise time is equal to 0 and the peak overpressure varies from a few bars to several tens bars. The shock wave diffraction on the lining panel sides is neglected.

Local approach. The study of the dynamic lining panel response first requires a model of the liner crush behavior. The honeycomb crushing effect is investigated by modelling the skin and the core with respectively volumic and shell finite elements as figure 8 shows [6, 7]. For the dynamic loading levels considered here, the crushing stresses induced within the core were found lower than the buckling loads; then, the core crushing effect on the lining panel behavior was neglected in the global approach.

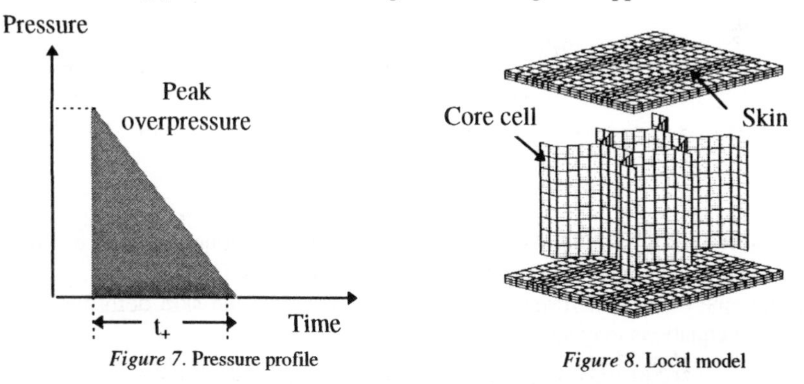

Figure 7. Pressure profile

Figure 8. Local model

Global approach. The numerical device is presented in figure 9.

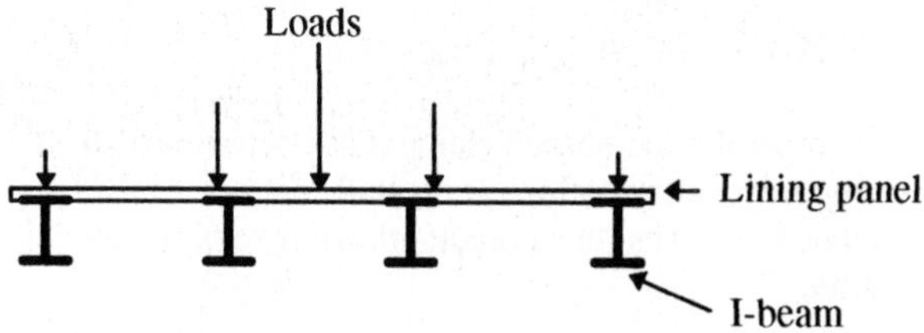

Figure 9. Lining panel attached on I-beams

Only one quarter of the structure was modelled because of the existence of two symmetry planes. The structural damping was represented with a pressure that is opposed to the liner displacement.

The disbonding of the lining panel was presumed to be caused by the shearing rupture of the interface skin-honeycomb [8, 9, 10]. This behaviour mode was approximated by means of membrane elements linking the composite skins to the core. The tensile rupture stress of these elements was assumed to be equivalent to the shearing rupture stress of the interface (figure 10).

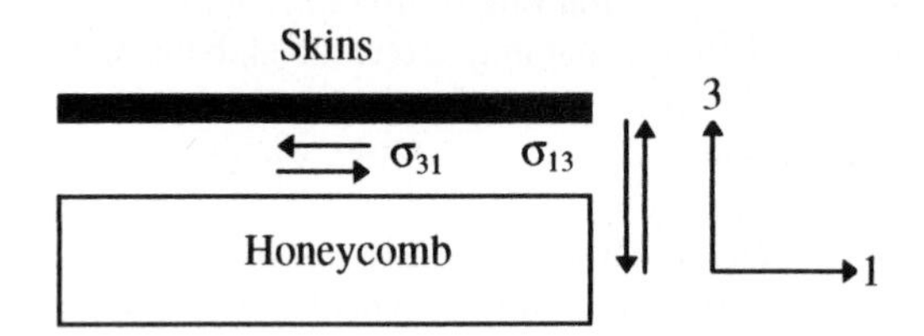

a) Interlaminar stress in the lining panel

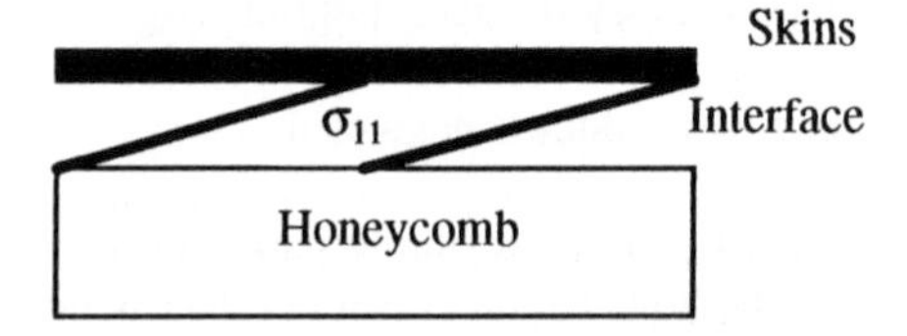

b) Tensile stress within the membrane element

Figure 10. Delamination Modelling

One can write :

$$\sigma_{31}{}^{r}.b.l.S = \sigma_{11}{}^{r}.b.h \tag{1}$$

$$\sigma_{11}{}^{r} = \frac{\sigma_{31}{}^{r}.l.S}{h} \tag{2}$$

where l, b and h are the length, width and thickness of a membrane element, S is the core surface that is in contact with the composite skin and $\sigma_{IJ}{}^{r}$ is the failure stress.

If membrane element deformation is much smaller than the skin deformation, then, the following equations may also be written :

$$\sigma_{11}{}^{r} / E_{11} << \sigma_{alu} / E_{alu} \tag{3}$$

$$E_{11} >> \sigma_{11}{}^{r} . E_{alu} / \sigma_{alu} \tag{4}$$

The characteristics of the membrane element are determined from equations 1, 2 and 4 as well as by using the results obtained from static tests.

Figure 11 on page 6 gives the computational residual deformation for the high loading configuration.

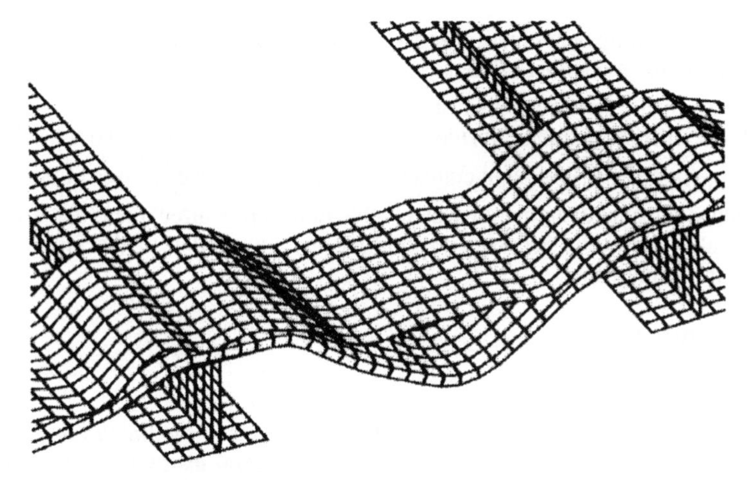

Figure 11. Numerical residual deformation of the liner under high loads

3.2.3. *Comparison between computations and trials (global approach)*

A good qualitative agreement was found between experimental and computional liner residual deformations (figures 6 and 11). By comparing the strain gage signals taken on I-beam webs, numerical efforts transmitted by the liner to the beams appeared to be higher than experimental ones (figure 12).

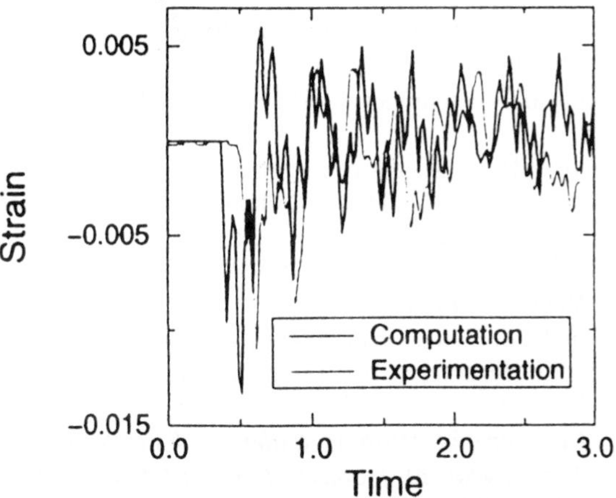

Figure 12. Comparison between experiments and computations (stiffener strain (%) in function of time (ms))

Crush tests realized on experimented lining panels didn't reveal any modification of the crushing modulus; this result validated the result found from the liner local modelling.

3.3. CONCLUSION

The works herein described allowed the forecast of the experimental results in a qualitative manner. They shows that the effort profile applying to the skin depends on the liner deterioration by delamination as well as skins and core failures and the requirement for improving the modelling of the :

- Out of plane shear response of the core because it contributes to reduce the liner global stiffness and the effort levels transmitted to the frames and the skin,
- Liner-frames and liner-skin contacts that may modify the stresses in the skin,
- Sandwich panel damping of which value may highly affect the effort levels.

They also shows the necessity to improve our knowledge of the failure modes and levels of the lining panel.

4. Future topics

Future topics will be to :

- Identify the main dynamic failure modes and levels of the lining panel submitted to loads of which rise times can vary from 0 to several ms by realizing both dynamic tests instrumented correctly and computer simulations,
- Develop a specific detailed model of the damage mechanisms defined experimentally.
- Develop a simplified model of the lining panel behaviour to be used to calculate the response of aeronautical structures made up of several thousand elements.

5. References

1. Gay, D. : *Composite materials*, Hermes publishing.
2. Favre, J.P., and Vautrin, A.: *CMO/caractérisation expérimentale et modélisation,* Comptes rendus des neuvièmes journées nationales sur les composites, AMAC (1994).
3. Allix, O.; Favre, J.P.; and Ladeveze, P.: *Essais et identification*, Comptes rendus des huitièmes journées nationales sur les composites, AMAC (1992).
4. Miravete, A.: *Impact and dynamic*, Proceedings of the ninth international conference on composite materials (ICCM/9) Vol.V : Composite behaviour, Univ. Of Zaragoza Woodhead Publ. (1993).
5. Guangyu Shi and Pin Tong : *Equivalent transverse shear stiffeness of honeycomb cores*, Int.J. Solids Structures Vol. 32 n° 10 pp 1383-1395, Elsevier Science Ltd (1995).
6. Porter, J.H.: *Utilizing the crushing under load properties of polypropylene and polyethylene honeycomb to manage crash energy*, Journal of materials and manufacturing, Vol. 103, n° section 5, pp. 659-666, SAE Transactions (1994).
7. Burgio, R.B.: *Modeling and computer simulation of the dynamic response and failure of composite structures subjected to shock loading*, Master's thesis, Naval postgraduate school Monterey, California (1995).
8. Pilling, M. And Fishwick, S.: *The application of progressive damage modelling techniques to the study of delamination effects in composite materials*, 6th europeen conference on composite materials, Bordeaux (FR, 1993).
9. Martin, R.H.: *Structural damage characterization*, Composite materials : Fatigue and fracture - Fifth volume, Atlanta (US), American society for testing and materials (1993).
10. Interfacial bond strength of composite materials, Published search, U.S. Department of Commerce, NTIS (1997).

IDENTIFICATION OF THE DYNAMIC MATERIAL PROPERTIES OF COMPOSITE SANDWICH PANELS WITH A MIXED NUMERICAL EXPERIMENTAL TECHNIQUE

I. PEETERS, H. SOL
Free University Brussels (VUB), Faculty of Applied Sciences
Pleinlaan 2, 1050 Brussels

1. Introduction

Since the eighties a mixed numerical/experimental technique (MNET) for the determination of the complex moduli of thin plates is developed at the department of Structural Analysis of the Free University Brussels (VUB) ([1],[2],[3]). It is a user-friendly and fast measuring technique based on the measurement of resonant frequencies and modal damping ratios of a free vibrating plate. This innovative test method determines the elastic constants and the damping characteristics of the plate material and is called 'the Resonalyser procedure'. The results obtained for the material data can then be used as input data for the analysis (e. g. in a finite element model) of more complex structures made out of the tested plate material.
The aim of the recent work is to develop a similar measuring technique for the estimation of the material characteristics of sandwich panels. Therefore a suitable model that can calculate the needed material data starting from a good chosen value of the parameters of that model has to be established. In this paper the latest work done to obtain an appropriate numerical model that can describe the dynamic behavior of a sandwich in an accurate but relatively simple way is presented. The model accounts for shear deformation because this effect can not be neglected here as in thin plate theory. In 1993 Hua H. started with the development of a sandwich model [4] and this model is now further investigated.
The sandwich model is used for the dynamic analysis of well-known self made panels. At first instance sandwiches with an isotropic core and ditto faces are treated. In the near future sandwiches with orthotropic and anisotropic composite faces will be studied. The frequencies obtained with the numerical model are compared with the experimental ones. The next work to be done is to calculate the modal damping ratios of the considered eigenmodes by making use of the energy method ([6],[7]). The work strategy is already explained in this paper. Again, a comparison between the calculated and the experimentally determined quantities must be carried out. The comparison between calculations and experiments is necessary to see whether the considered sandwich model is suitable to be used in a MNET for sandwich panels.

A. Vautrin (ed.), Mechanics of Sandwich Structures, 215–222.

2. Numerical Model

For the further understanding of the sandwich theory, figure 1 gives an image of the considered sandwich and the used coordinate system. In the formulas, index c is used for the core characteristics and index f for the properties of the faces.

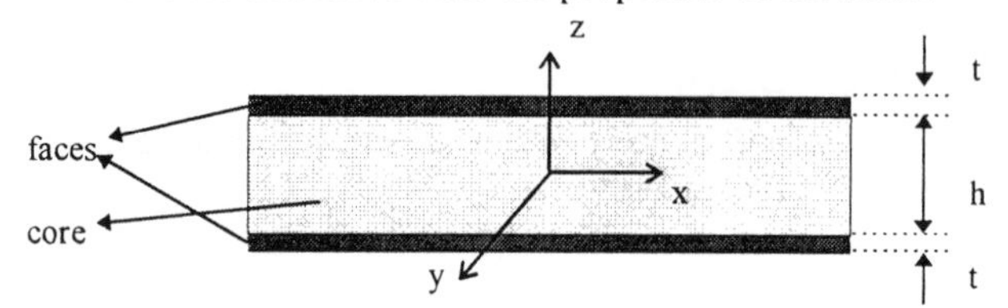

Figure 1. View of the considered sandwich and the coordinate system

2.1 EXPRESSION OF THE DISPLACEMENT COMPONENTS

The investigated theory is based on the following assumptions :

1. The material properties of the faces and the core are symmetric to the z = 0 plane;
2. The stresses σ_x, σ_y and τ_{xy} are uniformly distributed over the thickness direction of the faces (thin faces);
3. In the core $\sigma_x = \sigma_y = \tau_{xy} = 0$ and the core is only subjected to transverse shear stresses τ_{xz} and τ_{yz};
4. Under flexural deformation (anti-symmetric with respect to the mid-surface) $\varepsilon_z = 0$ in both faces and core;
5. The stress σ_z can be neglected.

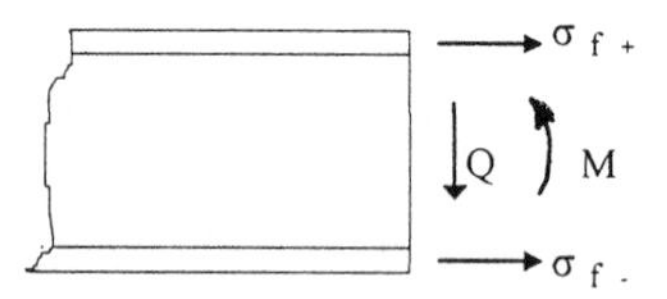

Figure 2. Moment and force distribution and sign convention

With assumption 1, the coupling of in-plane and bending deformation no longer exists. Assumption 3 together with the equilibrium equation gives for the core :

$$\frac{\partial \tau_{xz}^{c}}{\partial z} = \frac{\partial \tau_{yz}^{c}}{\partial z} = 0 \tag{1}$$

So τ_{xz}^{c} and τ_{yz}^{c} are uniformly distributed along the thickness direction. The shear stresses should be [5] :

$$\tau_{xz}^{c} = G_{xz}^{c}\gamma_{xz}^{c} = \frac{h+t}{h} G_{xz}^{c}\gamma_{xz} \tag{2a}$$

$$\tau_{yz}^{c} = G_{yz}^{c}\gamma_{yz}^{c} = \frac{h+t}{h} G_{yz}^{c}\gamma_{yz} \tag{2b}$$

The shear forces are given by :

$$Q_x = \tau_{xz}^{c} h \quad \text{and} \quad Q_y = \tau_{yz}^{c} h \tag{3}$$

Combining (2) and (3) gives :

$$\gamma_{xz} = \frac{Q_x}{G_{xz}^{c}(h+t)} \quad \text{and} \quad \gamma_{yz} = \frac{Q_y}{G_{yz}^{c}(h+t)} \tag{4}$$

The shear strains are also defined by the following expressions (u, v and w are the displacements in the x, y and z direction respectively).

$$\gamma_{xz} = \frac{\partial w}{\partial x} + \frac{\partial u}{\partial z} \quad \text{and} \quad \gamma_{yz} = \frac{\partial w}{\partial y} + \frac{\partial v}{\partial z} \tag{5}$$

As the mid-surface displacements u_O, v_O are zero (pure bending), the next expressions for the displacements in the x and y direction can be found after integration :

$$u = -z\left(\frac{\partial w}{\partial x} - \frac{Q_x}{G_{xz}^{c}(h+t)}\right) \quad \text{and} \quad v = -z\left(\frac{\partial w}{\partial y} - \frac{Q_y}{G_{yz}^{c}(h+t)}\right) \tag{6}$$

Setting

$$\theta_x = \left(\frac{\partial w}{\partial x} - \frac{Q_x}{G_{xz}^{c}(h+t)}\right) \quad \text{and} \quad \theta_y = \left(\frac{\partial w}{\partial y} - \frac{Q_y}{G_{yz}^{c}(h+t)}\right) \tag{7}$$

the displacements can be written as :

$$u = -z\theta_x \quad ; \quad v = -z\theta_y \quad ; \quad w = w(x,y) \tag{8}$$

With assumption 4 and equation (8) it can be seen that θ_x and θ_y are independent of z. From equations (4), (7) and (8) follows that a straight line normal to the middle plane of the core before deformation remains straight after deformation, but will rotate over angles θ_x and θ_y. This is an identical assumption as in the Mindlin theory.

2.2 EXPRESSIONS FOR MOMENTS AND SHEAR FORCES

For the upper face (z>0) the displacements in the x and the y direction are given by :

$$u^{+} = -\frac{h+t}{2}\theta_x \quad \text{and} \quad v^{+} = -\frac{h+t}{2}\theta_y \tag{9}$$

The expressions of u^- and v^- (lower face, $z < 0$) are the same, but with an opposite sign. The stresses in the faces can be written as follows (small displacements) :

$$\begin{Bmatrix} \sigma_x \\ \sigma_y \\ \tau_{xy} \end{Bmatrix} = \begin{bmatrix} c_{11} & c_{12} & c_{16} \\ c_{12} & c_{22} & c_{26} \\ c_{16} & c_{26} & c_{66} \end{bmatrix} \begin{Bmatrix} \varepsilon_x \\ \varepsilon_y \\ \gamma_{xy} \end{Bmatrix} = \begin{bmatrix} c_{11} & c_{12} & c_{16} \\ c_{12} & c_{22} & c_{26} \\ c_{16} & c_{26} & c_{66} \end{bmatrix} \begin{Bmatrix} \partial u / \partial x \\ \partial v / \partial y \\ \partial u / \partial y + \partial v / \partial x \end{Bmatrix} \tag{10}$$

The c_{ij} (i,j = 1, 2, 6) are the plane stress reduced stiffness components of the anisotropic material. By substituting (9) into (10) the stresses in the upper face are found. The stresses in the lower face can be calculated in a similar way.
The moments are given by :

$$\begin{aligned} M_x &= \frac{1}{2}(h+t)t\left(\sigma_x^+ - \sigma_x^-\right) \\ M_y &= \frac{1}{2}(h+t)t\left(\sigma_y^+ - \sigma_y^-\right) \\ M_{xy} &= \frac{1}{2}(h+t)t\left(\tau_{xy}^+ - \tau_{xy}^-\right) \end{aligned} \tag{11}$$

Substitution of (9) and (10) into (11) gives :

$$\{M\} = \begin{Bmatrix} M_x \\ M_y \\ M_{xy} \end{Bmatrix} = \begin{bmatrix} D_{11} & D_{12} & D_{16} \\ D_{12} & D_{22} & D_{26} \\ D_{16} & D_{26} & D_{66} \end{bmatrix} \begin{Bmatrix} -\partial\theta_x / \partial x \\ -\partial\theta_y / \partial y \\ -\left(\partial\theta_x / \partial y + \partial\theta_y / \partial x\right) \end{Bmatrix} = \left[D^f\right]\{\kappa\} \tag{12}$$

with

$$D_{ij} = \frac{c_{ij}(h+t)^2 t}{2} \qquad i, j = 1, 2, 6 \tag{13}$$

and $\{\kappa\}$: vector with plate curvatures

$\left[D^f\right]$: matrix with bending stiffnesses determined by the faces

The expressions for the shear forces can easily be found from equations (4) :

$$\{Q\} = \begin{Bmatrix} Q_x \\ Q_y \end{Bmatrix} = \begin{bmatrix} D_{44} & 0 \\ 0 & D_{55} \end{bmatrix} \begin{Bmatrix} \gamma_{xz} \\ \gamma_{yz} \end{Bmatrix} = \left[D^c\right]\{\gamma\} \tag{14}$$

with

$$D_{44} = G_{xz}^c(h+t) \quad \text{and} \quad D_{55} = G_{yz}^c(h+t) \tag{15}$$

and $\{\gamma\}$: vector with shear strains

$\left[D^c\right]$: matrix with shear stiffnesses determined by the core

Equations (13) and (15) are important to keep in mind as they give the expressions for the parameters of the sandwich model.

2.3 ENERGY FUNCTIONAL, DIFFERENTIAL EQUATIONS AND BOUNDARY CONDITIONS

The derivation of the energy functional is given in [5]. Assuming the time-displacement separable solution

$$w(x,y,t) = w(x,y)e^{i\omega t} \quad ; \quad \theta_x(x,y,t) = \theta_x(x,y)e^{i\omega t} \quad ; \quad \theta_y(x,y,t) = \theta_y(x,y)e^{i\omega t}$$

it can be written as follows (strain energy - kinetic energy) :

$$\pi = \frac{1}{2}\int_A \left({}^\tau\{M\}\{\kappa\} + {}^\tau\{Q\}\{\gamma\} \right) dA - \frac{\omega^2}{2}\int_A \left[pw^2 + I\left(\theta_x^2 + \theta_y^2\right) \right] dA \tag{16}$$

with $(p,I) = \int_{-h/2}^{+h/2}\left(1,z^2\right)\rho_c dz + 2\int_{h/2}^{t+h/2}\left(1,z^2\right)\rho_f dz$

in which ρ_f and ρ_c are the mass densities of the faces and the core separately. The expression for the energy functional is the same as in the Mindlin theory.

The corresponding governing differential equations and associated boundary conditions can be obtained by writing the forces and moments equilibrium. The expressions are the same as for a Mindlin plate, but the plate rigidity definitions are different. There are three differential equations, with three unknowns, namely θ_x, θ_y and w.

2.4 NUMERICAL SOLUTION FOR PLATE VIBRATIONS

For the identification of the material properties of rectangular thin plates using free vibrating data, Sol [1] used one finite element with higher order Lagrange polynomials as shape functions. The method proved to be useful in the calculation of resonant frequencies and associated mode shapes for rectangular thin plates.

Hua [4] extended this method to sandwich panels. The Lagrange shape functions $N_i(x,y)$ are to be used as trial functions. By the conventional displacement-based finite element method, the discretization procedure starts from the interpolation of the displacement field variables w, θ_x and θ_y by their nodal values (index i).

$$\begin{Bmatrix} w \\ \theta_x \\ \theta_y \end{Bmatrix} = \sum_{i=1}^{n} \begin{bmatrix} N_i & 0 & 0 \\ 0 & N_i & 0 \\ 0 & 0 & N_i \end{bmatrix} \begin{Bmatrix} w_i \\ \theta_{xi} \\ \theta_{xi} \end{Bmatrix} = \sum_{i=1}^{n} [N_i]\{d_i\} = [N]\{d\} \tag{17}$$

By minimizing the energy functional π (formula 16) with respect to the nodal values the eigenvalue equations are obtained.

$$([K] - \lambda[M])\{d\} = 0 \quad \text{with } \lambda = \omega^2 \tag{18}$$

3. The energy method

The prediction of the modal damping ratios will be carried out with the energy method as explained by Ungar and Kerwin[6]. They showed that the modal damping ratio can be written as a function of two energy quantities : the modal deformation energy P_j and the modal damping energy D_j. Expression (19) gives the exact relation.

$$\xi_j = \frac{D_j}{4\pi P_j} \tag{19}$$

The two energy quantities are defined as follows [3] :

$$P_j = \frac{1}{2} {}^{\tau}\{W\}_j [K'] \{W\}_j \tag{20}$$

$$D_j = 2\pi \frac{1}{2} {}^{\tau}\{W\}_j [K''] \{W\}_j \tag{21}$$

where {W} : real eigenvector (obtained from the dynamic analysis of the considered sandwich using the real part of the stiffness matrix)

[K'] : real part of the stiffness matrix

[K'']: imaginary part of the stiffness matrix

By substituting equations (20) and (21) in (19) a new expression for the damping ratio ξ associated with the j-th eigenmode can be found :

$$\xi_j = \frac{{}^{\tau}\{W\}_j [K'] \{W\}_j}{2\,{}^{\tau}\{W\}_j [K''] \{W\}_j} \tag{22}$$

In the near future a program will be written to carry out this calculation.

4. Testing of the sandwich model and discussion of the results

The developed sandwich model is verified by the calculation of the first eigen-frequencies f_j of sandwiches with polycarbonate faces and a PUR core. Two panels are studied : respectively with 1 and 3 mm thick faces. The total thickness of the panels is approximately 2 cm and the length and width are 30 cm.
The dynamic analysis of the considered testpanels with the developed model gives the needed eigenfrequencies and eigenmodes. The parameters of the model are the bending stiffnesses D_{ij}. These inputdata are calculated from the measured material properties of the faces and the core which are determined separately with the Resonalyser procedure (MNET for thin plates). The sandwich model also needs the dimensions of the panel and the mass density of the faces and the core.
The eigenfrequencies and the modal damping ratios are measured experimentally on the rectangular panels with free boundary conditions.
Table 1 and 2 give the results for the first three eigenmodes of each panel.

TABLE 1. Results obtained for sandwich 1
(1 mm thick faces)

	mode 1	mode 2	mode 3
f_j calculated (Hz)	201.0	314.4	374.2
fj measured (Hz)	216.2	337.9	427.0
δf_j (%)	7	7	12
ξ_j measured	0.00533	0.00472	0.00685

TABLE 2. Results obtained for sandwich 2
(3 mm thick faces)

	mode 1	mode 2	mode 3
f_j calculated (Hz)	213.4	335.1	384.1
fj measured (Hz)	253.9	395.2	484.9
δf_j (%)	16	15	21
ξ_j measured	0.00480	0.00474	0.00556

When the calculated (dynamic analysis with Mindlin model) and the experimental measured eigenfrequencies of the first three eigenmodes of the panel are compared, the following differences were noticed :
- the experimental eigenfrequencies are remarkably higher than the calculated ones ;
- the difference is greater for the sandwich with thicker faces ;
- the difference tends to be higher for the higher eigenmodes.

These discrepancies can have different reasons. First of all the testmaterial has to be fabricated correctly and the used material properties of the constituents of the sandwich need to be correct. More experiments are needed to see if the noted differences could be due to this. There is of course the possibility that the sandwich model should be adapted. Further investigation of the developed theory will be necessary.

5. Conclusion

A model for the dynamic analysis of a free vibrating sandwich panel was studied. It is similar as the Mindlin thick plate theory and gives a relatively simple description of the behaviour of a sandwich panel. The validity of the model is tested by calculating the eigenfrequencies and in the near future also the modal damping ratios (making use of the energy method) of two self made sandwich plates. It is seen that the measured and the calculated quantities do not match very well. Further research and correction of the model is necessary.

6. Acknowledgments

This work has been carried out with the financial support of the Flemish institute for the benefit of the scientific-technological research in the industry (IWT).

7. References

(1) Sol, H. (1986) *Identification of anisotropic plate rigidities using free vibration data*, Ph.D. dissertation, Department of Structural Engineering, Free University Brussels (VUB)

(2) Vantomme, J. (1992) *A parametric study of material damping in glass fiber reinforced thermosetting plastics*, Ph.D. dissertation, Department of Structural Engineering, Free University Brussels (VUB)

(3) De Visscher, J. (1995) *Identification of the complex stiffness matrix of orthotropic materials by a mixed numerical experimental method*, Ph.D. dissertation, Department of Structural Engineering, Free University Brussels (VUB)

(4) Hua, H. (1993) *Identification of Plate Rigidities of Anisotropic Rectangular Plates, Sandwich Panels and Circular Orthotropic Disks Using Vibration Data*, Ph.D. dissertation, Department of Structural Engineering, Free University Brussels (VUB)

(5) Plantema, F. J. (1966) *Sandwich Construction*, John Wiley & Sons, Inc., New York

(6) Ungar, E. E. and Kerwin, E. M. (1962) Loss factors of viscoelastic systems in terms of energy concepts, *The Journal of the Acoustical Society of America* **34**(7), 954-957

(7) Rikards, R. (1993) Finite Element Analysis of Vibration and Damping of Laminated Composites, *Composite Structures* **24**, 193-204

(8) Sol, H. and Hua, H. (1995) *Effect of the shear deformation on the first ten resonant frequencies of Poisson's plates*, Internal Report, Department of Structural Engineering, Free University Brussels (VUB)

NEW ELASTODYNAMIC SOLUTIONS TO FORCED VIBRATION PROBLEMS OF LAYERED RECTANGULAR VISCOELASTIC PLATES

S. KARCZMARZYK
Institute of Machine Design Fundamentals
Warsaw University of Technology
Narbutta 84, 02-524 Warszawa, POLAND

1. Introduction

Nowdays one can find in the literature a new approach to the investigations of layered plates which is called individual-layer analysis (Cho *et al.*, 1990). The approach has been developed to apply to the multilayered (laminated) thick plates problems. The characteristic feature of the approach is that the system of equations of the plate boundary problem consists of 11p differential equations, where p denotes the number of layers of the plate. Thus, to obtain numerical results for a three-layer plate one has to solve 33 partial differential equations. It is noted that the individual-layer approach is simplified since the equations of motion of an individual layer are derived the same way as in the classical theory of plates that is in terms of stress resultants and displacements. An alternative approach to the plate problems one can find in papers of Srinivas and Rao (1970), Levinson (1985) or Taylor and Nayfeh (1994). Solutions given in the papers are exact within the linear theory of elasticity. The authors use directly the linear elasticity equations of motion expressed in terms of stresses and displacements. To solve the equations they use the semi-inverse technique. Within the exact approach the cross-sectional warping and all the surface boundary conditions and the continuity of stresses and displacements conditions through the plate thickness have been taken into account. However, all the exact three-dimensional solutions have been derived only for the simply supported plates.

In this paper one can find an extension of the exact approach enabling the author to solve a few boundary problems of layered structures namely: plates, plate stripes (bands) and beams with various boundary (edge) conditions (Karczmarzyk 1992, 1995, 1996). All the author solutions have been derived by applying the semi-inverse technique and the method of superposition. The same method was applied to solve the boundary problem for the C-F-C-F layered plate where C denotes the clamped edge whereas F denotes the free edge. The new, closed-form, two-dimensional, exact within the linear elasticity theory solution has been proposed in section 4. All the solutions outlined here are limited to the structures composed of isotropic layers. Including anisotropy of the layers is possible. Application of the approach to the eigenvalue problem of the orthotropic plate stripe stiffened by parallel isotropic beams has been shown by Karczmarzyk (1995). Besides, one can show that the Levinson solution (1985) for the isotropic plate is the particular case of that one of Srinivas and Rao (1970).

A. Vautrin (ed.), Mechanics of Sandwich Structures, 223–230.

2. A Solution to the Linear Theory of Elasticity Equations of Motion

Most of the expressions given in this section, necessary to facilitate understanding of the contents, have been published by Karczmarzyk (1995,1996). The equations of motion of j-th layer can be written in the well known form

$$\left(\sigma_{kl,l}\right)_j = \rho_j\left(u_k\right)_j . \tag{1}$$

Displacements for the isotropic layer are assumed in the form:

$$(u_x)_j = -g_j(z)\frac{dX(x)}{dx}Y(y)e^{i\omega t}, \quad (u_x)_j = -g_j(z)\frac{dY(y)}{dy}X(x)e^{i\omega t},$$

$$(u_z)_j = f_j(z)X(x)Y(y)e^{i\omega t}, \tag{2}$$

It has been assumed, as in the cited papers of the author, that the functions of space variables x,y satisfy the equations

$$\frac{d^2X}{dx^2} = \theta_x^2, \qquad \frac{d^2Y}{dy^2} = \theta_y^2 . \tag{3}$$

It is noted that both θ_x^2 and θ_y^2 are unknown and can either be positive or negative for the purely elastic plate and the coefficients are in the complex domain for the viscoelastic plate. After taking into consideration the Hooke low and expressions (2), (3) one can transform equations (1) to the expanded form:

$$-\mu_j\frac{d^2g_j}{dz^2} - \left(\lambda_j + 2\mu_j\right)_j\left(\frac{X_{,3x}}{X_{,x}} + \frac{Y_{,yy}}{Y} + \frac{\rho_j\omega^2}{\lambda_j + 2\mu_j}\right)g_j + \left(\lambda_j + \mu_j\right)\frac{df_j}{dz} = 0,$$

$$-\mu_j\frac{d^2g_j}{dz^2} - \left(\lambda_j + 2\mu_j\right)_j\left(\frac{X_{,xx}}{X} + \frac{Y_{,3y}}{Y_{,y}} + \frac{\rho_j\omega^2}{\lambda_j + 2\mu_j}\right)g_j + \left(\lambda_j + \mu_j\right)\frac{df_j}{dz} = 0,$$

$$\left(\lambda_j + 2\mu_j\right)\frac{d^2f_j}{dz^2} + \mu_j\left(\frac{X_{,xx}}{X} + \frac{Y_{,yy}}{Y} + \frac{\rho_j\omega^2}{\mu_j}\right)f_j$$
$$-\left(\lambda_j + \mu_j\right)\left(\frac{X_{,xx}}{X} + \frac{Y_{,yy}}{Y}\right)\frac{dg_j}{dz} = 0 \tag{4}$$

In equations (4) the notation is used:

$$X_{,x} = \frac{dX}{dx}, \qquad X_{,xx} = \frac{d^2X}{dx^2}, \qquad X_{,3x} = \frac{d^3X}{dx^3}. \tag{5}$$

Number of equations in set (4) can be reduced provided that,

$$\frac{X_{,xx}}{X} = \frac{X_{,3x}}{X_{,x}} = \theta_x^2, \qquad \frac{Y_{,yy}}{Y} = \frac{Y_{,3y}}{Y_{,y}} = \theta_y^2. \tag{6}$$

Using (6) one obtains directly from (4) the set of two equations:

$$-\mu_j \frac{d^2 g_j}{dz^2} - (\lambda_j + 2\mu_j)_j \left(\widetilde{\beta}_2^2\right)_j g_j + (\lambda_j + \mu_j)\frac{df_j}{dz} = 0,$$

$$(\lambda_j + 2\mu_j)\frac{d^2 f_j}{dz^2} + \mu_j \left(\widetilde{\beta}_1^2\right)_j f_j - (\lambda_j + \mu_j)\left(\theta_x^2 + \theta_y^2\right)\frac{dg_j}{dz} = 0. \tag{7}$$

Factors $\left(\widetilde{\beta}_1^2\right)_j$, $\left(\widetilde{\beta}_2^2\right)_j$ are defined as follows:

$$\left(\widetilde{\beta}_1^2\right)_j = \theta_x^2 + \theta_y^2 + \rho_j\omega^2 \frac{1}{\mu_j}, \quad \left(\widetilde{\beta}_2^2\right)_j = \theta_x^2 + \theta_y^2 + \rho_j\omega^2 \frac{1}{\lambda_j + 2\mu_j}. \tag{8}$$

Functions g_j, f_j fulfiling (7) can be written in the form:

$$g_j(z) = g_{1j}\left(z, \widetilde{\beta}_{1j}\right) + g_{2j}\left(z, \widetilde{\beta}_{2j}\right), \quad f_j(z) = f_{1j}\left(z, \widetilde{\beta}_{1j}\right) + f_{2j}\left(z, \widetilde{\beta}_{2j}\right). \tag{9}$$

They are dependent each other i.e.,

$$f_{1j} = -\frac{\theta_x^2 + \theta_y^2}{\widetilde{\beta}_{1j}^2}\frac{dg_{1j}}{dz}, \qquad f_{2j} = -\frac{dg_{2j}}{dz}, \tag{10}$$

and can be obtained by using the following equations:

$$\frac{d^2 g_{1j}}{dz^2} = -\widetilde{\beta}_{1j}^2 g_{1j}, \qquad \frac{d^2 g_{2j}}{dz^2} = -\widetilde{\beta}_{2j}^2 g_{2j}. \tag{11}$$

It is noted that displacements defined by (2), (3), (6), (10), (11) satisfy the Saint-Venant continuity conditions.

3. Forced Vibrations of the Layered Cantilever Plate

It is well known that the cantilever and C-F-C-F plates can be considered as the corresponding plate stripes provided that the ratio length/width is greater than three. Therefore the solution given by Karczmarzyk (1995) enables accurate numerical results for the layered cantilever plates vibrating under transverse force uniformly distributed along the free edge parallel to the clamped one. The author has also proposed (1996) the extension of the two-dimensional solution including the base excitation. Displacement and stress vectors in both the solutions are composed of two ingredients

$$(\underline{u})_j = (\underline{u^I})_j + (\underline{u^{II}})_j \, , \qquad (\underline{\sigma})_j = (\underline{\sigma^I})_j + (\underline{\sigma^{II}})_j \, . \tag{12}$$

The ingredients with superscript I one obtains assuming in (8) that $\theta_y^2 = 0$ and $\theta_x^2 = \tilde{\alpha}^2$ is unknown and, for the purely elastic structure, less than zero. The ingredients with superscript II can be derived under assumption in (8) that $\theta_y^2 = 0$ and $\theta_x^2 = \tilde{\gamma}^{2*}$ is unknown and, for the purely elastic structure, greater than zero. It is noted that final form each of the solutions of the problems consists of one set of the linear algebraic nonhomogeneous equations and two sets of the linear algebraic homogeneous equations implying the two characteristic equations:

$$F_\alpha(\tilde{\alpha}) = 0 \, , \qquad F_\gamma(\tilde{\gamma}) = 0 \, . \tag{13}$$

The non-homogeneous set results from the boundary conditions at the plate edges. The homogeneous equations one obtains after equating to zero on the plate outer surfaces both the normal and the shear stresses as well as satisfying the continuity (stresses and displacements) conditions between adjoining layers - separately for each of the components I and II. The details have been published by Karczmarzyk (1995, 1996).

4. Forced Vibrations of Layered C-F-C-F Plate Under Uniform Pressure

The elastodynamic solution to the problem has been proposed as an extension of that one in section 3. The displacement and stress vectors consist now of three components:

$$(\underline{u})_j = (\underline{u^I})_j + (\underline{u^{II}})_j + (\underline{u^{III}})_j \, ,$$

$$(\underline{\sigma})_j = (\underline{\sigma^I})_j + (\underline{\sigma^{II}})_j + (\underline{\sigma^{III}})_j \, . \tag{14}$$

The components with superscripts I and II are the same as in the above case. The third component, denoted by superscript III, has been proposed in the form of the plane wave. The plate considered is composed of p isotropic layers and one of the layers is attached to a rigid supports as it is shown in Fig. 1.

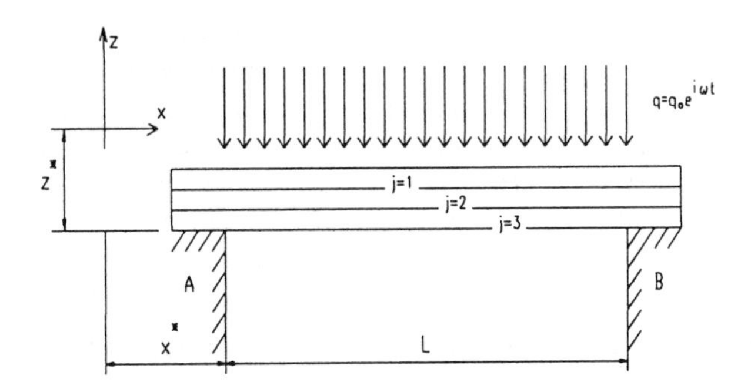

Figure 1. The C-F-C-F layered plate and location of the co-ordinate system

The plane wave problem is well known in the literature however it is outlined here to present some details of the solution expressed by (14). The plane wave displacements and the corresponding strains are defined as follows:

$$(u_z^{III})_j = f_j^{III}(z)\exp(i\omega t) = [\widetilde{Z}_{1j}\cos(\Omega_j z) + \widetilde{Z}_{2j}\sin(\Omega_j z)]\exp(i\omega t),$$

$$\Omega_j^2 = \frac{\rho_j \omega^2}{\lambda_j + 2\mu_j}, \qquad (u_x^{III})_j = (u_y^{III})_j \equiv 0, \tag{15}$$

$$(\varepsilon_{zz})_j = \frac{df_j^{III}}{dz}\exp(i\omega t), \quad (\varepsilon_{xx})_j = (\varepsilon_{yy})_j \equiv 0,$$

$$(\varepsilon_{xy})_j = (\varepsilon_{yz})_j = (\varepsilon_{zx})_j \equiv 0. \tag{16}$$

Symbols μ_j, λ_j denote the Lame parameters. The normal stresses corresponding to the displacements are as follows:

$$(\sigma_{zz}^{III})_j = (\lambda_j + 2\mu_j)\frac{df_j^{III}}{dz}\exp(i\omega t) = \frac{1-\nu_j}{\nu_j}\lambda_j \frac{df_j^{III}}{dz}\exp(i\omega t),$$

$$(\sigma_{xx}^{III})_j = (\sigma_{yy}^{III})_j = \lambda_j \frac{df_j}{dz}\exp(i\omega t) = \frac{\nu_j}{1-\nu_j}(\sigma_{zz}^{III})_j. \tag{17}$$

It is evident that for the plane wave the equation of motion is now of the form

$$\frac{d^2 f_j^{III}}{dz^2} + \frac{\rho_j \omega^2}{\lambda_j + 2\mu_j} f_j^{III} \equiv \frac{d^2 f_j^{III}}{dz^2} + \Omega_j^2 f_j^{III} = 0. \tag{18}$$

In order to calculate $\widetilde{Z}_{1j}$, $\widetilde{Z}_{2j}$ one has to use the boundary and continuity conditions for the normal stresses and deflections. For the structure composed of p layers the boundary conditions on the outer surfaces of the plate are of the form:

$$\left[\sigma_{zz}^{III}(\hat{z}_1)\right]_1 = 0, \qquad \left[\sigma_{zz}^{III}(\hat{z}_p)\right]_p = q. \tag{19}$$

q denotes the pressure sinusoidally varying in time and uniformly distributed on the surface of p-th layer - see Fig. 1. Continuity conditions of stresses and displacements between adjoining layers are now reduced to the form:

$$\left[\sigma_{zz}^{III}(\hat{z}_j)\right]_j = \left[\sigma_{zz}^{III}(\hat{z}_j)\right]_{j+1}, \qquad \left[u_z^{III}(\hat{z}_j)\right]_j = \left[u_z^{III}(\hat{z}_j)\right]_{j+1}. \tag{20}$$

It is noticed that for the structure of non-rectangular cross-section the condition of continuity of the normal stresses has to be replaced by the appropriate condition of continuity of the normal forces - see Karczmarzyk (1995). The boundary and continuity conditions can be written as the matrix equation

$$A_3 Z = B_3. \tag{21}$$

When the plate is composed of two layers the matrix A_3 and vector B_3 are follows:

$$A_3 = \begin{bmatrix} a_{11}^{III}\, a_{12}^{III} & - - - - \\ a_{21}^{III}\, a_{22}^{III}\, a_{23}^{III}\, a_{24}^{III} \\ a_{31}^{III}\, a_{32}^{III}\, a_{33}^{III}\, a_{34}^{III} \\ - - - - & a_{43}^{III}\, a_{44}^{III} \end{bmatrix}, \qquad B_3 = \begin{bmatrix} 0 \\ 0 \\ 0 \\ q \end{bmatrix}. \tag{22}$$

The minus signs denote that some elements of the matrix A_3 are equivalent to zero. After solving equations (21) that is calculating the constants $\widetilde{Z}_{1j}$, $\widetilde{Z}_{2j}$ for component III one can transform the end conditions of the plate to a nonhomogeneous set of the linear algebraic equations. Assuming that the origin of the co-ordinate system is placed within the plane equally distant from the supports the boundary conditions at the ends of the C-F-C-F plate for the solution composed of the three ingredients can be written:

$$(u_z(-L/2,z_j^*,t))_j = (u_z(L/2,z_j^*,t))_j = 0,$$

$$(\phi_{zx}(-L/2,z_j^*,t))_j = (\phi_{zx}(L/2,z_j^*,t))_j = 0, \tag{23}$$

The second equation means that the slopes at the ends are equal to zero. It is noticed that subscript j in (23) denotes the layer directly attached to the rigid supports of the plate. In the case shown in Fig. 1 the co-ordinate $z_j^* \equiv z_3^* = z^*$. First of the conditions can be rewritten in the form:

$$(u_z^I(L/2,z_j^*,t))_j + (u_z^{II}(L/2,z_j^*,t))_j = -(u_z^{III}(L/2,z_j^*,t))_j, \tag{24}$$

Since the component denoted by superscript III has been defined after solving (21) one can see that (23) is the non-homogeneus set of the linear algebraic equations. In order to obtain the corresponding solution for the cantilever plate one must follows the way presented in this section and in the papers by Karczmarzyk (1995, 1996). The solution of the boundary problem for the C-F-C-F structure is simpler than that one for the clamped-free (cantilever) structure.

5. Numerical Results and Conclusions

To assess the solution proposed here some calculations were made for the C-F-C-F and cantilever two-layer plates. In Table 1 one can see the maximum deflections calculated for the plates of stiffness-comparable layers. In Table 2 are listed the maximum deflections for the purely elastic and viscoelastic cantilever two-layer plate composed of a stiff and a compliant layer. To facilitate comparisons the numerical results have rather been obtained for the layered beams i.e. for the case when the Poisson ratios are equal to zero. The input data to obtain results in Table 1 are as follows. The Young modulus, mass of the cubic metre (mass density) and thickness for the first layer are equal to 200000 MPa, 7860 kg, 5 mm, respectively. The parameters for the second layer are equal to 16000 Mpa, 1750 kg, 20 mm, respectively. The maximum deflections at the midspan of the clamped-clamped plate are denoted by u_{CC} while the deflections at the free edge of the cantilever plate of the same length are denoted by u_{CF}.

TABLE 1. Numerical results for the C-F-C-F and cantilever two-layer plate for a range of L; amplitude of the exciting pressure $q_0 = 0.2 \cdot 10^6 \ N/m^2$, $\omega = 500$ rad/s.

L [m]	0.15	0.20	0.25	0.30	0.35	0.40
$u_{CF}(L)$ [mm]	0.2411	0.8068	2.1462	5.2171	13.3184	52.8045
$u_{CC}(L/2)$ [mm]	0.0052	0.0165	0.0403	0.0839	0.1562	0.2686
u_{CF}/u_{CC}	46.286	48.954	53.242	62.212	85.276	196.584

It is seen in Table 1 that for the short plate which eigenfrequency is much higher than the frequency of the excitation the ratio is close to 46. Within the strength of materials the $u_{CF}(L)$, $u_{CC}(L/2)$ and the ratio for the homogeneous beam are as follows:

$$u_{CF}(L) = \frac{1}{8}\frac{q_0 L^4}{EJ}, \quad u_{CC}(L/2) = \frac{1}{384}\frac{q_0 L^4}{EJ}, \quad u_{CF}/u_{CC} = 48. \qquad (25)$$

In Table 2 are the maximum deflections of the two-layer plate composed of the stiff and compliant layers. The Young modulus, mass density and thickness for the first layer are equal to 200000 MPa, 7860 kg, 2 mm, respectively. The parameters for the second layer are equal to 200 Mpa, 1400 kg, 18 mm, respectively. The calculations were made for the purely elastic structure and the corresponding viscoelastic one. It is assumed that the loss factor η_2 of the second (compliant), viscoelastic, layer depends on frequency.

TABLE 2. Numerical results for the cantilever two-layer plate for a range of ω; the pressure amplitude $q_0 = 0.1 \cdot 10^4 \;\; N/m^2$, L=0.35 m, $\eta_2 = 0.0005 \cdot \omega + 0.025$.

ω [rad/s]	70	80	90	100	110	120
$(u_{CF})_{p-e}$ [mm]	5.447	6.968	10.176	20.872	-133.47	-14.735
$(u_{CF})_{v-e}$ [mm]	5.418	6.899	9.936	18.754	-21.605	-13.803
ξ [%]	6.938	7.675	2.416	11.294	517.764	6.752

As expected the deflections of the viscoelastic structure are lower than the purely elastic one. The percentage difference ξ increases when the exciting frequency is close to the resonant frequency. It is finally concluded that the two-dimensional solution has been derived, exactly within the linear theory of elasticity, in the closed-form i.e. without expanding any functions in the Fourier series. Its final form consists of four sets of the linear algebraic equations while two of the sets are homogeneous and imply the characteristic equations. These are the evident features of the solution novelty.

6. References

Cho, K.N., Stritz, A.G. and Bert, C.W. (1990) Bending analysis of thick bimodular laminates by higher-order individual-layer theory, *Composite Structures* **15**, 1-24.

Karczmarzyk, S. (1996) An exact elastodynamic solution to vibration problems of a composite structure in the plane stress state, *Journal of Sound and Vibration* **196** , 85-96.

Karczmarzyk. S. (1995) New elastodynamic solutions to forced and free vibrations problems of plane viscoelastic composite structures, *Mechanique Industrielle et Materiaux* **48**, 107-110.

Karczmarzyk, S. (1992) A new linear elastodynamic solution to boundary eigenvalue problem of flexural vibration of viscoelastic layered and homogeneous bands, *Journal of Theoretical and Applied Mechanics* **30**, 663-682.

Levinson, M. (1985) Free vibrations of a simply supported rectangular plate: an exact elasticity solution, *Journal of Sound and Vibration* **98**, 289-298.

Srinivas, S. and Rao, A.K. (1970) Bending, vibration and buckling of simply supported thick orthotropic rectangular plates and laminates, *International Journal of Solids and Structures* **6**, 1463-1481.

Taylor, T.W. and Nayfeh, A.H. (1994) Natural frequencies of thick, layered composite plates, *CompositesEngineering* **4**, 1011-1021.

THEORY FOR SMALL MASS IMPACT ON SANDWICH PANELS

Robin Olsson
The Aeronautical Research Institute of Sweden
P.O. Box 11021, S-161 11 Bromma, Sweden

1. Introduction

This paper presents a theory for small mass impact on sandwich panels. Small mass impact response of plates is governed by wave propagation and is observed when the impactor/plate mass ratio is significantly smaller than unity. The present method is limited to impact times where through-the-thickness waves may be neglected so that the response is governed by flexural and shear plate waves. In contrast, large mass impact response is governed by the corresponding static deflection mode and is observed when the impactor/plate mass ratio is larger than unity. The different response types and their dependency on impactor/plate mass ratio for monolithic plates were discussed by Olsson (1993). In the present paper we discuss the specific conditions for small mass impact response of sandwich panels.

A number of instrumented impact experiments on sandwich panels were reviewed by Olsson & McManus (1996) but a study by Davies et al. (1995) is the only one which at least partially satisfies small mass impact conditions.

A model for small mass impact on sandwich panels was presented by Koller (1986) who neglected the through-the-thickness compliance of the core in the contact analysis. However, experimental evidence and a recent model by Olsson & McManus (1996) has shown the major influence of face sheet deflection and inelastic core crushing in indentation cases of concern in sandwich structures. A model for small mass impact on orthotropic Kirchhoff plates was given by Olsson (1992). A model for small mass impact on monolithic, transversely isotropic Mindlin plates with a Hertzian contact behaviour was given by Mittal (1987). However, sandwich panels have a much lower stiffness in transverse shear and indentation and a different, more or less linear, load-indentation relation.

In the present analysis the indentation model by Olsson & McManus (1996) is used to modify Mittal's (1987) solution to apply for specially orthotropic sandwich panels with a linear load-indentation relation. The

A. Vautrin (ed.), Mechanics of Sandwich Structures, 231–238.

response history is obtained by step wise solution of an integral equation. Closed form solutions are given for panels having infinite shear stiffness.

2. Indentation behaviour of sandwich panels

As shown by Hill et al. (1989) the load-indentation relation between an elastic half-sphere and a solid can be expressed by the Meyer power law

$$F = k_\alpha \alpha^q \tag{1}$$

where k_α is the contact stiffness, α the indentation and where q ranges from $q = 3/2$ for an elastic contact to $q = 1$ for a fully plastic contact. The applicability of Eq. (1) to impacts is shown by Andersson & Nilsson (1995).

The apparent indentation α in sandwich panels is composed of Hertzian elastic indentation α_H of the face sheet and local face sheet deflection α_F:

$$\alpha = \alpha_H + \alpha_F \tag{2}$$

The Hertzian elastic indentation as given by Olsson (1992) is

$$\alpha_H = \left(F/k_H\right)^{2/3} \quad \text{where} \quad k_H = \tfrac{4}{3} Q_H \sqrt{R} \quad \text{and} \quad 1/Q_H = 1/Q_{zf} + 1/Q_{zi} \tag{3}$$

where the face sheet and indentor out-of-plane stiffnesses Q_{zf} and Q_{zi} are

$$Q_z = E_z / \left(1 - \nu_{rz}\nu_{zr}\right) \tag{4}$$

Solutions for isotropic plates may be extended to specially orthotropic plates by use of an effective plate stiffness D^* derived by Olsson (1992)

$$\begin{aligned}
D^* &= \sqrt{D_{11}D_{22}}\left[\tfrac{\pi}{2}\Big/K\left(\sqrt{(1-A)/2}\right)\right]^2 && 0 \le A \le 1 \\
D^* &= \sqrt{D_{11}D_{22}}\left[\tfrac{\pi}{2}\Big/K\left(\sqrt{(A-1)/(A+1)}\right)\right]^2 (A+1)/2 && 1 < A \\
&\text{where} \quad A = \left(D_{12} + 2D_{66}\right)\Big/\sqrt{D_{11}D_{22}} && K(s) = \int_0^{\pi/2} \left(1 - s^2 \sin^2\theta\right)^{-1/2} d\theta
\end{aligned} \tag{5}$$

Here $K(s)$ is the complete elliptic integral of the first kind.

For small deflections and an elastic core the local face sheet deflection under a concentrated load is given by Olsson & McManus (1996) as

$$\alpha_F = F/k_F \qquad \text{where} \quad k_F = 8\sqrt{k_c D_f^*} = 8\sqrt{D_f^* Q_{zc}/h_c^*} \tag{6}$$

Here D_f^* is the face sheet plate stiffness, k_c is the core elastic foundation stiffness and Q_{zc} is the core out-of-plane stiffness given by Eq. (4). The effective thickness h_c^* of elastic cores given by Olsson & McManus (1996) is

$$h_c^* = \min\{h_c, h_c^{3D}\} \qquad \text{where } h_c^{3D} = h_f \tfrac{64}{27}\left[\tfrac{1}{3}(1-\nu_{rzc})(1-\nu_{zrc})Q_f^* Q_{zc}/G_{rzc}^2\right]^{1/3} \quad (7)$$

where $Q_f^* = 12 D_f^*/h_f^3$ for faces of thickness h_f and where the relation $(1-\nu_c)^2$ for isotropic cores has been replaced by $(1-\nu_{rzc})(1-\nu_{zrc})$ for transversely isotropic cores. In most cases the effective core thickness is given by h_c^{3D}.

In practice crushing of core cells occurs at low loads and the resulting behaviour is described by the model by Olsson & McManus (1996) which shows that softening due to core crushing and stiffening due to face membrane effects result in a more or less linear load-deflection relation

$$\alpha_F = (F/k_F)^{1/q} \quad \text{where} \quad q \approx 1 \;\; \text{and} \;\; k_F \le 8\sqrt{k_c D_f^*} \tag{8}$$

It is easily verified that the indentation compliance is dominated by local face sheet deflection for loads of concern in impact on sandwich panels.

The core crush radius a given by Olsson & McManus (1996) is

$$a = \bar{a}\sqrt{F/(\pi p_0)} \tag{9}$$

where $\bar{a}$, which is close to unity, was obtained by matching face boundary conditions between the crushed inner region and the elastic outer region.

3. Validity of small mass impact response models

The following expression by Olsson (1992) gives the position of the leading edge of the dominating transient flexural wave in orthotropic plates

$$r_{11}(\theta) = \sqrt{\pi}\left[2D_r(\theta)/m\right]^{1/4}\left(D^*/\sqrt{D_{11}D_{22}}\right)^{1/8}\sqrt{t} \tag{10}$$

Small mass impact response is satisfied as long as no boundary is reached.

For the dynamic impact problem we introduce dimensionless variables

$$\bar{F}(\bar{t}) = \bar{\alpha}^q(\bar{t}) \qquad \text{where} \quad \bar{t} = t/T \;,\; \bar{\alpha} = \alpha/(TV_0) \quad \text{such that } \dot{\bar{\alpha}}(0) = 1 \tag{11}$$

The time constant T for an impactor of mass M and initial velocity V_0 is

$$T = \left[M/\left(k_\alpha\sqrt{V_0}\right)\right]^{2/5} \quad \text{for } q = 3/2 \qquad \text{and} \quad T = \sqrt{M/k_\alpha} \quad \text{for } q = 1 \tag{12}$$

where q=3/2 corresponds to a Hertzian contact and q=1 to a plastic contact.

A minimum impact time is obtained by considering an immovable plate

$$t_{imp}/T \ge \pi \tag{13}$$

The minimum is exact for a linear contact law while the minimum for a Hertzian impact is 2% larger. The impact time on a movable plate is longer due to the additional compliance caused by deflection of the plate.

Koller (1986) relies on a serial solution which only is convergent for

$$2\varepsilon t/\pi < 1 \quad \text{where} \quad \varepsilon = \left(h_c + h_f\right)^2 G_{rzc} \Big/ \left(2h_c\sqrt{2mD_f^*}\right) > h_c G_{rzc} \Big/ \sqrt{8mD_f^*} \tag{14}$$

where m is the panel weight per unit area. Combination of Eqs (6), (7) and (11) to (14) shows that the solution by Koller (1986) is divergent for

$$3^{-1/12}(\bar{t}/\pi)\sqrt{2M/(mh_f^2)}\left[(1-\nu_{rzc})(1-\nu_{zrc})G_{rzc}Q_{zc}/Q_f^{*2}\right]^{1/3} h_c/h_f > 1 \tag{15}$$

To calculate the peak load it is necessary to consider cases where $\bar{t} \approx \pi/2$. Koller's (1986) model was verified by dropping 4 g steel balls on panels with rubber cores. It is easily verified that Koller's (1986) solution diverges for core materials and masses of concern when considering impact damage.

4. Response model

A model of more practical use than Koller's (1986) solution may be based on the solution by Mittal (1987) who derived the following dimensionless integral equation for indentation of a rigid mass on an infinite shear deformable plate

$$\bar{\alpha} = \bar{t} - \int_{\bar{t}_0}^{\bar{t}} \bar{F}(\bar{\tau})\left[(\bar{t}-\bar{\tau}) + \lambda \tfrac{2}{\pi}\left\{\arctan\left[(\bar{t}-\bar{\tau})/\beta\right] + 2\beta/(\bar{t}-\bar{\tau})\right\}\right] d\bar{\tau} \tag{16}$$

The time $\bar{t}_0$ is a small time increment to account for the action of surface loads over a finite area, which is necessary to maintain finite stresses. However, rather than using the arbitrary definition of Mittal (1987) we use the definition by Olsson (1992), originally derived by Sneddon (1945) when analysing forced motion of large plates under action of a uniform load within a small radius c. For sandwich panels we assume that surface loads are applied within the radius of the crushed core, a, as given by Eq (9).

$$\begin{aligned} &\text{Monolithic plates } (q = 3/2): && \bar{t}_0 = c^2\sqrt{m/D^*}\,/T = \tfrac{1}{4}\bar{\alpha}RV_0\sqrt{m/D^*} \\ &\text{Sandwich panels } (q = 1): && \bar{t}_0 = a^2\sqrt{m/D^*}\,/T \approx \tfrac{1}{4\pi}\bar{\alpha}\,k_\alpha V_0\sqrt{m/D^*}\,/p_0 \end{aligned} \tag{17}$$

where D^* is the panel plate stiffness, p_0 is the core crush stress and R is the impactor tip radius. For simplicity it was assumed that $\bar{a}=1$. The dimensionless inelasticity parameter λ and shear parameter β are given by

$$\lambda = M\Big/\left(8T\sqrt{mD^*}\right) \qquad \beta = \sqrt{mD^*}\Big/\left(KG_{rz}hT\right) \quad \text{Sandwich: } KG_{rz}h = G_{rzc}h_c \tag{18}$$

where K is the shear factor, G_{rz} and G_{rzc} are out-of-plane shear moduli of the panel and core and h and h_c are the corresponding thicknesses. The shear stiffness for sandwich panels is defined in agreement with the common assumptions of thin face sandwich theory.

Following Olsson (1992) the plate centre deflection given by Mittal (1987) is redefined on the following dimensionless form

$$\overline{w}_0 = w_0 8\sqrt{mD^*}\Big/\left(MV_0\right) = \tfrac{2}{\pi}\int_{\bar{t}_0}^{\bar{t}} \overline{F}(\overline{\tau})\left\{\arctan\left[(\bar{t}-\overline{\tau})/\beta\right] + 2\beta/(\bar{t}-\overline{\tau})\right\}d\overline{\tau}$$

$$\text{Kirchhoff plate, } \beta = 0: \qquad \overline{w}_0 = \bar{I} = \int_0^{\bar{t}} \overline{F}(\overline{\tau})d\overline{\tau} \tag{19}$$

Integral expressions for the flexural moment in Mindlin plates are given by Mittal (1987) and on closed form for Kirchhoff plates by Olsson (1992).

Equation (16) is now combined with Eq. (11), discretized and solved stepwise by use of the piecewise integration method by Timoshenko (1913)

$$\overline{\alpha}_{N+1} = \bar{t}_{N+1} - \sum_{i=0}^{N} \overline{F}(\overline{\tau}_i)\int_{\bar{t}_{i-1}}^{\bar{t}_i}\left[(\bar{t}-\overline{\tau}) + \lambda\tfrac{2}{\pi}\left\{\arctan\left[(\bar{t}-\overline{\tau})/\beta\right] + 2\beta/(\bar{t}-\overline{\tau})\right\}\right]d\overline{\tau} \tag{20}$$

The indentation can be determined stepwise from the resulting expression:

$$\begin{aligned}\overline{\alpha}_0 = 0 \qquad \overline{\alpha}_{N+1} &= (N+1)\Delta\bar{t} - \tfrac{1}{2}\Delta\bar{t}^2\sum_{i=0}^{N}\overline{\alpha}_i^{\,q}(2N-2i+1) \\ &- \lambda\tfrac{2}{\pi}\sum_{i=0}^{N}\overline{\alpha}_i^{\,q}\Bigg[\Delta\bar{t}(N-i+1)\arctan\left\{\tfrac{N-i+1}{\beta}\right\} - \Delta\bar{t}(N-i)\arctan\left\{\tfrac{N-i}{\beta}\right\} \\ &- \tfrac{1}{2}\beta\ln\left\{\frac{\beta+(N-i+1)^2\Delta\bar{t}^2}{\beta+(N-i)^2\Delta\bar{t}^2}\right\} + 2\beta\ln\left\{\frac{(N-i+1)\Delta\bar{t}+\bar{t}_0}{(N-i)\Delta\bar{t}+\bar{t}_0}\right\}\Bigg]\end{aligned} \tag{21}$$

The deflection is obtained by dividing the last sum in Eq. (21) with λ.

For Kirchhoff plates, β=0, we may write Eq.(16) as a differential equation

$$\ddot{\overline{\alpha}} + \lambda q\overline{\alpha}^{q-1}\dot{\overline{\alpha}} + \overline{\alpha}^q = 0 \qquad \overline{\alpha}(0) = 0 \qquad \dot{\overline{\alpha}}(0) = 1 \tag{22}$$

For a linear contact law, q =1, solutions were given by Olsson (1993)

$$\begin{aligned}\lambda < 2 \quad & \overline{F} = \overline{\alpha} = \sin\left(\bar{t}\sqrt{1-\lambda^2/4}\right)e^{-\lambda\bar{t}/2}\Big/\sqrt{1-\lambda^2/4} \\ \lambda = 2 \quad & \overline{F} = \overline{\alpha} = \bar{t}e^{-\bar{t}} \\ \lambda > 2 \quad & \overline{F} = \overline{\alpha} = \sinh\left(\bar{t}\sqrt{\lambda^2/4-1}\right)e^{-\lambda\bar{t}/2}\Big/\sqrt{\lambda^2/4-1}\end{aligned} \tag{23}$$

5. Application and discussion

As an application example we consider experiments by Davies et al. (1995) with a 0.8 m x 0.8 m x 0.025 m sandwich panel with quasi-isotropic glass/epoxy faces on a PVC core. All data are from Davies et al. (1995).

Figure 1 shows the indentation behaviour predicted by the model of Olsson & McManus (1996) together with results from indentation tests on smaller beam specimens. Solutions are given for a fully elastic core and for a yielding core where the face is modelled using either a membrane model or a nonlinear (NL) large deflection plate model with full bending stiffness of the face. The behaviour of a face sheet with delaminations is bounded between the latter solutions while fibre failure will result in additional softening. Fibre fracture was observed in the plotted beam experiments but not in the plate impact experiments. The predicted linearized indentation stiffness was estimated to 1.2 kN/mm.

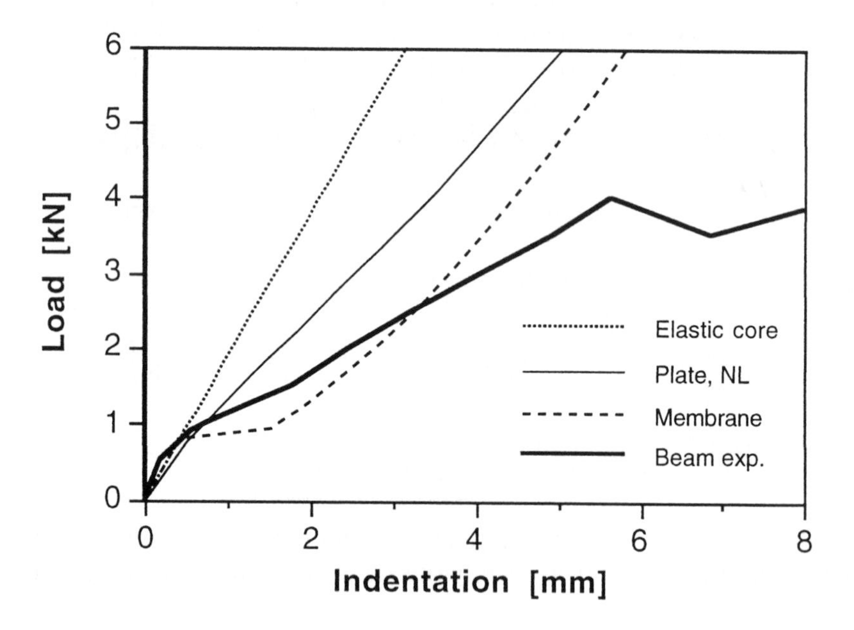

Figure 1. Predicted and observed indentation behaviour

Figure 2 shows the predicted and observed indentation histories of the 6.4 kg panel during impact at V_0=7.67 m/s by a 2.5 kg impactor with 12.5 mm tip radius. Good agreement with the experimental results is obtained within a few seconds of calculation on a personal computer, while the three-dimensional finite element analysis presented by Davies et al. (1995) required 30 h on an IBM RS6000 work station.

According to Eq. (10) the dominating flexural wave reaches the edges after 1.2 ms and after this predictions become increasingly inaccurate.

In the unloading phase the experimentally observed contact load and deflection are lower than predicted, mainly due to the neglected effect of permanent indentation on the unloading behaviour. This effect was addressed for monolithic plates by Chattopadhyay & Saxena (1991). The unloading phase is however of limited interest for damage prediction. If shear effects are neglected the peak force is obtained by combination of Eqs (1), (11) and (23) which in the present case results in a 50% overestimation of the impact force.

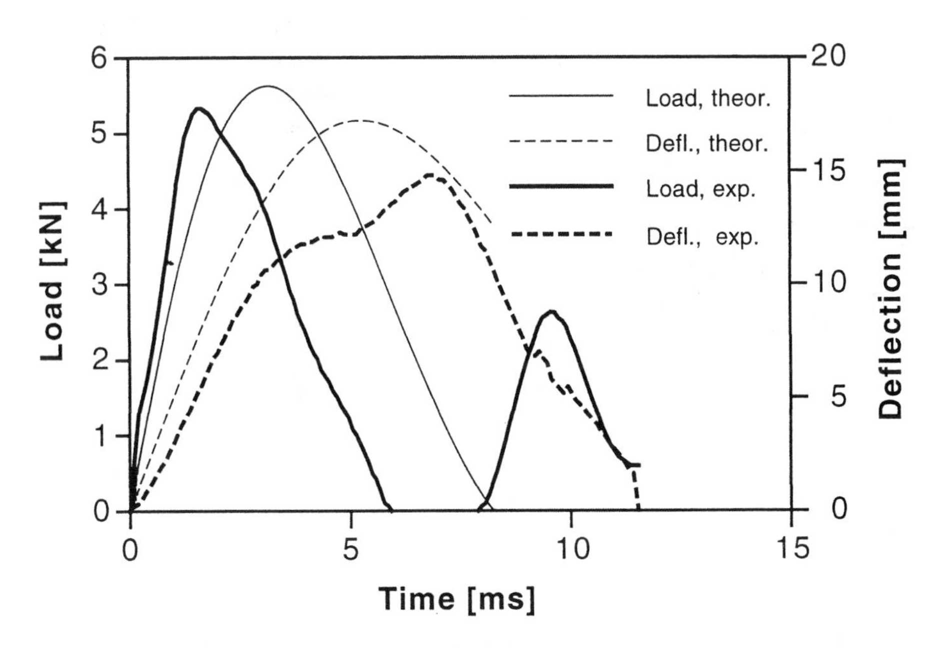

Figure 2. Predicted and observed impact histories

Face sheet damage influences the indentation behaviour, but has a small influence on the global response of the panel since damage in sandwich panels is more localised than in monolithic laminates. Thus the predicted peak load may be used to predict damage according to the methods suggested by Olsson (1996). The predicted core crush radius of 30 mm, as given by Eq. (9), provides an upper bound for the delamination radius which was reported to 15 mm by Davies, Choqueuse & Pichon (1994). Note that the crushed core material may cause premature face sheet buckling.

Finally we observe that sandwich indentation with significant face sheet membrane effects obeys an apparent Hertzian power law rather than a linear contact law, which can be modelled after modification of Eqs (8) and (17b).

6. Conclusions

A previous solution by Koller (1986) was found inapplicable for most cases expected to cause damage in sandwich panels. It was shown that a solution for shear deformable plates by Mittal (1987) is applicable after appropriate modifications to account for the different indentation behaviour and out-of-plane shear stiffness of sandwich panels. The method provides speedy results comparable to those of time consuming 3D dynamic FE-analyses.

7. Acknowledgements

This research was jointly funded by The Aeronautical Research Institute of Sweden and The Swedish Defence Materiel Administration.

8. References

Andersson, M. & Nilsson, F. (1995) A perturbation method used for static contact and low velocity impact, *Int. J. Impact Engng*, **16** (5/6) 759-775.

Chattopadhyay, S. & Saxena, R. (1991) Combined effects of shear deformation and permanent indentation on the impact response of elastic plates, *Int. J. Solids & Struct.*, **27** (13) 1739-1745.

Davies, P., Choqueuse, D. & Pichon, A. (1994) Influence of the foam core on composite sandwich static and impact response, In *Composites Testing and Standardisation,* ECCM-CTS2, Woodhead Publishing Ltd, Abington Hall, England, 513-522.

Davies, P., et al. (1995) Impact behaviour of composite sandwich panels, In *Impact and Dynamic Fracture of Polymers and Composites,* ESIS 19, MEP, London, 341-358.

Hill, R., Storåkers, B. & Zdunek, A.B. (1989) A study on the Brinell hardness test, *Proc. Royal Soc. Lond.*, **A423** 301-330.

Koller, M.G. (1986) Elastic impact of spheres on sandwich plates, *J. Appl. Math. & Phys. (ZAMP)*, **37**, 256-269.

Mittal, R.K. (1987) A simplified analysis of the effect of transverse shear on the response of elastic plates to impact loading, *Int. J. Solids & Struct.*, **23** (8) 1587-1596.

Olsson, R. (1992) Impact response of orthotropic composite laminates predicted from a one-parameter differential equation, *AIAA J.*, **30** (6) 1587-1596.

Olsson, R. (1993) Impact response of composite laminates – a guide to closed form solutions, *FFA TN 1992–33*, The Aeronautical Research Institute of Sweden, Bromma.

Olsson, R. (1996), Prediction of impact damage in sandwich panels, *Proc. Third Int. Conf. on Sandwich Construction*, EMAS, Solihull, England, 1996, 659-668.

Olsson, R. & McManus, H.L. (1996) Improved theory for contact indentation of sandwich panels, *AIAA J.*, **34** (6) 1238-1244.

Sneddon, I.N. (1945) The symmetrical vibrations of a thin elastic plate, *Proc. Cambridge Phil.Soc.*, **41** (1), 1945, 27–43.

Timoshenko, S.P. (1913) Zur Frage nach den Wirkung eines Stoßes auf einen Balken, *Z. Math. Phys.*, **62**, 198-209.

EFFECT OF BALLISTIC IMPACT ON THE COMPRESSION BEHAVIOUR OF CARBON/GLASS HYBRID COMPOSITE SANDWICH PANELS FOR MILITARY BRIDGING STRUCTURES.

N. PERCHE, B. HILAIRE, J.C. MAILE, J.M. SCANZI
Délégation Générale pour l'Armement.
Direction des Centres d'Expertise et d'Essais
Centre de Recherches et d'Etudes d'Arcueil.
16 bis avenue Prieur de la Côte d'Or
94 114 Arcueil CEDEX (F)

1. Introduction

Sandwich materials are evaluated as robust and low weight alternatives for military bridging structures. In service damage tolerance is obviously a key parameter for material selection in this kind of application. However, despite the amount of publications available in literature dealing with low velocity impact on sandwich structure, there is very little done on the study of ballistic impact.

The main objectives of this study are firstly to understand how sandwich structures behave when subjected to ballistic impact and secondly to assess the influence of the generated damages on sandwich mechanical properties. In the first step, presented hereafter, compression after ballistic impact has been carried out on particular structures. Damage analysis and post impact behaviour clarified some key points enhancing our understanding of the phenomena.

2. Material Description

The skins of the sandwich panels are made of hybrid glass/carbon woven plies composite. The plies are 0.367 mm thick and stacked in the panel length direction. The core material is a 10 mm thick Divinycell H80 PVC foam. Two different stacking sequences have been evaluated, the sandwich structures are denominated as H1 and H2, as shown in Figure 1.

A mechanical characterisation of the sandwich constituents has been carried out. The results, in terms of moduli and Poisson's ratios are given is table 1 [1][2].

A. Vautrin (ed.), Mechanics of Sandwich Structures, 239–246.

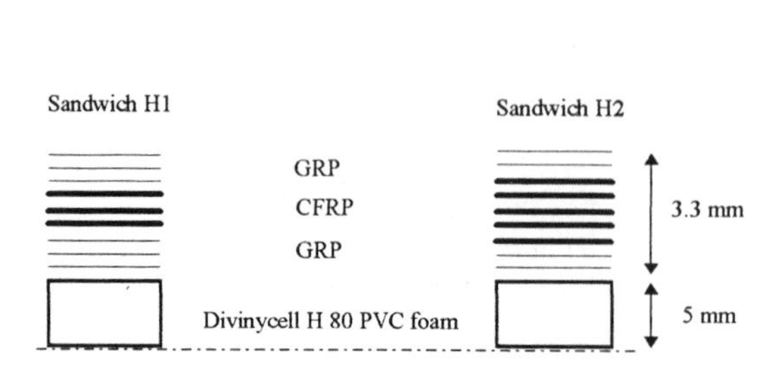

Figure 1. Stacking sequences description.

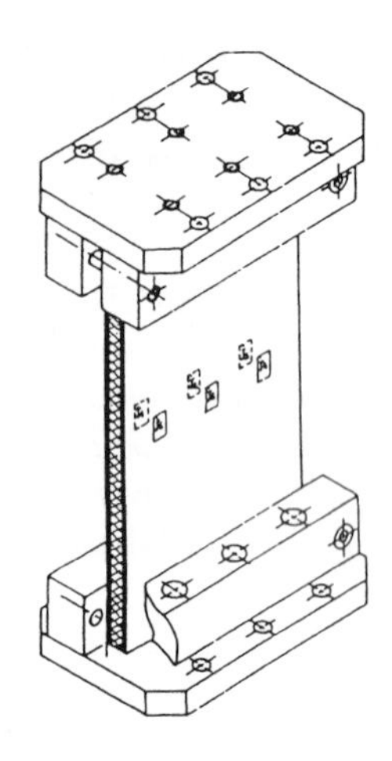

Figure 2. Compression specimen.

TABLE 1. Mechanical characteristics of the sandwich constituents.

	E_L (MPa)	E_T (MPa)	G_{LT} (MPa)	$G_{LT'}$ (MPa)	$G_{TT'}$ (MPa)	ν_{LT}
GRP	21000	21000	3300	2300	2300	0.14
CFRP	53200	53200	3500	2500	2500	0.06
PVC Foam	120	120	35	35	35	0.32

3. Compression Behaviour of Plain Panels.

3.1. EXPERIMENTAL SET UP

Fully built-in boundary conditions have been chosen for in plane compression testing of the panels. A drawing of the mounting is given in figure 2. An out of plane compression pressure of 1 MPa is applied to the specimen by means of two screws in each block. This pressure was sufficient to avoid slipping of the sample during testing. Furthermore, the end blocks are vertically screwed to the compression machine in order to prevent any transverse displacements. The panel length between end blocks was adjusted to 250 mm, their width is of 175 mm and six strain gauges are stuck to the sandwich skins (figure 2). Two sandwich panels of each type have been tested.

3.2. RESULTS AND DISCUSSION

Figure 3 shows typical response of the two centred gauges on opposite skins. Global buckling of the structure in the first mode is observed for all the samples. As expected from its higher carbon content, the compression modulus of H2 type panels is the higher.

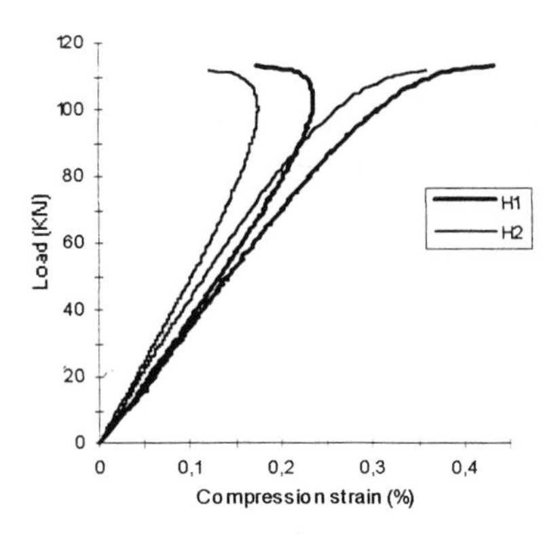

Figure 3. Compression behaviour of plain panels.

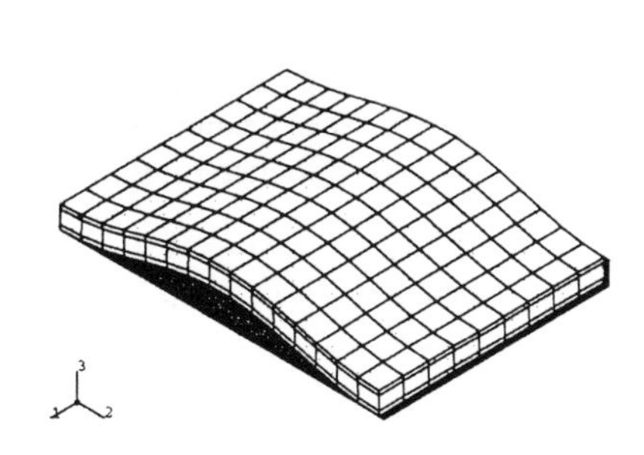

Figure 4. F.E. simulation

Despite the change in compression modulus, the buckling loads are equivalent, in the order of 100 KN. This is due to the particular configuration of those sandwich panels, as shown below.

The reduced buckling load (per unit of width) in the fundamental mode of a wide fully built-in on both ends orthotropic sandwich panel of length L can be expressed as [3]

$$N_{cr} = \frac{4\pi^2 D_{11} / L^2}{1 + 4\pi^2 D_{11} / L^2 H_{55}} \tag{1}$$

where D_{11} and H_{55} are obtained from the classical constitutive equations of composite materials with transverse shear. D_{11} is the reduced flexural rigidity of the panel and H_{55} its transverse shear rigidity, taking into account transverse shear correction factors [3]. L is the panel length.

From the mechanical characteristics given in table 1, and the stacking sequences described in figure 1, the values of flexural and shear rigidity are given in table 2. The reduced buckling loads are then calculated from equation 1 and the apparent buckling loads calculated as

$$Fcr = w.Ncr \tag{2}$$

w being the width of the panel.

TABLE 2. Rigidities and buckling loads of the sandwich panels.

	D_{11} (N.mm)	H_{55} ($N.mm^{-1}$)	Ncr ($N.mm^{-1}$)	Fcr (kN)
H1	9.51 10^6	634	573.5	100.4
H2	11.6 10^6	634	583.5	102.1

Figure 5 is a 3D plot (Maple V package) of the reduced buckling load as a function of flexural and shear rigidity in the range of the determined values for H1 ± 20 %, for a panel length of 250 mm. It clearly shows that buckling load of those kind of panels is mainly governed by the shear rigidity. Indeed, in equation 1, $4\pi^2 D_{11} >> H_{55}L^2$, and this equation simply reduces to $Ncr \approx H_{55}$. This would not be true for longer panels, as shown in figure 6, for a panel length of 750 mm, for which both flexural and shear rigidity have a clear influence on the buckling load.

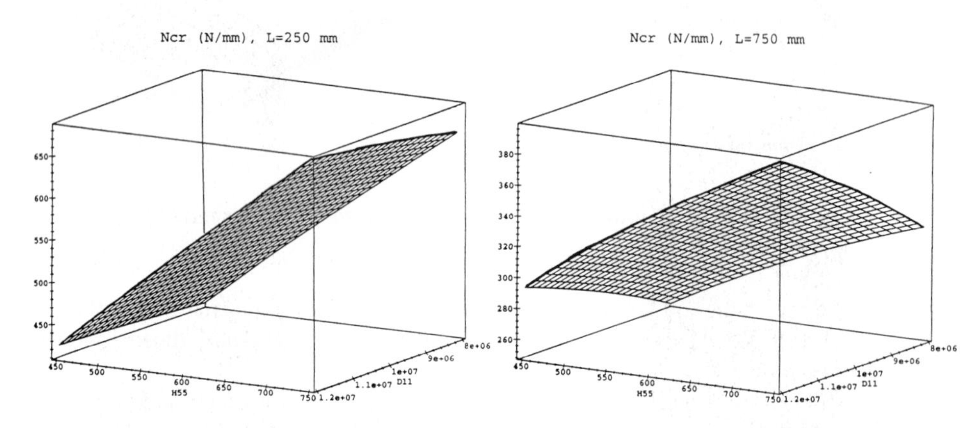

Figure 5. panel length = 250mm *Figure 6.* panel length = 750mm
Influence of flexural and shear rigidity on the sandwich panel buckling load.

In order to get a better understanding of the observed behaviour, a 3D numerical simulation of the sandwich panels has been carried out on Abaqus finite elements code. The foam is meshed with eight nodes elements and the skins are meshed with eight nodes reduced integration elements (one integration point per element) described as a composite material. The buckling loads are in accordance with experimental and analytical results, respectively 102.7 and 104.3 KN for H1 and H2 type panels. The shape of the deformed panel is given in figure 4.

4. Effect of a Ballistic Impact.

4.1. EXPERIMENTS

The sandwich panels, simply supported on a frame, have been subjected to ballistic impacts. The projectile is a 7.62 mm NATO bullet, the incident velocity is of about 800 $m.s^{-1}$. Two angles of incidences have been chosen: 0° (normal to the plane) and 70°. Impacts are centred in the sandwich structure and in the width direction for the 70° impact tests. Two sets of H1 and H2 samples have been tested in both conditions.

4.2. NORMAL TO THE PLANE BALLISTIC IMPACT

Normal to the plane ballistic impact only generated small circular delaminated areas around the impact hole, the radius being a bit larger on the back face than on the front face (10 mm against 6 mm), as shown on figure 7. After compression testing, the cutting of the panels did not reveal any skin / core debonding nor foam failure.

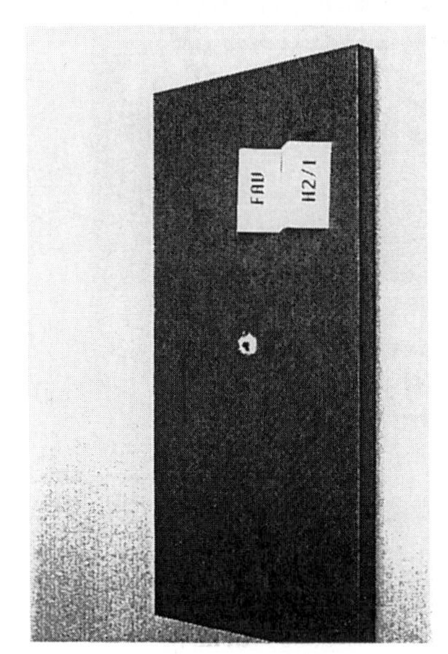

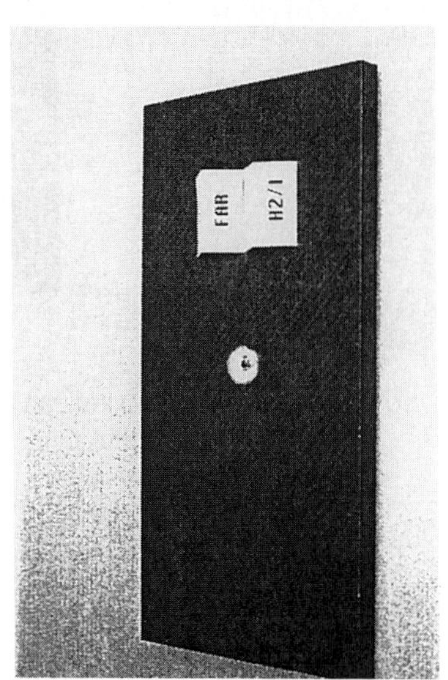

Figure 7. Photographs of a sample subjected to a normal impact.
(left: front face, right: back face)

The damages being very small compared to the panels dimensions, the compression after impact behaviour of those panels is equivalent to that of plain panels. A global buckling of the structure at a load of about 100 KN is observed.

4.3. 70° BALLISTIC IMPACT.

More dramatic damages were observed for the other set of samples subjected to ballistic impact, those damages were equivalent for H1 and H2 type panels. Figure 8 shows typical photographs of the samples before compression testing. Figure 9 is a projection onto the sandwich plane of the observed damages.

In addition to elliptical delamination around impact holes on front and back faces, a large debonding zone can be observed between the top skin and the foam core, as well as a small delaminated area on the front face. Those damages are attributed to a rebound of fragments of the projectile brass casing on the back skin. Those fragments come then back into contact with the top skin, their kinetic energy being high enough to generate delamination and debonding. Those metallic fragments are 'captured' in-between the two skins, they can then rebound one more time or slip against the top skin. Figure 10 details this scenario.

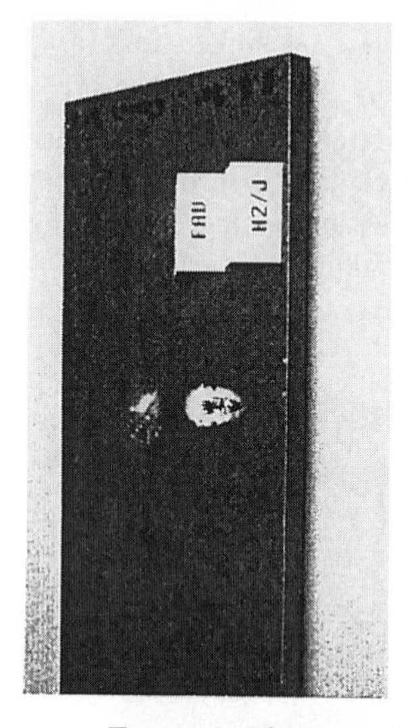

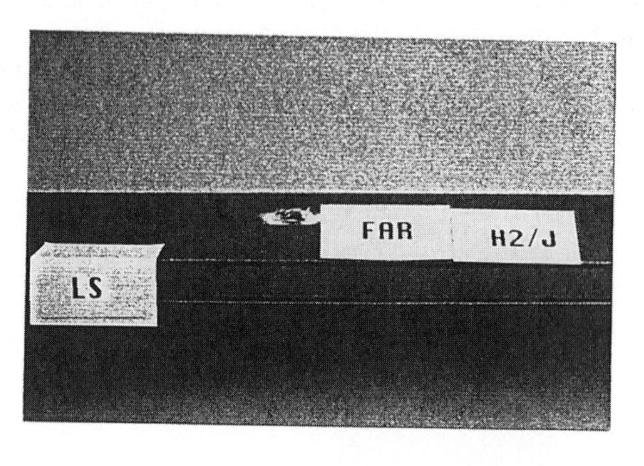

Figure 8: Photographs of sample subjected to a 70° ballistic impact
(left: front face, right: back face showing apparent debonding)

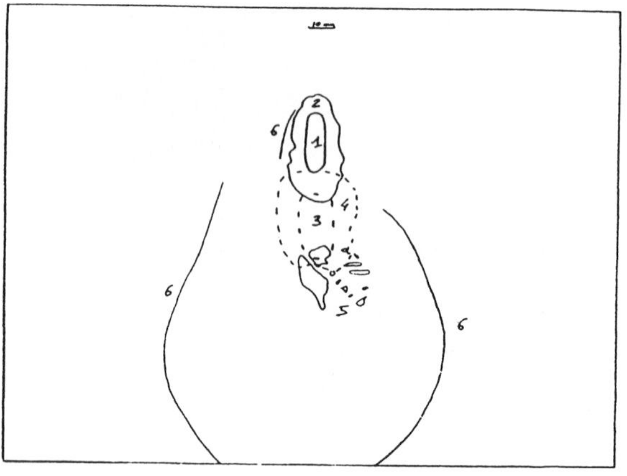

Figure 9. Projection of the damages onto the panel plane

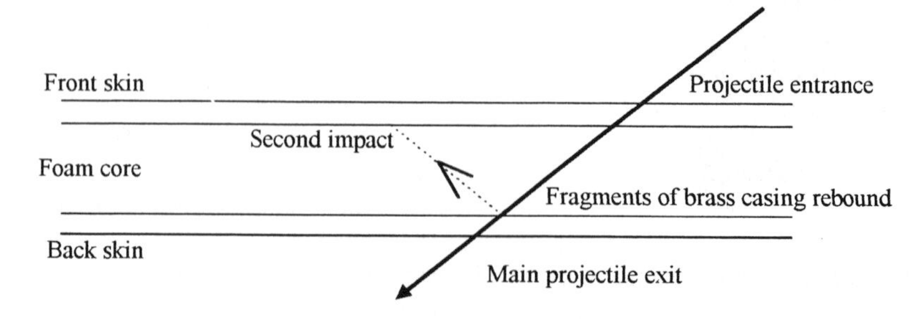

Figure 10. Projectile - panel interaction scenario.

Typical compression after impact behaviour of those panels is shown on figure 11 (H2 panel), figure 12 describes the strain gauges positions on the front face. The gauge B3 is on the back skin, directly opposite to gauge F3.

A numerical simulation of damaged panels has been carried out. The incident hole, as well as a triangular debonded zone have been meshed. In this zone, the elements of the top skin are simply disconnected from the foam elements. The maximum debonded length is of 105 mm.

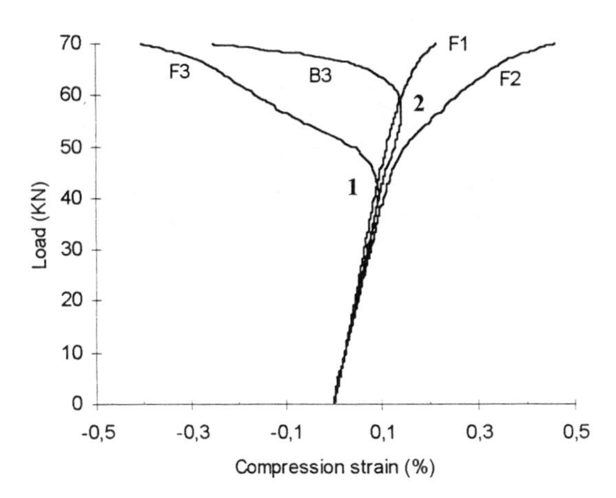

Figure 11. Compression after impact behaviour.

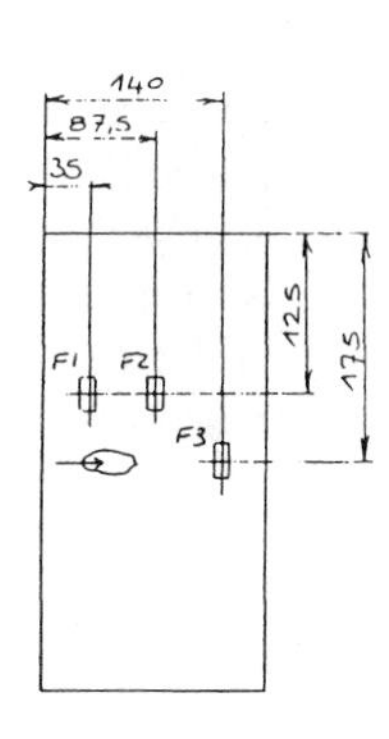

Figure 12. Strain gauges position on front face.

At an experimental load level of about 40 KN (mark **1** on Figure 11), a local buckling of the debonded skin is observed and gauge F3 goes back into tension. This is numerically reproduced on figure 13 at a load of 53 KN.

Between 40 and 60 KN, cracks propagate into the foam. Then, global buckling of the structure in an opening mode is observed (mark **2** on Figure 11) at about 60 KN. The numerical equivalent behaviour is shown on figure 14. The buckling load expected from the simple finite element model is of 76 KN.

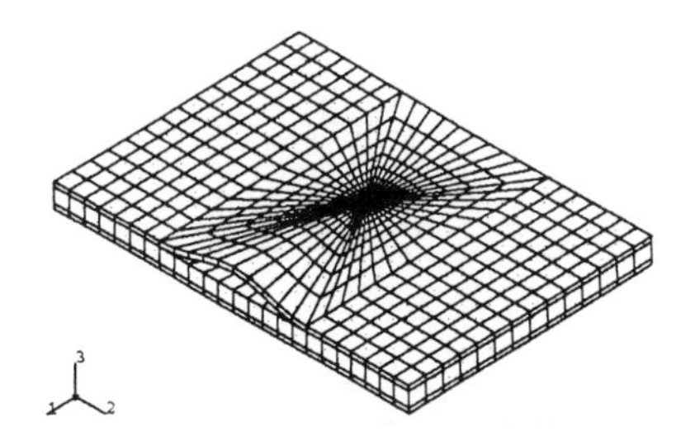

Figure 13. Local buckling of the front skin

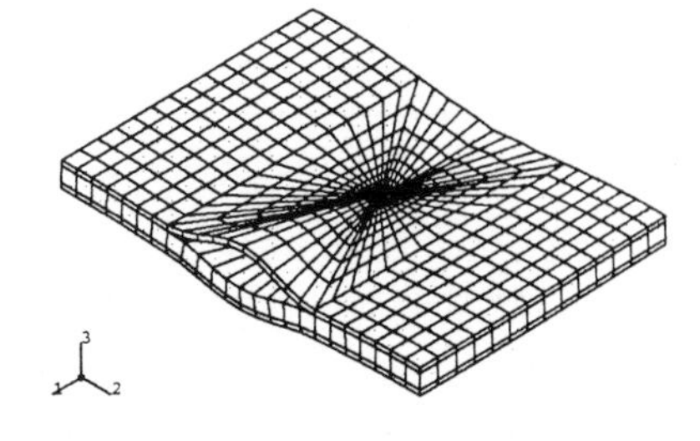

Figure 14. Global buckling of the structure

Finally, the ultimate load bearing capability of the structure is about 70 KN against 110 KN for undamaged panels, which is an acceptable result considering the amount of damage.

5 Conclusion

The principle of sandwich structures made of hybrid glass/carbon woven plies composite and foam core provides interesting compromises between rigidity, damage tolerance, weight and cost. However, this study showed that the interaction between the projectile and the sandwich structure during a ballistic impact at high angle of incidence is difficult to assess. Indeed, in this configuration, the damages are not directly generated by the perforation mechanisms but by the rebound of fragments of brass on the back skin. From this, there is clearly a need for further experimental investigation and work is still in progress in order to assess the influence of the nature and thickness of the foam as well as the nature and angle of incidence of the projectile.

References

1. Mailé, J.C. and Hilaire, B.: *Matériaux composites appliqués aux structures de franchissement, Résultats sur éprouvettes monolithiques*, rapport ETCA 93R111, 1993.
2. Mailé, J.C., Hilaire, B.: *Matériaux composites appliqués aux structures de franchissement, Résultats sur éprouvettes sandwichs*, rapport ETCA 94R007, 1994.
3. Berthelot, J.M.: *Matériaux composites, Comportement mécanique et analyse des structures*, MASSON, 1992
4. Habbit, Karlson, Sorensen: *ABAQUS V5.5, Theory Manual*, Habbit, Karlson & Sorensen Inc, 1995.

PENETRATING IMPACT STRENGTH OF SANDWICH PANELS - MEANINGFUL TEST METHOD AND SIMPLIFIED PREDICTION

MARTIN HILDEBRAND
VTT Manufacturing Technology
P.O. Box 1705, FIN - 02044 VTT, Finland
tel. +358 9 456 6211, fax. +358 9 456 5888, Martin.Hildebrand@vtt.fi

1. Abstract

This paper proposes a test method which is comprehensive enough to be able to quantify the impact strength, according to various definitions, of a sandwich and, most importantly, be applicable to sandwich panels of various types and scales. These findings are based on the results of a penetrating impact test series using three different test methods: The standardised ISO 6603 method and two non-standardised methods. The first non-standardised method uses a pyramid-shaped impactor instead of the cylindrical impactor used in the ISO 6603 method and in the second the impact test is performed quasi-statically using a cylindrical impactor.

The results obtained with the three test methods lead to a different ranking in impact strength of the panels. Hence, impact test results obtained with different test methods are not even qualitatively comparable.

The pyramid-shaped impactor is able to generate clearly more failure modes than the cylindrical impactor in the ISO 6603 method. Therefore, it is considered to be of more practical value for determining the impact strength of FRP-sandwich structures.

Additionally, a simplified approach to predict the impact strength of FRP-sandwich panels in respect to full penetration is presented. The method is semi-empirical and allows one to predict the absorbed energy at penetration of the inner face without the need for exotic input values. The empirical part of the method lies in the fact that certain damage modes are assumed to occur during the impact. These assumed damage modes have been observed in many earlier experiments. The accuracy of the method is reasonably high, in most cases better than ±30%.

The semi-empirical prediction method is particularly useful during the design stage of sandwich structures due to its reasonable accuracy and the fact that the needed input values are usually known at this stage.

A. Vautrin (ed.), Mechanics of Sandwich Structures, 247–254.

2. Introduction

During recent years, considerable research activities have been focused on the issue of FRP-sandwich impact strength. References [1 - 11] show for example that the research activities in the Nordic countries have been manifold and concerned different industrial branches. However, a common impact-testing method is not in use.

In eight of the above-mentioned references, impact tests have been performed. It is interesting to note that within these references, also eight different test methods have been applied. As the comparison of test results obtained with different methods is nearly impossible, it is obvious that the general knowledge and thereby also the predictability of sandwich impact strength will remain poor as long as no common method is in use.

The need of a common, reliable and comprehensive experimental method is even more emphasised if one takes into account the obvious lack of reliable analytical or numerical prediction methods for sandwich impact strength, specifically if full penetration of the sandwich is of major interest.

3. Meaningful test method

A test method has been developed [12, 13] which is comprehensive enough to be able to quantify the impact strength, according to various definitions, of a sandwich and, most importantly, be applicable to sandwich panels of various types and scales.

Compared to the standardised ISO 6603 method [14], there are two main differences: At first, the impactor is pyramid-shaped. In a penetrating impact through a sandwich, the projected contact area between impactor and sandwich grows with increasing indentation as opposed to the ISO 6603 impactor, in which the projected contact area remains after an initial growth constant.

The second difference concerns the post-processing of the results. The response of the test specimen during the test is calculated, and hence, the elastic part of the absorbed energy can be separated from the part related to the indentation. This is important, if flexible test specimens are tested or if panels with different stiffness are to be compared. The post-processing requires the knowledge of a set of elastic properties of the sandwich specimen.

The specimen size in this method depends on the sandwich total thickness: it is at least 250 × 250 mm, the size of the specimen support being at least 180 × 180 mm. A larger specimen and support size is used for specimens thicker than 40 mm. The specimens are not clamped (Figure 1).

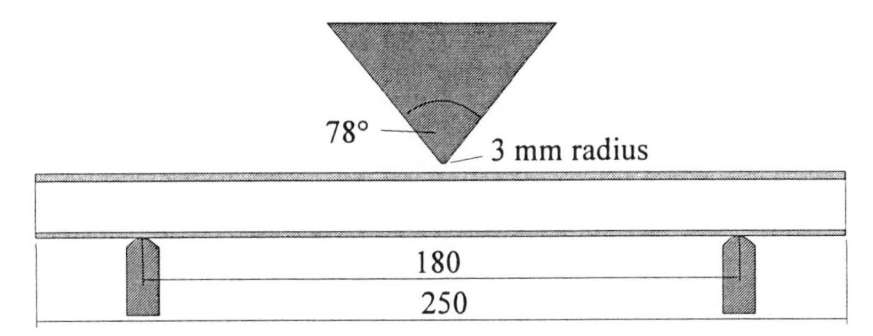

Figure 1. Geometry of the pyramid-shaped impactor, specimen and support.

The effect of the impactor geometry on the force and absorbed energy during the impact can be seen comparing the results of the 'pyramid' method with the standardised ISO 6603 method (Figure 2).

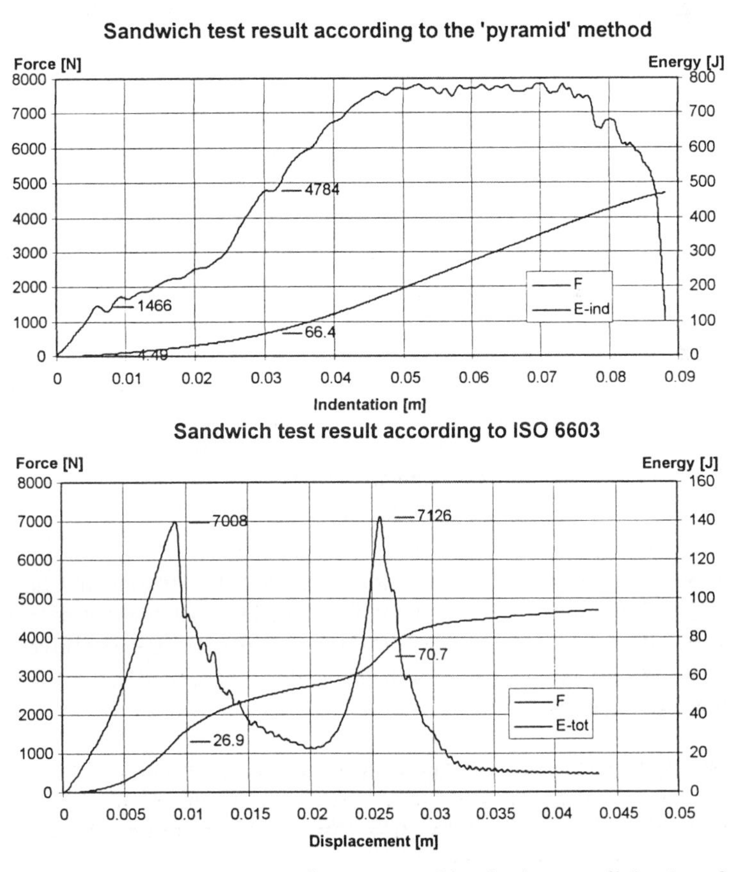

Figure 2. Comparison between results generated by the 'pyramid' (top) and the ISO 6603 (bottom) method. Force and corresponding absorbed energy values are shown at penetration of outer and inner face, respectively.

The failure produced by cylindrical impactors (as in the ISO 6603 method) in a sandwich panel is very local and the predominant failure mode is in shear. The test with the pyramid shaped impactor generates different failure modes, such as shear, core crushing and bending of the faces.

Since the 'pyramid' method is able to provoke a multitude of different failure modes, and additionally is able to provide consistent results at penetration of both outer and inner face [15], it is considered to be of most practical and general value for determining the impact strength of FRP-sandwich structures.

4. Simplified semi-empirical prediction method

A method is developed [16] in order to predict the impact strength as measured in the 'pyramid' test method. The force-indentation behaviour of sandwich panels is typically as shown in Figure 3. The penetration of the outer and of the inner face can be seen as a force peak.

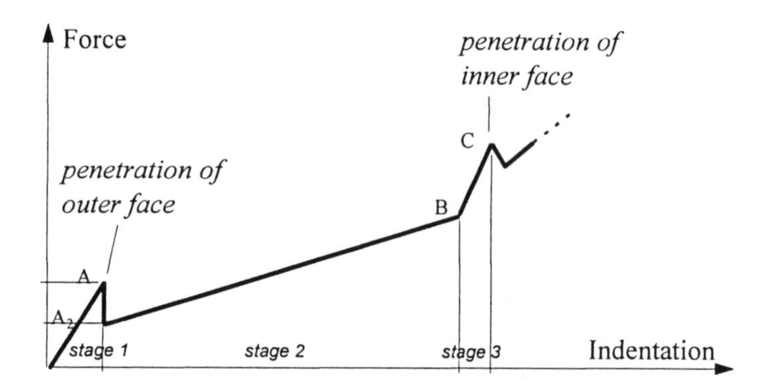

Figure 3. Typical force-indentation behaviour of a sandwich panel in the 'pyramid' impact test method.

The prediction method is semi-empirical, because the predominant failure modes occurring during the impact are assumed. These assumptions are based on the failure modes observed in experiments of over 70 different panels, representatives of typical boat, railway and ship applications. For the majority of these panels, the assumptions do agree with the actual behaviour.

The predominant failure modes during the three stages are as follows:

1) Penetration of outer face through outer face transverse shear failure
2) Bending of the outer face + core crushing in vertical and horizontal direction
3) Bending of the outer face + core crushing in vertical and horizontal direction + penetration of inner face through inner face transverse shear failure

Stage 1) Penetration of outer face through outer face transverse shear failure

The tip radius of the pyramid is rather small (R3) compared with the thickness of the outer face, therefore it causes local loading which leads to a transverse shear failure. It is assumed that the most critical stress is the transverse shear stress. Experiments performed with a low energy [13] lead to a local dent in the outer face confirming that the failure mode is in shear.

The force at penetration of the outer face is (see also Figure 3)

$$A = \tau_o A_o \tag{1}$$

where τ_o transverse shear strength of outer face
A_o area of shear failure, for geometrical reasons = $6.44\ t_o^2$
where t_o outer face thickness

$$\rightarrow \quad A = 6.44\ \tau_o\ t_o^2 \tag{2}$$

Stage 2) Bending of the outer face + core crushing in horizontal and vertical direction

After having penetrated the outer face, the pyramid locally bends the faces at the edge of the damaged area and the force correspondingly drops to the level A_2.

$$A_2 = \sigma_{bend\text{-}o}/6\ A_{bend} \tag{3}$$

where $\sigma_{bend\text{-}o}$ flexural strength of the outer face
A_{bend} flexural damage area = $s\ t_o$
s width of pyramid, based on the pyramid geometry $s(t_o) = 6.44\ t_o$

Due to the foundation of the core the bent laminate fails in tension. Therefore, the relevant strength value for $\sigma_{bend\text{-}o}$ is the tensile strength of the laminate.

$$\rightarrow \quad A_2 = 1.07\ \sigma_{bend\text{-}o}\ t_o^2 \tag{4}$$

When the impactor is progressing through the panel, the force level is determined both by outer face bending and by core crushing. Shortly before reaching the inner face, the force level at point B is

$$B = \sigma_{bend\text{-}o}\ A_{bend} + \sigma_{crush\text{-}horiz}\ A_{horiz} + \sigma_{crush\text{-}vert}\ A_{vert} \tag{5}$$

where $\sigma_{crush\text{-}horiz}$ core horizontal crushing strength
A_{horiz} horizontally crushed area
$\sigma_{crush\text{-}vert}$ core vertical crushing strength
A_{vert} vertically crushed area

The impactor geometry determines the crushed areas

$$A_{horiz} = 3.22\ z^2 \quad (6)$$
$$A_{vert} = 0.648\ z^2 \quad (7)$$

where z indentation

$$\rightarrow B = 1.07\ \sigma_{bend\text{-}o}\ (t_o + t_c)\ t_o + 3.22\ \sigma_{crush\text{-}horiz}\ t_c^2 + 0.648\ \sigma_{crush\text{-}vert}\ t_c^2 \quad (8)$$

where t_c core thickness

Stage 3) Bending of the outer face + core crushing in horizontal and vertical direction + penetration of inner face through inner face transverse shear failure

It is assumed that the penetration of the inner face is at the beginning based on the same failure type than that of the outer face, which is transverse shear. This can be observed in the experiments in panels which have a strong bond between core and inner face. In some cases, if the bond is weak, the inner face delaminates before being penetrated. In such cases, this simplified model is clearly too simple.

$$C = \sigma_{bend\text{-}o}\ A_{bend} + \sigma_{crush\text{-}horiz}\ A_{horiz} + \sigma_{crush\text{-}vert}\ A_{vert} + \tau_i\ A_i \quad (9)$$

where τ_i transverse shear strength of inner face
A_i area of shear failure, for geometrical reasons (pyramid geometry)
$A_i = 6.44\ t_i^2$
where t_i inner face thickness

$$\rightarrow C = 1.07\ \sigma_{bend\text{-}o}\ (t_o + t_c + t_i)\ t_o + 3.22\ \sigma_{crush\text{-}horiz}\ (t_c + t_i)\ t_c + 0.648\ \sigma_{crush\text{-}vert}\ (t_c + t_i)\ t_c + 6.44\ \tau_i\ t_i^2 \quad (10)$$

Throughout the three stages of penetrating the panel, it is assumed that the indentation corresponds to the thickness of the respective part. This means that local bending of the damaged area is not taken into account. With this assumption, the absorbed energy at penetration of the inner face is

$$\rightarrow E_{inn} = A\ t_o/2 + A_2\ t_c + (B\text{-}A_2)\ t_c/2 + B\ t_i + (C\text{-}B)\ t_i/2 \quad (11)$$

The correlation between the simplified prediction method and experimental results is shown in Figure 4a and 4b for the absorbed energy at penetration of the inner face. The linear regression of the experimental values is added into Figure 4a, assuming that the y-intercept is at zero.

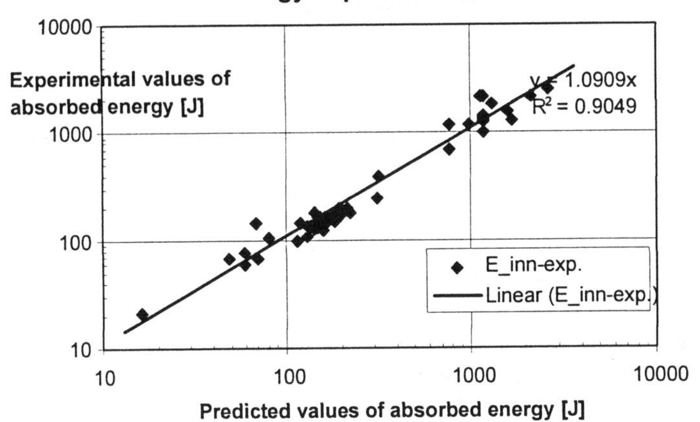

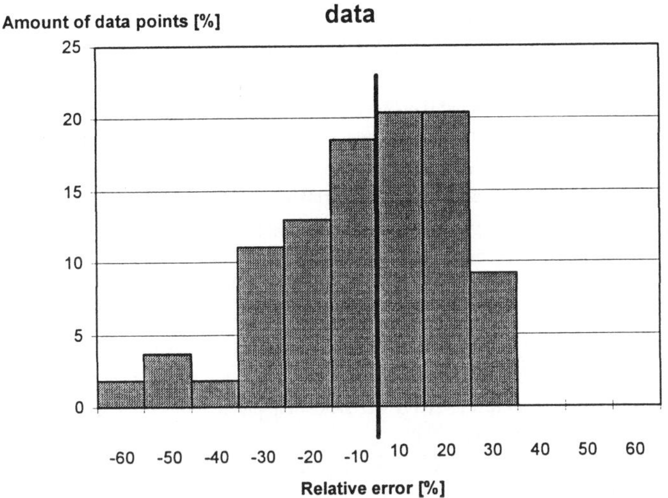

Figure 4a (above) and 4b (below). Correlation between simplified prediction method and experiments.

The overall correlation of the simplified model is quite reasonable throughout the whole strength range (Figure 4a). Looking at individual points, the predicted impact strength at penetration of the inner face lies in most cases within ±30% (Figure 4b).

Most importantly, for its use in the design stage of sandwich products, there is no need for exotic input values. The input values are sandwich geometry, tensile and transverse shear strength of the face laminates and core crushing strength both in panel and through-thickness direction.

It is well known that the strength values of many core materials and also of the face laminates are strain-rate dependent. Therefore, if material values at high strain rates - corresponding to impact loading speed - are known, these should be used as instead of the usually better known quasistatic values.

It is clear that the presented method has several shortcomings. The method will be further developed in order to incorporate a better foundation model for the outer face, a core material toughness parameter and other impactor geometries.

5. References

1 Gaarder, R. H. 1995. Methods to predict impact performance of sandwich structures. Espoo: Nokos Second Composites and Sandwich Meeting, 4.-5. 4. 1995. Pp. 25-27.

2 Brevik, A. F. 1995. Sandwich constructions for subsea applications. Presented at the third international conference on sandwich construction. 9 p.

3 Grenestedt, J. & Kuttenkeuler, J. 1995. Slow impact on Sandwich panels. Espoo: Nokos Second Composites and Sandwich Meeting, 4.-5. 4. 1995. Pp. 35-37.

4 Olsson, R. 1995. Prediction of impact damage in sandwich panels. Presented at the third international conference on sandwich construction. 10 p.

5 Lönnö, A. & Håkanson, P. 1995. Development of a super tough FRP-sandwich concept for the 10.8 m combat craft 90E. Presented at the third international conference on sandwich construction. 12 p.

6 Hildebrand, M. 1994. The effect of raw-material related parameters on the impact strength of sandwich boat-laminates. Espoo, Technical Research Centre of Finland, VTT Publications 211. 36 p. + app. 19 p.

7 Auerkari, P. & Pankakoski P. H. 1995. Strength of sandwich panels with impact defects. Espoo: VTT Technical Report VAL B 75. 19 p. + app. 4 p.

8 Burchardt, C., et al. 1993. Slagpåvirkning af sandwichplader. Aalborg: M.Sc. Thesis, Aalborg Universitets Center. 90 p. + app. 82 p.

9 Aamlid, O. 1995. Oblique impact testing of aluminium - and composite panels. Hovik: Det Norske Veritas Research, Technical Report No. 95-2042. 36 p.

10 Marum, S.E. 1993. Impact testing of GRP/PVC sandwich panels (In Norwegian). Mandal: Kvaerner Mandal test report EO 094. 23 p. + app. 49 p.

11 Nilsen, P.E., Moan, T., Gustafson, C.-G. 1992. Dynamic and quasi-static indentation of PVC/GRP sandwich plates. Warley: Proceedings of the second international conference on sandwich construction. EMAS. Pp. 121 - 137.

12 Kivelä, J. 1992. The determination of local impact strength for small-craft sandwich structures (in Finnish). Espoo: Helsinki University of Technology. M.Sc. Thesis. 62 p. + app. 19 p.

13 Hildebrand, M. 1996. Improving the impact strength of FRP-sandwich panels for ship applications. Technical Research Centre of Finland, VTT Technical Report VAL B 138.

14 ISO 6603/2-1989. Plastics - Determination of multiaxial impact behaviour of rigid plastics - Part 2: Instrumented puncture test.

15 Hildebrand, M. 1996. A comparison of FRP-sandwich penetrating impact test methods. Espoo, Technical Research Centre of Finland, VTT Publications 281. 33 p. + app. 1 p.

16 Hildebrand, M. 1996. Penetrating impact strength of FRP-sandwich panels - empirical and semi-empirical prediction methods. Technical Research Centre of Finland, VTT Technical Report VAL B 154.

CONTRIBUTION OF THE RESPONSE SURFACE METHODOLOGY TO THE STUDY OF LOW VELOCITY IMPACT ON SANDWICH STRUCTURES.

F. COLLOMBET*, S. PECAULT*, P. DAVIES**, D. CHOQUEUSE**, J.L. LATAILLADE* and A.TORRES MARQUES***

*LAMEF- ENSAM Bordeaux, FRANCE

**Lab. Matériaux Marins, IFREMER Centre de Brest, FRANCE

***I.N.E.G.I., Rua do Barroco, S. Mamede Infesta, PORTUGAL

1. Introduction

The behaviour of sandwich structures under low velocity impact loading is a multi-parameter subject. The main question is : how can we treat this problem to determine which parameters are critical ? The materials which can be used as skin or core of sandwich are numerous. The impact load can be defined in terms of mass and velocity of the striker, size of the sandwich plate, boundary conditions, geometry of the impactor [1], [2], [3], [4], [5], [6]. The effects of the plate dimensions are rarely studied. Researchers do not always define everytime the impact by the couple (mass,velocity) but often retain as the critical parameter the kinetic energy of the striker [7].

2. Response Surface Methodology

We propose a Response Surface Methodology to contribute to the study of the influence of mass and velocity of the striker and the size of the sandwich plate on the response of sandwich panels. The square sandwich structures investigated are composed of glass/polyester skins on PVC cores.

2.1 PRINCIPLES OF THE METHOD

The investigation of the experimental area is performed by a limited number of tests. As there is no physical background for modelling the influence of the three factors on the responses, the Response Surface Methodology is based on the hypothesis that the measured responses can be locally interpolated by an empirical polynomial $\hat{Y}$. The fundamental result is that the model variance is expressed as the product of two terms; σ^2 the experimental variance and A depending only on the tests distribution in the experimental area as a function of the postulated polynomial where :

$$\mathrm{var}(\hat{Y}) = A\,\sigma^2 \tag{1}$$

σ^2 is estimated by the repetition of tests. This means that the model variance can be minimized by an efficient choice of the experimental distribution. This optimization is independent of the experimental variance.

A. Vautrin (ed.), Mechanics of Sandwich Structures, 255–262.

2.2 INTERPOLATION AND EXPERIMENTAL DISTRIBUTION

We choose a second order polynomial to interpolate the responses which assumes that responses are non linear. The mass **m** and the velocity **v** of the striker, the size **s** of the sandwich panels are the variables of this polynomial which is defined by the following equation (where I is the mean value) :

$$\hat{Y}(m,v,s) = I + \mathbf{am} + \mathbf{bv} + \mathbf{cs} + \mathbf{dm^2} + \mathbf{ev^2} + \mathbf{fs^2} + \mathbf{gmv} + \mathbf{hms} + \mathbf{ivs} \quad (2)$$

We retain the experimental distribution proposed by Doehlert [8], [9]. It allows the identification of the coefficients (I,..., i) of the second order polynomial $\hat{Y}$. We use normalized parameters X_1, X_2 and X_3 in the range from minus one to plus one. In this form, the Doehlert matrix is valid for many applications :

TABLE 1. Doehlert matrix

number of tests	X_1	X_2	X_3
1	1	0	0
2	-1	0	0
3	0.5	0.86	0
4	-0.5	-0.86	0
5	0.5	-0.86	0
6	-0.5	0.86	0
7	0.5	0.28	0.816
8	-0.5	-0.28	-0.816
9	0.5	-0.28	-0.816
10	0	0.57	-0.816
11	-0.5	0.28	0.816
12	0	-0.57	0.816
13	0	0	0

The thirteen experiments imposed by the Doehlert design were applied with falling weight set-ups. For this kind of experimental set-up X_2 is taken as projectile speed since this is the easiest to vary. To minimise the number of strikers, X_3 is chosen as mass of the projectile. For the independent parameters, we consider five levels for span of sandwich panels (the square panels are simply supported on two sides), seven levels for velocity of the projectile and three levels for mass.

3. Experimental Set-ups and Sandwich Panels

Three falling weight set-ups were used at LAMEF-ENSAM in Bordeaux (France), the Marine Materials Laboratory of IFREMER-Brest (France) and I.N.E.G.I. of Porto (Portugal) [10]. Each set of tests leads to a distinct polynomial identification. For each impact test, we measure seven dependent responses which are maximum contact force, panel deflection, maximum projectile displacement, damage area in upper skin, local crushing rate, energy restored to the projectile, time for panel to return to its original position. The ranges of the parameters investigated are given below :

TABLE 2. Ranges of parameters

laboratory	impact energy (J)	mass (kg)	velocity (m/s)	span (cm)
LAMEF-ENSAM	11 to 100	4 to 7	2 to 6	30 to 80
IFREMER-Brest	30 to 300	9 to 24	2 to 6	30 to 80
INEGI-Porto	30 to 300	4 to 6	4 to 11	30 to 80

However there are overlapping areas in terms of isoenergy conditions with different couples (mass,velocity) and others for the same couples (mass,velocity).

The square panels tested are simply supported on two sides. Two kinds of sandwich structure are investigated with the same glass polyester facings (two millimeters thickness) and two different foam cores of the same density (80 kg/m^3 for twenty millimeters nominal thickness). The cells of P.V.C. foams are closed. The constitutive material consists of thermoset and thermoplastic copolymers which are crosslinked (rigid foam) or linear (ductile foam).

4. Polynomial Identifications

For each set of tests, impacts at one condition were repeated three times to evaluate experimental scatter (variance) which is for instance about 10 % for the maximum contact force. From the fifteen tests of each experimental work (13 plus 2 repeats), ten coefficients of polynomial are identified. Using Snedecor's criterion the significance of each response is evaluated. For example, for the LAMEF work and the rigid P.V.C. foam, two responses are not statistically significant (energy restored to the projectile, time for panel to return to its original position) as the uncertainty of the model (second order) is greater than the repeatability of the measurements.

5. Two Examples from IFREMER-Brest Work

The thirteen experiments imposed by the Doehlert design (defined in Table 3.) were applied with the IFREMER of Brest falling weight set-up :

TABLE 3. Surface methology experiments at IFREMER-Brest

test number	span (cm)	velocity (m.s^{-1})	mass (kg)	impact energy (J)
01	80	4	16.58	133
02	30	4	16.58	133
03	67.5	6	16.58	298
04	42.5	2	16.58	33
05	67.5	2	16.58	33
06	42.5	6	16.58	298
07	67.5	4.6	23.92	253
08	42.5	3.3	9.21	50
09	67.5	3.3	9.21	50
10	55	5.3	9.21	129
11	42.5	4.6	23.92	253
12	55	2.6	23.92	81
13	55	4	16.58	133

Iso-level line graphs are presented (here for a given span of fifty five centimeters) in order to emphasize the influence of each factor on the plate responses.

Fig.1. shows the iso-level lines of the damaged area on the upper skin of a 55 cm * 55 cm sandwich plate. In the same way for a given span of fifty five centimeters, Fig.2. shows that three distinct regions can be identified for the maximum contact force.

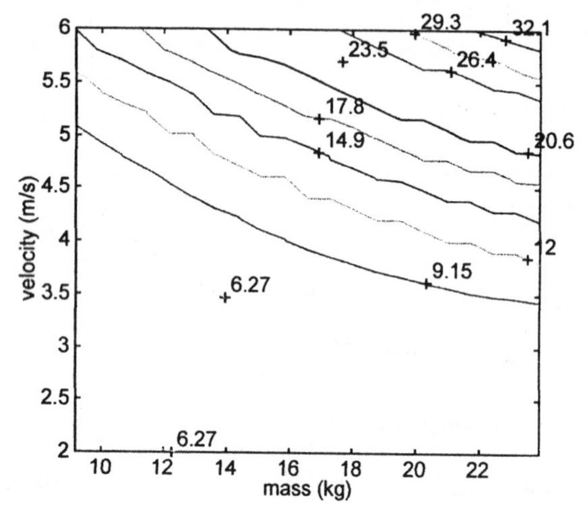

Fig.1. : Iso-level lines of the polynomial modelling the damaged surface (cm^2) on the upper skin of a 55 cm * 55 cm sandwich plate

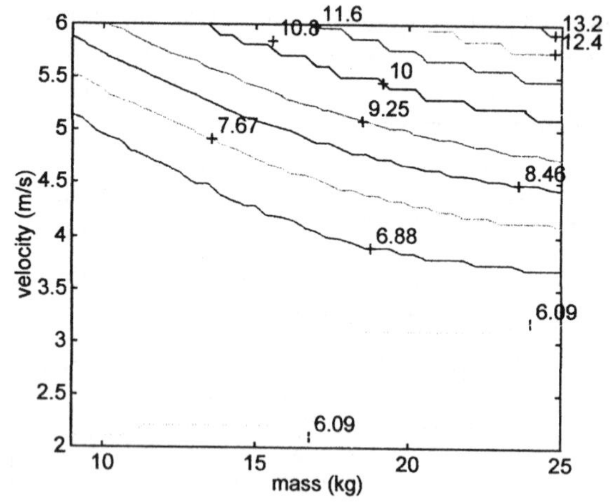

Fig.2. : Iso-level lines of the polynomial modelling the maximum force (kN) on the 55 cm * 55 cm sandwich plate

In Fig.1. for low mass and velocity levels, the distance between the iso-level lines is wide. This means that the influence of mass and velocity on the damaged area is weak. The greater the mass or the velocity, the closer the isolevel lines. The influence of the mass and the velocity becomes stronger at the higher values of the parameters in the range 9-25 kg * 2-6 ms^{-1}. On the other hand, the iso-level lines are almost perpendicular to the velocity axis. This shows that the velocity is the main factor of influence of the damaged area. The lines of iso-level are not straight lines but present a curvature.

This indicates that the influence of the two terms $\mathbf{v}^2$ and **mv** are significant. A second degree polynomial is necessary to model the responses in an efficient manner.

In Fig.2. for low velocities, the mass has very little influence on the contact force, whereas at high velocities the mass dominates force response. There is an intermediate region where neither dominates. For other experimental conditions these regions may vary.

The evolution may be related to the damage area of the upper skin which has a very similar shape. However, the striker mass is the preferential parameter of the contact duration as [10].

6. Nature of Damage

Fig.3. shows several damage mechanisms are acting. First, the upper skin is damaged with cracking then fiber breakage and perforation. Second, the foam is dynamically compressed and shear cracking appears for the rigid foam only (not for ductile foam) and then lower skin debonding and perforation.

Fig.3. : Cross section in the plane of impact point of sandwich structure with rigid foam

The influence of foam can be studied using local crushing measured during LAMEF work presented in Table 4 :

TABLE 4. Surface methology experiments at LAMEF-ENSAM

test number	span (cm)	velocity ($m.s^{-1}$)	mass (kg)	impact energy (J)
01	80	4	5.56	44.5
02	30	4	5.56	44.5
03	67.5	6	5.56	100
04	42.5	2	5.56	11.1
05	67.5	2	5.56	11.1
06	42.5	6	5.56	100
07	67.5	4.6	6.76	71.5
08	42.5	3.3	4.36	23.7
09	67.5	3.3	4.36	23.7
10	55	5.3	4.36	61.2
11	42.5	4.6	6.76	71.5
12	55	2.6	6.76	22.8
13	55	4	5.56	44.5

Fig.4 shows that there is no difference between the ductile and rigid foams for the local crushing values of impacted sandwich structures. Once the skin is pierced, these foams offer no resistance to indentation (tests with complete crushing of foams).

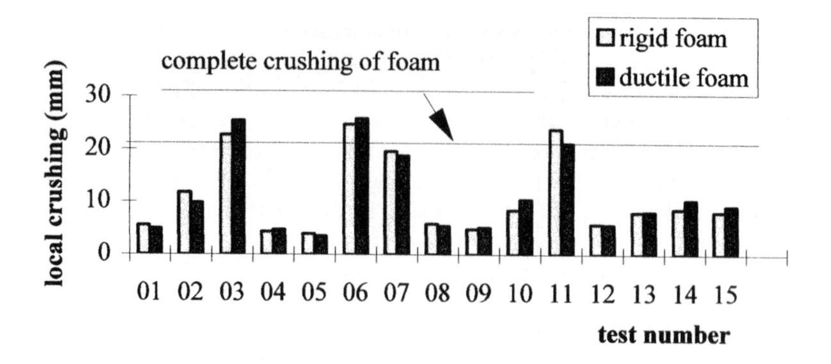

Fig.4. : Comparaison of local crushings of sandwich structures measured during the LAMEF work between ductile and rigid foams.

The I.N.E.G.I. of Porto work is presented in Table 5 :

TABLE 5. Surface methology experiments at I.NE.G.I. of Porto

test number	span (cm)	velocity (m.s^{-1})	mass (kg)	impact energy (J)
01	80	7.45	4.76	133
02	30	7.45	4.76	133
03	67.5	11.2	4.76	298
04	42.5	3.7	4.76	33
05	67.5	3.7	4.76	33
06	42.5	11.2	4.76	298
07	67.5	8.7	5.76	216
08	42.5	6.2	3.76	73
09	67.5	6.2	3.76	73
10	55	9.9	3.76	185
11	42.5	8.7	5.76	216
12	55	4.9	5.76	71
13	55	7.45	4.76	133
14	55	7.45	4.76	133
15	55	7.45	4.76	133

In Fig.5a. and Fig.5b., the sequence of photos shows the lower skin damage for three impact conditions for increasing energies (70 up to 220 J) for two types of foam.

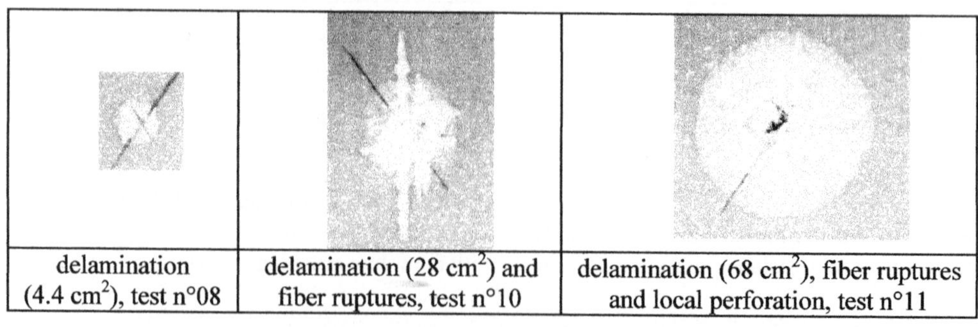

delamination (4.4 cm^2), test n°08	delamination (28 cm^2) and fiber ruptures, test n°10	delamination (68 cm^2), fiber ruptures and local perforation, test n°11

Fig.5a. : Lower skin damage of sandwich structures for the ductile foam (work of I.N.E.G.I. of Porto).

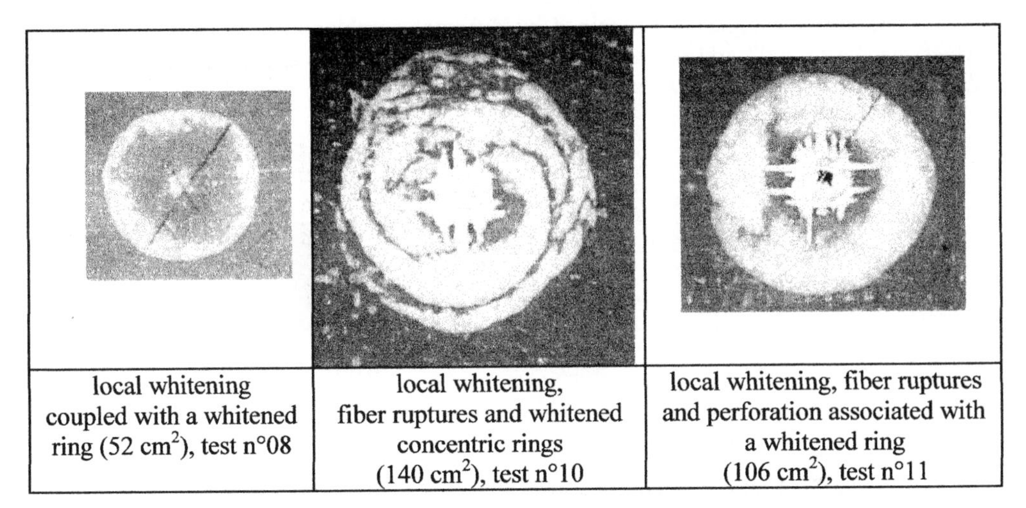

local whitening coupled with a whitened ring (52 cm^2), test n°08	local whitening, fiber ruptures and whitened concentric rings (140 cm^2), test n°10	local whitening, fiber ruptures and perforation associated with a whitened ring (106 cm^2), test n°11

Fig.5.b. : Lower skin damage of sandwich structures for the rigid foam (work of I.N.E.G.I. of Porto).

From the results shown in Fig.5a. and Fig.5b., it is clear that the foam has a significant influence on damage development. Moreover the mechanisms are not the same for both P.V.C. foams, and shear cracks are not observed in the ductile foam.

The different role of each type of foam is emphasized by Fig.6. which shows a comparison of the areas of damage of the lower skin for the rigid and ductile foams for equivalent impact energies.

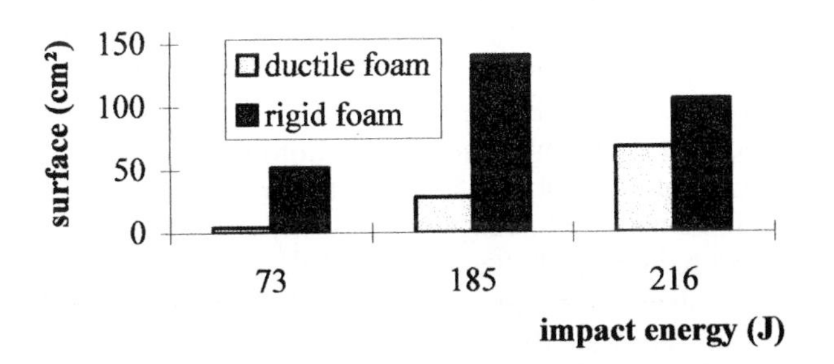

Fig.6. : Comparison of the lower skin damage of sandwich structures between the rigid and ductile foams (work of I.N.E.G.I. of Porto).

7. Conclusion

One of the main conclusions from this study is the importance of both mass and velocity in defining the severity of impact. From either energy sets, we are able to compare responses. One example is shown Fig. 7. in the case of two impacts for the same energy of about 300 J.

The 300 J energies are obtained by different combinations of mass and velocity which are the tests 03 and 06 in the works of I.N.E.G.I.-Porto and IFREMER-Brest. The damage extent is shown for two spans :

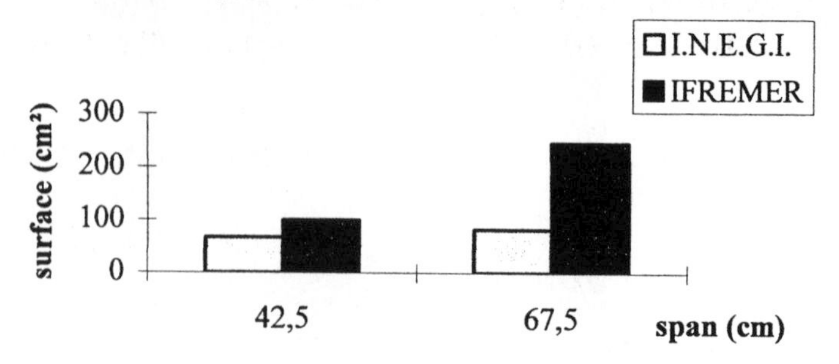

Fig.7. : Comparison of the lower skin damage for the rigid foam of sandwich structures at isoenergy (tests 03 and 06, works of I.N.E.G.I. and IFREMER).

Fig.7. clearly shows that the definition of an impact event in terms of an impact energy is ambiguous, both mass and velocity must be specified. The use of the Response Surface Methodology approach enables us efficient test programs to be run, so that critical test parameters can be identified and focussed on for the study of sandwich or composite structures under impact loading.

8. References

1. Bernard M.L. and Lagace P.A., 1989, "Impact resistance of composite sandwich plates", Journal of Reinforced Plastics and Composites, vol 8, 432-445.
2. Chun-Gon K. and Eui-Jin J, 1992, "Impact resistance of composite laminated sandwich plates", Journal of Composite Materials, Vol 26, n°15, p 2247-2261.
3. Davies P., Choqueuse D. and Pichon A., 1994, "Influence of the foam core on composite sandwich static and impact response", European Conference on Composites Testing and Standardisation, Sept 13-15, Hamburg Germany, 513-521.
4. Nemes J.A. and Simmonds K.E., 1990, "Low velocity impact response of foam-core sandwich composites", Journal of Composite Materials, Vol 26, n° 4.
5. Mines R.A.W., Worrall C.M. and Gibson A.G.,1994, "The static and impact behaviour of polymer composite sandwich beams", Composites, Vol 25, n° 2, pp 95-110.
6. Shih W.K. and Jang B.Z.,1989, "Instrumented impact testing of composite sandwich panels", Journal of reinforced plastics and composites, vol 28, 270-298.
7. Robinson P. and Davies G.A.O., 1992, "Impactor mass and specimen geometry effects in low velocity impact of laminated composites", Int. J. Impact. Engng, Vol 12, n°2, pp 189-207.
8. Doehlert D.H.,1970, "Uniform shell designs", Applied Statistics, Vol 19, n°3, pp 231-239.
9. Doehlert D.H. and Klee V.L.,1972, "Experimental designs through level reduction of the d-dimensional cuboctahedron", Discrete Mathematics 2, pp 309-334.
10. Pécault S., 1996, "Tolérance au choc de structures sandwichs et effets d'échelle", PhD thesis (in french), N°96.37, ENSAM-Bordeaux.

ENERGY ABSORPTION CHARACTERISTICS OF A FOAM-CORED SANDWICH PANEL WITH FIBRE-REINFORCED PLASTIC INSERTS

J.J. CARRUTHERS*, M.S. FOUND** & A.M. ROBINSON*

* *Advanced Railway Research Centre*
The University of Sheffield, Regent Court, 30 Regent Street,
Sheffield, S1 4DA (UK)

** *Department of Mechanical Engineering*
The University of Sheffield, Mappin Street,
Sheffield, S1 3JD (UK)

Abstract

This paper describes an experimental investigation into the energy absorption properties of a foam-cored sandwich panel with integral fibre-reinforced plastic tubes and frusta. The panels were tested under both quasi-static flatwise and edgewise compression. Under flatwise compression, it was found that the panels with inserts which failed by stable progressive brittle fracture exhibited the best energy absorption characteristics. Under edgewise compression, the panels with inserts which prevented separation of the face plates offered the most predictable and repeatable performance. Particular insert geometries which achieve these ideals are identified.

1. Introduction

The topic of crashworthiness has received much attention over the last decade. Structural design philosophies based principally on proof strengths have been superseded by a realisation of the importance of carefully designed energy absorbing crush zones. These ensure a controlled collapse force which brings colliding vehicles to rest without transmitting damaging accelerations to their occupants. Considerable research interest has been shown in the use of composite materials for crashworthiness applications because it has been demonstrated that they can be designed to provide energy absorption capabilities which are superior to those of metals when compared on a weight-for-weight basis [1,2]. It has been found that, in general, fibre-reinforced plastics (FRPs) do not exhibit the ductile failure processes associated with metals. Instead, the brittle nature of most fibres and resins tends to generate a brittle mode of failure. Provided that the crushing mechanisms can be controlled so that the FRP fails in a stable progressive manner, very high levels of energy can be absorbed.

A. Vautrin (ed.), Mechanics of Sandwich Structures, 263–270.

In recent years, some progress has been made in establishing the influence of material, geometric and experimental parameters on the energy absorption capability of FRP tubes, with perhaps Hull [3] providing the definitive work in this field. However, the practical application of this knowledge has been limited by the simplistic nature of the geometries investigated. Although tubes can be considered structurally representative up to a point, the question still largely remains of how best to reproduce the high energy absorptions demonstrated in the laboratory within real applications.

This paper describes one approach to the development of structural crashworthy composites. It is based on the use of a foam-cored sandwich panel with integral energy absorbing FRP inserts. A sandwich panel design was chosen as the basis for the study in order to obtain the necessary strength and stiffness for use in structural applications. The function of the FRP inserts, which were in the form of tubes and hollow conical frusta, was to control the failure loads (and hence the energy absorption capability) of the panels.

2. Material Design & Specification

The sandwich panel specimens (Figure 1) were manufactured using resin transfer moulding (RTM). Each consisted of a rigid closed-cell polyurethane foam of nominal density 120 kg/m^3 surrounded by facings of glass-reinforced polyester. Incorporated within the core of each specimen were four FRP inserts fabricated from [±45°] tubular braided glass fibres in a polyester resin. The fibres at the ends of each braided insert were merged with those of the face plate laminate so as to provide a mechanical tie between opposing facings. It has previously been shown [4] that this arrangement inhibits separation of the face plates, even after core debonding. Furthermore, it has been found [4] to enhance the mechanical properties of panels, particularly with respect to shear stiffness and strength.

A number of different insert geometries were tested, although there was no variation within a given specimen. Details are provided in Table 1. The geometries of

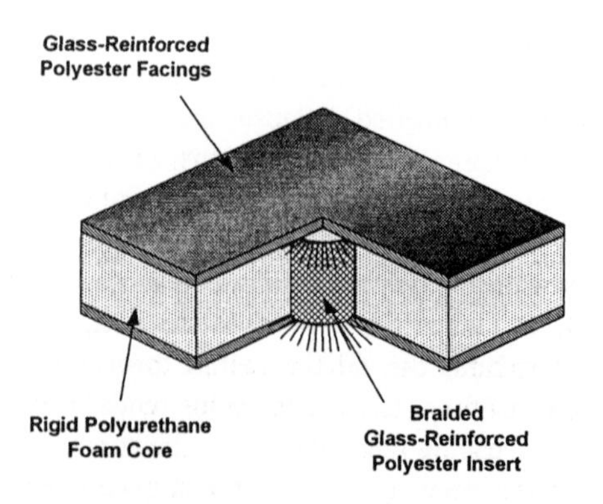

FIGURE 1. General structural design of the energy absorbing composite sandwich panels.

the conical frusta were chosen in accordance with Mamalis *et al.*'s [5] recommendations for designing FRP frusta which fail by high energy progressive crushing. Similar mean diameters were then selected for the tubular inserts. A conventional sandwich panel, without any inserts, was also tested for comparison purposes.

TABLE 1. The different geometries of FRP insert investigated

Insert Type	Semi-Apical Angle	Length (mm)	Inside Diameter (mm)	Outside Diameter (mm)	Fibre Volume Fraction (%)
None	-	-	-	-	-
Tube	-	25	14	15	59
Tube	-	25	14	16	29
Frustum	10°	25	Wide End = 19.4	Wide End = 20.6	49
Frustum	10°	25	Wide End = 19.4	Wide End = 21.5	24

3. Testing & Analysis Procedures

The specimens were compressed in the flatwise direction between two parallel rectangular platens using a Mayes 100 kN capacity servo-electric test machine. All the specimens were crushed at a uniform rate of 5 mm/min and the load-displacement characteristic was recorded in each case.

4. Results

The behaviour of the sandwich panel specimens was found to be very different under flatwise and edgewise compression. The core generally controlled failure in the former instance, the face plates in the latter. Each is therefore considered separately below.

4.1. FLATWISE COMPRESSION

4.1.1. Failure Mechanisms

Under flatwise compression, all specimens exhibited a linear-elastic response at small crush distances ($< \approx 1$ mm). This was due to the bending and stretching of the cells within the rigid polyurethane foam and the elastic compression of the FRP tubes and frusta. Those specimens with FRP inserts displayed a stiffer elastic response than those without. At large crush distances ($> \approx 15$ mm), the collapse load of all specimens increased rapidly with displacement. The cores of the sandwich panels had become fully crushed, and their useful energy absorption capability was exhausted. Core densification had been reached. Between these two extremes (i.e. for crush distances of $\approx$1-15 mm), the load-displacement response of the sandwich panels was found to vary considerably with insert geometry. In general, three different types of behaviour were observed: a uniform response, a uniform response with an initial peak load, and a non-uniform response (Figure 2). Each is considered in turn below.

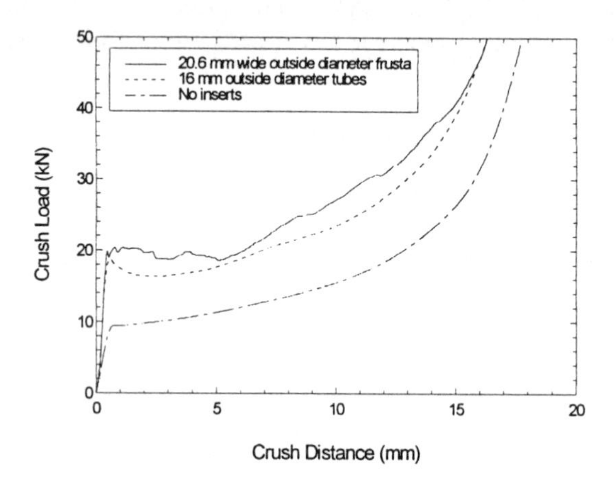

FIGURE 2. A comparison of the load-displacement characteristics of representative specimens from each of the three general failure categories.

Uniform Response. This was the characteristic behaviour of the sandwich panels with no GRP inserts. Between the elastic and densification regions there was a long collapse plateau in which the load increased only slowly with displacement. Such a response is typical of polymer foams [6], and is associated with the gradual collapse of the cells within the foam. The implication is that, in the absence of any inserts, the properties of the sandwich panels under flatwise compression are dictated by the bulk core material.

Uniform Response With an Initial Peak Load. This was the characteristic behaviour of the sandwich panels with 16 mm outside diameter tubular inserts or 21.5 mm wide outside diameter conical inserts.

These specimens all exhibited a pronounced peak load at the end of the elastic region. Audible cracking accompanied the termination of the peak load, suggesting failure of the FRP inserts. The load then dropped and assumed a largely uniform response, similar to that typical of foams, but at a higher average crush level.

These observations would tend to indicate an initial catastrophic failure of the FRP inserts rather than the onset of controlled progressive crushing. X-ray analysis supported this hypothesis. Figure 3 shows the sequence of failure for a 16 mm outside diameter tube. The images clearly show the development of a central circumferential fracture. This then generates a telescopic collapse mode in which the bottom half of the tube slides up inside the top half. Apart from the initial fracture, there is no further evidence of brittle failure. This explains the uniformity in the load-displacement characteristic.

It is believed that the majority of the specimens which failed in a catastrophic manner did so because of inconsistencies in the fibre distribution within the FRP tubes

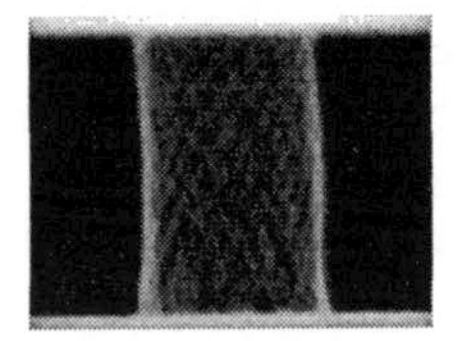 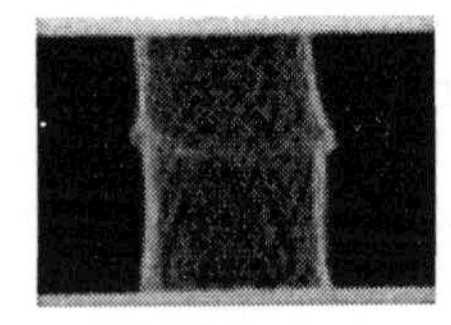 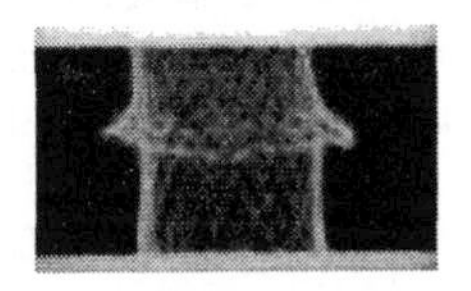 

FIGURE 3. X-ray images showing the sequence of failure for a 16 mm outside diameter tube. Images are shown at crush distances of 0 mm, 2 mm, 4 mm and 13 mm.

and frusta. Unstable failure was initiated from a local non-uniformity in the material or geometry as a result of fibre displacement during lay-up or injection.

Non-Uniform response. This was the characteristic behaviour of the remainder of the specimens, i.e. those with 15 mm outside diameter tubes and 20.6 mm wide outside diameter. Following initial failure, the load-displacement characteristics of these specimens exhibited pronounced serrations. Furthermore, there was audible cracking throughout the crush event.

These observations are consistent with the progressive crushing of the inserts. Figure 4 depicts the sequence of failure for a 15 mm outside diameter tube which clearly shows *controlled* brittle failure from one end of the insert. The serrations in the load-displacement characteristic are caused by the stick-slip nature of the brittle fracture processes.

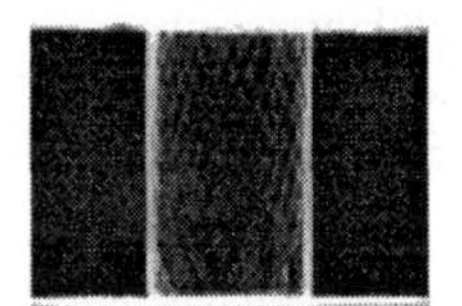  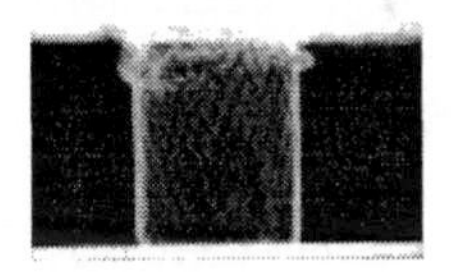 

FIGURE 4. X-ray images showing the sequence of failure for a 15 mm outside diameter tube. Images are shown at crush distances of 0 mm 2 mm, 4 mm and 13 mm.

4.1.2. Energy Absorption Capability

Having established the failure mechanisms of the different types of sandwich panel under edgewise compression, their energy absorption capabilities were then assessed. The amount of energy absorbed by each specimen was calculated from the area under its load-displacement characteristic. These absolute energy absorptions were then normalised by mass in order to allow for a direct comparison between specimens.

Figure 5 shows the variation of mass specific energy absorption with insert geometry. The columns represent mean values for specimens with a given type of insert, and the error bars indicate recorded extremes. It can be seen that only two of the insert geometries provided significant improvements over the basic sandwich construction. In other words, for many specimens, any increase in energy absorption by virtue of the inserts was offset by their higher mass. The two insert geometries which did show improvements were the 15 mm outside diameter tubes and the 20.6 mm wide outside diameter. Sandwich panels with these types of insert showed increases in mean specific energy absorption of 12% and 34% respectively. It should be recalled that these were the only geometries which were found to crush in a stable, progressive manner. Therefore, the energy absorption potential of the brittle fracture processes was exploited very efficiently. Those specimens which failed in a catastrophic manner were much less efficient and resulted in correspondingly lower values of specific energy absorption.

It can also be seen from Figure 5 that the specimen with the 20.6 mm wide outside diameter frusta offered a high level of repeatability. This is significant because repeatability is an important aspect of crashworthy design. A minimum level of performance must be guaranteed and, on a larger structural scale, there will be a need to predict and ensure a preferred sequence of collapse.

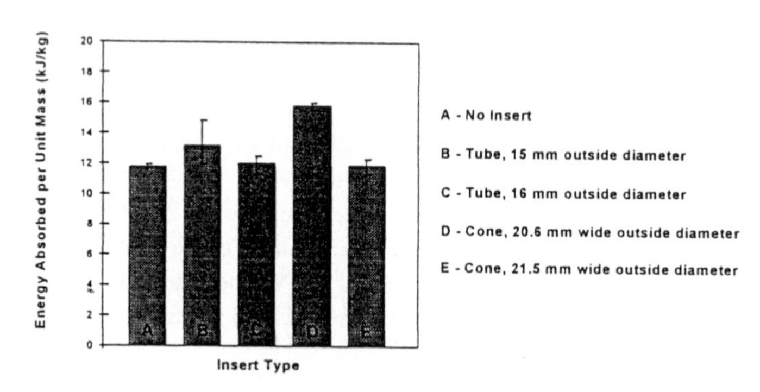

FIGURE 5. Comparison of mass specific energy absorptions for the different types of sandwich panel.

4.2. EDGEWISE COMPRESSION

Under edgewise compression, there was a tendency for the sandwich panel specimens to fail by unstable buckling. Progressive crushing with useful energy absorption did not generally occur. This was because the majority of the loading was taken by the thin facings rather than by the core material.

The load-displacement characteristics of all specimens tested were of the same general form (Figure 6). Each exhibited an initial linear-elastic region which terminated in a very high peak load before dropping. This peak load coincided with the failure of one of the face plates. All specimens then proceeded to buckle, with the outside of one of the facings bent into a concave profile, and the outside of the other into a convex. After this initial failure, crushing continued at low load until the specimens rotated or slipped out of the grips.

More specifically, the test specimens typically exhibited one of two types of load-displacement response. The source of this variation was face plate separation (see Figure 6). Those specimens which did exhibit face plate separation had a lower initial peak load than those which did not. Furthermore, catastrophic face plate separation was accompanied by a very sharp drop in crushing load and then a somewhat uneven load-displacement characteristic. Specimens whose facings remained intact exhibited a less dramatic drop in peak load and a subsequently much more uniform response.

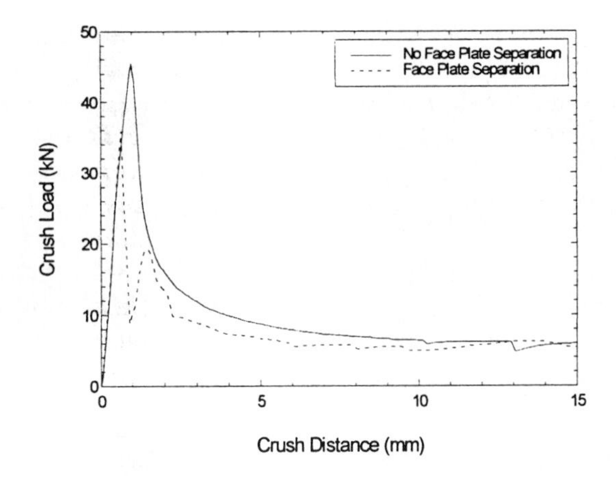

FIGURE 6. Comparison between the load-displacement characteristics of a sandwich panel which exhibited face plate separation under edgewise compression and one which did not.

The post-failure response and structural integrity of the sandwich specimens was found to be greatly influenced by the FRP inserts. These were supposed to inhibit face plate separation by effectively tying opposing facings together. In practice, this was achieved with varying degrees of success.

Specimens with no GRP inserts and those with conical inserts showed the complete separation of one facing (Figure 7). With the conical inserts, this always involved the facing attached to the narrow ends of the frusta where the forces acting to separate the facings were concentrated over a relatively small interface region.

Specimens with tubular inserts showed no face plate separation, with the panels crushing quasi-progressively from one end (Figure 8). Clearly, these specimens exhibited the best failure characteristics. They retained their structural integrity and demonstrated the most predictable and repeatable load-displacement characteristics.

5. Conclusions

Foam-cored sandwich panels with integral FRP tubes and frusta have been tested under quasi-static flatwise and edgewise compression.

Under flatwise compression, the sandwich panels which failed by stable progressive crushing exhibited the best energy absorption characteristics. The main prerequisite for ensuring this mode of collapse was consistency in the fibre distribution within the GRP tubes and frusta. Otherwise the inserts tended to fail catastrophically at a local non-uniformity in material or geometry. Conical inserts were generally found to offer the most predictable and repeatable performance as their geometry assisted in ensuring consistency of manufacture.

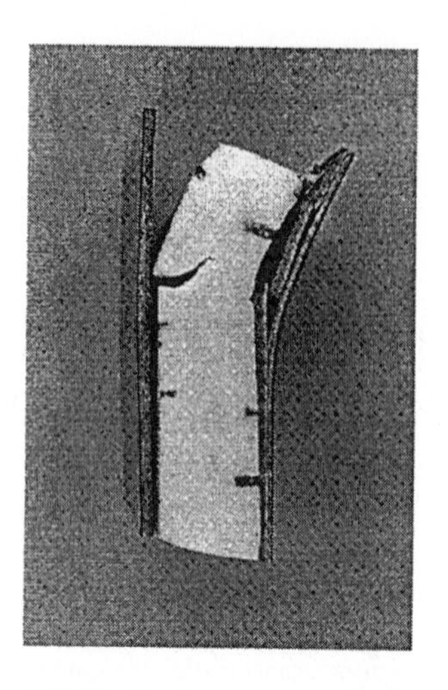

FIGURE 7. Face plate separation: a sandwich panel with no FRP inserts.

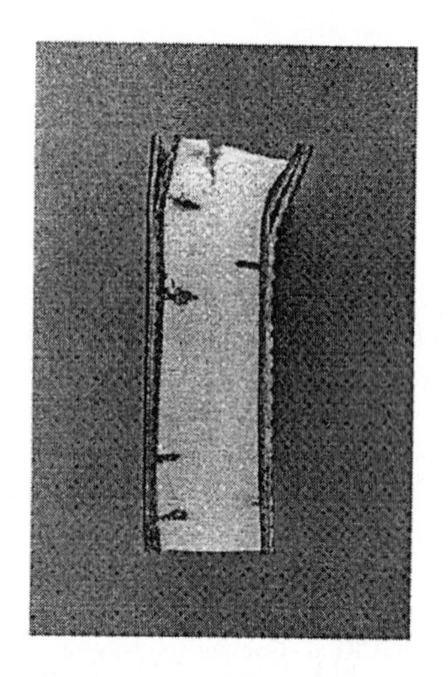

FIGURE 8. No face plate separation: a sandwich panel with tubular FRP inserts.

Under edgewise compression, the sandwich panels with inserts which prevented separation of the face plates had the preferred failure characteristics. These exhibited the most predictable and repeatable failure mechanism.

Overall, for the specimens tested, no one type of insert geometry was found to give the best performance under both loading conditions. The conical geometries were generally better under flatwise compression, and the tubes under edgewise. Therefore, in developing composite structures based on such material systems, consideration should be given to the local tailoring of insert geometries. Mechanical properties can then be optimised according to expected loading conditions.

6. References

1. Thornton, P.H.: Energy absorption in composite structures, *Journal of Composite Materials* **13** (1979), 247-262.
2. Schmueser, D.W. and Wickliffe, L.E.: Impact energy absorption of continuous fiber composite tubes, *Journal of Engineering Materials and Technology*, **109** (1987), 72-77.
3. Hull, D.: A unified approach to progressive crushing of fibre-reinforced composite tubes, *Composites Science & Technology* **40** (1991), 377-421.
4. Richardson, M.O.W., Robinson, A.M., Eichler, K. and Moura Branco, C.: Mechanical behaviour of a new stress dissipating composite sandwich structure, *Cellular Polymers* **13** (1994), 305-317.
5. Mamalis, A.G., Manolakos, D.E., Viegelahn, G.L., Sin Min Yap and Demosthenous, G.A.: On the axial crumpling of fibre-reinforced thin-walled conical shells, *International Journal of Vehicle Design* **12** (1991), 450-467.
6. Gibson, L.J. and Ashby, M.F.: *Cellular Solids: Structure & Properties*, Pergamon Press, Oxford, 1988.

THEORETICAL AND EXPERIMENTAL ANALYSIS OF DISSYMMETRICAL SANDWICH PLATES WITH COMPRESSIBLE CORE

B. Castanié [*] J.J Barrau [**] S. Crézé [***] J.P Jaouen [****]

** Professeur Agrégé. Université Paul Sabatier*
ENSAE, Laboratoire Structure 31055 Toulouse France
*** Professeur. Université Paul Sabatier, Bat 3PN 31062 Toulouse France*
**** Chef du laboratoire structure ENSAE 31055 Toulouse France*
***** Ingénieur Eurocopter France 13700 Marignanne France*

1-Introduction.

The design of some modern ultra-light weight aicraft or helicopter structures uses the dissymmetrical sandwich technology. In this kind of design, stresses are taken by the thick skin of the sandwich known as working skin and made of multi-layered carbon. The behaviour with global buckling is fulfilled by the core and a thin skin known as stabilizing skin made of two carbon or Kevlar layers. The certification of such structures rests partly on monoaxial tests in compression or shear with deformable square of representative test plates, or on tests under complex stresses of complete substructures. The purpose of this study is on one hand to build an original way of testing which would be able to test representative plates under uniaxial compression, shear and combined compression/shear loading, and on the other hand to build a specific non-linear theory with the associated program able to describe the behaviour of the real plates.

2-Experimental Study.

An original test machine was designed and built at the Sup'Aéro Structure Laboratory. The complete test installation is showed in Fig 1. The representative test plate (1) is bolted to the machine in its central part. This last one is made of one longitudinal box (2) and two crossing I-beams (3).

A. Vautrin (ed.), Mechanics of Sandwich Structures, 271–279.

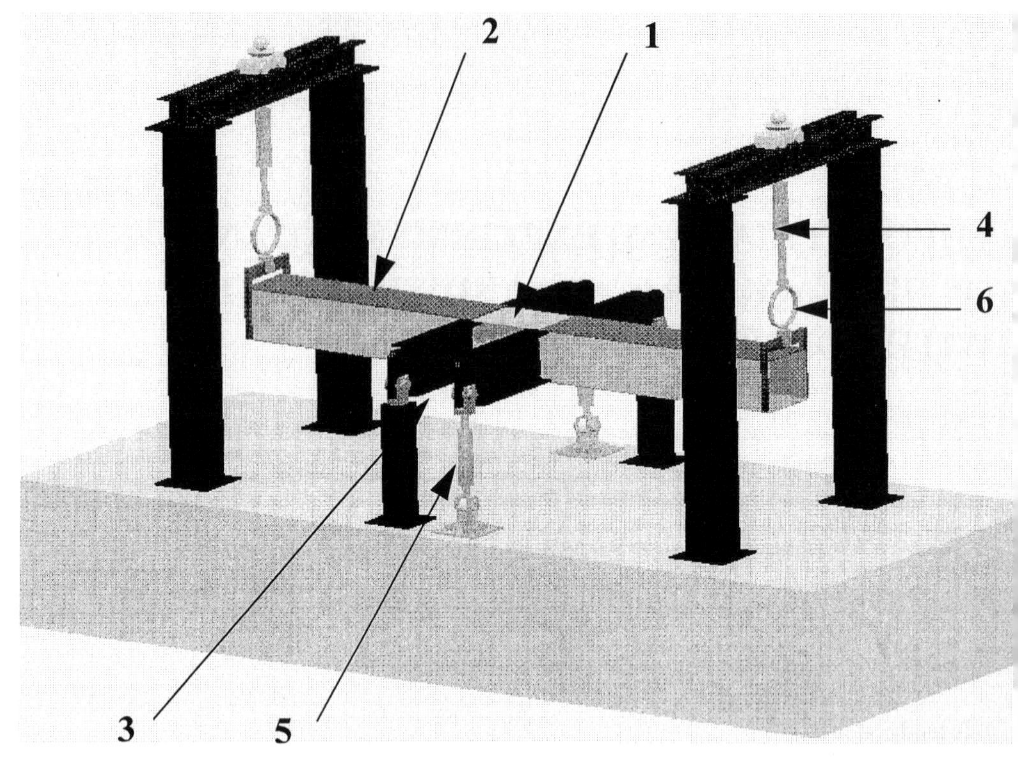

Figure 1. The testing machine

When the box is loaded with the actuators (4) at its two end points, then the test machine is subjected to four point bending and the test plate is loaded under compression or traction. When the I-beams are loaded with the actuators (5), an antisymmetric displacement of the end points of the I-beams is achieved so that a torque in the central part of the box is generated and the test plate is loaded under shear. If these two ways are simultaneously carried out the test plate is loaded under combined stresses. The loads are measured with the two load sensors (6) and specific strain gauges on the strap of the actuators (5). The test plate provided by Eurocopter France (Fig 2) comprises a peripheral zone due to bolted fixing and a central test zone of 200x200mm representative of the dissymetrical sandwich technology:

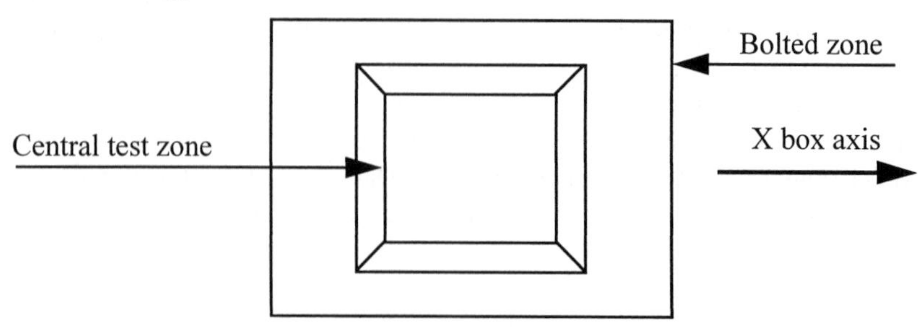

Figure 2. Test plate design.

At the present stage of the study, two tests under compression and two tests under combined loads have been performed. Every test plate is equipped with many strain gauges allowing a quantitative study. On half of the test plates a photoelastic plate is stuck allowing a qualitative study of the stress distribution for the lower stresses while the other half is painted in white in order to make a Moiré study of the deformation state. The strain values reached at the center of the working skin at the last force increment during the test in the principal direction and the angle with the X- box axis are given in the following table:

	COMPRESSION N°1	*COMPRESSION N°2*	*COMBINED N°1*	*COMBINED N°2*
Direction 1	-10460	-12650	-10018	-11470
Direction 2	1300	1304	3522	5360
Angle	2	2	28	60
Failure mode	WORKING SKIN	WORKING SKIN	WORKING SKIN	WORKING SKIN

Table 1. Maximun strains and failure modes for the tests performed.

In each case, the failure occurs when the maximum compression strain of the carbon is reached. Besides the way the sandwich plate breaks is always the same, a static failure of the working skin following the 90° direction of the compression principal direction. For example the breaking pattern of the compression n°1 and combined n°2 plates are showed in fig 3:

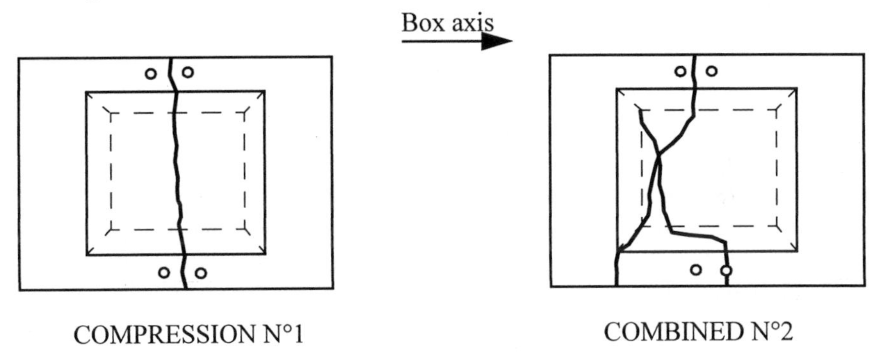

Figure 3. Breaking pattern examples.

As a conclusion to the first part of the study, it has been demonstrated that this original machine design is fully able to carry out tests under compression/shear loadings. The whole clamped boundary conditions of the test machine delays the emergence of the non-linear phenomena by comparison with the usual tests. However, the test machine is more representative of aircraft structures and gives a better idea of the static margin. Besides, the high strain values that have been obtained validate the technology of dissymmetrical sandwich structures. The study will continue with impacted plate tests to give the margin for this case. Tests under shear stresses will be conducted.

3-Theoretical study.

In this second part of this study, the aim is to make an industrial user-friendly program able to describe the behaviour of the real plates for the lowest computing cost. Because of the complexity of the peripheral zone, only the central part of the sandwich plate is described (fig4). The geometric parameters are also given in this figure and we give the subscript **1** to the working skin, **2** to the stabilizing skin and **c** to the core. The center of the working skin is the origin of the axis. Because of the large deflection but small rotation and strain, the Von Karman theory is used [3, 6].

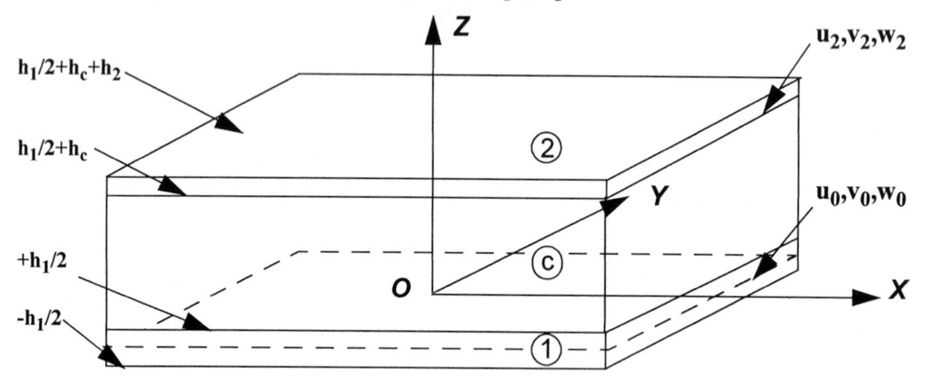

Figure 4. Sandwich plate geometry and displacement unknows.

3.1 HYPOTHESIS FOR THE WORKING SKIN

The working skin remains thin with this technology but is subjected to local flexural moment, consequently the Kirchoff plate theory is applied. This leads naturally to the choice of the displacement of the midplane (u_0,v_0,w_0) as unknown factors (see fig 4). So the displacement field is:

$$\bullet u_1(x,y,z) = u_0(x,y) - z \cdot \frac{\partial w_0}{\partial x} \quad . \tag{1}$$

$$\bullet v_1(x,y,z) = v_0(x,y) - z \cdot \frac{\partial w_0}{\partial y} \tag{2}$$

$$\bullet w_1(x,y,z) = w_0(x,y) \tag{3}$$

And the nonlinear Von Karman strains have the following expressions:

$$\varepsilon^1_{yy} = \frac{\partial v_0}{\partial y} - z \cdot \frac{\partial^2 w_0}{\partial y^2} + \frac{1}{2} \cdot \left(\frac{\partial w_0}{\partial y}\right)^2 \tag{4}$$

$$\gamma^1_{xy} = \frac{\partial u_0}{\partial y} + \frac{\partial v_0}{\partial x} - z \cdot \frac{\partial^2 w_0}{\partial x \partial y} + \frac{\partial w_0}{\partial x} \cdot \frac{\partial w_0}{\partial x} \tag{5}$$

$$\varepsilon^1_{xx} = \frac{\partial u_0}{\partial x} - z \cdot \frac{\partial^2 w_0}{\partial x^2} + \frac{1}{2} \cdot \left(\frac{\partial w_0}{\partial x}\right)^2 \tag{6}$$

Since the working skin remains thin and if the stacking sequence is built according to a mirror symmetry, there is no membrane-bending coupling and the skin is homogenized as an orthotropic skin. So the constitutive law is [2]:

$$\begin{bmatrix} \sigma^1_{xx} \\ \sigma^1_{yy} \\ \tau^1_{xy} \end{bmatrix} = \begin{bmatrix} E^1_{11} & E^1_{12} & 0 \\ E^1_{21} & E^1_{22} & 0 \\ 0 & 0 & G^1_{12} \end{bmatrix} \bullet \begin{bmatrix} \varepsilon^1_{xx} \\ \varepsilon^1_{yy} \\ \gamma^1_{xy} \end{bmatrix} \tag{7}$$

Where the E_{ij} are the equivalent orthotropic coefficients for the skin stacking sequence.

3.2. HYPOTHESIS FOR THE STABILIZING SKIN.

The stabilizing skin is generally made with one or two layers and therefore it is comparable to a membrane. The displacements are uniform throughout the thickness and we choose those of the interface with the core as unknown factors (u_2,v_2,w_2) (see fig 4). The skin is also assumed to be orthotropic. The expression of displacements, strains and the constitutive law are very similar with those of the working skin without the bending or curvature terms.

3.3. HYPOTHESIS FOR THE CORE

The compressive and shear strains are supposed to be uniform throughout the thickness of the core. The face-parallel stresses are negligible. Besides, coupling between compression and shear are also energetically negligible. So with the low rotation hypothesis, the displacement fields have the following expressions:

$$u_c(x,y,z) = \frac{1}{h_c} \cdot \left[\left(z - \frac{h_1}{2}\right) \cdot u_2 - \left(z - \left(\frac{h_1}{2} + h_c\right)\right) \cdot u_0 + \frac{h_1}{2} \cdot \left(z - \left(\frac{h_1}{2} + h_c\right)\right) \cdot \frac{\partial w_0}{\partial x}\right] \tag{8}$$

$$v_c(x,y,z) = \frac{1}{h_c} \cdot \left[\left(z - \frac{h_1}{2}\right) \cdot v_2 - \left(z - \left(\frac{h_1}{2} + h_c\right)\right) \cdot v_0 + \frac{h_1}{2} \cdot \left(z - \left(\frac{h_1}{2} + h_c\right)\right) \cdot \frac{\partial w_0}{\partial y}\right] \tag{9}$$

$$w_c(x,y,z) = \frac{z - \frac{h_1}{2}}{h_c} \cdot [w_2 - w_0] + w_0 \tag{10}$$

And the strain fields become after simplification:

$$\varepsilon^c_{zz} = \frac{w_2 - w_1}{h_c} \tag{11}$$

$$\gamma^c_{xz} = \frac{(u_2 - u_0)}{h_c} + \frac{1}{h_c} \cdot \left[\left(z - \frac{h_1}{2}\right) \cdot \frac{\partial w_2}{\partial x} - (z - (h_1 + h_c)) \cdot \frac{\partial w_0}{\partial x}\right] \tag{12}$$

$$\gamma^c_{yz} = \frac{(v_2 - v_0)}{h_c} + \frac{1}{h_c} \cdot \left[\left(z - \frac{h_1}{2}\right) \cdot \frac{\partial w_2}{\partial y} - (z - (h_1 + h_c)) \cdot \frac{\partial w_0}{\partial y}\right] \tag{13}$$

While the constitutive relation can be expressed as [4]:

$$\begin{bmatrix} \sigma^c_{zz} \\ \tau^c_{xz} \\ \tau^c_{yz} \end{bmatrix} = \begin{bmatrix} E^c_{zz} & 0 & 0 \\ 0 & G^c_{xz} & 0 \\ 0 & 0 & G^c_{yz} \end{bmatrix} \bullet \begin{bmatrix} \varepsilon^c_{zz} \\ \gamma^c_{xz} \\ \gamma^c_{yz} \end{bmatrix} \tag{14}$$

So the strain energy in the core Wc has got an explicit compression term which is not considered negligible with this approach :

$$Wc = \iiint(\sigma^c_{zz} \cdot \varepsilon^c_{zz} + \tau^c_{xz} \cdot \gamma_{xz} + \tau^c_{yz} \cdot \gamma^c_{yz})dV_c \tag{15}$$

3.4 RITZ UNKNOWN FACTORS

The displacement fields are approximated with a Ritz method using a polynomial basis. For numerical reasons, non dimensional coefficients ξ and η are used to give the x-y position of a current point of the sandwich plate (fig 5). Besides, the problem remain isostatic for u_0 and v_0 displacements and 3 degrees of freedom must be locked in the x-y plane (Fig 5).

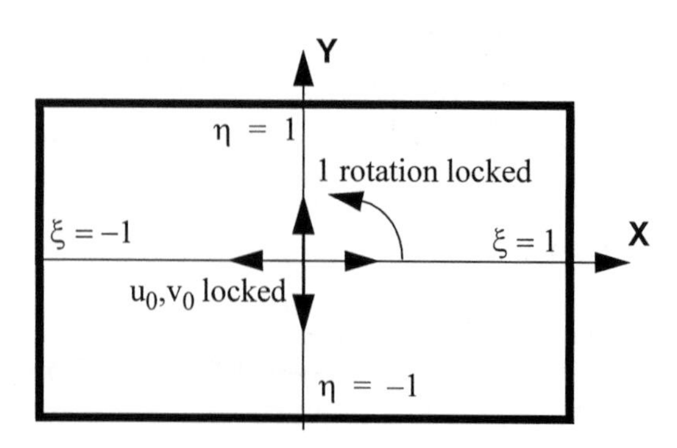

Figure 5. Degrees of freedom locked in the x-y plane.

The associated mathematical expressions are:

$$\bullet u_0(0,0) = 0\ ;\ v_0(0,0) = 0;\qquad \frac{\partial}{\partial\eta}u_0(0,0) = 0 \tag{16}$$

And the Ritz polynomials for the membrane displacements become:

$$u_0(\xi,\eta) = \sum_{k=1}^{M_{u_0}} B_{k0}\cdot\xi^k + \eta\cdot\sum_{k=1}^{M_{u_0}} B_{k1}\cdot\xi^k + \sum_{k=0}^{M_{u_0}}\sum_{l=2}^{N_{u_0}} B_{kl}\cdot\eta^l\cdot\xi^k \tag{17}$$

$$v_0(\xi,\eta) = \sum_{l=1}^{M_{v_0}} C_{0l}\cdot\eta^l + \sum_{k=1}^{M_{v_0}}\sum_{l=0}^{N_{v_0}} C_{kl}\cdot\eta^l\cdot\xi^k \tag{18}$$

For the deflection, the following function was chosen:

$$w_0(\xi,\eta) = \sum_{k=0}^{M_{w_0}}\sum_{l=0}^{M_{w_0}} A_{kl}\cdot\xi^k\cdot\eta^l\cdot(1+\xi)^{L_1}\cdot(1-\xi)^{L_2}\cdot(1+\eta)^{L_3}\cdot(1-\eta)^{L_4} \tag{19}$$

It can describe the usual boundary conditions: free, simply supported or clamped edges according to the values 0, 1 or 2 assigned to the L_i coefficients . The polynomials of the stabilizing skin displacement (u_2,v_2,w_2) are made with the basis $\xi^i.\eta^i$. Because of the geometric non linearity, the membrane displacements are dependent upon the deflection and the compatibility between the different polynomials must be achieved. For exemple:

$$M_{u_0} = 2\cdot(M_{w_0} + L_1 + L_2) - 2 \tag{20}$$

So, the number of unknown factors are dependent on the L_i boundary conditions and the degree of w_0. Table 2 gives the total number of unknows factors for 3 different problems and different degrees of w_0 function.

DEGREE	1	2	3
2 sides simply supported	57	149	277
4 sides simply supported	161	287	449
4 sides clamped	437	631	861

Table 2. Number of unknown factors.

3.5 SOLUTION METHOD

After writing the Principle of Virtual Works, to minimize the residual force vector, the tangent stiffness matrix is built. Then the standard Newton-Raphson method or the initial stress method are applied. The matricial formulation is borrowed from nonlinear finite element methods [3, 6]. This leads to the calculus and assembly of 81 elementary stif-

stiffness matrix. From a practical point of view, a two-sides simply supported problem takes 35" on a Pentium 90 micro-computer.

3.6 CORRELATION OF THE SECOND COMPRESSION TEST

To describe the behaviour of the test plates with the simplified geometry (fig4), according to the real design of the plates, a constant part N% of the stresses per unit length can enter the stabilizing skin directly and the loading plane can be different from the mid-plane of the working skin. The compression case is given in the fig 6 :

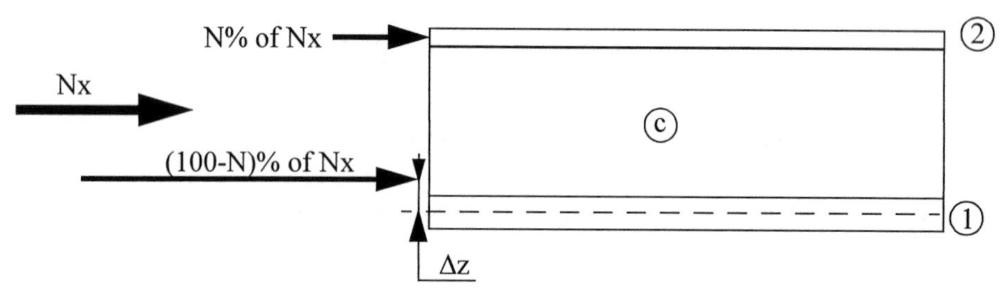

Figure 6. Loading of the dissymetrical sandwich plate in the compression case.

To correlate the second test under compression, all the edges are clamped and N = 4 %. This gives the following results (The strains in fig 6 are given at the center of the skins) :

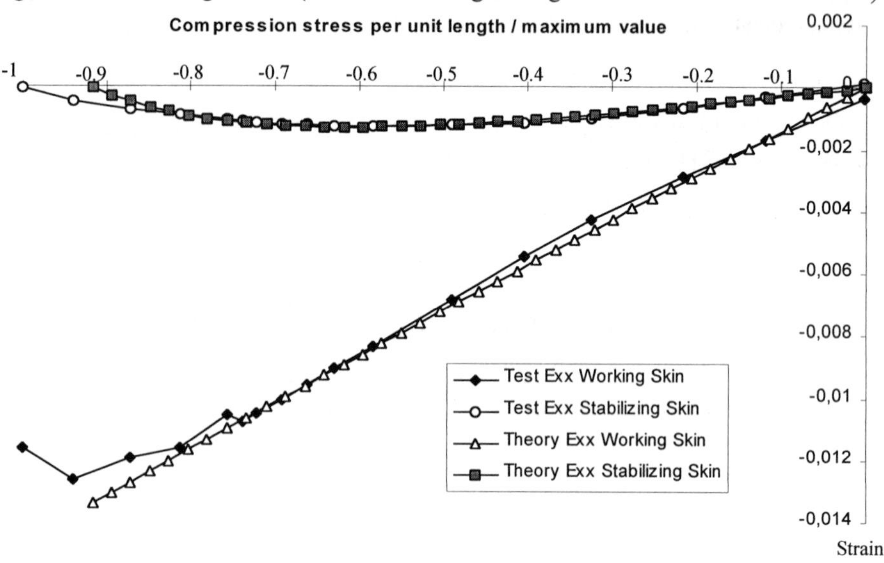

Figure 7. Correlation of the second compression test.

The calculus is made with a degree 1 for the w_0 polynomial and it appears to be enough for a good correlation of the compression strain at the center of the skin. The difference observed for the higher strain is due to the difficulty in obtaining the static equilibrium of the test machine with the actuators (5) and a break in linearity may occur. The N% coefficient is a coefficient representative of two physical phenomena. First the well-known fact that the real boundary conditions cannot be fully clamped. Second, the real geometry and the tapered transition zone are determining factors in the mechanical behaviour of the working skin for the relatively short dimensions of the test plates.

4- Conclusion:

As a conclusion, a non-linear theory able to describe the dissymmetrical sandwich behaviour with a compressible core has been developed. The hypothesis and the choice of the displacement unknowns leads to the simpliest required formulation. A good correlation between tests and theory is observed for the compression tests. The program enables the engineer to validate his design choices more quickly than the finite element method. The original test machine demonstrates the validity of the design to perform tests under combined stresses. It will now be used to carry out shear tests of the dissymmetrical sandwich plates and local buckling studies. The study will go on with testing of other kinds of polynomials in order to limit the number of unknown factors and further with the theory of the dissymetric sandwich shell and tests using the same technology.

5-References:

1. Allen, H.G., (1969) *Analysis and Design of Structural Sandwich Panels*, Pergamon Press, Oxford.

2. Barrau, J.J.,Laroze, S.,(1987) *Mécanique des structures, Tome 4, Calcul des structures en matériaux composites*, Eyrolles-Masson.

3. Cristfield, M.A., (1991) *Non linear finite element analysis of solids and structures*, Wiley.

4. Gay, D., (1991) *Matériaux composites*, Hermes.

5. Lee, L.J. and Fan, Y.J., (1996), Bending and vibration analysis of composite sandwich plates, *Computers and structures* **60**, 106-112.

6. NAFEMS,(1991), *Introduction to Nonlinear Finite Element Analysis*, E Hinton.

VISCOELASTIC PROPERTIES OF STEEL/POLYMER/STEEL SANDWICH STRUCTURES

S. BISTAC, M.F. VALLAT, J. SCHULTZ
Institut de Chimie des Surfaces et Interfaces
15, rue Jean Starcky 68057 Mulhouse cedex France

Abstract

Polymer lamined-steel sandwiches are used for their vibrations and sound damping properties in the automotive industry and household applications. The studied sandwich assemblies are constituted by two steel sheets and one thin polymer layer (ethylene vinyl acetate copolymer EVA) inserted between them. In this work, the influence of the thickness of the polymer on the damping properties of the sandwiches has been analysed. Dynamic mechanical measurements have been performed on steel/polymer/steel sandwiches. The mechanical transition temperature Tmech at which tanδ goes through a maximum is related to the glass transition of the polymer, and reflects the mobility of the polymer chains. The results show that for high polymer thicknesses, Tmech remains constant and corresponds to the glass transition temperature of the bulk copolymer. However, when the polymer thickness decreases, Tmech increases greatly. This increase can be explained by the reduced mobility of the polymer chains near the interface. These observations suggest the presence of an interphase, localized in the vicinity of the interface with the substrate, for which the polymer properties differ from that in the bulk. Migration and orientation of the polar acetate groups of the EVA towards the polar steel surface and consequently crystallinity modifications are able to explain the reduction of the chains mobility.

1. Introduction

Polymer laminated-steel sandwiches are used for their vibrations and sound damping properties in the automotive industry and household applications. The studied assemblies are constituted by two steel sheets and one thin polymer layer inserted between them. The viscoelastic behaviour of the polymer is considered to be the most important parameter to assure sufficient damping properties [1]. In this work, the influence of the thickness of the polymer on the damping properties of the sandwiches has been analysed. The relationship between the damping properties of the polymer/metal sandwiches and the bulk properties of the polymer has been examined. Viscoelastic properties of steel/ethylene vinyl acetate copolymer (EVA)/steel sandwiches have been studied by dynamic mechanical measurements. The variation of

A. Vautrin (ed.), Mechanics of Sandwich Structures, 281–286.

the loss factor tanδ with temperature is analysed as the function of the thickness of the polymer film inserted in the sandwiches.

2. Experimental procedure

The polymer used is a ethylene vinyl acetate copolymer (EVA) containing 28 wt % of vinyl acetate and grafted with about 1% maleic anhydride. This grafting improves adhesion to steel [2]. The substrate used is mild steel provided by Sollac (Usinor France), with a thickness equal to 0.35 mm. The surface is degreased by cleaning with trichloroethane in an ultrasonic bath.

The sandwiches are obtained under pressure at 180°C. The pressure of 1.5 MPa is applied during 5 minutes. Circulating water in the press platens ensures the cooling of the assemblies. Different thicknesses ranging from 20 to 350 μm are obtained by inserting spacers between the two steel sheets.

Dynamic mechanical measurements have been performed on steel/polymer/steel sandwiches using DMTA instrument from Rheometric Scientific. The global response of the sandwich submitted to a three points flexure test is recorded. A sample 10 mm wide and 45 mm long is clamped in its extremities and submitted in the middle to sinusoidal vibrations. Measurements are made at 1 Hz from -80 to +80°C with a scanning temperature of 2°C/min. This setup provides the real (E') and imaginary (E") parts of the dynamic modulus and the loss factor tanδ (= E"/E').

The variation of the loss factor tanδ is studied as a function of temperature and is related to the damping properties of the assembly. The mechanical transition temperature Tmech at which tanδ goes through a maximum is related to the glass transition of the copolymer, and reflects the mobility of the polymer chains.

3. Results

In figure 1 is given the variation of tanδ as a function of the temperature for steel/EVA/steel sandwiches, with polymer layers of thicknesses ranging between 40 and 350 μm.

The amplitude of the damping peak decreases, due to the fact that the relative amount of viscoelastic material (EVA) is smaller when the polymer thickness decreases.

But, more surprising and interesting, the mechanical transition temperature Tmech is shifted towards higher temperatures when the polymer thickness decreases.

Figure 2 presents the evolution of Tmech. as a function of the EVA thickness.

At high thickness values, Tmech remains constant and corresponds to the glass transition temperature of the bulk copolymer (-15°C).

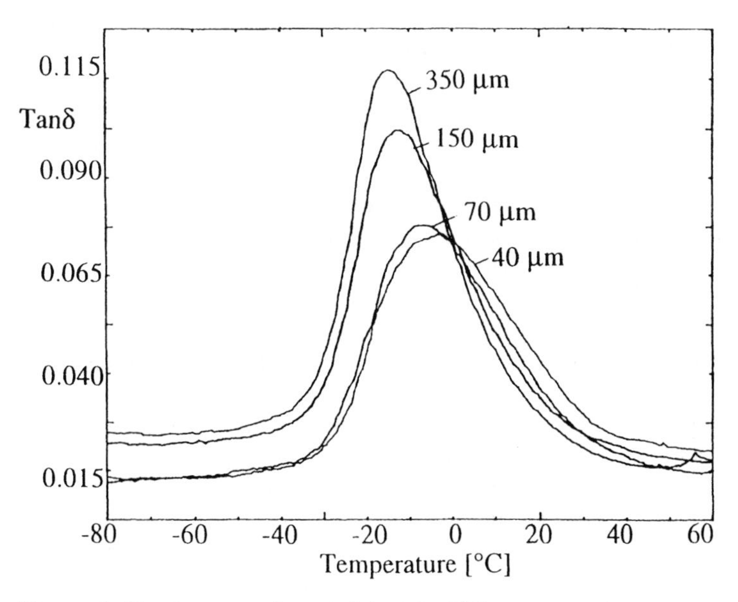

Figure 1. Tanδ curves of sandwiches for different EVA thicknesses.

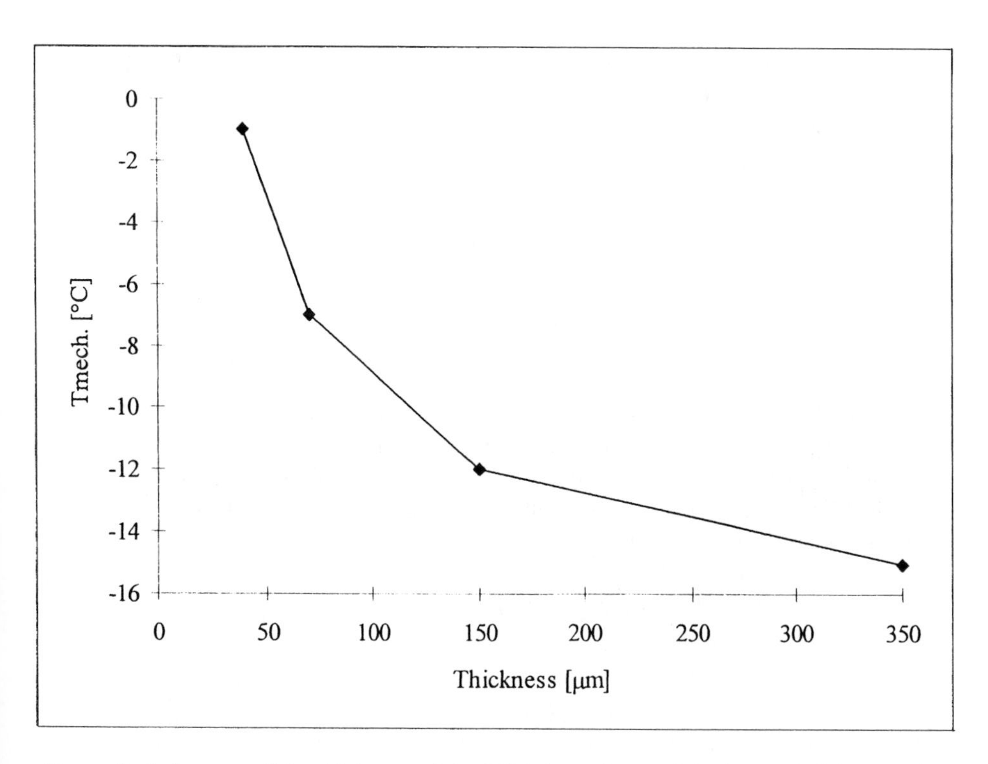

Figure 2. Influence of the thickness of the EVA layer inserted in the sandwiches on the Tmech value

In order to be sure that the observed variations cannot be attributed to mechanical coupling effects, other experimental work has been performed using other polymers such as styrene-butadiene rubber (SBR) or polyvinyl acetate (PVAc), which weakly adhere to the steel substrate, and polyamide 11 (PA11), a strongly adhering polymer. It has been shown that no significant evolution of Tmech occurs as a function of polymer thickness for these other studied systems, at least in the range of thickness studied. Table 1 summarises the results obtained for the steel/EVA/steel sandwiches and for the sandwiches using other polymers.

TABLE 1 . Tmech values of steel/EVA/steel sandwiches, steel/ SBR/steel sandwiches, steel/ PVAc/steel sandwiches and steel/PA11/steel sandwiches, for different thicknesses of polymer.

Sandwiches	Thickness [μm]	Tmech. [°C]
Steel/EVA/Steel	40	-1
	70	-7
	150	-12
	350	-15
Steel/SBR/Steel	50	-41
	100	-45
	300	-43
Steel/PVAc/Steel	40	41
	80	39
Steel/PA11/Steel	50	59
	120	56

4. Discussion

The increase of Tmech with decreasing thickness can be explained by the reduced mobility of the polymer chains near the interface. These observations suggest the presence of an interphase. An interphase is a zone in the polymer, localised in the vicinity of the interface with the substrate, for which the polymer properties differ from the bulk polymer properties. When the polymer thickness is changed, the relative proportion of the interphasial zone with respect to the bulk polymer varies and increases when the polymer thickness decreases.

For thin polymer thickness, the proportion of the interphase is higher and its properties are also detectable. For high thickness value, this interphase exists, but his viscoelastic properties are not detected and are concealed by the bulk properties

The results indicate that the interface with the substrate has an influence on the polymer properties, leading to the formation of an interphase with restricted mobility.

The increase of Tmech has already been reported in the literature for polymer/substrate assemblies. A similar evolution of Tmech has been that reported for similar assemblies by Braunisch [3]. Studies of assemblies with epoxy resins have also shown an increase

of Tmech when the polymer thickness decreases, attributed to the presence of an overcrosslinked interphase in contact with the substrate [4]. Interfacial layers with specific properties have been also evidenced for poly(methyl methacrylate) films of different thicknesses [5].
The observed phenomena for steel/EVA/steel sandwiches may be therefore attributed to the formation of an interphasial layer in which the amorphous chains are constrained differently from the bulk chains. The strong interfacial interactions between grafted EVA and the steel surface cannot explain the reduction of the chains mobility in the intephase. Results obtained for sandwich containing non-grafted EVA (with low adhesive properties) have in fact shown the same evolution of Tmech when the polymer thickness is decreased. A more plausible explanation is based on microstructural changes of the polymer, in the vicinity of the interface with the steel.
The EVA studied is a semi-crystalline copolymer, with a degree of crystallinity of about 21%. The melting peak is rather complex and widespread starting around room temperature up to 120°C, with two major components : one at low temperature (around 45-50°C) and a second one at higher temperature (around 90°C).
Previous studies by Differential Scanning Calorimetry (DSC) of the EVA films inserted in the sandwiches have shown some crystallinity organisation differences as a function of the polymer thickness [6]. For thin layers of EVA, the contribution of the low temperature peak, corresponding to small and disorganised crystals, is much more pronounced.
More perturbed crystals present in the interphasial zone can constitute physical ties reducing the mobility of the linking amorphous phase. These microstructural modifications are probably induced by the migration and orientation of the polar acetate groups of the EVA towards the polar steel surface, for interfacial energy minimisation reasons. An increase of the polar component of the surface energy when EVA copolymers have been moulded against a polar substrate has been observed [7] The increase of the polar component indicates that more vinyl acetate groups are present on the polymer surface. This orientation phenomena is able to induce microstructural changes, particularly crystalline modifications which perturb the mobility of the amorphous phase. Physical interactions are sufficient for these orientation effects, as shown by the results obtained for a non-grafted EVA.

5. Conclusion

The study of the viscoelastic properties of the steel/EVA/steel sandwiches have shown that the thickness of EVA in the sandwich can affect greatly the temperature of the mechanical transition Tmech. The increase of Tmech with decreasing thickness is the consequence of the reduced mobility of the polymer chains, in the vicinity of the interface with the substrate. These results indicate the formation of an interphase, whose properties are different from the bulk polymer properties. The decrease of the chains mobility in the interphasial zone is probably the consequence of a gradual change of the polymer properties near the interface with the steel. Migration and orientation of the polar acetate groups of the copolymer, inducing crystalline modifications can explain the reduction of the mobility of the polymer chains in the interphase. The variation of the mechanical transition temperature when the polymer thickness is varied can have some important consequences on the practical sound and vibrations damping properties of the assemblies. Moreover, it would be necessary to know precisely the thicknesss and the mechanical and viscoelastic characteristics of this interphase in order to make some mechanical calculations and simulations of the sandwich structure.

6. Aknowledgements

The authors wish to thank the SOLLAC Company for their technical and financial support, and specially Mrs Pascale Mercier and Mr Henri Guyon, from the LEDEPP Laboratory.

7. References

[1] Hartmann, B. (1990) *Sound and Vibration Damping with Polymers,* Corsaro, R.D & Sperling, L.H. Eds., ACS Symposium Series **424,** Washington.

[2] Bistac, S., Vallat, M.F. and Schultz, J. (1996) Adhesion in polymer/steel sandwiches, *J. Adhesion* **56,** 205-215

[3] Braunisch, H. (1969/1970) Schwingungsgedämpfte dreischichtige verbundsysteme, *Acustica* **22**, 136-144

[4] Bourhala, H., Chauchard, J., Lenoir, J. and Romand, J. (1990) Influence de l'épaisseur de l'adhésif et du vieillissement sur les propriétés mécaniques dynamiques d'un assemblage : adhésif structural/acier inoxydable, *Die Angewandte Makromoleculare Chemie* **178** (2941), 47-62

[5] Tretinnikov, O.N. and Zhabankov, R.G. (1991) The molecular structure and glass transition temperature of the surface layers of films of poly(methyl methacrylate) according to infrared spectroscopy data, *J. Mat. Sci. Letters* **10**, 1032-1036

[6] Bistac, S., Cheret, D., Vallat, M.F. and Schultz (1997) Interphases in ethylene vinyl acetate copolymer/ steel sandwiches, *J. Appl. Polym. Sci.*, in press

[7] Chihani, T., Bergmark, P. and Flodin, P. (1995) Surface modification of ethylene copolymers molded against different mold surfaces. Part 2 : Changes at the outermost surface, *J. Adhesion Sci. Technol* **9**(7) 843-857

BENDING FATIGUE BEHAVIOUR OF PUR-EPOXY AND PHENOLIC 3D WOVEN SANDWICH COMPOSITES

H. JUDAWISASTRA, J. IVENS, I. VERPOEST
Department of Metallurgy and Materials Engineering
Katholieke Universiteit Leuven, de Croylaan 2, B-3001 Leuven, Belgium

1. Abstract

In this paper the bending fatigue behaviour of 3D-woven sandwich composite panels is studied. Panels with different core properties were selected. During the fatigue test the displacement was monitored at a constant load to obtain the stiffness degradation. The Wöhler curves of all panels are presented. The results of the property degradation are correlated with the mechanical properties of the different sandwich panels. To obtain core shear failure, the fatigue set-up can be modified using panel with thicker skins.

2. Introduction

Sandwich materials have become very popular in the construction because of the light, cost-effective and durable structures they can generate. The properties of the sandwich structures are strongly influenced by the core material. This research focuses on a new type of integrated core sandwich panel based on 3D-woven sandwich fabrics.

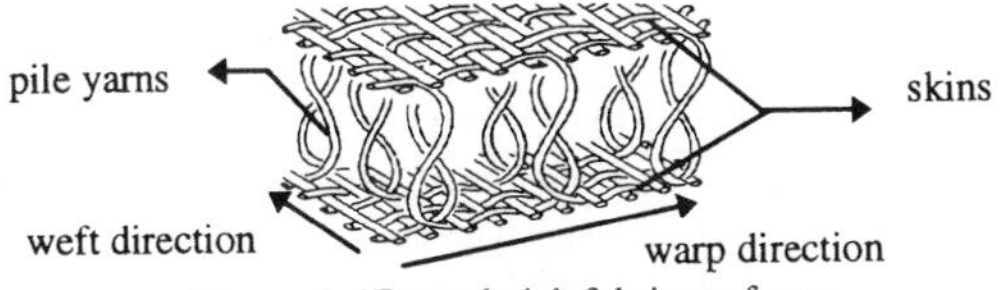

Figure 1. 3D sandwich fabric preform

The well established velvet weaving technique is used to produce the 3D sandwich fabric preform, but the cutting process of the piles is skipped (figure 1). Compared with other core materials (Honeycomb, PVC, and Balsa), 3D-woven sandwich fabrics show a cost advantage and also provide a higher delamination resistance. In this study, 3D sandwich fabrics made of glass fibre are impregnated with epoxy or phenolic resin. The empty core can be foamed up with a polyurethane or a phenolic foam to improve the shear resistance of the panels.

The 3D woven fabric panels have been applied in vehicle applications which are subjected to repetitive dynamic loads. The production process and the basic mechanical properties of the panels have been studied thoroughly at KU Leuven for several years.

A. Vautrin (ed.), Mechanics of Sandwich Structures, 287–294.

However, the fatigue data have not been available yet. Thus, it is necessary to study the long time behaviour of these 3D woven fabric panels. Since the shear resistance is a weak spot of the material [1], three point bending (3PB) fatigue tests will be performed.

3. Experimental Procedures

Four types of fibre glass-epoxy panels, 2 unfoamed and 2 PUR foamed panels, have been selected and produced at KU Leuven (table 1). On these four panels only the warp direction was tested as this is the weakest direction [1, 3]. The other panel, produced by Metalleido S.R.L (Italy), is a fibre glass-phenolic panel with phenolic foam injected into the core. This panel was tested in both warp and weft direction (figure 1). Additional layers of two dimensional fabric (2 layers on the top skin and 1 layer on the bottom skin) have been laminated on this phenolic panel.

TABLE 1. Materials

Panel	Skin Thickness (mm)	Nominal pile length (mm)	Foam	Foam density (kg/m^3)
Epoxy-89021	0.35	20	NO	-
Epoxy-89021	0.35	20	PUR	88
Epoxy-89018	0.35	10	PUR	76
Epoxy-86005	0.35	10	NO	-
Phenolic-89020	1.1 & 0.7	16	Phenolic	150

Static Three Point Bending (3PB) tests and compression tests were performed to determine the core properties and the static 3PB strength. The 3 point bending with combination of 2 different span lengths is carried out according to ASTM-standard C393. By this method it is possible to obtain the static ultimate load (P_u), core shear stress (τ_c) and the effective core shear modulus (G_c). The other 3PB test data can be seen in table 2. The flatwise compression test is performed, according to ASTM-standard C365 on a specimen of 50mm x 50 mm, to experimentally determine the compression strength, σ_c, of the sandwich panels.

TABLE 2. 3PB test data

3PB Static Test	Specimen length	Specimen thickness	Span width	Span length	Loading
Long Span	350 mm	Depend on the material used	50 mm	250 mm	till failure
Short Span	350 mm		50 mm	100 mm	elastically

Shear tests were also carried out to check the core shear strength and modulus, determined in the 3PB test. This test was carried out based on ASTM C-273 by which the core shear strength, τ_c, and the effective core shear modulus, G_c, are determined. The specimen width was kept the same (50 mm) while the specimen length was varied depending on the thickness of the panels.

Since large differences were found between the shear strength results from the shear test and the 3PB test, additional 3PB test with thicker skin were performed as a comparison test for the determination of the core shear properties. The dimensions of the specimens and set up of this test is the same as the previous 3PB test on long specimens and loading was applied up to failure. The only difference is that stainless steel skins, 1 mm thickness, are attached as extra skins on both sides of the panels .

Fatigue tests, with R = 0.1, were performed with the same set up as used for the static 3PB test on a long span. A frequency of 1-2 Hz was chosen. Different load levels, based on the 3PB static strength (80% and 65% of 3PB Pu), were chosen to determine the Wöhler curve. Tests were perfomed up to failure or to a maximum of 10^6 cycles. The machine will stop automatically if the displacement exceeds 20 mm either due to specimen failure or stiffness degradation. With the help of the Computer Aided Fatigue Testing (CAFT) program, the displacement is monitored at a constant load to obtain the stiffness degradation. The Stiffness Degradation (SD) is obtained from :

$$SD = \left[\left(\left(\frac{\Delta P}{\Delta \delta}\right)_{\text{final cycle}} - \left(\frac{\Delta P}{\Delta \delta}\right)_{\text{first cycle}}\right) \Big/ \left(\frac{\Delta P}{\Delta \delta}\right)_{\text{first cycle}}\right] \times 100\% \quad (1)$$

with : ΔP = difference between maximum and minimum load (N)
$\Delta\delta$ = displacement of the specimen due to ΔP (mm)

4. Results and Discussion

4.1. BASIC MECHANICAL CORE PROPERTIES

4.1.1. *3PB and Compression Tests Results*

Figure 2 shows a variation in core mechanical properties of the panels. This properties are dependent on the microstructure of the core (pile fibre density, pile fibre length, pile angles, pile fibre degree of stretching, resin content and foam support) [1, 2, 3].

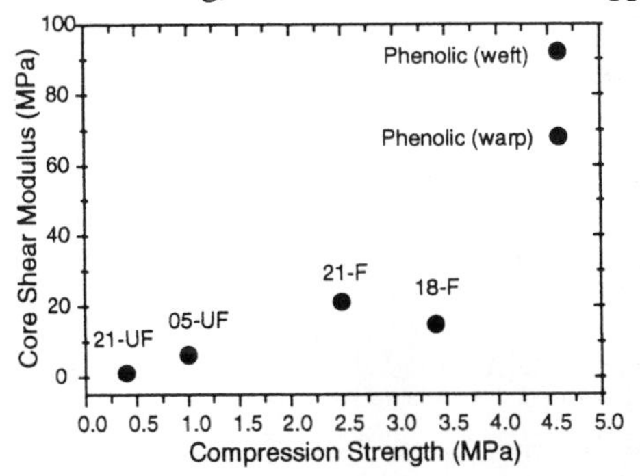

Figure 2. Core properties of panels

During loading, on the epoxy panel, pile failure initiates the panel failure. Further investigation revealed a mixed failure mode : a compression induced skin failure because of the local buckling of the skin into the local core failure area. On the phenolic foam panels, damage was initiated by foam and pile failure. In the warp direction, the core finally fails in shear, whereas in the weft direction the final failure is caused by compression failure of the skin.

4.1.2. *Shear Test Results*

The core shear moduli of panels obtained from the 3PB and shear tests (figure 3), show a very similar tendency with a small differences. This means that the 3PB test method with two different span lengths can be used as an easy alternative test method to obtain the core shear modulus instead of the more complex shear test method.

Due to the very large difference in "core shear strength" between the 3PB test and the shear test of the PUR-epoxy panels (figure 4), the shear stress is described as the "core shear stress at Pu (ultimate load)". The different stress condition in both tests is the reason for the results differences : the pure shear stress can be created in the core during the shear test while in the 3PB test, it also generates tension and compression stresses in the core which play an important role in lowering apparent "shear strength". This is also proved by the final failure mode : where a mixed mode failure occurs in the 3PB test, pure core shear failure is found in the shear test. This proves that the results of the shear test are more reliable than the shear properties obtained 3PB tests.

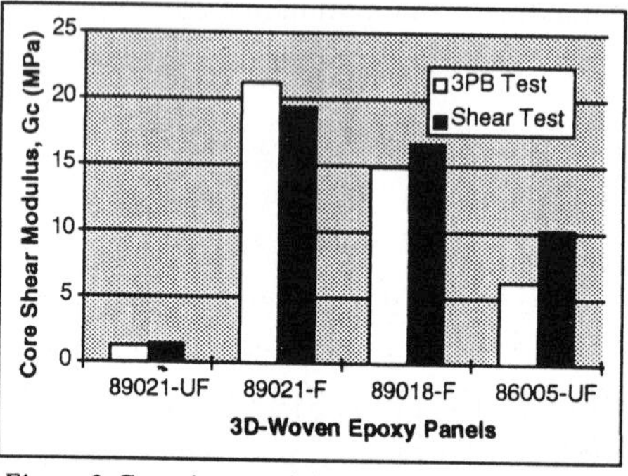

Figure 3. Core shear modulus of panels (from static 3PB test and shear test)

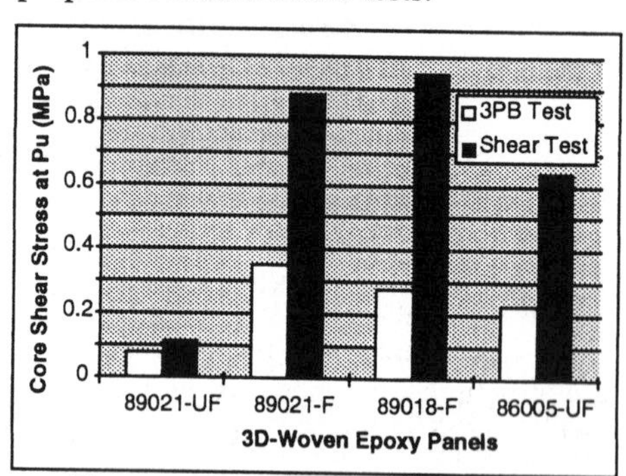

Figure 4. Core shear stress at Pu of panels (from static 3PB test and shear test)

4.1.3. *Comparison among 3PB Test, Shear Test and New 3PB Test Results*

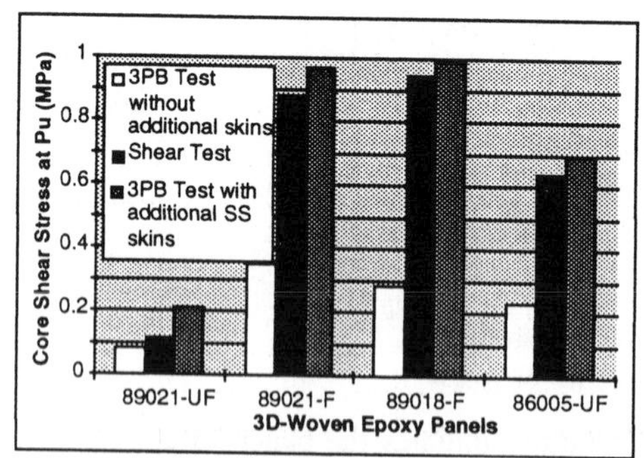

Figure 5. Core shear stress at Pu of 3D-woven epoxy panels resulting from 3 different tests

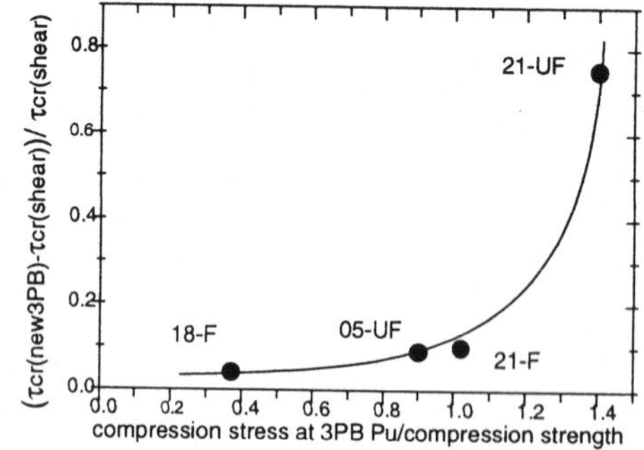

Figure 6. The importance of the compression stress in the higher core shear strength (from new 3PB test)

Figure 5 shows that the new 3PB test results are closer to the shear test results than to the previous 3PB test results since the same failure mode (i.e. core shear failure) was obtained. The higher core shear strength resulting from the new 3PB test than the shear test results seems to be caused by the compression stresses, that alter the shear state in the core. The importance of the compression stresses can be shown in figure 6. It is clear that the difference between the shear test and bending test with thick skins increases as the compression stresses approach the core compression strength.

Figure 5 also shows that the new 3PB test with additional skins results in a much higher core shear stress at Pu compared to the previous 3PB test. This was expected since during the new 3PB test, core shear failure was generated (using a thicker skin) instead of a mixed mode failure, obtained in the previous 3PB test. The stress state for these two different panels is schematically drawn in figure 7. It shows that by increasing the thickness of the skin (from the 0.35 mm to the 1.5 mm) the core shear stress only decreases marginally (approximately 10%). However, if the changes of the other stress components are considered (i.e.: skin compression and tension stresses, core tension and compression stresses), large differences are found.

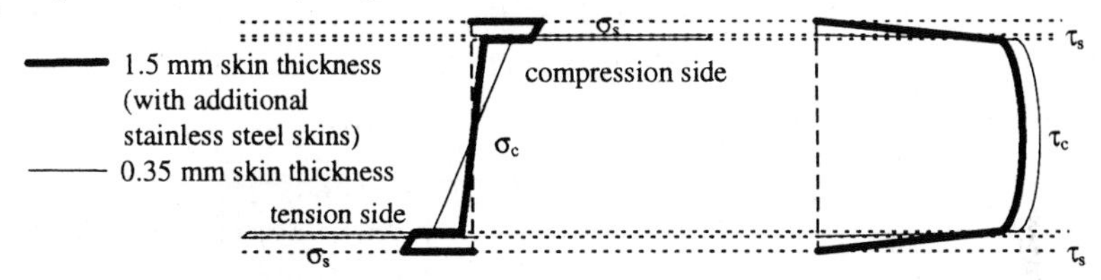

Figure 7. Stress state in the 3D-woven PUR epoxy panel with two different skin thickness

Finally, it can be concluded that the new 3PB test with additional stainless steel skins can be used to obtain core shear strength, as long as no big compression stresses are present. Furthermore, using thicker skins can be used as the future static and fatigue test set-up to obtain the correct core shear strength.

4.2. FATIGUE BEHAVIOUR

Table 3 shows the fatigue test results of all panels including the property degradation.

TABLE 3. Fatigue test results of 3D woven panels

PANEL	FATIGUE BEHAVIOUR	MAXIMUM LOADING			
		80% Pu	**65% Pu**	**50% Pu**	**35% Pu**
89021-UF	cycle to failure	32,000 ± 15,000	>995,000*	-	-
Epoxy	stiffness deg.(%)	42.3 ± 1.7	66.9	-	-
warp	failure mode	core, partly skin	core*	-	-
89021-F	cycle to failure	52,000 ± 20,000	$>10^6$	-	-
Epoxy	stiffness deg.(%)	2.9 ± 0.7	2.7 ± 0.1	-	-
warp	failure mode	skin	no failure	-	-
89018-F	cycle to failure	$>10^6$	-	-	-
Epoxy	stiffness deg.(%)	5.4 ± 1.6	-	-	-
warp	failure mode	no failure	-	-	-
86005-UF	cycle to failure	66,000 ± 7,500	$>10^6$	-	-
Epoxy	stiffness deg.(%)	10.3 ± 0.2	4.5 ± 0.4	-	-
warp	failure mode	core and skin	no failure	-	-
89020	cycle to failure	198 ± 62	44,000 ± 8,100	105,500 ± 25,000	$>10^6$
Phenolic	stiffness deg.(%)	21 ± 12.4	51 ± 8.7	55 ± 3.0	18.4 ± 3.4
warp	failure mode	core, delamination on some panels	core, delamination on some panels	core, delamination on some panels	no failure
89020	cycle to failure	40 ± 23	210 ± 11	231,000 ± 74,500	$>10^6$
Phenolic	stiffness deg.(%)	-	15 ± 1.3	36 ± 1.8	21 ± 6.2
weft	failure mode	skin	core, skin	core, skin	no failure

4.2.1. *Wöhler Curves of 3D Woven Sandwich Panels*

The Wöhler curves of the panels are plotted in figure 8. The core shear stress at one cycle to failure is the core shear stress at static 3PB Pu and 10^6 cycles is assumed to be the fatigue limit. The "asterisk" sign on the 89021-UF panel means that the panel did not show any visible failure when the machine stopped. For this panel, the maximum deflection limit of 20 mm was exceeded, due to the very high stiffness degradation (67%). It can clearly be seen that the fatigue limit for every panel is determined by the static strength : the lower the static strength, the lower the fatigue limit obtained.

To compare the relative fatigue behaviour, normalizing Wöhler curves is plotted as can be seen in figure 9. As fatigue tests were done on thin skin material, the data can not be normalized to the real core shear strength but is normalized to the core shear stress at static 3PB test failure, τ_{cr}.

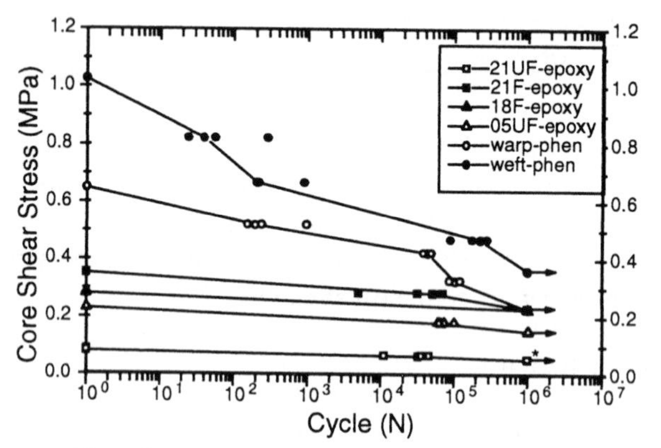

Figure 8. Wöhler curves of 3D-woven sandwich panels

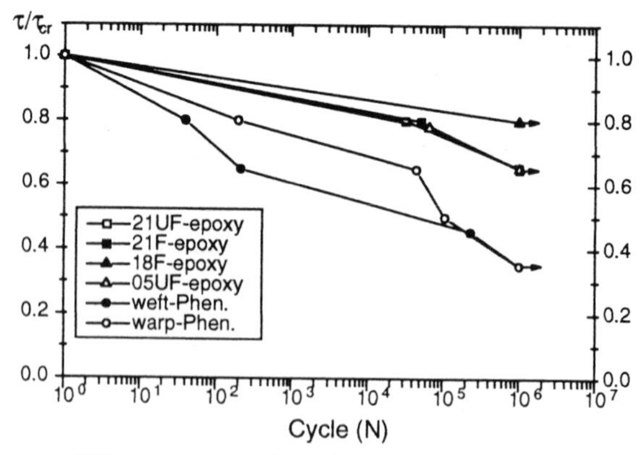

Figure 9. Wöhler curves of 3D-woven sandwich panels (normalized to core shear stress at S3PB Pu)

Figure 9 shows that the fatigue behaviour of the PUR-epoxy panels is better than the phenolic panels, eventhough the static properties are much lower. In the PUR-epoxy panels, the fatigue limit of 10^6 cycles was reached at 80% or 65% of the 3PB-Pu instead of 35% in the phenolic panels. The reason of the lower fatigue behaviour of the phenolic panels is due to the brittle behaviour of the phenolic foam. During the cyclic loading shear cracks are formed in the foam and reduce the foam to a loose powder that falls out of the core. It therefore provides a rapid decrease in the core performance and results in much lower fatigue properties.

The foamed epoxy 89018 panel shows superior fatigue behaviour compared to the other panels. This panel reaches the fatigue limit at 80% of the 3PB static strength (10^6 cycles) and has a small stiffness degradation (5.4 %). The very good performance of the 89018 foamed panels is most likely due to the higher properties of core structure by which the core prevents the skin from buckling on the compression side and resists the high shear stress during cyclic loading. The fatigue behaviour of the 89021-UF panel is similar to the 89021-F and the 86005-UF panels. However this panel shows a severe stiffness degradation (67%) due to the lower core properties of the weak panel.

The warp and weft direction of the phenolic foamed panels seem to have the same fatigue limit and the same high stiffness degradation (20%). However, it seems that the panel properties in the weft direction decrease more than in the warp direction. This behaviour is related to the deflection limit in the fatigue machine and the final failure mode of the panels. A high loading in the strong (weft core) panel tends to result in a skin buckling failure while a core shear failure is easy to be obtained in the soft (warp core) panel (but similar skin strength) with a rather low loading condition. If skin failure occurs during fatigue loading, the fatigue machine will stop since the deflection limit is exceeded suddenly. When core shear failure occurs during fatigue loading, the core degrades but the skins can still carry the load. The deflection limit will be reached slowly before the fatigue machine switches off automatically.

Investigations of the failure mode during and after loading revealed a mixed mode of failure almost identical to the static test failure, except for the 89018-F panels where no (clear) failure was observed. Pure core shear failure was obtained only in the warp direction of the phenolic panel since in the weakest (warp) direction, the phenolic foam cracked more severely during cyclic loading. The presence of a delamination on the warp direction panel, is possible since the skins consist of more than 1 layer. However the weaker skin properties on this direction [3] and the weaker core properties also induce this damage. On the weak (warp) skins, both compressive and shear stresses will act as an in plane and an out of plane stress raiser and create delamination.

4.2.2. *Influence of other Properties to the Property Degradation*

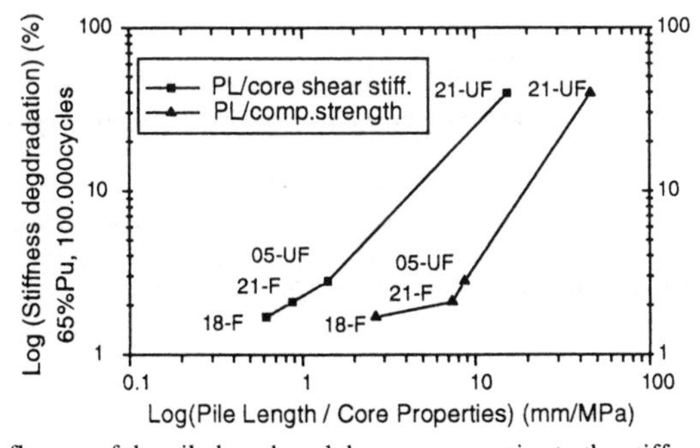

Figure 10. The influence of the pile length and the core properties to the stiffness degradation

To analyse more in detail the stiffness degradation of the panels, the data are taken only from the PUR-epoxy panels at 65% 3PB Pu with fatigue limit 10^5 cycles. From figure 10, a correlation between the stiffness degradation, the core shear modulus and the pile length can be seen clearly. The stiffness degradation is influenced by the pile length and the core properties : the stiffness degradation increases as a function of the ratio of the pile length over the core properties.

5. Conclusions

- The 3D woven glass fabric epoxy panels with polyurethane foam show excellent fatigue behaviour, particularly on 89018 fabric-PUR foam panel which shows the best results : fatigue life > 10^6 cycles with stiffness degradation < 6% at 80% Pu loading.
- The 3D woven glass fabrics phenolic panels with phenolic foam have rather low mechanical fatigue properties due to its brittle foam behaviour : fatigue life > 10^6 cycles with stiffness degradation ≈ 20% at 35% Pu loading.
- The stiffness degradation of the panel is proportional to the ratio of the pile length over core properties.
- The failure mode obtained during 3PB testing and flexural fatigue testing in the panels shows a compression induced skin failure instead of pure core shear failure. The test set up can be modified by increasing the skin thickness of the panels to reduce the skin buckling probability, hence, pure core shear failure can be obtained.

6. Acknowledgements

This text presents research results of the Belgian programme on Interuniversity Poles of Attraction, funded by the Belgian state, Prime Minister's Office, Science Policy Programming. The scientific responsibility is assumed by its authors. H. Judawisastra is financed through grants of the Governmental Agency for Cooperation with Developing Countries (ABOS-VLIR). J. Ivens is financed through grants of the Flemish Institute for the Promotion of the Scientific-Technological Research in Industry (IWT).

7. References

1. Van Vuure, A.W., Ivens, J., Verpoest, I.: *Sandwich panels produced from sandwich-fabric preforms*, Proceedings of the International Symposium on Advanced Materials for Lightweight Structures, ESTEC, 1994, 609-612.
2. Van Vuure, A.W., Ivens, J., Verpoest, I.: *Sandwich-fabric Panels*, Proceedings of 40th International SAMPE Symposium and Exhibition, Anaheim, U.S.A., May 8-11, 1995, .966-976.
3. Coenaerts, T.: *Influence of the pile bundle geometry on the core properties of a 3D sandwich fabric composites*, EUPOCO Thesis, KU-Leuven, 1993-1994.

CHARACTERIZATION OF THE MECHANICAL BEHAVIOR OF WOOD LAMINATES WITH THE HELP OF A FLEXION-COMPRESSION DEVICE

Mixed flexion compression tests on wood laminates

G. MUSSOT-HOINARD, G. FERRON
Université de Metz
Laboratoire de Physique et Mécanique des Matériaux
URA CNRS 1215 - ISGMP
Ile du Saulcy - 57045 Metz Cedex 01 - FRANCE

1. Introduction

Uniaxial tension or compression tests and three points bending tests are often not representative of the loadings undergone by structural elements used in building construction. Thus, a flexion-compression device, previously built for studying long fibers composites [1], has been designed with the help of ABAQUS/Standard for wood-based panels. After having described the principle of the device, results concerning the damage and failure phenomena of maritime pine plywood samples submitted to mixed flexion-compression loadings will be explained.

2. Experimental device

The principle of the flexion-compression device is based (figure1) on the conversion of the uniaxial tension force of a conventional testing machine, in a symmetrical system of normal compressive loads (N) and bending moments (C), acting on the ends of the samples.

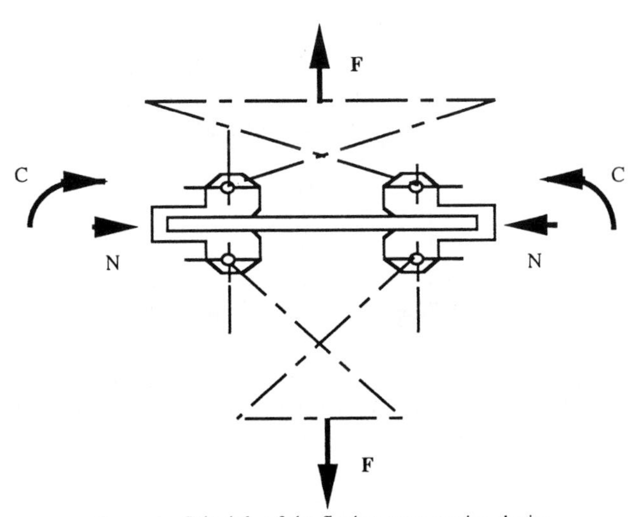

Figure 1 : Principle of the flexion-compression device

A. Vautrin (ed.), Mechanics of Sandwich Structures, 295–302.

This is obtained by means of a system of jointed arms. According to the arms arrangement, different M/N ratios can be obtained (M is the bending moment in the central section of the sample). The M/N ratio remains practically constant during the loading. Assuming a linear elastic behavior for a homogeneous material, the through-thickness stress distribution is the sum of the stress distribution due to the normal load σ_c such that

$$\sigma_c = -\frac{N}{S} \tag{1}$$

and that due to the bending moment σ_f with

$$\sigma_f = \frac{My}{I_z}. \tag{2}$$

Three types of tests can then be defined:

- a quasi-pure bending test (figure 2) (M/N = 250 mm) corresponding to a through-thickness stress distribution such that the neutral fiber is located at the mid-width of the sample .

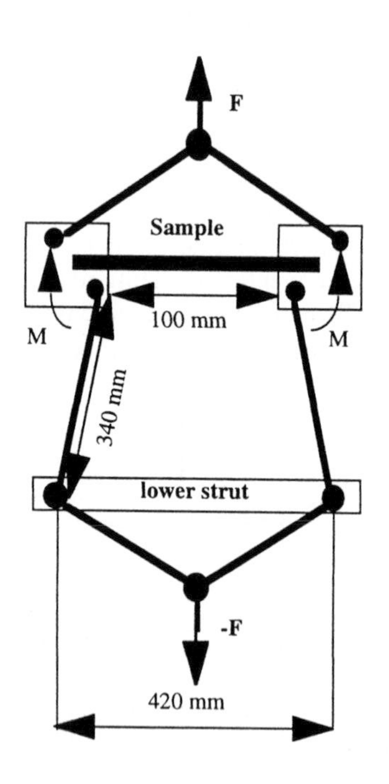

Figure 2 : Schematic view of the device in the case of quasi-pure bending test

- two mixed flexion-compression tests (figure 3), the first one with a M/N = h/3 ratio (Position 1 of the upper strut), the second one with a M/N = h/6 ratio (Position 2 of the upper strut). The neutral fiber is situated at h/4 of the upper face under tension in the first case while it coincides with this face in the second case.

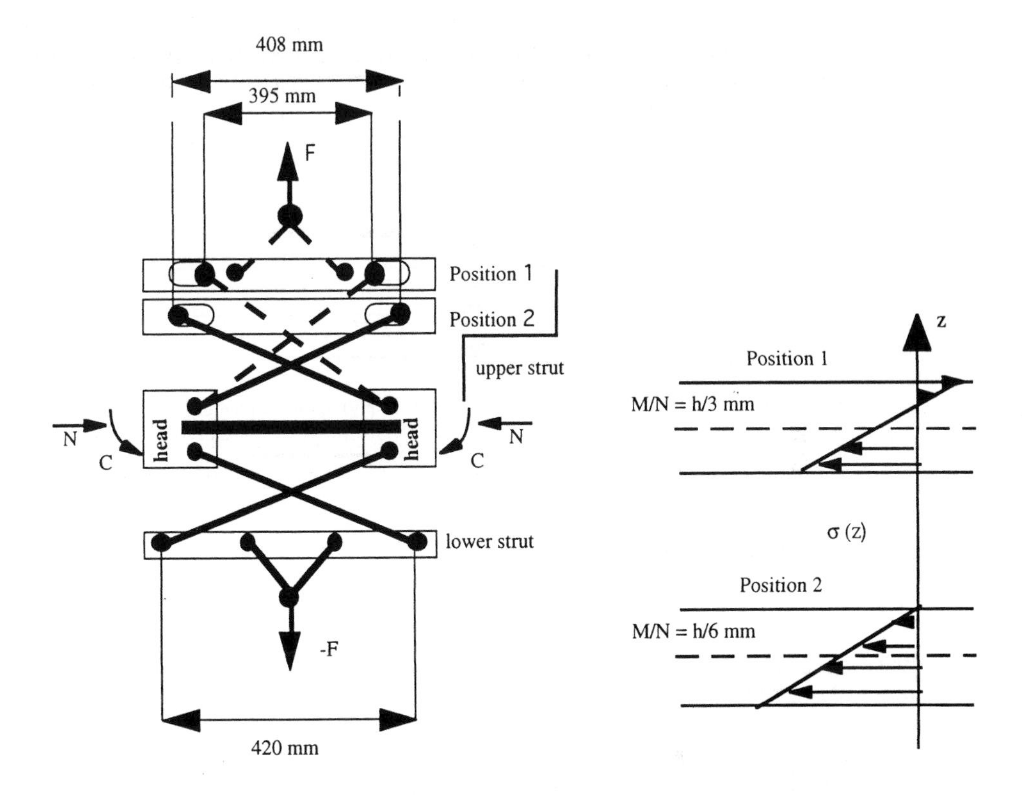

Figure 3 : Schematic view of the device and distribution of compressive and tensile stresses over the cross-section in the case of mixed flexion-compression test

3. Maritime pine plywood specimens

3.1. PLYWOOD MANUFACTURING

Samples are plywood specimens. They are manufactured according to a usual process. The first preparation consists in debiting wood logs into blocks of suitable lengths. Then a veneer (a thin layer of wood), obtained from the initial wood block by a high speed rotary-cutting, comes away in a continuous sheet before being cut at prescribed width and dried. Plywood panels result from the assembly by hot pressing of generally an odd number of elementary layers glued together with phenolic resin. Mostly, face grain directions between two adjacent plies are at 90° so as to reduce the anisotropy of mechanical properties of panels. The in-plane directions of the veneer are the longitudinal (**L**) and tangential (**T**) orthotropic directions of the tree trunk [2].

3.2 TESTED SAMPLES

The tested samples are made of maritime pine from France. Their average density is

590 kg/m³ at 12% of moisture. Specimens are plywoods arranged in a 5-layer (5-ply) configuration. Layers are arranged in a symmetrical 0°-90°-0°-90°-0° configuration of face grain directions (figure 4) in a width ratio of 1,5:2:2:2:1,5. Dimensions of tested samples are 360x100x9 mm³.

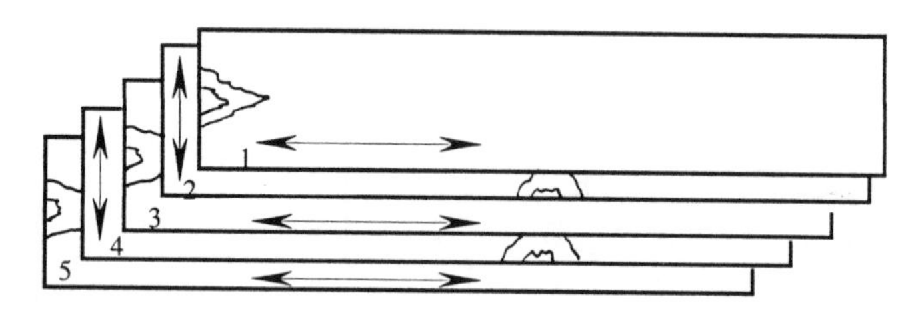

Figure 4 : Face grain direction arrangement of a 5-layer (5-ply) maritime pine plywood specimen.

4. Tests and results

4.1 TESTS AND SAMPLES INSTRUMENTATION

4.1.1. *Tests*

Quasi-static tests are carried out with the help of the flexion-compression device, previously described, installed on a conventional Zwick tension machine. An annexed stabilization system has been adapted to prevent any instability during loading. This system assures the perpendicularity between the vertical axis of the setting and the axis joining two symmetrical points on the heads transmitting the loading to the sample. From a kinematical point of view the deformation of the system is then necessarily symmetrical.
Experimental results are illustrated in the case of flexion-compression tests of ratio $M/N = h/3$. The crosshead speed is equal to 1 mm/mn.

4.1.2. *Samples instrumentation*

On the one hand, longitudinal local strains are measured on each face by means of longitudinal strain gages glued on the center of the upper and lower faces of samples. On the other hand, acoustic emission phenomena [3,4,5] are detected by a matched pair of piezo-electrical acoustic emission collectors supplied by Physical Acoustics Corpoaration (R15, 50-200 KHz, 18 mm diameter). They are coupled to the upper face of samples under tension by means of silicone grease, a rubber band ensuring the continuous contact during the tests. They have been chosen for their operating frequency range, in relation with the strong absorption power of wood due to its grainy and porous structure.

4.2. RESULTS AND DISCUSSION

The current $F(\Delta L)$ and $\varepsilon(\Delta L)$ mechanical responses and $E(\Delta L)$ physical response are described and analysed to get a better understanding of the damage phenomena.

4.2.1. *Results*

F(ΔL) and ε(ΔL) curves. The 5-layer (5-ply) maritime pine plywood goes to failure for mixed flexion compression tests in three stages according to the representative F(ΔL) and ε(ΔL) curves presented on figure 5.

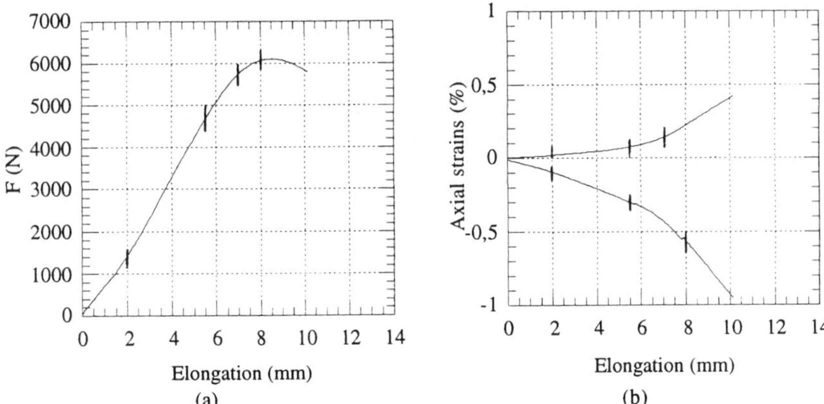

Figure 5 : Evolution of (a) the force on the setting and (b) the axial strains on the upper and lower faces as a function of the elongation of the device for maritime pine plywood submitted to mixed flexion compression tests

The first part between 0N and 1400N is affected by the putting under load of the device and thus is not representative of the material behavior. The different stages of loading are described in Table1 in terms of F and ΔL in relation with the tension and compression strain domains. Two stages can be defined :

- stage I is subdivided into two parts.

  stage I1 between 1400N and 4700N is characterized by a linear evolution of the force on the setting and a linear evolution of axial strains (with a lower slope in tension than in compression).

Stages	Tension	Compression
	0 < F(N) <1400 0 < ΔL(mm) < 2	
I-1	1400 < F(N) <4700 2< ΔL(mm) < 5.4	
I-2	4700 < F(N) <5800 5.4 < ΔL(mm) < 7	4700 < F(N) <6100 5.4 < ΔL(mm) < 8
II	F(N) : 5800 - 6100 - rupture ΔL(mm) : 7 - 8 - 10	F(N) : 6100 - rupture ΔL(mm) : 8 - 10

Table 1 : Description of the three stages in terms of F and ΔL in relation with the tension and compression strain domains

stage I2 is characterized by a transition phase for load and strains and goes from 4700N to 5800N for the tension domain and to 6100N for the compression one (figure 6).

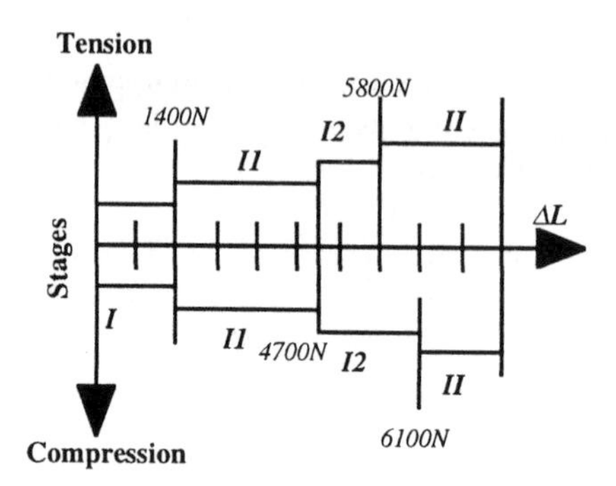

Figure 6 : Synoptic view of the stages in tension and in compression as a function of ΔL

- stage II is again characterized by a linear evolution of the axial tension and compression strains (with higher slopes than for stage I). It starts from 5800N in tension and from 6100N in compression, and spreads up to rupture.

E(ΔL) curves. Tests are accompanied by fairly large acoustic emission phenomena that manisfest continuously up to the end of stage I in tension. During the first part of stage II in tension, i.e. up to 6100N, 8 bursts of 0.9- 4.8s are emited at a gap of 1.4s to 5.3s. This phenomenon continues at the begining of stage II in compression when the value of the force remains around its maximum. It is characterized by 5 bursts of 1- 4.8s emited at a gap of 3.8 to 8s. For other samples submitted to the same test, the discontinuous burst emission can be observed up to rupture.

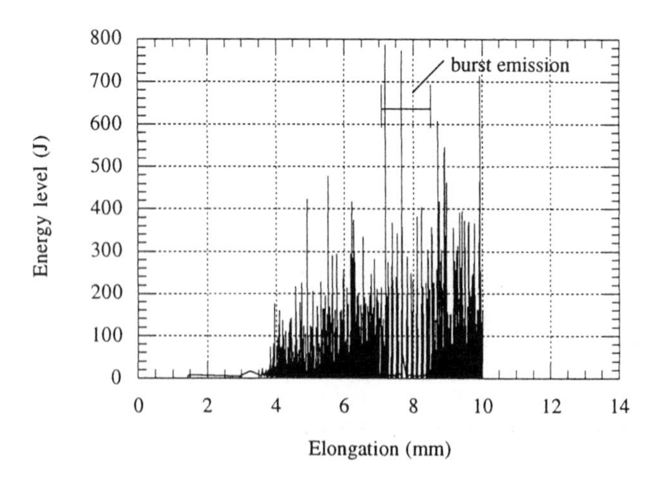

Figure 7 : Evolution of the energy emission acoustics level as a function of elongation in the case of maritime pine plywood submitted to flexion compression tests of M/N = h/3 ratio

4.2.2. *Discussion*

According to the evolution of strains on the external faces as a function of the elongation of the device (figure 5b), the ratio of upper tensile strain to lower compressive stain evolves from 1/3 (end of stage I1) to 1/2 (end of the test). In the assumption of Love-Kirchhoff plates, this evolution represents a displacement of the neutral fiber initially situated in the neighbourhood of the interface of ply1/ply2 in direction of the lower part of the sample up to the mid-width of ply 2. This strain distribution is described on figure 8a.

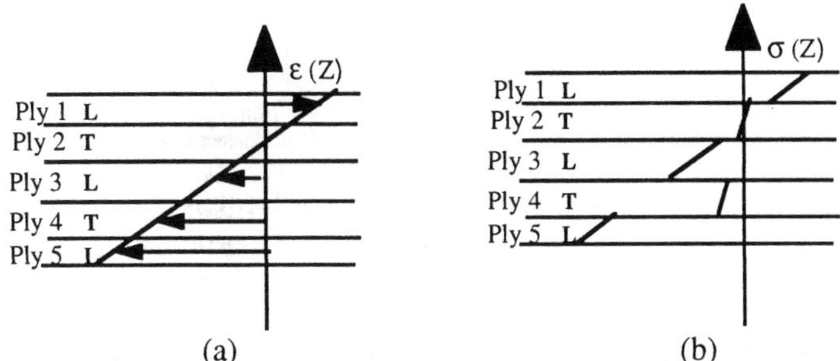

Figure 8 : Through-thickness distribution of (a) axial strains and (b) axial stresses at the end of the test

For discussion, the corresponding stress distribution is displayed on figure 8b by only taking into account the differences of Young's moduli in **L** and **T** directions.

The acceleration of straining, both in tension and in compression during these stages (stages I2 and II) is the geometrical manifestation of a localization of the damage events at the mid-length of the sample.

According to Meredith [6], Young's moduli of wood in tension and in compression are approximately equal, but the elastic limit is considerably lower in compression than in tension. According to Moehler [7], the admissible stress in tension (or in compression) is 20 to 25 times less perpendicularly to the face grain direction than parallel to it.

In accordance with Meredith and Moehler studies and with naked eye observations, the following damage scenario is proposed :
- longitudinal ply 5 that is the most constrained shows fine kinks on its free face.
- transversal ply 4 accommodates compression by simultaneous rolling and slip of fibers (see figure 9).

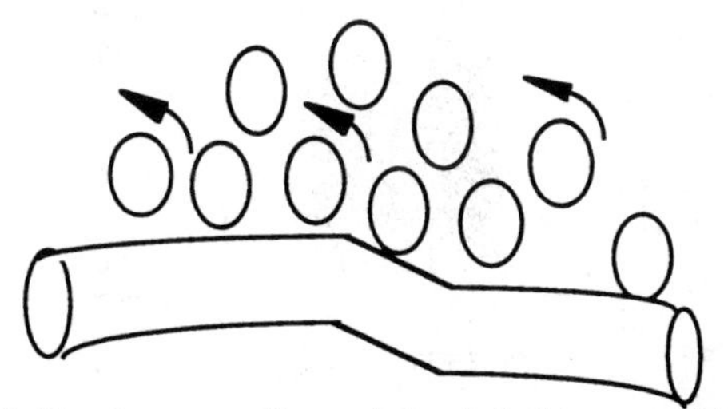

Figure 9 : Simultaneous rolling and slip of ply 4 transversal fibers and kinking of ply 5 longitudinal fibers under compression

- longitudinal ply 3 that is the second more constrained ply is forced to accommodate its deformation with interfibers slip under compression because of the presence of the adjacent plies (no free surface).
- transversal ply 2 on its upper part damages by decohesion of transversal fibers as described on figure 10.

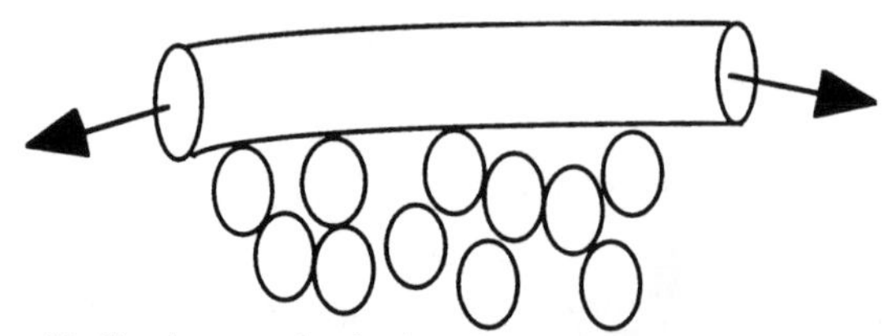

Figure 10: Simultaneous decohesion of second ply transversal fibers and elongation of first ply longitudinal fibers under tension

- longitudinal ply 1 fails according to the usual brittle mode observed for wood in tension.

5. Conclusion

A device has been designed for loading plywood panels under mixed flexion-compression to have a better understanding of the damage up to rupture. Two characteristic zones are identified from the mechanical and physical replies. Damage and fracture events have been proposed. Further studies involving numerical simulations are in progress to correlate the evolution of the overall stiffness of plywood with the evolution of damage within the plies.

6. References

1. Grandidier, J.C., Ferron, G., Potier-Ferry, M.: Microbuckling and strength in long-fiber composites, theory and experiments, *Int.J. Solids Structures* 29 (1992), 1753-1761.
2. AFNOR, NF B 51-002: *Caracrérisation physiques et mécaniques des bois*, 2ème édition de bois et liège, 1988.
3. Roget, J.: *Essais non destructifs: l'émission acoustique. Mise en oeuvre et apllication*, AFNOR-CETIM, 1988.
4. Beattie, A.G., Acoustic emission, principles and instrumentation, *Journal of Acoustic*
5. Ansell, M.P: Acoustic emission from softwoods in tension, *Wood Sci. Technol.* 16 (1982), 35-58.
6. Meredith, R.: *Mechanical propertiesof wood and paper*, North-Holland Publ.Co., Amsterdam, 1953.
7. Moehler, K.: *Zur Berechnung und tragender Sperrholz-Konstruktionen*, *VDI-Zeitschrift* 107 (1965) 17,729-738.

APPLYING THE GRID METHOD TO THE MEASUREMENT OF DISPLACEMENT AND STRAIN FIELDS THROUGH THE THICKNESS OF A SANDWICH BEAM

L. DUFORT, M. GRÉDIAC, Y. SURREL AND A. VAUTRIN
École des Mines de Saint-Etienne
SMS/Dept. of Mechanical and Materials Engineering
158, cours Fauriel
42 023 Saint-Etienne Cedex 2 (FR)

Abstract.

The present work is an experimental investigation. The final goal is to measure the displacement field through the thickness of a sandwich beam in order to check and compare the efficiency of different kinematics-based models. The first experimental results obtained on a sandwich beam under three-point bending are presented in this paper.

1. Introduction

A large number of beam or plate models have been developped to describe the displacement field through the thickness of sandwich structures. All of them are based on various kinematical assumptions concerning the displacement field through the thickness of the structures. The aim of this work is to measure this displacement field on the side of a sandwich beam under three-point bending. A suitable optical method is used to measure this field. It is then compared with different theories encountered in the literature.

2. Presentation of the different theories

First order theories, like the Timoshenko's theory [J.-M. Berthelot, 1992](Figure 1), assume that straight lines normal to the middle surface remain straight after deformation :

A. Vautrin (ed.), Mechanics of Sandwich Structures, 303–310.

$$u(x,y,z) = u_0(x,y) - z\frac{\partial w_0}{\partial x} + z\gamma_x^0(x,y) \tag{1}$$

However, these theories are often not sufficiently accurate to model correctly the strain or stress distribution through the thickness of sandwich structures.

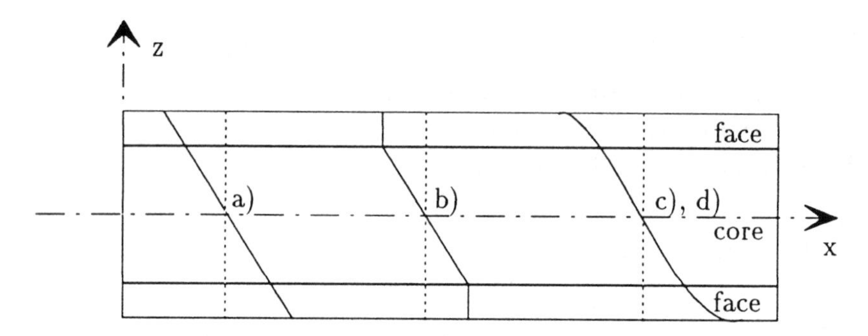

Figure 1. Profile displacement through the thickness for different models : a) Timoshenko, eq. 1; b) Sandwich theory, eq. 2; c) Reddy, eq. 3; d) Touratier, eq. 4.

The so-called "sandwich's theory" [J.-M. Berthelot, 1992] is based on the same assumption for the core but it assumes no shear deflection inside the skins :

$$u_{core} = u_0(x,y) + z\omega_x(x,y); \; u_{face} = u_0(x,y) \pm \frac{h}{2}\omega_x(x,y) \tag{2}$$

Higher-order theories proposed for instance by Reddy [J. N. Reddy and C. F. Liu, 1985] or Touratier [M. Touratier, 1991] have third degree or sine terms added to the first degree term and take into account the warping of any straight line perpendicular to the middle surface. The displacement through the thickness of the symmetric sandwich structure is then given by a parabolic or a sine expression :

$$u(x,y,z) = u_0(x,y) - z\frac{\partial w_0}{\partial x} + z\left[1 - \frac{4z^2}{3h_t^2}\right]\gamma_x^0(x,y) \tag{3}$$

$$u(x,y,z) = u_0(x,y) - z\frac{\partial w_0}{\partial x} + \frac{h_t}{\pi}\sin\left(\frac{z\pi}{h_t}\right)\gamma_x^0(x,y) \tag{4}$$

where :

$\gamma_x^0 = \left(\frac{\partial w_0}{\partial x} + \omega_x\right)$: transverse shear strain measured at the middle plane.
ω_x : total rotation.
h_t : total thickness.
u_0 : in-plane displacement.
$\frac{\partial w_0}{\partial x}$: rotation due to the bending (without shear).

3. Experimental setup

The experimental method used to measure the displacement field is an optical one developed in our laboratory and named " grid method" [Y. Surrel and N. Fournier, 1996] and [E. Alloba *et al.*, 1996].

A pattern of parallel black and white lines is marked onto the sample surface. This pattern acts as a spatial carrier. Let $\vec{R}(X, Z)$ be the position of a given material point in the initial state, and $\vec{r}(x, z)$ its position in the final state after loading. If a unidirectional grid is used, the reflected light intensity in the initial state at point $\vec{R}(X, Z)$ can be written as :

$$I_i(X, Z) = I_0\left[1 + \gamma \cos\left(\frac{2\pi X}{p}\right)\right] = I_0\left[1 + \gamma \cos\left(\phi(X, Z)\right)\right]$$

where I_0 is the local intensity bias, γ the contrast, p the grid pitch and $\phi(X, Z)$ the phase at $\vec{R}(X, Z)$.

The deformation of the grid can be interpreted as a phase modulation of the carrier. The inverse displacement field from the final to the initial state, $\vec{u}(u, v)$, is defined by :

$$\vec{R} = \vec{r} + \vec{u}$$

As the grid is supposed to follow the material displacements, the reflected intensity at point $\vec{r}$ in the final state is the same as at point $\vec{R}$ in the initial state, and so, in the final state :

$$I(x, z) = I_i(X, Z) = I_i(x + u, z + v)$$

$$I(x, z) = I_0\left[1 + \gamma \cos\left(\frac{2\pi(x + u(x, z))}{p}\right)\right] = I_0\left[1 + \gamma \cos\left(\phi(x, z) + \Delta\phi(x, z)\right)\right]$$

and so the phase modulation of the carrier becomes :

$$\Delta\phi(x, z) = 2\pi \frac{u(x, z)}{p}$$

The phase-shifting method is used here to evaluate the phase field of the grid. Three unknowns are involved: the bias I_0, the contrast γ and the phase ϕ, so at least three intensity values, given by the adjacent pixels on the same picture, are required to obtain the phase.

If two images are processed, the first in the initial state and the other one in the final state, the substraction of the corresponding phase fields gives a phase field which is proportional to the displacements during the test. Then a phase variation of 2π corresponds to a displacement equal to the grid pitch. The phase shifting method is used so as to enable the measurement of the local phases with an accuracy higher than $2\pi/122$. This value is the standard deviation of the measurement noise on the determined phase. This means that the displacement field accuracy is 1/122 of the pitch. Strain components are obtained through numerical differentiation of the displacements.

The sandwich beam used in this study is symmetrical and made of a balsa core and two glass/epoxy faces. One of the half lateral surfaces of the sandwich beam requires cleaning, sandpapering and coating with frush blaster and white paint. One vertical grid and one horizontal grid of 610 micron pitch each are then bonded, so as to obtain the displacement components in x and z directions. This beam rests on a three-point bending testing setup and the studied surface is homogeneously illuminated by four neon tubes (Figure-2).

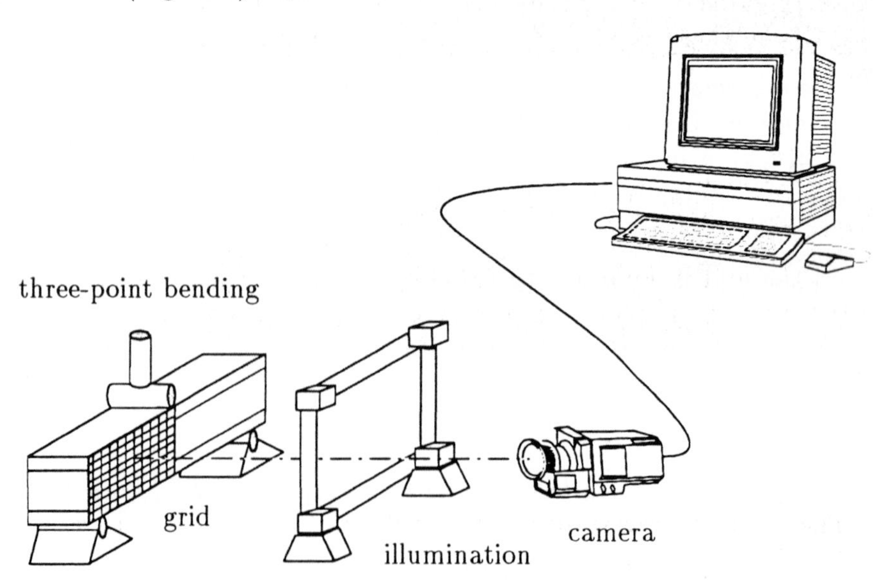

Figure 2. General view of the experimental setup.

4. Results and discussion

Both horizontal and vertical displacement fields are measured and depicted in Figure 3.

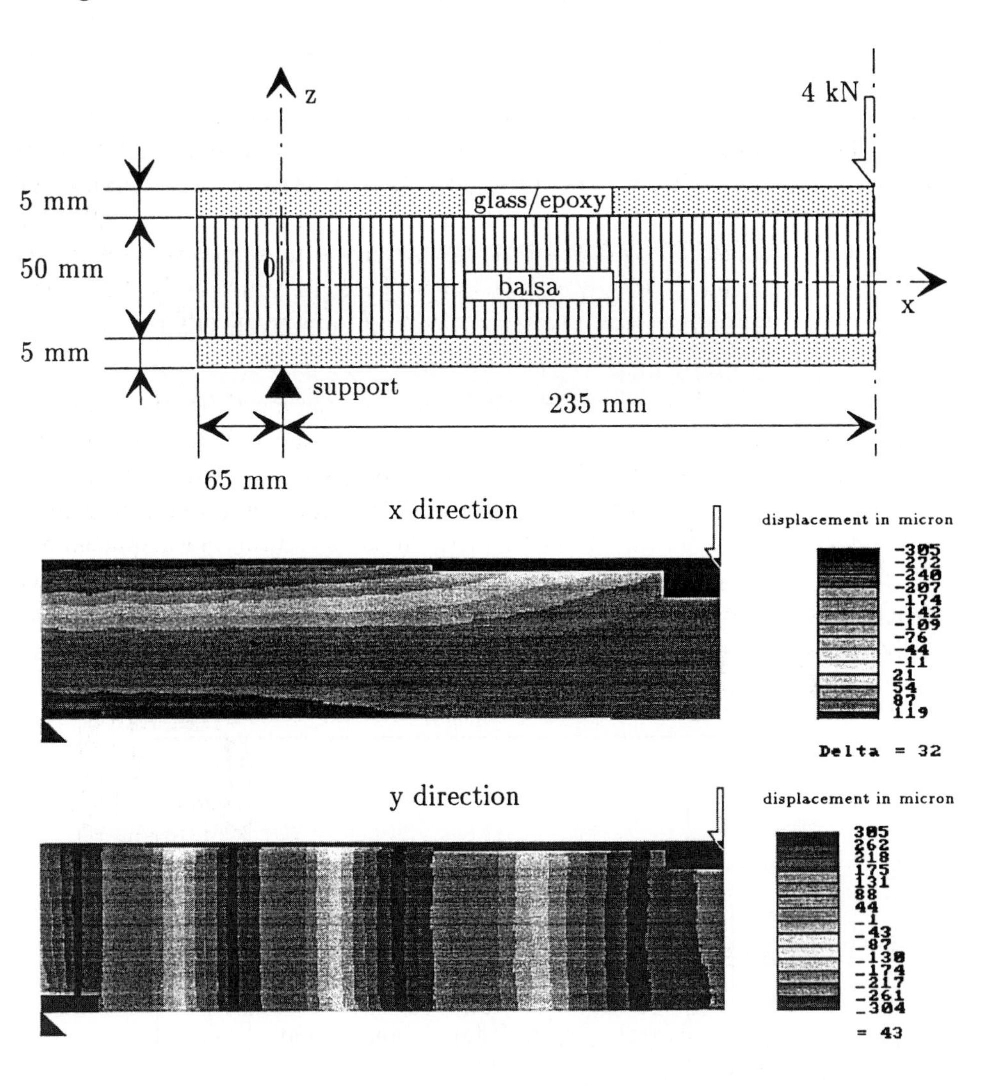

Figure 3. Experimentals displacement fields.

The different areas subjected to tensile and compressive stresses as well as the relative position of the neutral axis can clearly be distinguished.

The different dark areas are due, on one hand, to the shape of the support and the loading point which obscure the grid, and on the other hand to the optical method in itself. As the initial picture is shot with a fixed camera, the beam offset due to the loading results in a loss of information.

As the displacements along the vertical axis (z) are more than the value of the grid pitch, the scale of the values is modulo 610 microns.

A good agreement is found between the isodisplacement profile along the longitudinal axis and the one obtained from a 2D finite element analysis (Figure 4). However, due to a rigid body displacement of the beam, all the values are shifted and the neutral axis is offset above the middle plane.

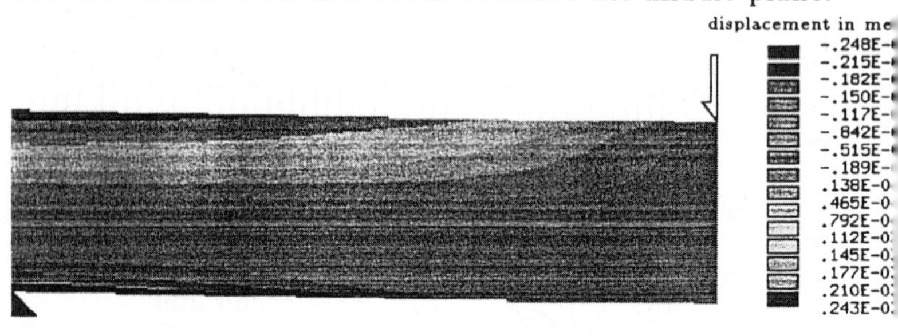

Figure 4. F. E. modelling of the displacement field along the longitudinal axis (x)

As far as the displacement profils through the thickness are concerned (with in-plane displacement equal to zero)(Figure 5), we noticed experimentally, for this type of sandwich beam, a horizontal displacement through the core and a small curvature in the faces.

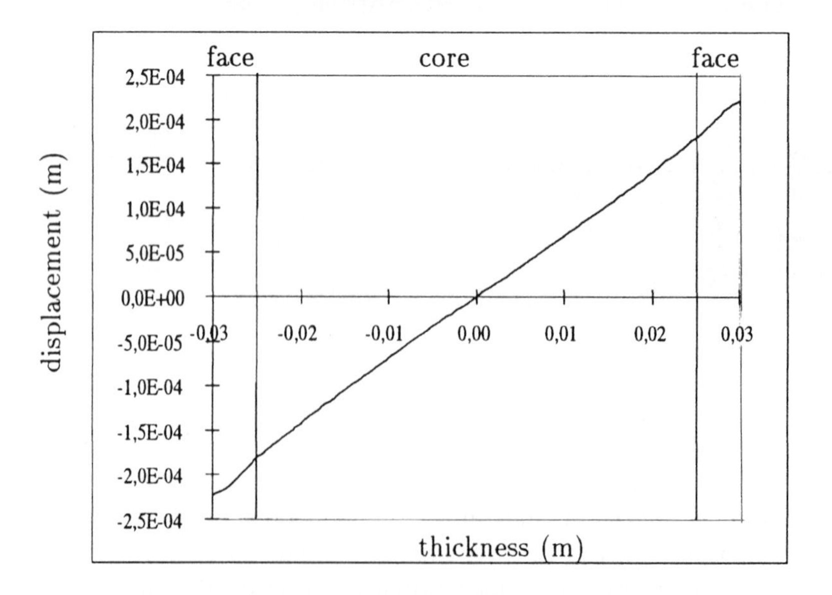

Figure 5. Experimental displacement field along the longitudinal axis, at $x = 25$ *mm*.

A slight warping of the section normal to the middle plane is spotted in finite element analysis (Figure 6) and at a lower scale in the Reddy and Touratier theories.

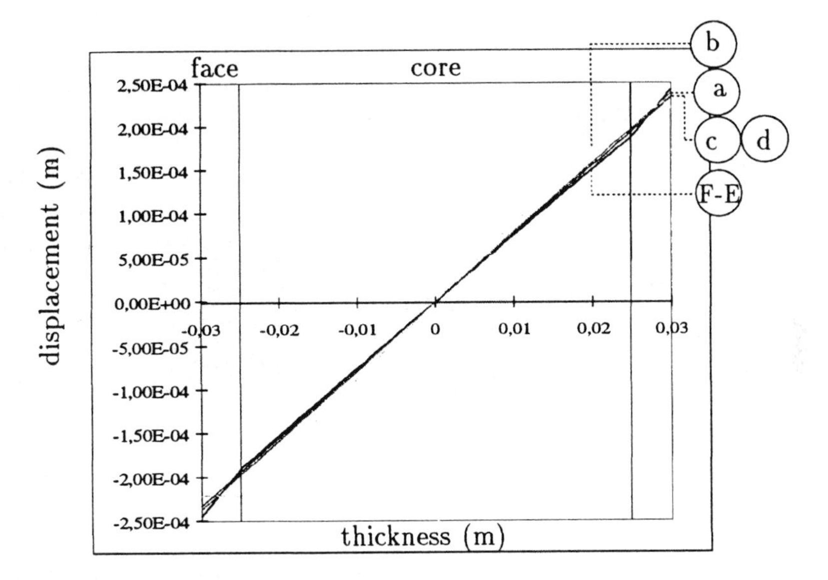

Figure 6. Displacement field along the longitudinal axis for various theories, at $x = 25$ mm : a) Timoshenko, eq. 1; b) Sandwich theory, eq. 2; c) Reddy, eq. 3; d) Touratier, eq. 4.

As a result, it seems that additional terms to the Timoshenko's theory proposed in the higher-order theories do not clearly appear in this first investigation. Further work will be carried out on different sandwich beams with different cores and faces so as to measure the displacement field in other configurations and to outline a significant warping through the beam thickness.

5. Conclusion

This paper shows the first measurements of a displacement field through the thickness of a sandwich beam with a suitable optical method. The objective was to observe the warping of sections which are taken into account in various higher-order theories. The constitutive materials of this first beam was however not well-suited as no warping was observed. Hence, further experiments will be carried out on different beams in order to emphasize this phenomenon.

References

Alloba, E., Pierron, F. and Surrel, Y. (1996) *Whole-field investigation of end-constraint effects in off-axis tests of composites, Abstract Proceedings of the VIII International Congress on Experimental Mechanics and Experimental/Numerical Mechanics in Electronic Packaging, SEM, Bethel USA*, pp. 405-406.

Berthelot, J.-M. (1992) *Matériaux composites : Comportement mécanique et analyse des structures*, Masson.

Reddy, J. N. and Liu, C. F. (1985) *A higher-order shear deformation theory of laminated elastic shells, Int. J. Engng. Sci.*, **23** (**3**), 319-330.

Surrel, Y. and Fournier, N. (1996) *Displacement field measurement in the nanometer range, Optical Inspection and Micromeasurements, SPIE, Washington USA*, **2782** pp. 233-242.

Touratier, M. (1991) *An efficient standard plate theory., Int. J. Engng. Sci.*, **29** (**8**), 901-916.

MOISTURE DETECTION IN SANDWICH PLATES USING INFRARED THERMOGRAPHY UNDER MICROWAVE EXCITATION

J.C. KNEIP, X.J. GONG, S. AIVAZZADEH
Laboratoire de Recherche en Mécanique des Transports
Institut Supérieur de l'Automobile et des Transports
49, Rue Mlle Bourgeois, BP 31, 58027 Nevers cedex, France

1. Introduction

The increasing use of sandwich materials implies more and more awareness about the long term effects by conceptors of environmental conditions. Among these, water absorption is of prime interest. Many studies (Cardon, 1996) have shown clearly that, by physicochemical interactions with material constituents, absorbed water could generate important modifications in mechanical properties. Furthermore, interactions can be accelerated through with the presence of damage in the materials. Certain investigatory means can be very useful in describing the state of a structure.

Therefore we have developed a non destructive testing technique to allow the detection of water absorbed in polymeric sandwich materials to be made (Kneip, 1997). The technique presented in this paper uses microwave excitation to create local heating in the material and an infrared thermography camera to detect it.

After describing the principles and the experimental process of the technique, we expose some early results obtained on sandwich plates.

2. Principles

2.1. INFRARED THERMOGRAPHY

The infrared thermography principle consists in observing the radiation emitted from a sample, and translating it into temperature. This is a very powerful method of investigation, firstly, because it enables observation of the thermal state of material or structure to be made very quickly, and secondly, because there is no contact or perturbation of the structure.

Primarily, the use of infrared thermography for non destructive testing of materials has been based on the use of photothermal techniques (Parker (1961), Balageas (1993)). Using these, we analyse the thermal response of the sample to an external thermal flux, with an infrared thermography camera. These thermal

A. Vautrin (ed.), Mechanics of Sandwich Structures, 311–317.

observations can be made in either reflection or transmission of the thermal excitation (figure 1). In every case, the principle of these techniques is based on the observation and interpretation of thermal diffusion delays during the transmission or the reflection of the thermal excitation. Defects have to slow down sufficiently the diffusion of the thermal spike to be detectable. Then, the higher the resistance of the defect versus the structure is, the more important the thermal contrast on the surface of the sample becomes. That's why these techniques are essentially used to detect and characterise delaminated zones in polymeric laminates.

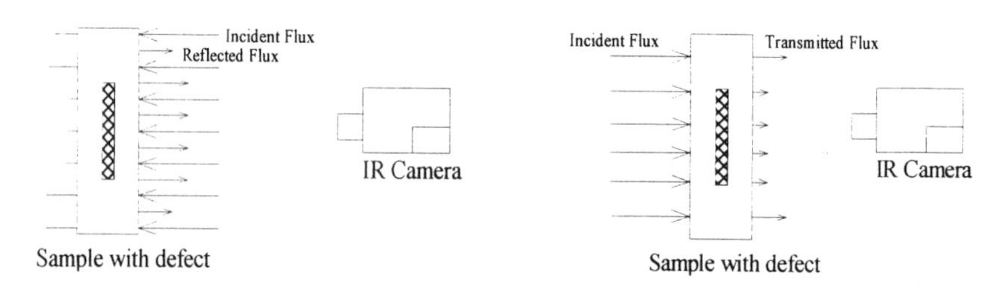

Figure 1. Principles of photothermal techniques

As regards its use in the field of moisture detection, there are very few published works. However we can point out Platonov's (1988) or Denel's (1989) works on honeycomb sandwich structures. These studies have shown the limits of these photothermal techniques for the detection of moisture. Similarly, we have shown that the detection of moisture was possible only when moisture concentrations were very high (several tens of percentage for porous materials) or with completely water-filled alveoli for honeycomb sandwich structures. Furthermore, sandwich materials are generally of large thickness (several tens of millimetres), in which we can find large cavities. All of these facts contribute to decrease the thermal flux, and then to create difficulties in detecting correctly the moisture concentration zone.

2.2. MICROWAVE TECHNIQUE

The principle action of microwaves on matter is based on the conversion of electromagnetic energy into thermal energy, due to the interaction between electrical charges which are present in the material under the influence of an external electrical field. This process, using dielectric hysteresis, is used with dielectric liquid or solid materials, containing polar molecules, and characterised by its permittivity ε' and its dipolar relaxation constant ε''. Dissipated energy in the material can be directly related to this two constants.

Unlike infrared heating (with photothermal methods, in our case), microwave heating takes place directly inside the sample, because of its wave-length (about λ=12.4 cm) and its important penetration power in mainly polymeric materials. Thus, thermal modelisation of heat become easier than in photothermal process. Moreover, because of its high asymmetrical configuration, the water molecule presents an exceptional polarity. This fact makes it the ideal material for this type of heating. These different aspects lead imagine that we could expect in quick and selective heating for our materials (Thuery, 1989).

The microwave heating technique uses wave-guide technology. A wave-guide enables electromagnetic energy to be carried and coupled to the sample. The difficulty is that each application needs the development of a specific applicator, as a function of the sample's geometry and the kind of heating required.

3. Experimental systems

3.1 MEASUREMENT AND EXCITATION PROCESS

The non destructive testing process involves two basic parts.

The first part, used to create heat in the water present in the material, is formed by an microwave heating system including a variable power microwave generator (0-2 kW at 2.45 GHz (λ=12.4 cm in void)). The wave applicator is a overdimensionned monomode wave-guide developed by CNRS (Bertaud, 1982), as we can see in figure 2. The dimensions of the microwave applicator were defined to cover the maximum heating surface.

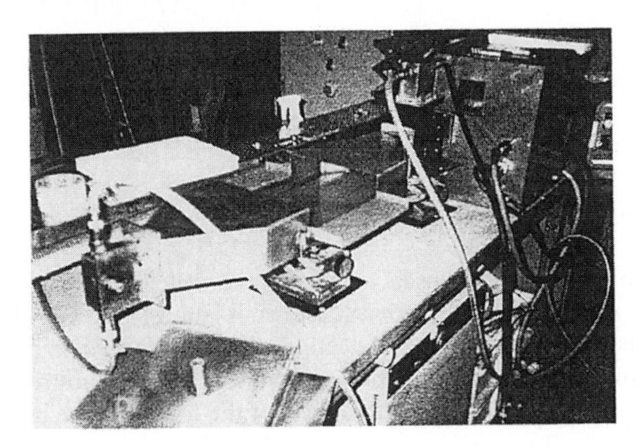

Figure 2. Photography of the micro-wave system

This system is used in the progressive wave state, which allows to gives less efficient but more homogeneous heating than in the stationary wave state. The absorption of the radiation at the end of the applicator is done through with a water load. This system works in the transverse electric 01 mode (TE_{01}). In the material the electric fields are influenced by the dielectric and geometrical characteristics of the sample.

The second part of the system is formed by an infrared thermography camera (AGEMA TH880 Swb), with quantic monodetector, liquid nitrogen-cooled (77 K). It deals with a short wave camera, using the infrared atmospheric transmission band 3-3.5 μm. An recording system enables data to be saved and viewed in real time on a PC compatible, with a resolution of 128 x 64 pixels, at 6.25 pictures/second.

3.2. TESTED MATERIALS

We have realised our tests on sandwich materials, similar to those presented by Denel in 1989, with dimensions of 70 x 70 mm², formed by a polypropylene honeycomb core and glass fibre mat skins (figure 3). The thickness of the core of these tested sandwich samples went from 10 to 40 mm.

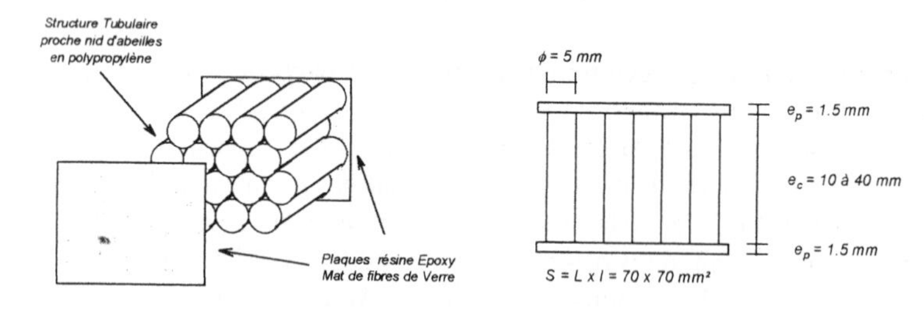

Figure 3. Schematic representation of tested sandwich samples

Humidity in tested materials was artificially injected with the help of a syringe. Thereby we had the possibilty to know the quantity of injected water and the place of the injection.

3.3. OPERATING PROCEDURES

The tested sample is firstly exposed to microwave solicitations. It is placed in the waves guide on a polypropylene support, which is almost transparent to microwave excitation. Immediately after the end of heating, sample is placed on a support to allow its observation by infrared thermography camera. An infrared mirror, placed behind the sample, can allow us to obtain thermal view of the front and the back of the sample simultaneously.

The necessary time of the microwave solicitation to obtain sufficient heating was determined to avoid any vaporisation of the moisture present. Then, we empirically have opted for a solicitation of 0.2 kW during 10 seconds.

4. Experimental Results and Discussion

For our tests, we tried to use almost similar samples as Denel had presented in his works (Denel, 1989). He had shown the possibilities of the photothermal technique, in transmission mode, for the detection of water in honeycomb structures. The main limitation of its method was the necessity to have almost completely water filled alveoli.

Our prime tests with similar conditions showed very good agreement to the Denel's results. As we expected, micro-wave solicitation combined to infrared thermography allows to observe localised heating source, with a very important thermal contrast with the rest of the sample. So, like Denel (1989), our technique allows us the detection of moisture. Furthermore, we observed that results were similar for all the thickness of samples, from 10 to 40 mm.

Then we tested samples with low quantity of contained water, up to 50 mg in one alveolus, injected with the help of a syringe, representing from 1/4 to 1/16 of its volume, in function of thickness of samples.

We expose here three results obtained for different thickness. S_1: 10 mm, S_2: 25 mm, S_3: 40 mm. The thermogramms exposed in figure 4 to 6 have been obtained 5 seconds after the end of the solicitation (0.2 kW during 10 seconds).

We clearly observe in each case a localised heating source, corresponding to absorbed water. These results are very interesting, firstly because, contrary to Denel's limitations, we showed that the detection of low quantities of absorbed water was possible. Secondly, because we showed that this technique was usable for thick structures investigation (up to 40 mm in our case). As we showed for laminates (Kneip, 1997), this last point constitutes an important advantage of this technique on photothermal techniques.

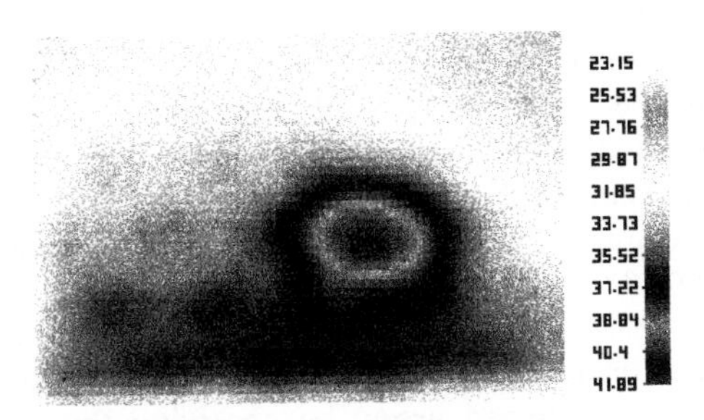

Figure 4. Thermogramm of sample S_1, at t_o + 5 seconds.

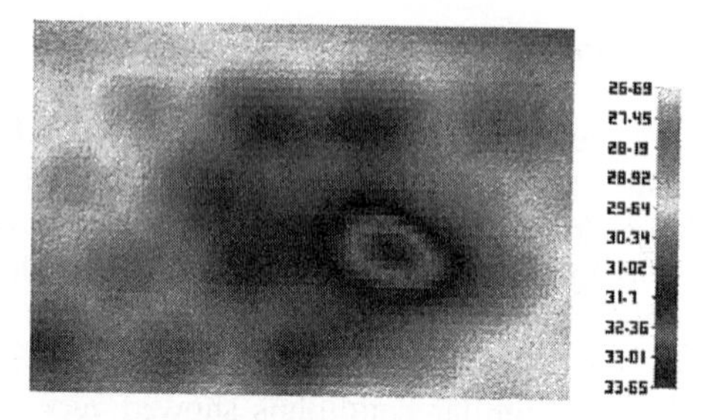

Figure 5. Thermogramm of sample S_2, at t_o + 5 seconds.

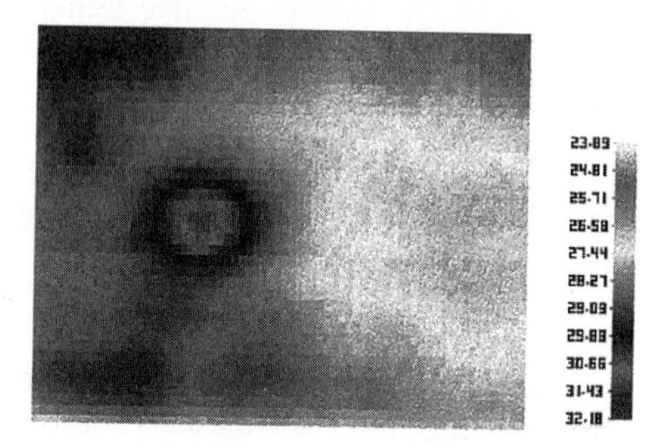

Figure 6. Thermogramm of sample S_3, at t_o + 5 seconds.

5. Conclusion

Our aim was to present this novel technique and to highlight of its possibilities in allowing the detection of absorbed water in materials to be performed. Thus we have shown that it is an easy-to-use technique in which a micro-wave solicitation is combined with an infrared thermographic observation.

In the case of water detection in honeycomb sandwich plates, we have shown that this technique was particularly well adapted in comparison of classically photothermal use techniques. The main advantages are the possibility of detecting low quantities of absorbed water and of quickly locating it. Furthermore it can be used to investigate thick structures, which constitutes an important drawback in photothermal techniques.

This technique can be use to localise moisture in this kind of structure, and also indirectly to detect damaged areas, by observing the anomalous presence of absorbed moisture, for example.

The limitations of this technique are mostly confined to the form of the wave-guide which has to be adapted to the geometry of the sample.

6. References

Balageas D. (1993) Contrôle non destructif des matériaux composites par thermographie IR, *Revue des Laboratoires d'essais*, N°35, mai 1993, p 12-18.

Cardon A.H., Fukuda H. & Reifsnider K.L. (1996) Progress in durability analysis of composite systems, *Proceedings of the international Conference Duracosys 95,* Edited by Cardon A.H., Fukuda H. & Reifsnider K.L., , Brussels.

Denel A.K. Golubev A.J. & Zhalnin A.P. (1989) Complex thermal testing of honey-comb panels, *Sov. J. of NDT (Defectoscopy)*, 1989, N° 9, p 55-56.

Kneip J.C. (1997) Application de la thermographie infrarouge à la détection de l'humidité dans les polymères et matériaux composites, *Thèse de l'Université de Bourgogne*, Janvier 1997.

Parker W.J., Jenkins R.J., Buttler G.P. & Abbott G.L. (1961), *J. of Applied Phys.*, vol. 32, N°9, 1961.

Platonov V.V. (1988) Ellipsometry in defectoscopy, *Sov. J. of NDT (Defectoscopy)*, N° 11, 1988, p 3-8.

Thuery J. (1989) Les micro-ondes et leurs effets sur la matière, CDIUPA, Tech&Doc, Ed. Lavoisier.

MODELLING OF THE CORE PROPERTIES OF SANDWICH-FABRIC PANELS WITH THE HELP OF FINITE ELEMENTS

J. PFLUG, A.W. VAN VUURE, J. IVENS, I. VERPOEST
Katholieke Universiteit Leuven
Department of Metallurgy and Materials Engineering
De Croylaan 2, B-3001 Heverlee, Belgium

1. Introduction

Sandwich-fabrics are produced with a velvet weaving technique. Due to the integrally woven nature of the fabric, panels and structures produced from the fabrics have a very high skin-core debonding resistance. Due to the one-step production of a sandwich preform the cost of sandwich panels and structures based on the fabric can stay limited.

Figure 1 shows several different sandwich-fabric panels obtained from sandwich-fabric preforms. Different fabric preforms for various core microstructures are available. After impregnation and curing, optionally a foam can be introduced into the core.

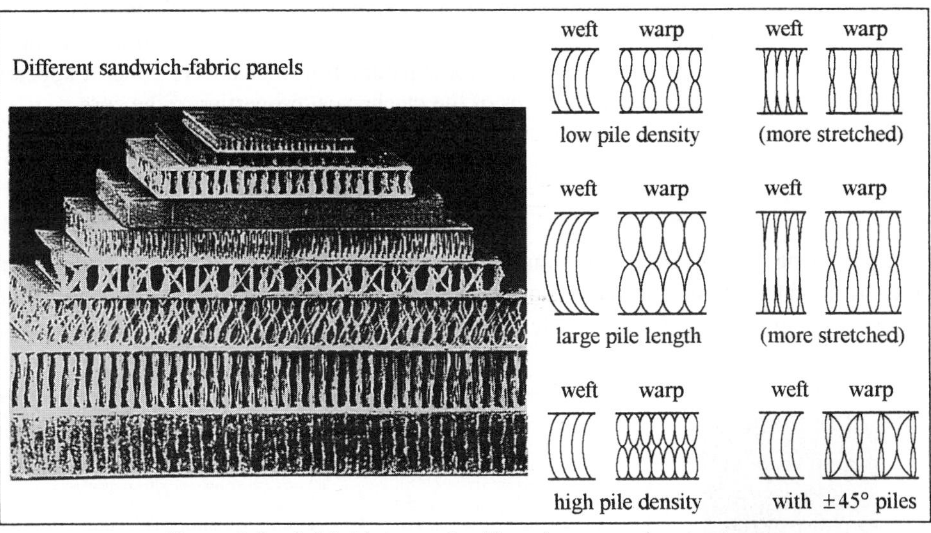

Figure 1. Sandwich-fabric panels with various core microstructures

The basic mechanical properties of the sandwich core (compression, shear) appear to have a strong dependence on the specific core microstructure. Parameters of interest are pile length, pile density and distribution, degree of stretching of the piles, woven angles of the piles, pile angles due to the production process, type of resin, resin content and distribution, type of foam and foam characteristics (especially foam density).

A. Vautrin (ed.), Mechanics of Sandwich Structures, 319–326.

The use of a foam is optional, but can be considered indispensable for panels of higher thickness (more than 15 mm) to reach sufficient properties for structural applications. There is a synergistic effect between piles and foam in the core.

These aspects have been reported in previous papers [3]. In this paper, the processing of the fabrics into sandwich panels and structures is also mentioned. Examples of applications are given in [4] and [5]. An overview of many aspects concerning sandwich-fabric panels (processing, mechanical properties including vibrational behaviour and structural damping) has appeared in [2].

Due to the large variation in properties depending on the specific and intricate core microstructure, there is a need for a design tool which can predict the panel properties as a function of the various parameters. A start was made in the development of such a tool by modelling the properties of unfoamed panels, with the help of finite elements. Until now, only elastic properties were evaluated.

A pre-processing program for a finite element code (in this case COSMOS/M), called 3DW-MODEL, was written, which models the pile shape and resin distribution [1]. Input parameters are data from the weaving company (fabric, weave and yarn type) and the panel producer (panel thickness, induced pile angles and resin content).

The pre-processing program produces a model of the unit-cell (repeating unit) of the material, including the (standard) skins of the sandwich-fabric panel. Possible resin connections with neighbouring unit cells are taken into account. Based on this an input file for the finite element code is prepared. Some parametric studies were done. It appears possible to make quantitative predictions of the mechanical properties of the core.

2. Microstructural modelling

The pile shape is calculated by assuming a sinusoidal shape of the piles in weft direction and the combination of a sinus- a tangent- and a linear function in warp direction. The model takes into account that usually two S-shaped piles couple and form an 8-shaped pillar together.

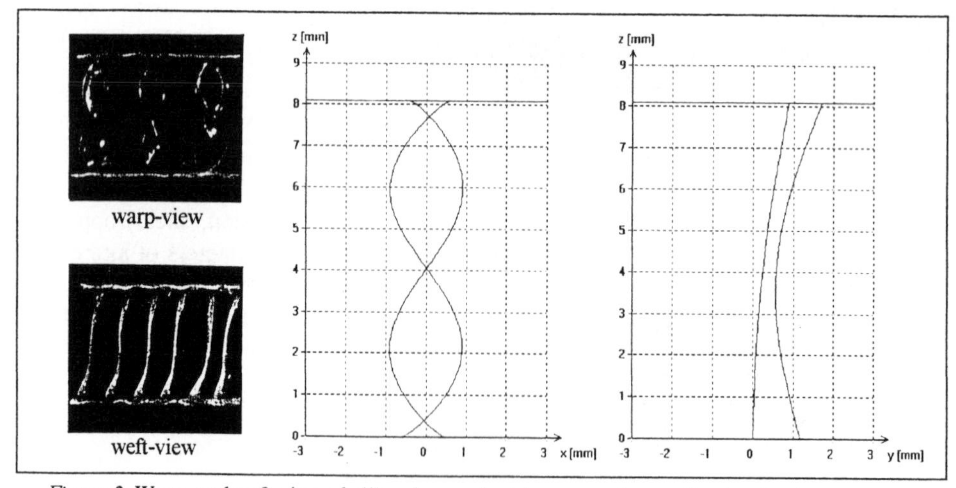

Figure 2. Warp- and weft-view of pillar shapes of a 10 mm panel according to pile shape program

Only at very high degrees of stretching the piles can uncouple again and a zig-zag formation of the piles is formed in the weft direction. The program is able to calculate the pile and pillar shape realistically for a range of degrees of stretching (thickness) based on the weaving scheme, the panel thickness and the induced pile angles during production.

In figure 2 the predicted pillar-shape is compared with the actually observed shape. The pile shapes of a 10 mm panel at 85 % of pile stretching are shown from the 2 principal material directions (warp = weaving direction and weft direction).

To model the resin distribution, in the first place estimations were made of the resin content in the skins and inside the pile bundles, based on experimental characterisations. The rest of the resin is found in the core. As an example figure 3 shows pile cross-sections at ¾ of the panel height with two different types of resin connections.

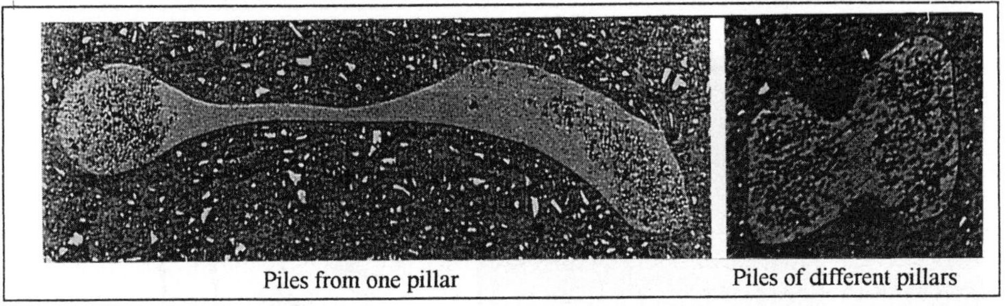

Piles from one pillar
Piles of different pillars

Figure 3. Pile cross-sections and resin connections between piles and pillars

The distribution of the excess resin is modelled with a very simple, but quite effective criterium: At all positions closer than a certain distance from glass fibre segments (skins and piles), resin is found.

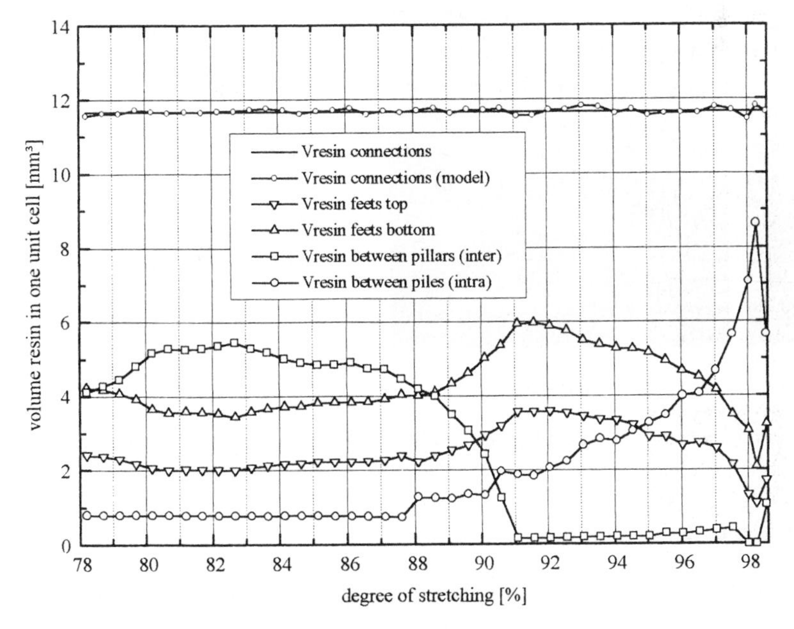

Figure 4. Resin concentrations near the skins and in connections as a function of the degree of stretching

The program does an iteration to match the volume of all the resin concentrations with the global resin content. The used criterium is a very simple model of a capillary effect.

The model is capable of predicting the resin concentrations at the pile fibre feet near the skins and the resin connections between piles and pillars quite realistically. The influence of the degree of stretching on the resin distribution in a 20 mm pile system is shown in figure 4. The degree of stretching has a strong influence on the type and the amount of resin connections.

In figure 5 the predicted resin distribution for a 10 mm and a 20 mm pile systems are given. The 3-dimensional views of the unit-cells of the materials are compared with actual cross-sections. Connections with the surrounding unit-cells are shown. In both panels the global resin content is 55 weight %. The resin connections are represented by rods.

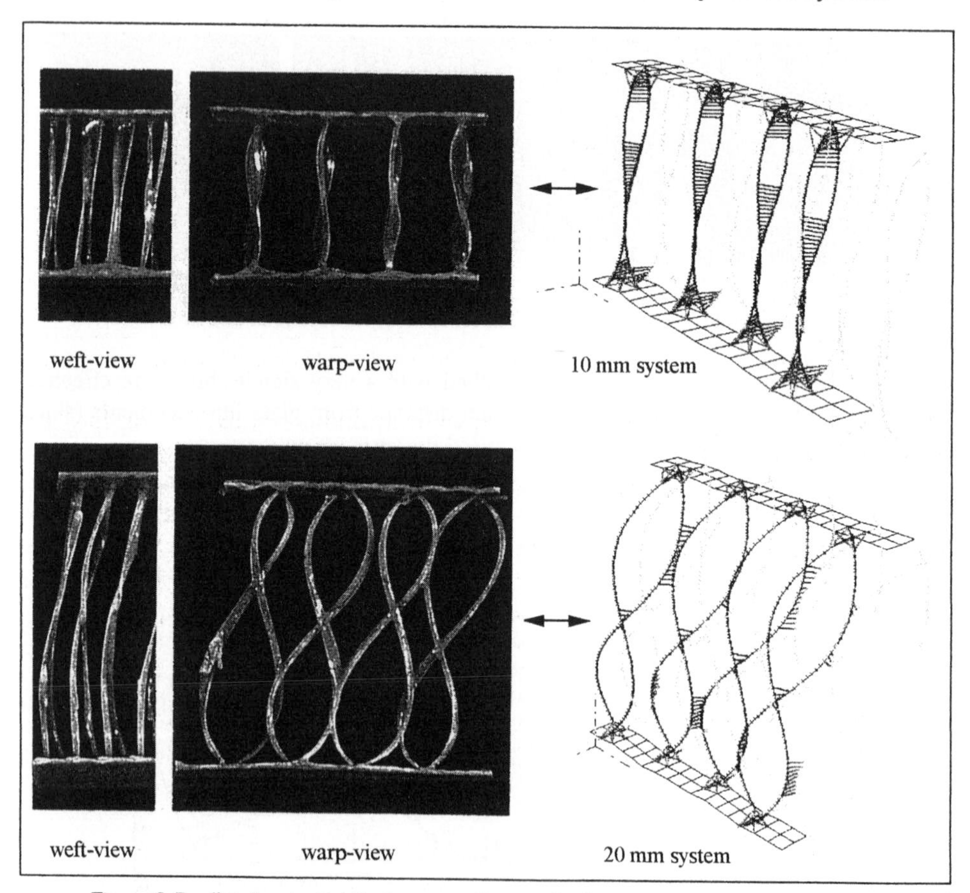

Figure 5. Predicted resin distribution according to pile shape and resin distribution model

3. Finite Element Calculations

The pile shape and resin distribution model was written in such a way that it can produce an input file for finite element programs [1].

For the skins plate (shell) elements are chosen. For the piles beam elements are taken, with a circular cross-section and a homogeneous distribution of the monofilaments assumed. Both assumptions were experimentally verified to be realistic. For the resin connections inside pillars and in between pillars beam elements with resin properties are taken. For the resin concentrations at the pile feet volume elements were chosen.

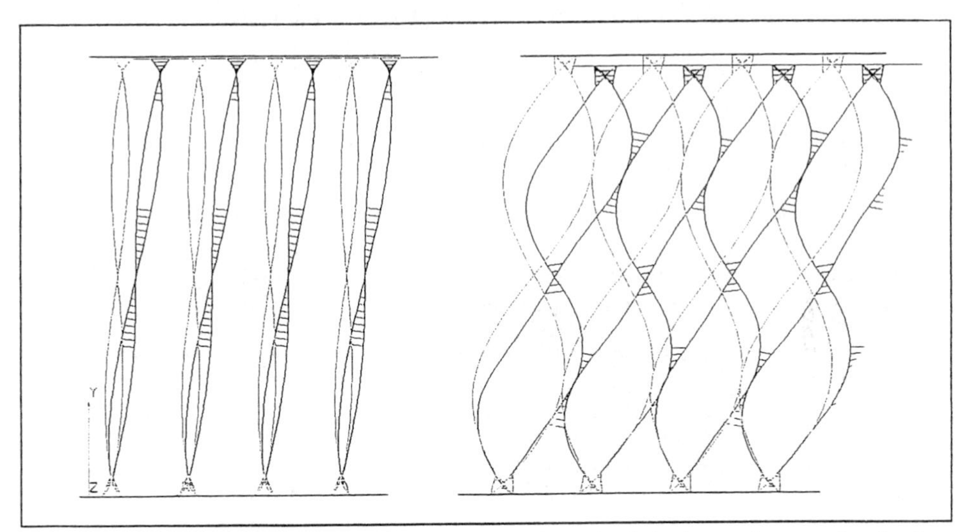

Figure 6. Influence of the pillar-connections on the deformation behaviour under warp shear loads

This finite element approach allows for detailed investigations on the deformations and stress concentrations in the core microstructure. In figure 6 the influence of the inter-connections between the pillars on the deformation behaviour under warp shear loads is shown. Figure 7 shows the influence of the inter-connections between the pillars on the stress distribution under warp shear loads.

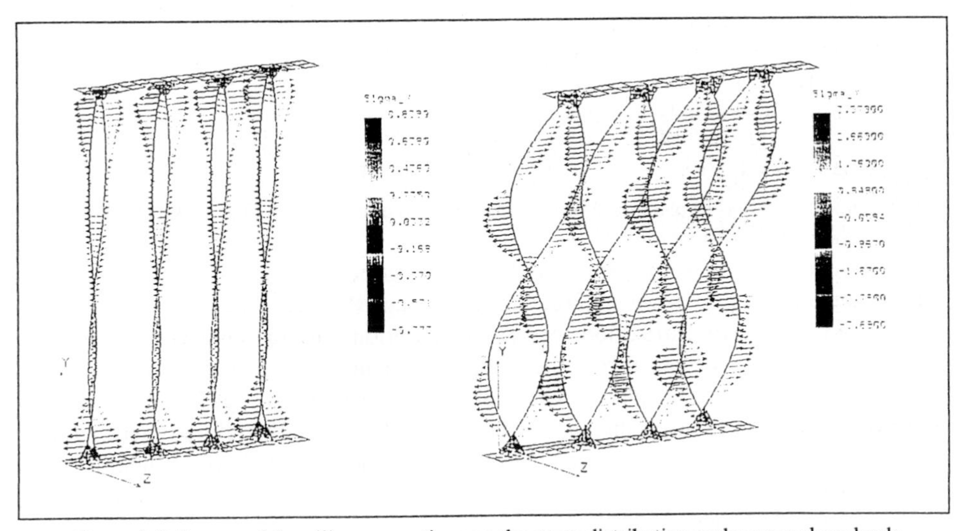

Figure 7. Influence of the pillar-connections on the stress distribution under warp shear loads

The stresses are better distributed in a panel with pillar-connections, since a diagonal reinforcement is formed with a connected pile fibre system (network).

In the finite element calculations the homogenisation principle was used. This principle allows the calculation of the properties of an infinite plate by the consideration of only one unit cell, by coupling the displacements and rotations at corresponding boundaries. In the framework of this work, also the effect of the choice of the boundary conditions was evaluated. The homogenisation theory proved to give reliable results for an infinite plate.

4. Results

Calculations on various sandwich-fabric cores show a good correspondence with experimental results. As an example in figure 8 a comparison is made between the predicted and measured core shear modulus in the warp and weft direction for 10 mm LD panels, as a function of the degree of pile stretching (panel thickness).

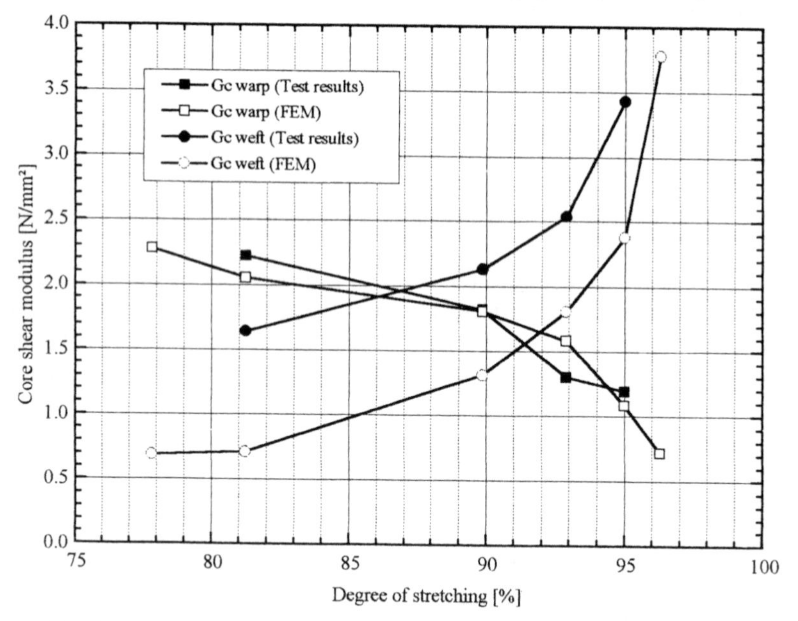

Figure 8. Predicted (FE) and measured core shear moduli as a function of the degree of stretching

The prediction for the warp shear modulus is very good. For the weft modulus too low values are predicted at lower degrees of stretching. Possibly this has to do with the prediction of the resin distribution. The simple criterion now simply puts resin in places closer than a certain distance from the composite segments. It is possible that in reality resin concentrates itself at places where other resin has already gathered. This could especially happen in the weft direction, where the pillars are rather close together for the 10 mm LD panels. If this were taken into account, the predicted weft modulus would increase.

The possible different resin distribution in reality can also explain the rather strong discrepancy which is found between the predicted weft shear modulus and experimental results for 10 mm LD panels at a range of resin contents, as shown in figure 9. The onset of resin wall formation (extensive inter-connections between the pillars in weft direction), which is thought responsible for the sharp increase in stiffness, is predicted at significantly higher resin contents than found in practice.

The predictions for the flatwise compression modulus are good for lower degrees of stretching (up to about 85%). At high degrees of stretching the predicted values are much too high. The compression modulus increases because at higher degrees of pillar stretching the pillars fill up with resin and the pile fibres are aligned in the load direction.

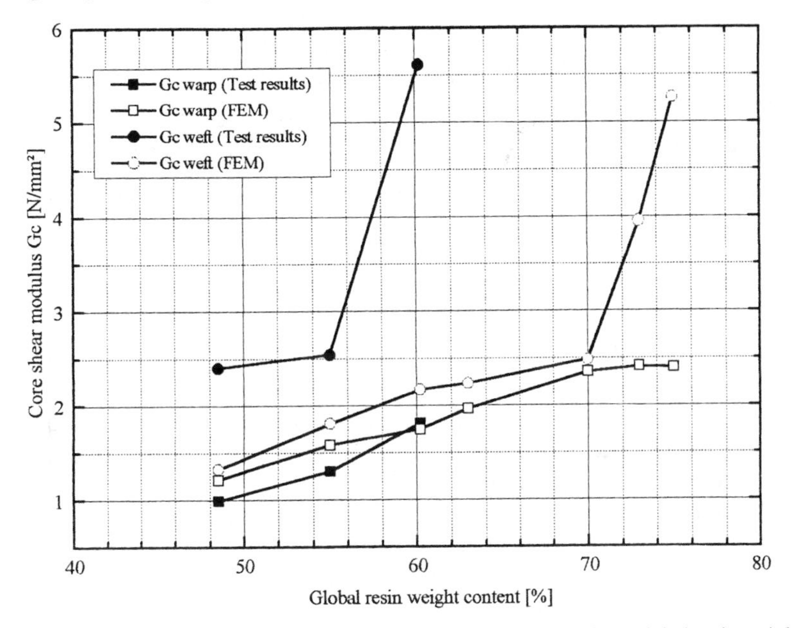

Figure 9. Predicted (FE) and measured core shear moduli as a function of the global resin weight content

The behaviour of sandwich-fabric panel cores in shear and flatwise compression can be understood by considering the flexural behaviour of the pile beams. At very high degrees of stretching the piles are almost straight and will have a very high compression stiffness. If due to loading the curvature of the piles increases somewhat, the compression stiffness of the system may drop rapidly. This is something to be verified with a non-linear analysis.

5. Conclusions

A start was made with the simulation of the core properties of sandwich-fabric panels with the help of finite elements. A special finite element preprocessing program was written for this purpose. It is based on a detailed model of the pile and pillar shapes and the resin distribution.

Strong point of the developed pile shape and resin distribution model is that it needs as input only data from the weaving company and the panel manufacturer. From these data a quite realistic estimation of the core microstructure is made.

Until now, the model has only been implemented for the basic weaving type from the Brite-Euram project Aficoss. In panels based on this fabric, usually 2 piles combine to form an 8-shaped pillar. For other weaving types modifications may be necessary. For instance in fabrics of higher pile densities groupings of 4 piles into pillars are possible.

For the finite element calculations the homogenization principle was applied. This appears to give very reliable results for the properties of an infinitely large plate, based on the consideration of only one unit-cell of the material.

Until now only elastic properties (warp and weft shear modulus and compression modulus) were calculated. The correspondence with the experimental results is fairly good for the shear moduli. Absolute predictions of the properties are possible.

The results indicate that improvement of the resin distribution model might be necessary, as certain accumulations of resin are not so well modelled in the present state of the model. It could be interesting to incorporate the effects of a certain randomness in the actual microstructures. This can be done by considering larger problem sizes in which the randomness has been introduced.

The predictions for the compression modulus are not good at high degrees of stretching. A first step to solve this problem might be to perform a non-linear analysis for this property.

A major addition which is necessary to model the behaviour of the full range of sandwich-fabric panels, is to expand the model to foamed cores. Question is how the foam can be introduced in the core. Solid elements with foam properties seem a solution, but apparently the creation of a mesh of solid elements in the highly irregular space in between the piles is not a simple problem.

It is possible to develop a design-tool for sandwich-fabric panels on the basis of the here presented work. Such a tool could e.g. be useful in the development of new fabrics. The properties of panels produced from newly conceived fabrics could be estimated before actual production and testing.

6. Literature

1. J. Pflug, Prediction of the core properties of sandwich-fabric composites with Finite Element Methods, Msc. thesis Eupoco program, KU Leuven, 1996
2. A.W. van Vuure, Composite panels based on woven sandwich-fabric preforms, PhD Thesis KU Leuven, 1997
3. A.W. van Vuure, J. Ivens, I. Verpoest, Sandwich-fabric panels, Proc. Int. SAMPE Conf., Anaheim, May 1995, 12 p.
4. A.W. van Vuure, J. Ivens, I. Verpoest, K. Swinkels, M. Fantino, Sandwich-fabrics as core materials for light-weight structures and in flat panel applications, Proc. Eur. SAMPE Conf., Salzburg, May 1995, p. 93-103
5. A.W. van Vuure, J. Ivens, I. Verpoest, M. Fantino, G. Clavarino, G. Fantacci, Sandwich-fabrics as core materials and in flat panels for marine applications, Proc. 3d Int. Conf. on Sandwich Construction, Southampton, Sept. 1995, p. 171-182

DETERMINATION OF MATERIAL PROPERTIES FOR STRUCTURAL SANDWICH CALCULATIONS: FROM CREEP TO IMPACT LOADING.

P. DAVIES*, R. BAIZEAU*, A. WAHAB*,
S. PECAULT**, F. COLLOMBET**, J-L LATAILLADE**

** Marine Materials Laboratory*
IFREMER, Centre de Brest, 29280 Plouzané, France

*** LAMEF, ENSAM Bordeaux,*
Esplanade des Arts et Métiers, 33405 Talence, France

Introduction

In order to use sandwich materials efficiently designers require reliable material properties. The properties commonly available are values given by material suppliers determined by standard quasi-static tests. These are used in initial material selection and then more detailed design calculations are performed. The loadings encountered are rarely quasi-static and predictions of the behaviour of composite sandwich structures under slow (creep) or rapid (impact) loadings can be made by various methods, from simple analytical models to complex numerical codes. Common to all these approaches is the need for appropriate material input data.

This paper presents experimental results from recent studies in which sandwich structures were considered for marine applications where both long term loads and impact were of concern. The materials considered are glass rovimat (woven rovings and mat) reinforced polyester facings and closed cell PVC foam cores (denoted "ductile" and "rigid" here, both of nominal density 80 kg/m^3). Such materials are widely used in marine applications and are manufactured by hand lay-up. This fabrication route can result in considerable variation in properties so experimental verification of properties is needed.

The work is presented in three parts. First, the results from constant load creep tests on composite facings and core foams are presented. Then the efforts made to quantify the influence of loading rate on the response of facings and cores are described, and the use of temperature to accelerate tests is shown. The third part of the paper is devoted to a brief correlation of the data generated with the behaviour of sandwich beams under creep conditions, and of sandwich panels under falling weight impact loading.

A. Vautrin (ed.), Mechanics of Sandwich Structures, 327–336.

1. Creep tests.

1.1 COMPOSITE FACINGS UNDER TENSILE LOADING

Creep behaviour of composites has attracted much attention and several authors have presented results showing the viscoelastic nature of glass and carbon reinforced composites, particularly when loaded off-axis [1]. However, few data are available for the type of compsites tested here which have low fibre content and are produced by hand lay-up. A series of tests was therefore performed in which 2, 5 and 8-ply specimens were strain gauged and subjected to different constant tensile load levels for periods up to 12 months. A large database was established allowing creep response to be predicted. An example of creep data is shown in Figure 1, results from longer tests have been presented previously [2]. One important aspect of these tests is the influence of fibre content. Thinner laminates, such as those used for sandwich facings (2-ply), tend to have higher resin contents when produced by hand lay-up and this must be taken into account when making predictions for sandwich beams. The creep behaviour is linear viscoelastic in the stress range of interest here (up to 57 MPa).

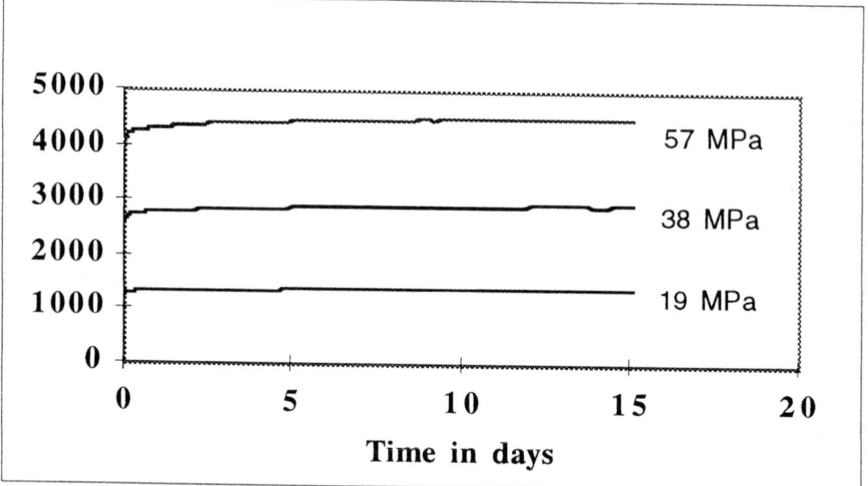

Figure 1. Tensile creep data, microstrain v time for 5-ply laminates loaded in fibre weave direction.

1.2 PVC FOAM CORES UNDER SHEAR LOADING

There have been several previous studies of the creep of foams, particularly polystyrene and polyurethane. Results relevant to the present study were obtained by Huang & Gibson [3,4] who analysed shear creep results for polyurethane foams and proposed models for linear and non-linear viscoelastic response based on micro-mechanics and using the solid polymer properties of the cell walls. They found that a power law of the form ($\gamma = \gamma_0 + mt^n$) described the creep of those materials well. In this equation γ is the shear strain, t the time, γ_0,the initial strain and m and n are the creep parameters. Chevalier [5] has used three-point flexure to characterize creep behaviour of a range of core materials including PVC, for times up to 8 weeks. Caprino et al. [6] presented

compression creep data for a PVC foam obtained at temperatures from 40 to 90°C for loading times up to 3 days. These were used to construct a master curve to extrapolate to longer times. Shenoi et al. [7] have tested sandwich beams subjected to ten point loading to study shear creep response of a ductile linear PVC foam. They concluded that the linear viscoelastic range for this material is limited to about 30% of the shear yield stress. Weissman-Berman also discusses viscoelastic behaviour of beams and uses measured creep compliance to predict 30 day beam behaviour [8].
In the present work shear creep tests on foam samples were performed on a universal test machine under force control, at different stress levels. Tests were quite short (a few hours) and temperature was increased to accelerate creep for the rigid foam (see ref. [2]) The ductile foam showed considerable creep even at 20°C as shown in Figure 2.

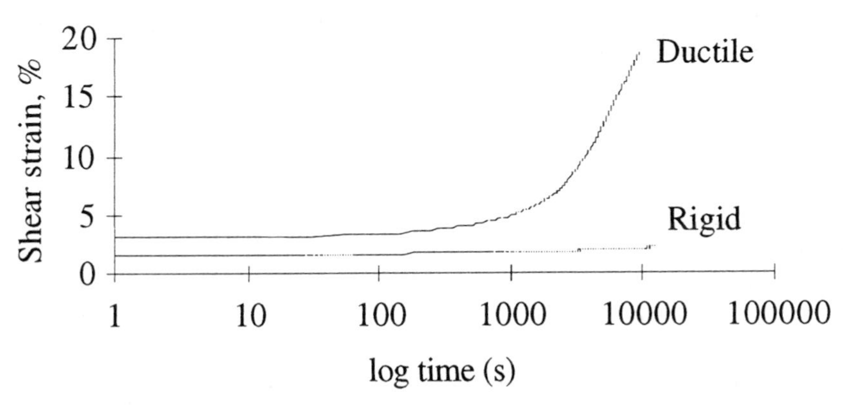

Figure 2. Shear creep results for rigid and ductile PVC foams for applied shear stress of $\tau = 0.4$ MPa

2. Influence of loading rate

2.1. COMPOSITE FACINGS UNDER TENSILE LOADING

For the drop weight loadings described below in Section 3 the maximum strain rate in the facings was measured by strain gauges to be in the range 1-5/s (10.9 kg dropped from 1 to 3 metres). The standard quasi-static tensile test gives a value corresponding to 0.0005 /s or slower so it was necessary to determine how loading rate affects the modulus. Several authors have investigated loading rate effects on tensile behaviour of composites and Barré et al. have reviewed published data recently [9]. For woven glass/epoxy composites Harding & Welsh suggest that moduli increase by 17 and 33% going from quasi-static to intermediate rates (23/s) for specimens tested in the 0° and 45° directions, [10]. It was not possible to perform fast tests here but a series of tests was performed at different loading rates on 0° and 45° specimens, to see if moduli values were consistent with such an increase. Results are shown in Figure 3, together with those from reference [10]. The trends in increasing moduli appear to be similar for these woven materials.

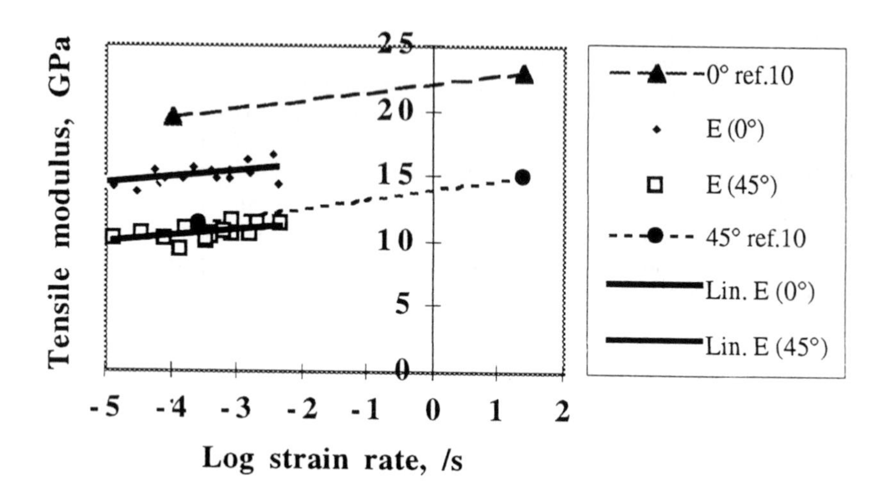

Figure 3. Effect of loading rate on glass/polyester composite tensile moduli, and results for woven glass/epoxy in ref. [10] .

2.2.PVC FOAM CORES

2.2.1. Shear loading

Most of the work published on the influence of rate effects on core behaviour has been initiated by material suppliers. For example, Gellhorn & Reif [11] present extensive data from compression and shear tests on foams at rates up to 500/s. Feichtinger gives results from shear tests on balsa, foams and honeycombs up to 385/s [12]. Lonno & Olsson present a rapid tensile test method and discuss some of the difficulties in measurements of high rate tensile data for foams [13]. All indicate increases in foam moduli with increasing strain rate but it is not always easy to analyse the data as different test set-ups are used for different test rates

In the present work the influence of loading rate on the shear behaviour was first examined directly, through a series of relatively slow tests performed on standard plate shear specimens (ASTM C273) at different loading rates on a screw-driven test machine. Examples of curves obtained are shown in Figure 4. Figure 4 indicates that there are significant differences in the knee point as rate is increased. Moduli increases over three decades are 12 % for the rigid foam and 20 % for the ductile foam. Such tests can be performed at cross head rates up to 100 mm/min., using a fast data acquistion system, and this corresponds to a test lasting a few seconds (initial shear strain rate of 0.01/s). However, the tests rates are still rather slower than the maximum shear rate predicted for the drop weight impact tests described in Section 3 (1-10/s).

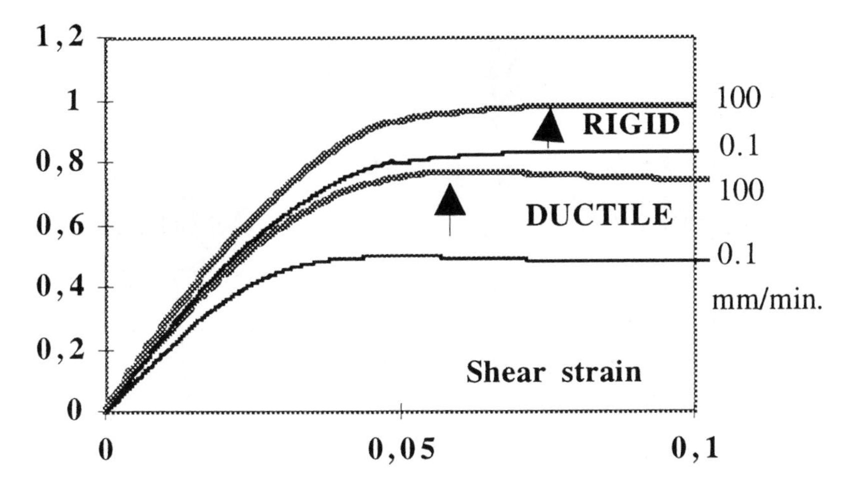

Figure 4. Shear response of PVC foam cores at 2 loading rates, initial parts of curves.

In order to extend the range of rates, tests were then performed at low temperatures, down to -40°C, to investigate the applicability of a time-temperature equivalence shift to data similar to that described for compression tests in reference [6]. Some results are shown in Figure 5.

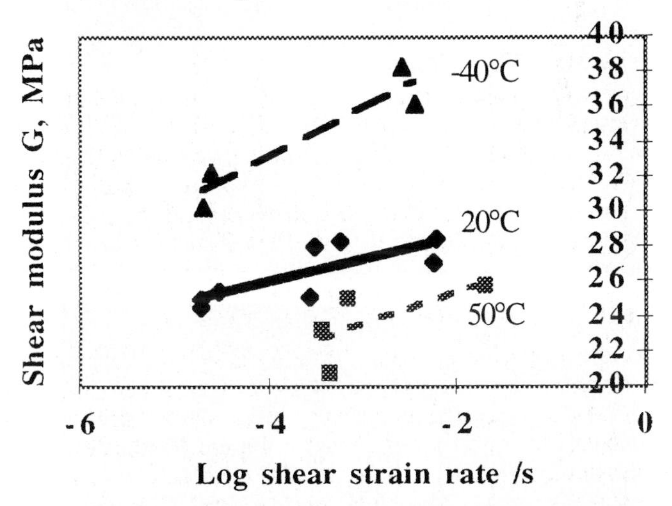

Figure 5. Shear modulus v strain rate for tests at different temperatures, rigid foam

The forms of the plots for each temperature indicate that a horizontal shift of the data from cold tests may be possible to extend the strain rate range, but it was felt that data at higher rates was necessary to validate such a shift.

A third approach was therefore employed, involving a drop weight impact set-up (Figure 6) similar to that used by Feichtinger [12].

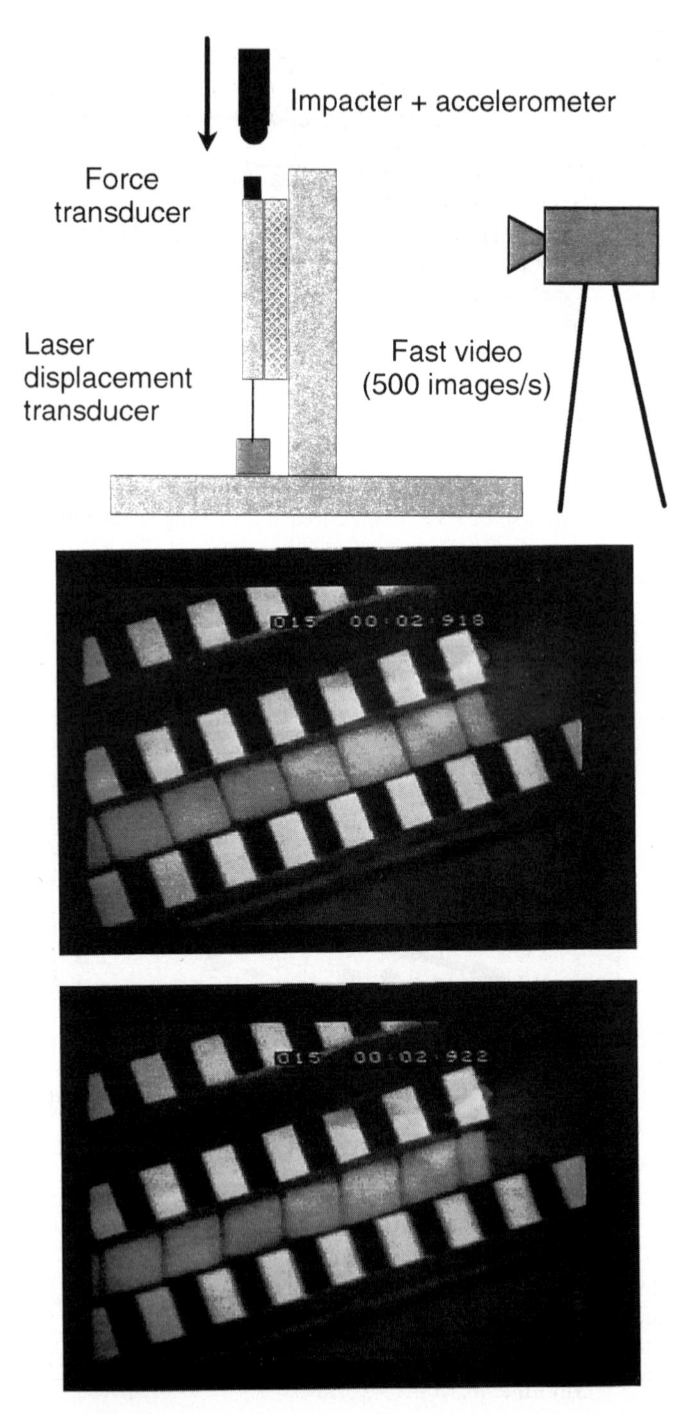

Figure 6. Drop weight shear test set up and fast video images before and during impact

A high speed video was employed, both to check measurements and to examine how the specimen deformed during the test. The recorded images shown in Figure 6 are separated by 4 ms and indicate a reasonably uniform shear strain state in the specimen. A damper was used between the impacter and the load cell, as initial tests without damping gave very noisy force signals. This damper reduced the loading rate but it was still in the range required (around 10/s). The shear rate could also be varied to some extent by varying the drop height. Preliminary results from these tests suggest that at these rates moduli values of both foams are about twice the quasi-static values. Figure 7 shows that such values are compatible with extrapolated values obtained at low temperatures (Figure 5) and using a temperature-loading rate equivalence.

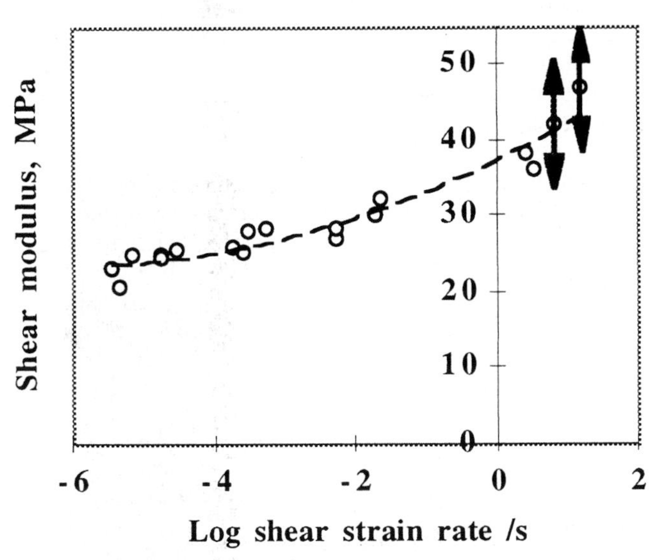

Figure 7. Shear modulus v log strain rate. Test results at different temperatures shifted onto one curve, and drop weight shear data ranges (arrows)

It should be emphasized that these are preliminary results and more tests are needed but they do suggest that the shear moduli values of both foams at the strain rates of interest for the drop weight impact tests on sandwich panels are significantly higher than those measured in quasi-static tests. Failure behaviour is not discussed here but that also evolves with rate [14].

2.2.2. *Compression loading*

In parallel with the shear tests run at IFREMER, compression tests were performed using a Hopkinson bar set-up at ENSAM. Specimens were 20mm diameter cylinders cut in the through-thickness direction of the foam. Strain gauged cylindrical PMMA bars were used to transmit the load to the specimen, Figure 8.

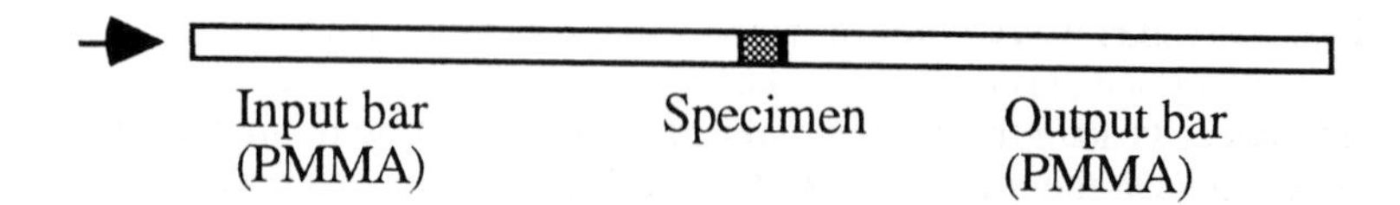

Figure 8. Hopkinson bar set-up to measure high rate compression behaviour.

This set-up allows much higher strain rates to be obtained, around 250/s, and this is the order of magnitude of crushing rates measured during drop weight impact tests performed recently with small diameter impacters [15,16]. The difficulty with these tests is ensuring a uniform stress field throughout the specimen. Measurements suggested that such a stress state was not achieved. The uniform stress state is reached in about 0.4 ms but before this input and output stresses differ by 100 %. The use of an average stress/strain relationship then has no physical sense so it is hazardous to quote moduli values. Nevertheless by using a rheological model (a viscoelastic-plastic model proposed by Sokolowski was applied) it is possible to fit the measured input and output data. The hypothesis of uniform stress can thus be avoided. When this is applied to the rigid foam a good global agreement is obtained but the local representation of what happens in each section is not in agreement with the crushing behaviour: the model suggests that the stress inducing cell wall buckling is achieved in all sections of the specimen but the tests show that irreversible collapse only occurs in the region in contact with the input bar. However, this technique does allow qualitative studies to be made. For example, figure 9 presents the maximum stress measured on the output bar corresponding to the collapse of cell walls of both foams as a function of temperature at a compression strain rate of 250/s It is interesting to note that at room temperature the ductile foam shows a higher crushing stress than the rigid foam. These results also indicate the higher sensitivity of the ductile foam to temperature.

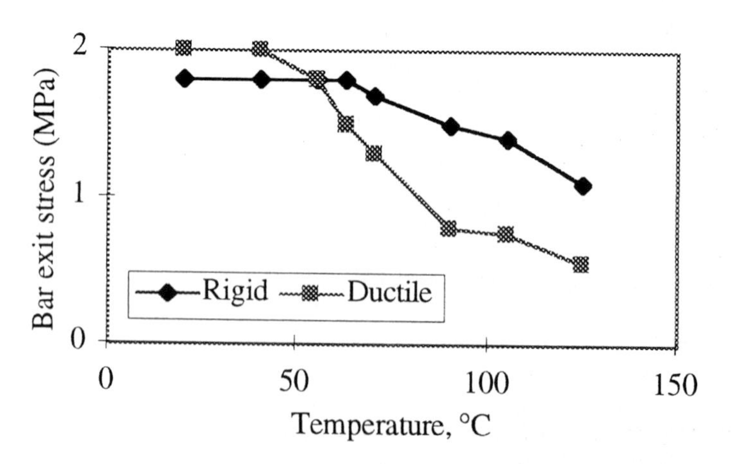

Figure 9. Output bar stress v test temperature, Hopkinson bar.

3. Tests on Sandwich structures

The sandwich structures described here consist of 2 mm thick facings made up of two layers of glass rovimat (a woven roving layer lightly bonded to a mat layer) in polyester, laminated directly onto 20 mm thick PVC foam cores by hand lay-up.

3.1. CREEP TESTS ON BEAMS

Sandwich beams were loaded in four point bending on a creep test bench. Load point displacement transducers and lower facing central strain gauges were monitored continuously. The applied loads were chosen to keep facing strains low (around 0.1%), and shear stresses were 0.32 MPa for both sets of beams. Figure 10 presents results for both materials in terms of measured upper load point displacements. The larger creep observed in the ductile foam in these tests reflects the results from the shear tests.

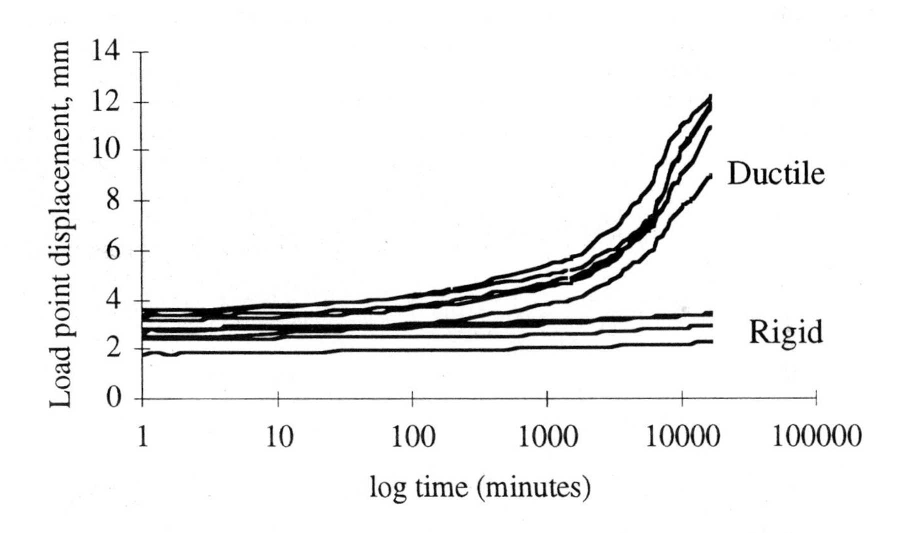

Figure 10. Four point flexure creep test results (5 specimens of each foam)

3.2. DROP WEIGHT IMPACT TESTS ON PANELS

A large number of impact tests have been performed [14,17,18] but only one example will be described here. Sandwich panels simply supported on four edges with 300 mm x 300 mm free area were impacted by a 100 mm diameter 10.9 kg steel hemispherical impacter dropped from different heights. The impact response of these panels, in terms of contact force, displacement and lower surface strains, was calculated using an analytical model based on the superposition of local indentation response and global vibration modeshapes, developed recently [19]. This model takes into account through thickness shear and uses a modal acceleration method to reduce the calculation time by first integrating the static component of the response. For the rigid PVC sandwich this gave a good correlation using quasi-static shear properties, not only for force and displacement but also for facing strains (within 15%). As the material fails by core shear and the model allows core shear strains to be calculated it may be possible to develop a failure criterion for this failure mechanism. For the impact test configuration described above the model predicts that at the drop height at which shear failure is observed experimentally (2.25 metres) the maximum shear strain is around 4%. This corresponds approximately to the knee on the curves in Figure 4 but more work is required to validate this approach. For the ductile foam the model did not give a good correlation with force and displacement using quasi-static properties. In this case improved test-analysis correlation was obtained by using a higher shear modulus, (28 GPa instead of 21), suggesting that rate dependence of properties must be taken into account.

4. Conclusions

In order to use sandwich materials with confidence the appropriate material properties must be used in predictive models. The results presented here show that basing design for creep and impact on quasi-static properties can lead to errors in predictions and hence to the use of inappropriate materials. The loading rate must be considered when the choice of core material is made, as a foam may show superior properties at high rate but not satisfy a stiffness requirement under long term loading or vice-versa. The characterization of these materials is not trivial but it is essential if their use is to be optimized.

References

[1] Dillard DA, ch. 8 in "Fatigue of Composite Materials", Elsevier 1990 pp339-84.
[2] Davies P, Craveur L, Proc. ECCM-CTS3, London 1996, Woodhead, pp 367-76.
[3] Huang JS, Gibson LJ, J. Mat. Sci., 26, 1991, pp637-47.
[4] Huang JS, Gibson LJ, J. Mats. in Civil Eng., 2, 3, August 1990 pp171-182.
[5] Chevalier JL, Proc, 3rd IFREMER conference, Paris Dec. 1992, pp107-116.
[6] Caprino G, Teti R, Messa M in Proc. 3rd Int. conf. on Sandwich Construction, ed Allen HG, EMAS 1996 pp813-24.
[7] Shenoi RA, Clark SD, Allen HG, Hicks IA, Proc. 3rd Int. conf. on Sandwich Construction, ed Allen HG, EMAS 1996 pp789-99.
[8] Weissman-Berman D, Proc. Sandwich Constructions 2, 1992.
[9] Barré S, Chotard T, Benzeggagh ML, Composites A, 1996 pp1169-81.
[10] Harding J, Welsh LM, J. Mat. Sci. 18, 1983 pp 1810-26.
[11] Van Gellhorn E, Reif G, Proc Sandwich Constructions 2, 1992 pp541-57
[12] Feichtinger KA, Composites (French) 1, jan-fev 1991, Vol 31 pp37-47.
[13] Lonno A, Olsson K-A, Proc. 2nd Int. conf. on Sandwich Construction, 1992.
[14] Davies P, Choqueuse D, Verniolle P, Prevosto M, Genin D, Hamelin P, ESIS19 "Impact and Dynamic fracture of polymers & composites", 1995 pp341-58
[15] Pecault S, PhD thesis ENSAM Bordeaux, 1996 (in French).
[16] Collombet et al., presentation at this conference
[17] Davies P, Choqueuse D, Pichon A, Proc. ECCM-CTS2, Woodhead Publishers, Hamburg 1994, pp513-522
[18] Davies P, Choqueuse D, Bigourdan B, Proc. 3rd Int. conf. on Sandwich Construction, ed. Allen HG, EMAS 1996, pp647-58.
[19] Wahab A, PhD thesis ENSAM Bordeaux, 1997, (in French).

MECHANICAL MODELLING AND CHARACTERISATION OF A NEW STRUCTURAL FILLER MADE OF SMALL SPHERES ASSEMBLY

K. IOUALALEN, Ph. OLIVIER & J.P. COTTU
Laboratoire de Génie Mécanique de Toulouse - IUT P. Sabatier
50 Chemin des Maraîchers - 31077 Toulouse Cedex 4 France

1. Introduction

This paper outlines a part of our ongoing research to develop a new structural filler that consists of small spheres (5,4mm diameter) and is designed to have similar applications to honeycombs. This material developed by ATECA RDM comes in the form of an assembly of small shallow spheres (figure 1). A such a material and is much more easier to put in shape than honeycombs. In fact, for the manufacturing of sandwich panels and shells which have complex shapes (i.e. double curvature) none cutting and machining operations are required for this structural filler to be put in shape. Furthermore, the use of an assembly of small shallow spheres as a structural filler enables the classical problems of honeycomb deformations to be avoided. When compared to polymeric foams used as structural filler, the sphere's assemblies offers higher dimensional and thermal stability. From a manufacturing point of view, once the skin panels of the sandwich structure are produced, the shallow spheres just have to be poured between those latter. The spheres assembly is then made by simply adding a liquid resin which once crooslinked will ensure the cohesion between the spheres. It is worthy of note that the resin which assembles the spheres is exactly the same as the one which constitutes the spheres wall. The aims fixed by our industrial partner (ATECA RDM) were firstly the improvement of the manufacturing process of spheres wall material - which is in fact a particulate composite material (Figure 1) - and secondly the determination of the mechanical characteristics of sphere assemblies.

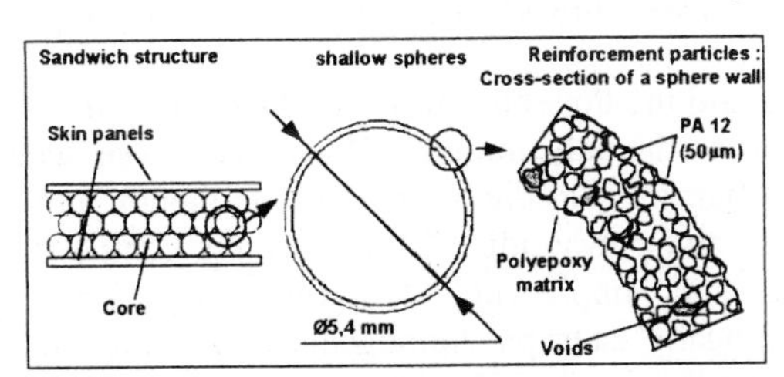

Figure 1. Material presentation.

To these ends, a threefold study has been performed on the spheres wall material characteristics, on a single sphere mechanical behaviour and lastly on the sphere assemblies characterisation. Finally at the end of this work, a simple prediction method has been settled. This method enables the mechanical characteristics of sphere assemblies to be determined directly from those of the spheres wall material.

A. Vautrin (ed.), Mechanics of Sandwich Structures, 337–344.

2. Determination of the Mechanical Characteristics of Spheres Wall Material

The spheres wall material is a particulate composite material. As is can be seen in figure 1, this composite material is made of a polyepoxy (EP) matrix produced by Ciba-Geigy and of polyamide 12 (PA12) particles produced by Hülls. Consequently, the spheres wall material cannot be considered as a homogeneous material and in addition, as it is shown in figure 1, the manufacturing process of the shallow spheres induces non negligible void contents. This is why the spheres wall material is considered as a composite material, it is to say as a heterogeneous structure stemming from the association of a great number of particles embedded in an polymeric matrix. In order to determine the mechanical properties of this particulate composite, two ways can be used. Experimental tests can be carried out directly on composite plates previously manufactured. Alternatively, the mechanical properties of the particulate composite can be predicted from those of its constituents on the basis of homogenisation modelling.

Several composite plates (EP-PA12) as well as unfilled polyepoxy (EP) and polyamide 12 (PA12) plates have been manufactured by ATECA RDM according to the cure cycle used for the spheres production. Note that for composites the mass fraction of polyamide 12 particles has been set once and for all at 60%, exactly as in the spheres themselves. A series of classical tensile tests has been carried out on samples cut-out from the composite, the polyepoxy and the polyamide 12 plates according to French standards. The obtained results are shown in figure 2.

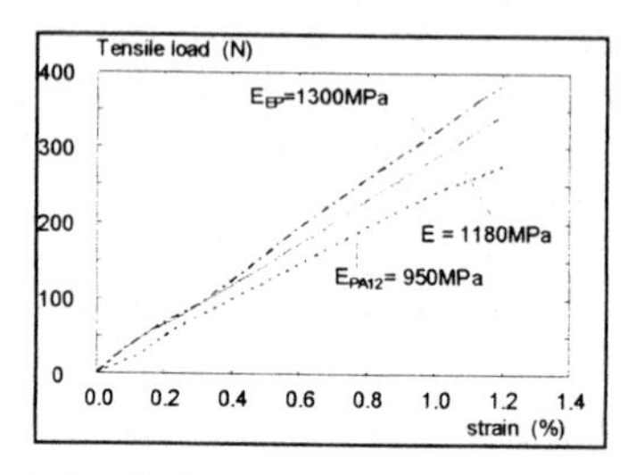

Figure 2. Tensile load-Strain curves for all components.

The elastic moduli E and the Poisson ratios ν of the composite (EP-PA12), the unfilled polyepoxy resin (EP) and the polyamide 12 (PA12) have been determined directly from load/strain curves in figure 2. On the basis of elastic moduli and Poisson ratios values of unfilled polyepoxy and polyamide 12, those of the composite material have been calculated by using both simple rule of mixtures and homogenisation modelling. Among all the statistically isotropic homogenisation modelling, we have used here those of Kerner [1] and Halpin-Tsai [2]. The obtained results are shown in table 1. E is the elastic modulus of the particulate composite, while G and K are respectively its Coulomb and compression moduli. As it can be seen in table 1, the moduli values given by the homogenisation modelling are very close to those obtained by using a simple rule of mixture. In all cases, compared to experimental values, these modelling result in an overestimate of moduli values say by 10%. This small difference between experimental and predicted values can be due to the physical bonding which develop between the constituents of the particulate composite. Effectively, as shown in a previous study [3], there are physical bonds of hydrogenous type between the polyamide 12 particles and the polyepoxy matrix. Furthermore, it is of importance to note that the modulus of the polyamide 12 particles is very close to the one of the polyepoxy matrix. In this case, when the moduli of the constituents are very close, Hashin [4] has shown that the homogenisation modelling can reduce to simple rule of mixtures. This can explain the small difference between the moduli values predicted with rule of mixture and those obtained from the homogenisation modelling.

TABLE 1. A comparison of the particulate composite (spheres wall material) characteristics obtained by using different methods

Method	E (MPa)	G (MPa)	K (MPa)
Experimental tests	1180	x	x
Rule of mixtures	1080	419	908
Kerner modelling	1078	415	896
Halpin-Tsai modelling	1087	419	896

3. Characterisation of Shallow Spheres

The mechanical tests used to perform the characterisation of the shallow spheres, must meet the spheres manufacturer (ATECA RDM) specifications of simplicity and low cost. Two types of mechanical loading have been thus selected : hydrostatic compression and compression of a single sphere between two parallel plates. This latter test is currently used by the spheres manufacturer as quality inspection. It is important to mention that these compression tests on single sphere between parallel plates have been carried out by spheres manufacturer request. In fact ATECA RDM manufactures only spheres and do not produce any spheres assembly, since the assembly is directly made once the spheres have been poured between the skins of the sandwich structure to manufacture.

The compression test between parallel plates gives relatively complex load/deflection curves (figure 3). As it can be seen in figure 3, at the beginning of the compression test, the load/deflection curve remains approximately rectilinear. This shows that all the strains that the sphere is undergoing remain reversible. The elastic modulus of the spheres wall material can be determined from this curve segment. Beyond this elastic linear or pre-buckling behaviour, the sphere undergoes plating which appear on the curve as a complex stretch. During the very beginning of their plating the spheres undergo a dynamic buckling phenomenon called snapping [5]. This dynamic buckling phenomenon has been modelled by Brodland and Cohen [5] which have performed a theoretical study of spheres snapping under a concentrated load. These authors have shown thus that the snapping essentially depends on the sphere wall thickness and on the type of loading. The compression test between parallel plates only enables few mechanical characteristics of spheres to be determined. Effectively, as it can be seen in figure 3, only the buckling limit which marks the end of the spheres elastic linear behaviour and the spheres ultimate compressive strength can be determined. Only the measurement of the buckling limit and the ultimate compressive strength proved to be reproducible from one tested sphere to an other.

Beyond mechanical tests, various approaches can be used for the spheres characteristics to be determined.

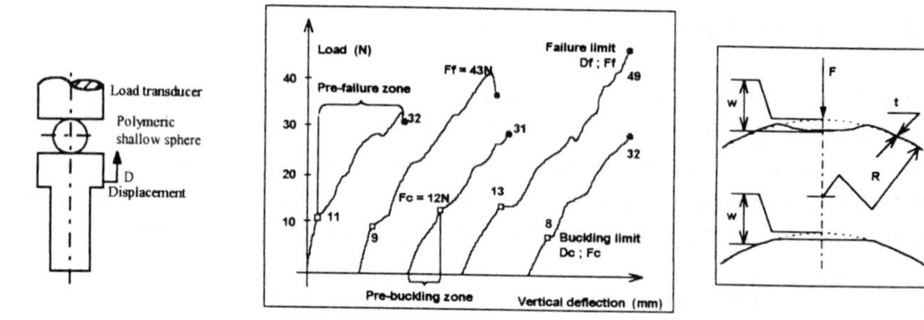

Figure 3. Experimental compression device and (Load/Vertical-deflection) curves corresponding, and deflection w before and after buckling.

3.1. Analytical Modelling

First of all the analytic modelling offered by Timoshenko and Woinowsky-Krieger [6] for spherical shells have been used. These modelling enable the value of the deflection w at a sphere apex (see figure 3) to be calculated from the elastic modulus E and the Poisson ratio ν of the sphere wall material (equation (1)). In this case, an shallow sphere of radius R and thickness t is submitted to an axisymmetric concentrated load F applied to its apex (figure 3). On the other hand, Pogorelov [7] has settled a relation (equation (2)) which allows for the buckling of a sphere under the effects of an axisymmetric load F concentrated on its apex (figure 3).

$$w = \frac{\sqrt{3\cdot(1-\nu^2)}}{4}\cdot\frac{F\cdot R}{E\cdot t^2} \tag{1}$$

$$w = 0.308\cdot\left(1-\nu^2\right)^{3/2}\cdot\frac{F^2\cdot R^2}{E^2\cdot t^5} \tag{2}$$

3.2. Empirical Modelling

It is the accurate analysis of shallow spheres buckling curves which has enable several authors [5,8] to strikingly identify the parameters which govern the snapping. Furthermore, on the basis of buckling curves analysis a modelling of the snapping phenomenon has been settled. Thus Brodland and Cohen [5] as well as Kaplan [8] have studied the snapping of a sphere and obtained a relation (equation (3)) which enables the critical buckling load Fc to be calculated from the sphere elastic constants E and ν. In relation (3), the ki are numerical constants.

$$Fc = \frac{E\cdot t^3}{R}\cdot\left[k_1 - k_2\cdot\left(1-\nu^2\right)^{1/4}\cdot\sqrt{\frac{R}{t}} + k_3\cdot\left(1-\nu^2\right)^{1/2}\cdot\frac{R}{t}\right] \tag{3}$$

3.3. F.E.M. Modelling

Finally the compression test of a single sphere between two parallel plates has been studied by use of finite elements method (FEM) [9]. A series compression tests has been thus simulated by using various values of sphere elastic modulus, while the compressive concentrated load applied to the shallow sphere apex has been set once and for all at F = 10N. For each simulation, the calculations carried out by the finite elements code have enable the deflection w (see figure 3) to be determined as a function of the modulus value of the sphere wall material E. The obtained results are plotted in figure 4. The modulus/deflection curve in figure 4 is then used to determined the value of the sphere elastic modulus. In fact, the experimental compression tests (cf. figure 3) show that under a 10N concentrated load F applied on its apex, the shallow sphere undergoes a deflection w = 0.14 mm. This latter value is then applied to the modulus/deflection curve in order to determine the elastic modulus of the sphere (see figure 4). This gives E = 1050 MPa.

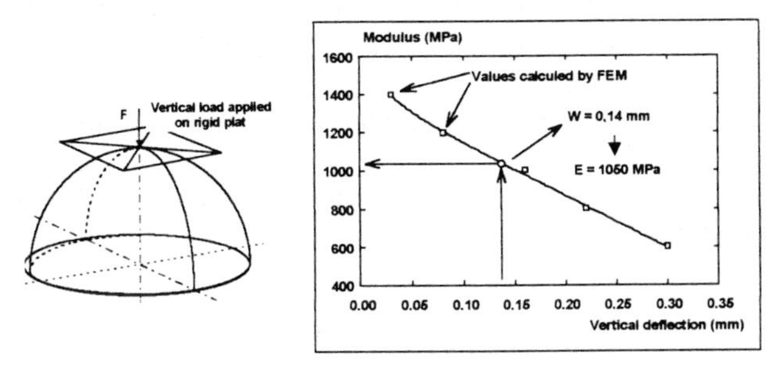

Figure 4. Sphere modelling and modulus determination from FEM values.

4. Characterisation of Spheres Assembly

In order to determine the mechanical characteristics of spheres assembly, two ways have been studied. The first way is the direct determination of the spheres assembly characteristics by submitted an assembly to experimental tests. The second one is based on the analyse of a single sphere behaviour in an assembly. This approach is to submit a single sphere to a mechanical loading for which its behaviour is identical to the behaviour of a sphere in an assembly. It is of importance to mention that in an assembly the spheres have not any specific laying out. There are neither conformation and nor cubical structure, the spheres stand at random locations and the assembly structure can be called a bulk structure.

4.1. Experimental Determination of an Assembly Characteristics

Compression tests have been carried out on a series of sphere assemblies according to French standard NFT54602 which is usually employed for foam testing. Figure 5 shows the experimental device used to carry out these tests. This means that now the spheres assembly is considered as a material. The obtained results are shown in figure 5. The load/deflection curve in figure 5 shows several stages. From the beginning of compression test up to the minimum limit (see figure 5) the spheres assembly begins to compress little by little. Effectively, all the spheres of the upper and lower sides of the assembly make contact with the parallel plates of the compression device. From the minimum limit up to the maximum limit (see figure 5) all the spheres undergo an elastic deflection. During that short period no sphere failure or buckling have been recorded. Beyond the maximum limit some spheres are undergoing buckling immediately followed by failure while some others are still submitted to an elastic linear deflection. This means that in a spheres assembly the failure of some spheres occurs even before some other ones have buckled. It is clear that all the spheres do not carry the same loading. It is obvious that in an assembly the load transmission is made possible only by the contacts between the spheres. Since in a spheres assembly the spheres have random locations, everything leads one to believe that each one of the spheres of an assembly has not the same number of contact points with the adjoining spheres.

The spheres assembly elastic modulus E_{sa} (the assembly is considered as a material) has been determined on the load/deflection curve (figure 5) between the minimum and maximum limits. In fact, the spheres assembly shows an elastic linear behaviour during this curve segment. This gives E_{sa} = 23.5 MPa.

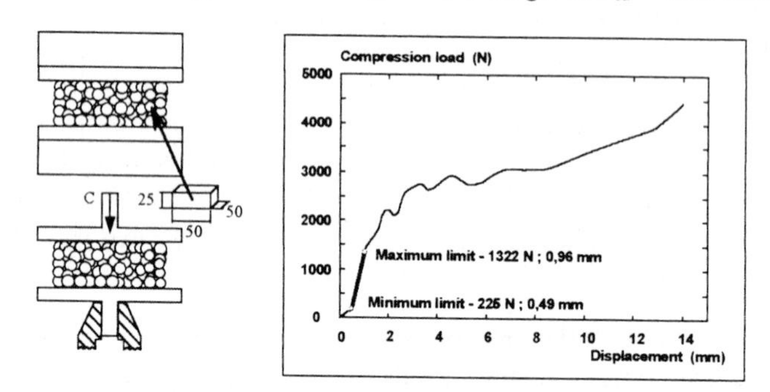

Figure 5. Compression/displacement curves on spheres assembly.

4.2. Behaviour of a Single Sphere in the Assembly

Single spheres are going to be submitted to a mechanical loading such that their behaviour remains the same of which they have in an assembly. This procedure should enable the assemblies characteristics to be determined from those of a single sphere. More precisely, the spheres assembly characteristics should be determined from the study of the behaviour of a single sphere loaded exactly as if it was in an assembly.

At the end of the mechanical tests carried on spheres assembly, it turns out that an evaluation of the number of contact points between the spheres is more than necessary. In order to perform this evaluation, the compactness of various bulk structure assemblies has been experimentally determined. The number of contact points between spheres in these assemblies is determined from the compactness values. Numerous work has been undertaken on the subject of number of contact points in disk and cylinder assemblies [10]. These studies have shown that only 15% of the disks or cylinders do not carry any load. In our case, for a three-dimensional assembly of spheres, 28% of the spheres do not carry any load. This give 8 contact points per sphere.

As a result, a shallow sphere with 8 contact points has been studied by using the finite elements method. Eight rigid plates in contact with a sphere have been used to simulate the 8 contact points through which the loading is transmitted (figure 6). Each one of the 8 plates carries the same concentrated load located on the contact point. It is important to mention that from a modelling point of view, once the sphere and the 8 plates are meshed, each contact between the sphere and plate is simulated by using 16 contact elements.

Moreover, compression tests using 8 contact points have been carried out on several shallow spheres. These compression tests have been performed by using low displacement values in order not to give rise to any buckling. The obtained results are shown in figure 6 together with those resulting from the FEM study. As it can be seen, for loading lower than 1N, the deflection value given by the FEM modelling remains much more lower than the experimental one. This difference might be due to the small number of contact elements used in the FEM modelling. In fact, a small number of contact element results in a error in the deflection value which becomes higher than the deflection itself. For loading higher than 1N, the deflection values get closer to the experimental ones as fast as the load increases. When the loading is higher than 8N the deflection values provided by the FEM rise dramatically and pass the experimental load/deflection curve. This is essentially due to the compression experimental device which is made of 8 small metallic spheres located around the shallow sphere to test. In fact, as soon as the first load is applied, the 8 metallic spheres induce a local deflection of the tested sphere by triggering a premature local buckling which is not considered by the FEM modelling.

The elastic modulus of the spheres assembly E_{sa} has been determined first on the load/deflection curve stemming from the experimental multicontacts compression test and secondly from the results provided by the FEM simulation. This gives E_{sa} = 9.5 MPa for the multicontacts experimental test and E_{sa} = 24.5 MPa for the FEM simulation. It is obvious that the simulation by FEM of the multicontacts compression test on a single sphere gives results very close to those obtained with the compression tests carried out directly on assemblies. Remember that in this FEM simulation the sphere is loaded exactly as if it was in an assembly. This proves that the compressive behaviour of a sphere assembly can be predicted with a very good accuracy by simulating the behaviour of a single sphere in the assembly. The experimental multicontacts compression tests which have been carried out on single spheres give a modulus value E_{sa} which is far from being acceptable. The large difference between the assembly modulus value thus obtained and the one given by the test performed directly on the assembly is certainly caused by the 8 metallic spheres above mentioned.

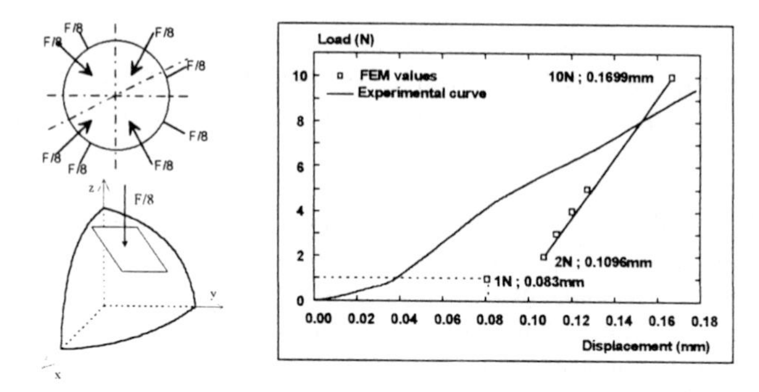

Figure 6. Experimental and theoretical (Load/Vertical deflection) curves for multicontact tests.

5. Concluding Remarks

We have shown that the spheres wall material characteristics (i.e. elastic modulus and Poisson ratio) can be easily determined with enough accuracy from those of the components, by using a simple rule of mixtures. Furthermore, we settled a FEM modelling which enables the characteristics (i.e. elastic modulus of a spheres assembly) of a spheres assembly to be determined from the behaviour of a single sphere considered as if it was in the assembly. This procedure gives very encouraging results in very good accordance with those directly obtained from compression tests on sphere assemblies. In the FEM modelling the input parameters are the sphere dimensions and the characteristics of the spheres wall material. So, it is obvious that this modelling enables the characteristics of a spheres assembly to be predicted from those of the spheres wall material. This means that the spheres manufacturer (ATECA RDM) which produces only single spheres, is now able to predict the characteristics of the assemblies without further manufacturing step.

6 References

1 Kerner, E.H. : The elastic and thermo-elastic properties of composite media, *Proceed. Phys. Soc.*, **B 69** (1956), 808-813.

2 Halpin, J.D., Kardos, J.L. : The Halpin-Tsaï equations : A review, *Polym. Eng. Science*, **16** (1976) 344-352.

3 Ioualalen, K. : Etude et développement d'un nouveau matériau de remplissage structural constitué de sphères creuses à paroi polymère chargée, Ph. D. Thesis. Toulouse (France) 1996.

4 Hashin, Z., Strickman, S. : A variational approach to the theory of the elastic behavior of multiphase materials, *Journal of Mechanics and Physics of Solids* **11** (1963),127-140.

5 Brodland, G.W., Cohen, H. : Deflection and snapping of spherical caps, *Int. Journal Solids Structures* **23** (1987) 1341-1356.

6 Timoshenko, S., Woinowshy-Krieger, S. : *Théorie des plaques et coques*, Editeur Polythechnique C. Béranger 1961.

7 Pogorélov, A. : *Théorie des enveloppes en déformations trans-critiques*, Edition russe 1965.

8 Kaplan, A. : *Buckling of spherical shells - In thin-shell Structures : Theory, Experiment and Design*, Edited by C. Fung and E.E. Sechler, Prentice-Hall Englewood Cliffs New Jersey 1974..

9 Huddleston J.V. : Finite deflections and snap-throught of high circular arches, Journal of applied mechanics decembre (1968),763-769.

10 Weber, J. : Recherches concernant les contraintes intergranulaires dans les milieux pulvérulents, *Cahier du Groupe Français de Rhéologie*, **Tome I-3** (1966), 161-170.

EXPERIMENTAL DETERMINATION OF THE DYNAMIC PROPERTIES OF A SOFT VISCOELASTIC MATERIAL

C. REMILLAT, F. THOUVEREZ, J. P. LAINE and L. JÉZÉQUEL
Laboratoire de Tribologie et Dynamique des Structures
Ecole Centrale de Lyon
36 Av. Guy de Collongue
69131 Ecully Cédex
FRANCE

1 Introduction

Diminishing the vibration level in industrial structures becomes an essential point, both to avoid fatigue failure and to improve the comfort, from a vibratory or an acoustical point of view . This goal is generally fulfilled through the use of damping treatments. These can be single layer or constrained layer damping treatments. In both case, the efficiency of these methods is closely related to the knowledge of the dynamical characteristics of the applied material. Indeed, for successful results, the damping treatment should suit the vibratory problem. For a reliable prediction of the vibratory response of the damped system, one needs accurate values of the damping material physical characteristics. These are the complex moduli, which are temperature and frequency dependent ([3]).
Several techniques exist for the experimental determination of the complex moduli as a function of frequency and temperature. Actually, it is well known that the temperature and frequency variables can be combined in a single one called the reduced frequency. From that point, the master curves of the material, which consist in the modulus and loss tangent curves of the complex moduli as a function of the reduced frequency, and the shift factor evolution as a function of temperature, completely characterize the material. Moreover, as it is difficult to obtain the frequency variation of the dynamic moduli at a given frequency in a broad frequency range, this frequency-temperature correspondence is often exploited ([4], [5], [7]). In this study, assuming measurement at room temperature will enable focusing only on the frequency dependence of the complex moduli. Measurement methods for determining these moduli as a function of frequency, can be classified following different criteria. One can oppose resonance to non-resonance methods ([1], [8]), or, as proposed here, direct to indirect methods. The second type regroups all the experimental sets using flexural vibrations of composite beams, as the well known Oberst's method for example. These are easy to implement, and seem at first sight peculiarly suitable to test soft viscoelastic materials. Indeed, these materials can flow under their own weight, and need a support structure to stiffen the tested structure.

A. Vautrin (ed.), Mechanics of Sandwich Structures, 345–352.

However, they can lead to identification difficulties in case of thin layer, or really very soft materials ([6]). The material effect becomes then difficult to separate from the one of the base structure or other added layers. Unlike indirect methods, the direct ones are specially designed to only measure the damping material effect. This is certainly an advantage for the later extraction of the moduli values from the measured data. However, the material softness makes that kind of method delicate to bring into operation, from a technical point of view. Moreover, the experimental set is surely material dependent. This is the reason why very few attention has been paid to the dynamic characterization of mastic like materials. Nashif [2] did propose an adaptation of the classical resonance technique, associated with the frequency-temperature correspondence. Yet, this can not be used for materials with Young's modulus below $10^6 Pa$, and not either for a full characterization at room temperature only.
This paper presents a complete step, from the building of a relevant experimental setting, to the obtaining of the material rheological law. The dynamic characteristics variations with respect to temperature were not considered here to focus only on the frequency aspect of the problem. Consequently, all experiment are made at room temperature. The studied material is isotropic, so a priori, two different measurements should be made to have a complete description of it. However, if the mastic is assumed to be incompressible, the Poisson's coefficient is known. This leads us to consider, in a first approach, the measurement of the shear modulus only. Two different types of boundary conditions, namely clamped and free, are applied to the experimental set for higher accuracy. Estimation errors due to uncertainty in measurement are investigated. Then, the shear modulus observability through that peculiar experimental set is checked. Finally, a generalized Maxwell model is used to fit the experimental curve, providing a rheological model for the material in the frequency range of interest.

2 Experimental setting

The experimental setting was especially built for the viscoelastic material to be excited in pure shearing. The mastic is sandwiched between two cylindrical and coaxial parts. A punctual force is applied axially to the inner part (see fig. 1), so that the material is experimenting only shear stresses. This is true if the inner part can be assumed to be perfectly rigid. The inner part dimensions must then be chosen accordingly to that hypothesis, and so as to be as light as possible.
The mastic thickness must also be carefully selected. This is a compromise between two effects. A small mastic thickness will lead to a high equivalent rigidity, and then high force levels should be employed to properly excite the material. On the other hand, if the mastic thickness is too large, the flow effect will become important.
Two different configurations, shown in fig. 1, are considered. The outer part can be either clamped or not. As we will see later, the clamped configuration can provide a better evaluation of the mastic rigidity in the low frequency range than the free one. However, it can not be employed in the whole frequency range of interest, as the clamped boundary conditions is difficult to achieve. This fact is highlighted by the emergence of undesirable structural modes. To avoid the appearance of these

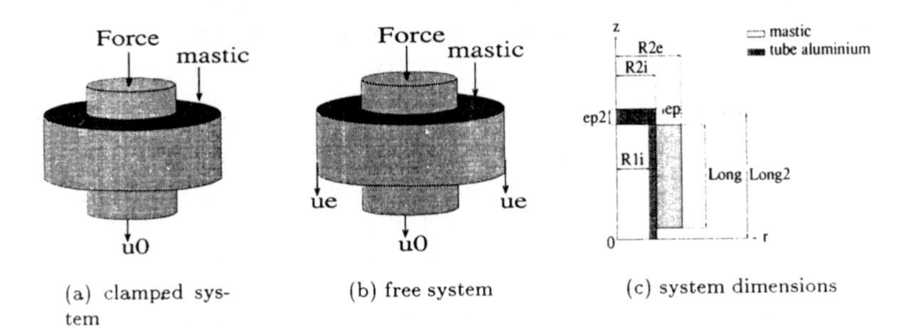

(a) clamped system (b) free system (c) system dimensions

Figure 1: Principle of the experimental setting for shear measurement

modes, a significant effort in designing the outer part should be made, as pointed out in [9] in an equivalent study about traction-compression experimental settings. For the sake of simplicity, we decide to mainly use a free experimental set on the main frequency range of study, and to later correct the low frequency range results by means of measurementsusing the clamped system.

3 Experimental set modeling

The shear modulus experimental determination is closely linked to a theoretical model of the measured function, which is the acceleration-force transfer function at the same point. Taking advantage of both the geometrical simplicity of the experimental set, and of the mastic stress state, a spectral method is used. The true displacement field is then approximated by means of a trial function, which must satisfied the kinematics conditions. The minimum trial polynomial function is :

$$\begin{cases} u_r(r,\theta,z) = 0 \\ u_\theta(r,\theta,z) = 0 \\ u_z(r,\theta,z) = a_1 r + a_0 \end{cases} \tag{1}$$

The chosen displacement must satisfied the boundary conditions, which are, for the clamped configuration :

$$\begin{cases} u_z(r = R_{1e}) = 0 \\ u_z(r = R_{2i}) = u_0 \end{cases} \tag{2}$$

If the system is free, the mass of the outer part is part of the whole movement. This is introduced through the following boundary conditions :

$$\begin{cases} u_z(r = R_{1e}) = u_e \\ u_z(r = R_{2i}) = u_0 \end{cases} \tag{3}$$

where u_e represents the outer part velocity, which is assimilated to a punctual mass (see fig. 1). This simple model can easily be improved, taking a higher polynomial degree in the trial function. Yet, it happens to be enough for that case. The model parameters are classically determined by applying the principle of virtual work :

$$\int_{R_{1e}}^{R_{2i}} \int_0^{2\pi} \int_0^{Long} \partial w_e + \partial w_i r dr d\theta dz + \partial W_{icomp} + \partial W_{ext} = 0 \tag{4}$$

The elementary work done by the elastic forces can be expressed by :

$$\partial w_e = \sigma_{rz} \partial \epsilon_{rz} \tag{5}$$

where $\sigma_{rz} = 2G^* \epsilon_{rz}$ is the shear stresses in the material, G^* represents the mastic shear modulus, and ∂W_{ext} is the elementary work associated to the punctual force F_{ext} :

$$\partial W_{ext} = F_{ext} \partial u_0 \tag{6}$$

The elementary work due to the inertia forces in the mastic is :

$$\partial W_i = -\rho_{visc} \ddot{u}_z \partial u_z \tag{7}$$

ρ_{visc} being the mastic density. Complementary terms are added for taking into account the elementary works associated to inertia forces :

- in the inner part

$$-m_{tub} \ddot{u}_0 \partial u_0 \tag{8}$$

- due to various sensors

$$-m_{correc} \ddot{u}_0 \partial u_0 \tag{9}$$

- in the outer part appearing in the free system case

$$-m_{sys} \ddot{u}_e \partial u_e \tag{10}$$

with m_{sys} mass of the outer part. The material shear modulus is then obtained by equating the theoretical and measured transfer function $-\omega^2 u_0 / F_e xt$. The proposed model relevance was checked by comparing to a three dimensional FE model given by the commercial code ANSYS, and was found to be in good agreement.

4 Sensibility and observability analysis

4.1 SENSIBILITY ANALYSIS

Two types of parameters should be distinguished : measurement parameters, and those related to the experimental setting. Among them, we will take a peculiar interest to the influence of the outer part mass and of the modulus and phase of the measured transfer function, for the free system case. All related curves are given

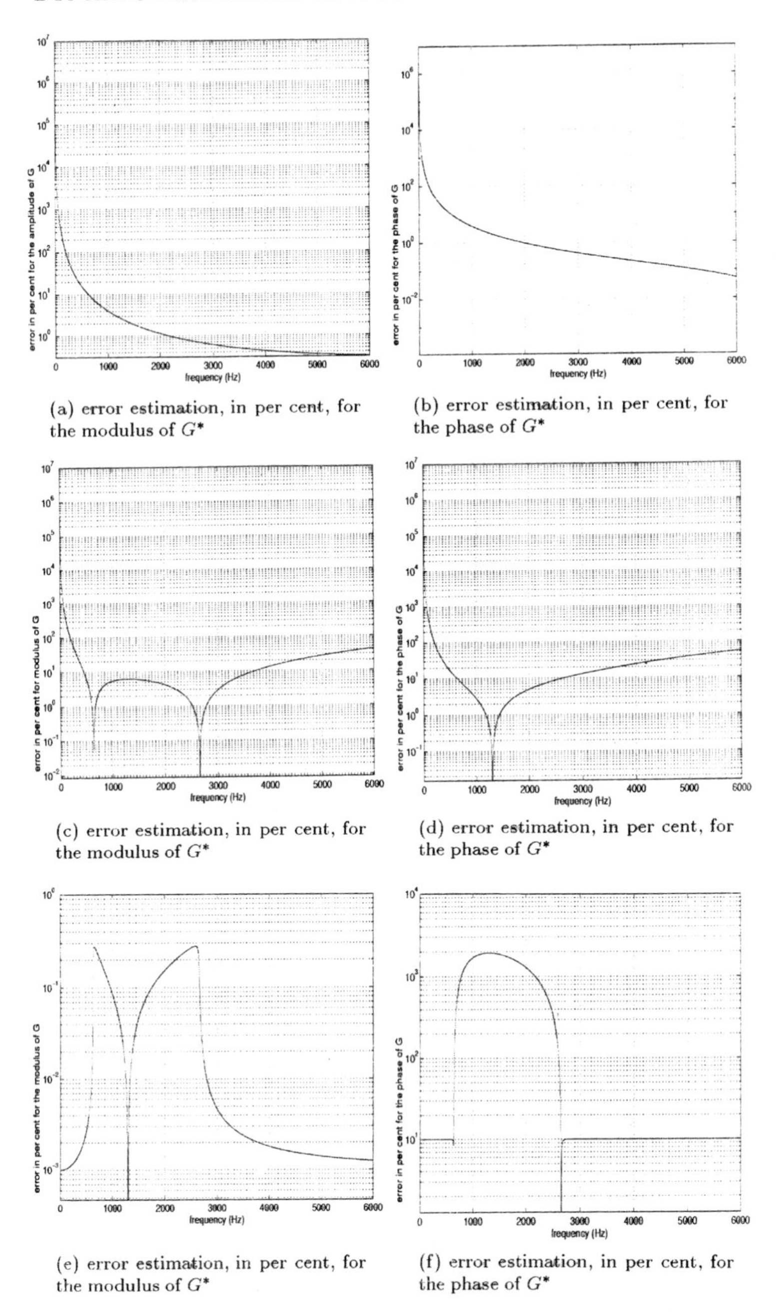

(a) error estimation, in per cent, for the modulus of G^*

(b) error estimation, in per cent, for the phase of G^*

(c) error estimation, in per cent, for the modulus of G^*

(d) error estimation, in per cent, for the phase of G^*

(e) error estimation, in per cent, for the modulus of G^*

(f) error estimation, in per cent, for the phase of G^*

Figure 2: Sensibility of G^* to a 10% uncertainty about the outer part mass (fig. a and b), about the transfer function modulus (fig. c and d) or the transfer function phase (fig. e and f)

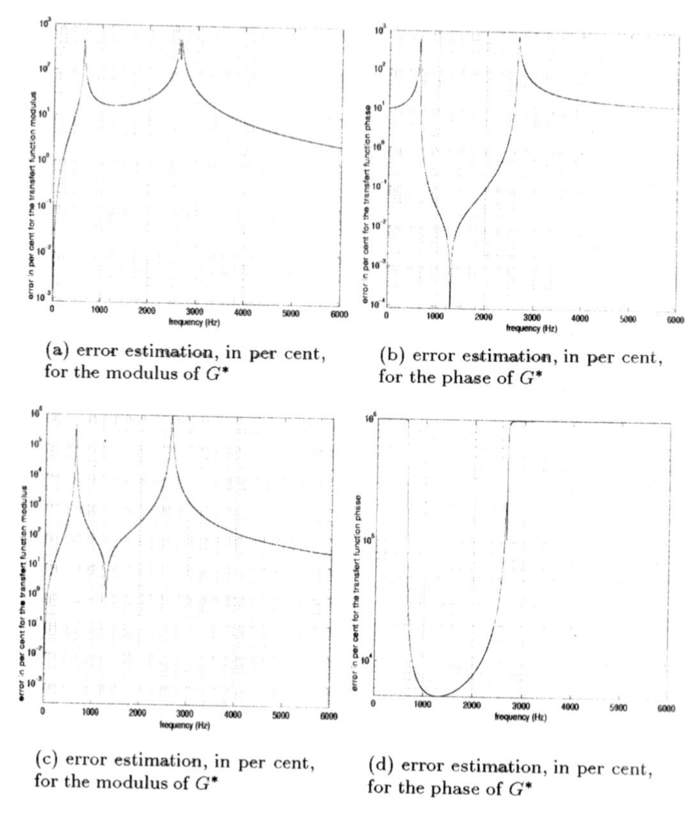

(a) error estimation, in per cent, for the modulus of G^*

(b) error estimation, in per cent, for the phase of G^*

(c) error estimation, in per cent, for the modulus of G^*

(d) error estimation, in per cent, for the phase of G^*

Figure 3: Observability of G^* : sensitivity of the transfer function to a 10% uncertainty about the modulus of G^* (fig. a and b), about the phase of G^* (fig. c and d)

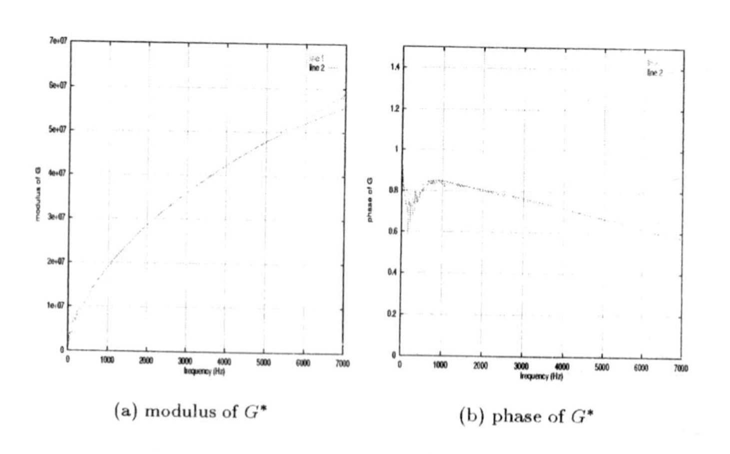

(a) modulus of G^*

(b) phase of G^*

Figure 4: Comparison between experimental results and a six elements Maxwell model

in fig. 2. It can be seen that errors upon the outer part mass is extremely unfavorable in the low frequency range, justifying the use of complementary measurements by means of the clamped system. Taking a look at errors measurement enables to conclude that the modulus of G^* is not sensible to uncertainties on the phase of the transfer function. Yet, error measurement for the modulus of the transfer function can greatly compromise the evaluation of both the amplitude and phase of the shear modulus. A bad phase estimation for the transfer function has essentially consequences upon the shear modulus phase.

4.2 OBSERVABILITY ANALYSIS

The observability of G^* through a peculiar experiment can be defined by the ability that one has to obtain accurate frequency values or not for G^*, using that measurement method. It is computed using the derivatives of the transfer function with respect to G^*. The curves in fig. 3 show the observability of both the phase and amplitude of the shear modulus upon the phase and modulus of the transfer function. It can be seen that the mastic stiffness can not be accurately measured in the low frequency range, using the free system. The observability maxima are located near the resonance and antiresonance frequencies. A priori, the influence of the amplitude of the shear modulus is mostly sensible between these two frequencies The mastic damping, which is related to the shear modulus phase, is mostly linked to the transfer function phase.

5 Experimental results

The measurements were performed on a broad frequency band (0-7000 Hz), using the free system. An impact hammer was employed for the excitation. Complementary measurements were made at a lower frequency range (0-400 Hz), by means of the clamped system, using a random excitation. This will greatly improve the low frequency results, as there are important uncertainties coming from the outer part moving mass that made the previous measurements not accurate enough. Equating the theoretical and experimental transfer functions leads to a linear relation between G^* and the measured transfer function, enabling the extraction of the frequency values of G^*. For a later use in a computing code as for a better physical understanding, a generalized Maxwell model is employed to fit the experimental curves for G^*. This rheological model is taken following the expression :

$$G^* = K_0 + \sum_{i=1}^{n} \frac{K_i j\omega C_i}{K_i + j\omega C_i} \tag{11}$$

with K_0 representing the static stiffness, K_i and C_i being the stiffness and viscosity of the ith Maxwell element.
A means square minimization process is employed to fit the experimental values of G^* using this rheological model. Six Maxwell elements can provide a good approximation, as shown in figure 4. The static stiffness value is estimated to be $5.4\ 10^5\ Pa$, though this result should be taken with care, as only a quasi-static measure could provide the real value.

6 Conclusion

An experimental setting for measuring the shear modulus of a soft viscoelastic material was presented. The result of the sensitivity and observability analyses enables to conclude that this is well suited for that type of material. A rhelogogical model was then used to fit the obtained experimental curves. This should be useful for later computing of both the frequency or time vibratory response of a damped structure. An estimate of the static stiffness, which is difficult to measure, was obtained.

References

[1] T.-K Ahn and K.-J. Kim. Sensitivity analysis for estimation of complex modulus of viscoelastic materials by non-resonance method. *JSV*, Vol. 176(4):pp. 543–561, 1994.

[2] C. M. Cannon, A. D. Nashif, and D. I. G. Jones. Damping measurements on soft viscoelastic materials using a tuned damper technique. *Shock and Vibration Bulletin*, Vol.:pp. 151–163, 1968.

[3] J. D. Ferry. *Viscoelastic properties of polymers*. John Wiley & sons, 1980.

[4] D. I. G. Jones. Temperature-frequency dependence of dynamic properties of damping materials. *JSV*, Vol. 33(4):pp. 451–470, 1974.

[5] D. I. G. Jones. A reduced-temperature nomogram for characterization of damping material behavior. *Shock and Vibration Bulletin*, Vol.:pp. 13–22, 1978.

[6] D. I. G. Jones and M. L. Parin. Technique for measuring damping properties of thin viscoelastic layers. *JSV*, Vol. 24(2):pp. 201–210, 1972.

[7] A. D. Nashif. Materials for vibration control in engineering. *Shock and Vibration Bulletin*, Vol. 43:pp. 145–151, 1973.

[8] S. O. Oyadiji and G. R. Tomlinson. Determination of the complex moduli of viscoelastic structural elements by resonance and non-resonance methods. *JSV*, Vol. 101(3):pp. 277–298, 1985.

[9] S. O. Oyadiji and G. R. Tomlinson. Characterization of the dynamic properties of viscoelastic elements by the direct stiffness master curve methodologies, part1: design of load frame and fixtures. *JSV*, Vol 186(4):pp. 623–647, 1995.

RECENT DEVELOPMENTS IN 3D-KNITTINGS FOR SANDWICH PANELS

DIRK PHILIPS, IGNAAS VERPOEST
Department of Metallurgy and Materials Engineering,
Katholieke Universiteit Leuven, de Croylaan 2, B-3001 Leuven, Belgium
JORIS VAN RAEMDONCK
IPA NV, Zoomstraat 6, B-9160 Lokeren, Belgium

0. Abstract

In this paper, the concept and production of 3D-knitted fabrics and composites is explained. So far, most effort has been spent on the processing of the fabric [1] into a composite. Finally, a few preliminary results are shown and some future applications are mentioned.

1. Concept of 3D-knittings

3D-knitted fabrics are basically very simple textile fabrics. They consist of two layers of flat knitting that are simultaneously knitted on a double-bed rashell knitting machine (Fig.1). The skins are connected by pile fibers on a regular basis. One can imagine the 3D-knit (Fig.2) constructed from two grids of knitted beams connected by the vertical pile fibers.

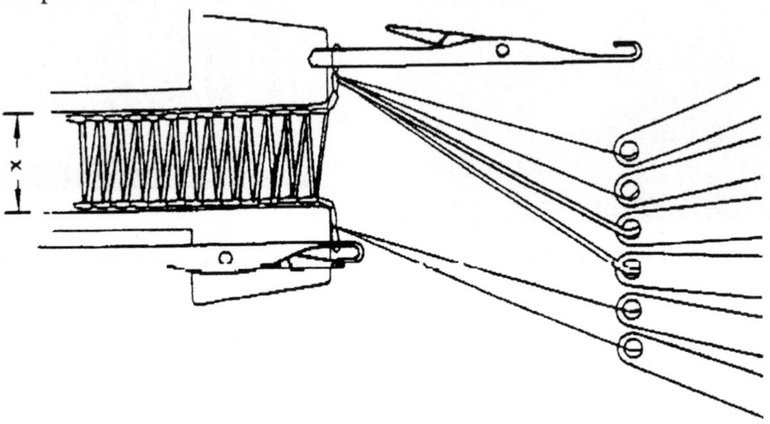

Figure 1: Production of 3D-knitted fabrics

A. Vautrin (ed.), Mechanics of Sandwich Structures, 353–360.

During the production process, one step is dedicated to knit a loop for the top skin of the fabric. In this step, the needle takes the pile fiber along with the top skin yarn and produces one knitting loop. This can be clearly seen in Fig.3: each loop in the knitted beam consists of two yarns[1]. The inner yarn is the one that's being used to knit the skin, while the outer yarn is a pile fiber that is co-knitted into the skin.

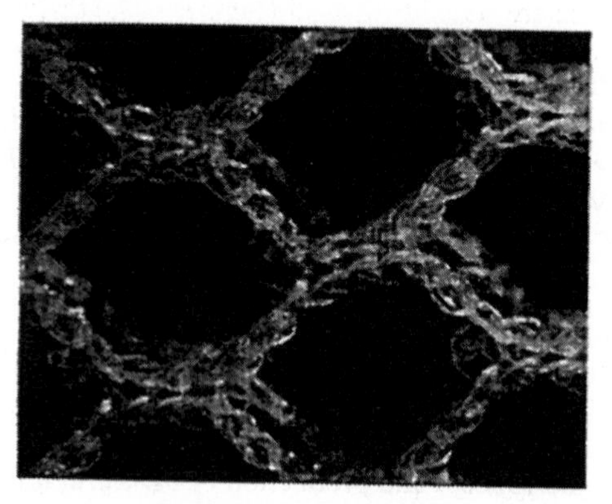

Figure 2: 3D-knitted fabric

In a second knitting step, the pile fiber is moved close to the opposite skin. During the third step, a new stitch is produced. This time with the bottom skin yarn and the pile fiber at the bottom skin of the fabric. Finally, the whole cycle is repeated. From this, it is obvious that the pile fibers make up an integral part of the skins (Fig.4) will result in a higher skin/core delamination resistance of the composite.

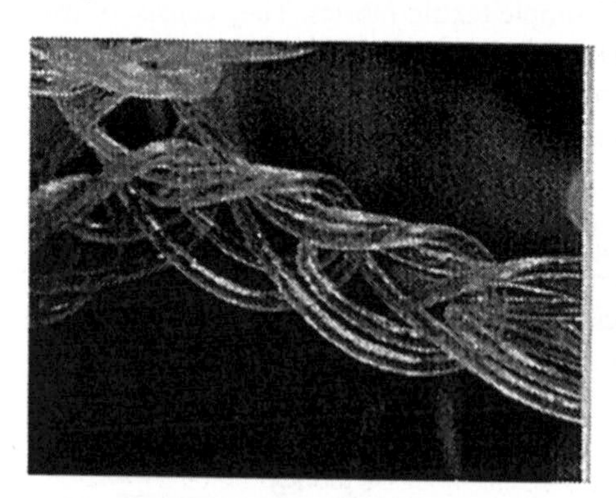

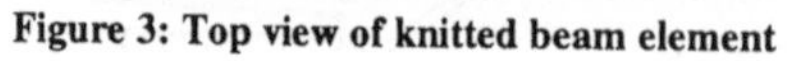

Figure 3: Top view of knitted beam element

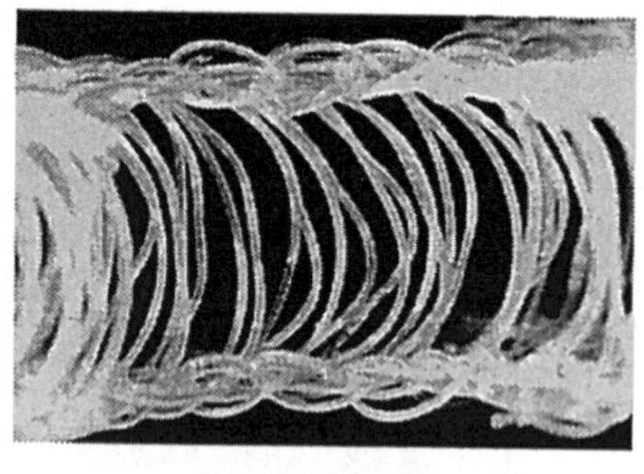

Figure 4: Side view of 3D-knitted fabric

The final product is a sandwich textile structure which is very deformable because of the knitted loops and because of the open structure. This extreme drapability

[1] In this case two monofilaments

opens up possibilities to deep-draw a 3D-knitted prepreg and to make complex shaped sandwich structures without a lot of manual labour.

3D-knitting technology allows a wide variety of architectures to be produced. For the moment only the hexagonal knits are being used for making composites although rhombic fabrics are possible as well (Fig.5,6). The difference depends only on the number of stitches used for the connection between the knitted beam elements.

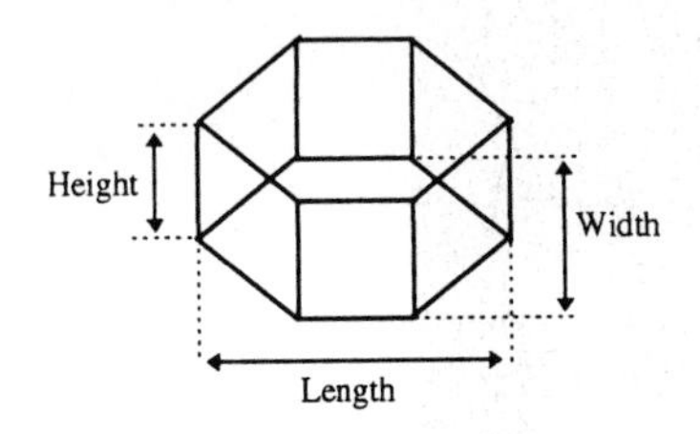

Figure 5: Hexagonal cell structure

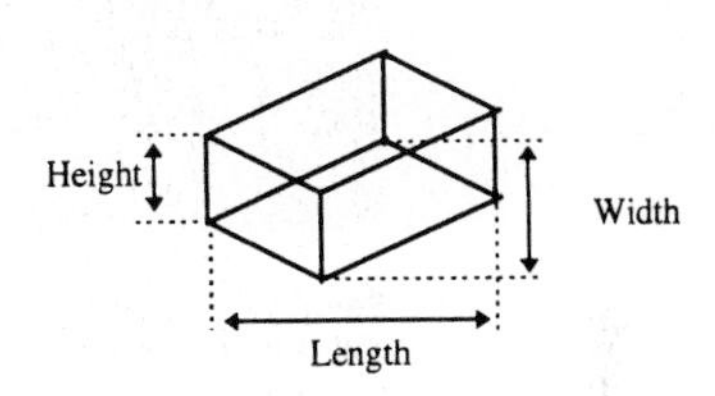

Figure 6: Rhombic cell structure

Another attractive aspect of the architecture is the orientation of the pile fibers. Normally the pile fibers stand vertical, but to increase the stability of the prepreg and to increase the shear stiffness of the composite a portion of the piles can be knitted under an angle [2]. This clearly will be of interest for varying and optimising the mechanical properties such as bending or impact.

2. 3D-Knitted composites

Up to now, some basic research has been performed on 3D-knittings for composite structures. All 3D-knits used so far are existing types that have originally been used for textile applications. Preliminary tests showed very promising results. However, the potential of this new material is much bigger when the textiles are specially designed or adapted for the composite applications. Our present research focusses on increasing the stability of the knits and on the use of special yarns.

One of the current concerns is the correct impregnation of the knits. This difficulty is related to the PET-monofilaments in the core of the sandwich fabric. The PET-yarns which are necessary to separate the two skins do not absorb resin and are difficult to impregnate homogenously (Fig.7). To get a better impregnation with the resin and to obtain better mechanical properties, multifilaments can be spun around the PET-monofilaments. So far, natural fibers (ramie-viscose) and polyester multifilaments have been used for this purpose [3]. Multifilaments can sometimes absorbe resin (i.e. ramie-viscose) or they can trap the resin in the cavities between the filaments. In this way the processing becomes easier and the composite is more homogenous (Fig.8). The nice thing about this, is that the amount and the type of fibers covering the central PET-monofilament can be adapted according to needs.

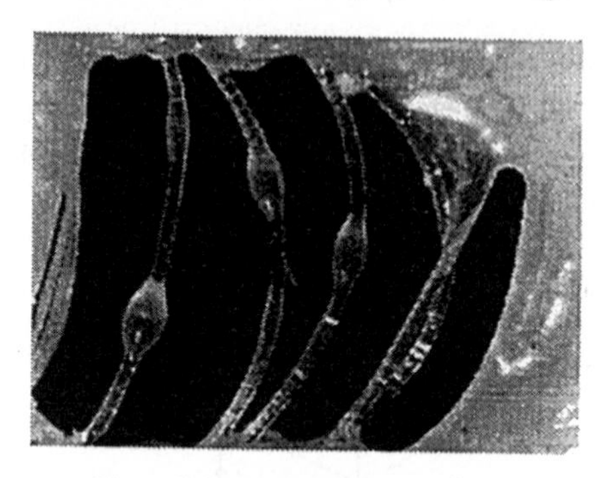

Figure 7: Bad impregnation of plain PET-monofilament pile fibers

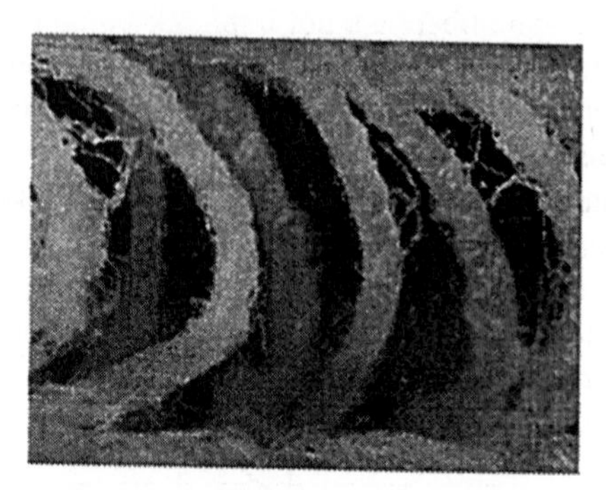

Figure 8: Homogenous impregnation of combined PET/ramie-viscose pile fibers

Another topic being investigated is the influence of the unit cell geometry on the composites properties. All kinds of hexagonal and rhombic patterns can be produced to modify the stiffness of the sandwich plates. These patterns determine the orientation of the knitted beams in the skins and hence the stiffness of the composite. The orientation is also a function of the amount of stretching of the fabric (Fig.10,11). Changes in orientation make a big difference for the final stiffness of the composite. The influence of the variation can be calculated with analytical formulas such as for classical honeycombs.

3. Bending stiffness

The major parameters determining the sandwich stiffness are the cell geometry of the skins, the type of yarns, the resin content and distribution and the degree of stretching.

All three-point-bending tests have been carried out with an Instron 1196 or 4467 testing machine using a load cell of 1kN. The span length between the supports for the sample was 140 or 100 mm. No extensometer was used since the measured force was already quite small. Time was used for calculating the displacement.

3.1. SOME RESULTS

The influence of the first three parameters is demonstrated in Fig.9. In this figure two types of 3D-knitted composites are presented. All samples are about 7mm thick. The first four samples are plain PET-monofilament knits which cannot be impregnated properly, especially the pile fibers. H10 has hexagonal unit cells while R7 is rhombic. The resin content and especially the distribution are important parameters [4]. When the fabric is underimpregnated the stiffness is really low as can be seen in the samples H10,47w% and R7,45w%. The flexural stiffness of sample H10,70w% is already somewhat higher because the skins have been impregnated more. However, the PET-pile fibers are not impregnated at all. This leads to a very low shear stiffness and thus a low bending stiffness. When the textile is overimpregnated (R7,78w%) it becomes much stiffer but the homogeneity of the composite is not good. In this case, the stiffness is much higher because of the presence of resin pockets, especially between the pile fibers[2]. Also the excess resin in the corners between the knitted beam elements limit the deformation of the grid structure and hence stiffens the composite a lot. The second group of samples has PET/ramie-viscose pile fibers which have a much better and more homogenous impregnation quality. The flexural stiffness of these samples is already higher. The main reasons for this are the better impregnation of the piles and the fact that ramie-viscose fibers are stiffer than polyester fibers.

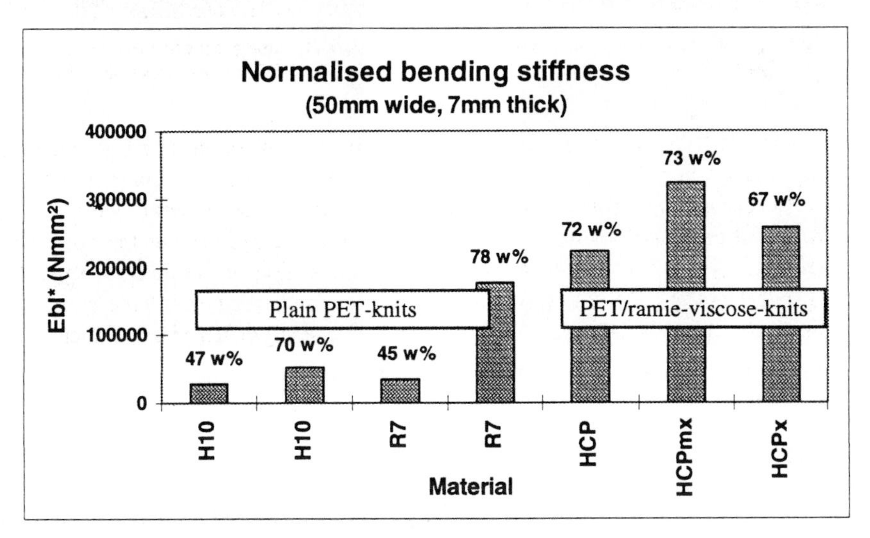

Figure 9: Flexural stiffness of 3D-knitted composites

3.2. INCREASING COMPOSITE DENSITY

Mechanical properties can be improved if the amount of material inside the skins is increased. This is absolutely required because the largest part of the skins is just holes and they do not contribute to the stiffness. A typical material has around 10% of the skin area filled with material.

A first way to increase the skin density is to reduce the cell size. In practice this is not easy because knitting with combined monofil/multifilament piles leads to bigger knitting loops and hence larger unit cells. The only way to reduce cell size is by stressing the knitting yarns more which introduces new problems such as breaking glass fibers in the skins and shearing of the fabric.

[2] formation of resin "walls"

A second way to increase skin density is by introducing extra fibers into the skins so that more resin will be absorbed into the skins. Fibers bundels can be inserted or they can be knitted into the beam elements. The knitted beams will become much stiffer when glass fibers or other stiff fibers are used. The problem here is that the glass breaks because of the highly curved loops, because the local stresses are high. The newest knits all have glass in their skins.

3.3. STRETCHING THE FABRIC

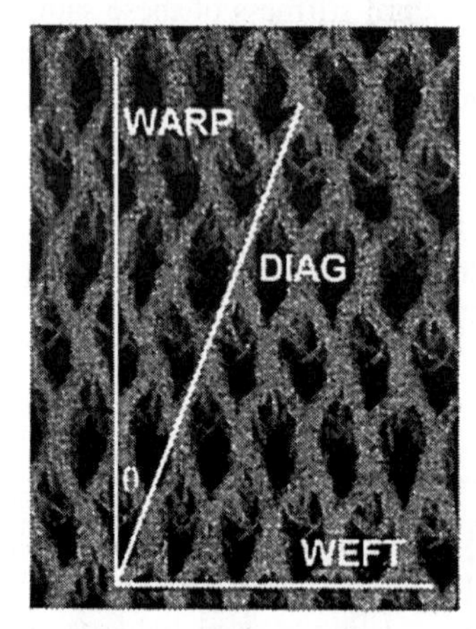

Figure 10: The fabric is stretched to 18°

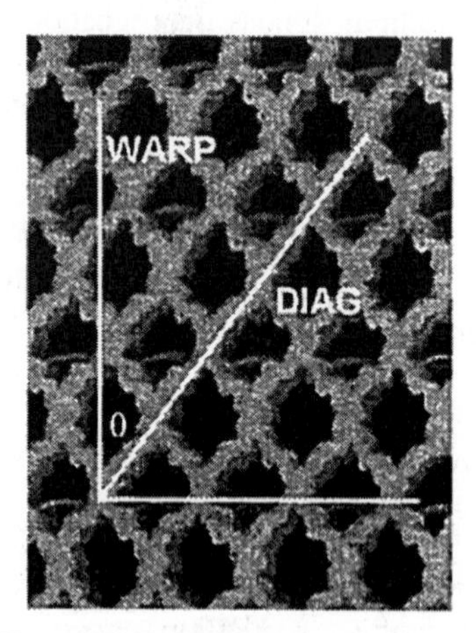

Figure 11: The fabric is stretched to 32°

In Fig.10,11,12, the relation between the degree of stretching the fabric and the final composites properties is being examined. In fig.11 the measured flexural stiffness of one type of composite with three different stretching degrees is presented. All samples are around 5mm thick. The more the fabric is stretched[3] in the weft direction, the weaker the warp direction because less beams are oriented in that direction. The same is true for the diagonal direction because the beam elements are not arranged in a straight line any more, but they become zigzagged. The flexural stiffness of the second sample is somewhat too high, but this effect comes from the higher resin content: 61w% instead of 56w%. Bending stiffness in the weft direction is always low.

---

[3] the bigger the angle

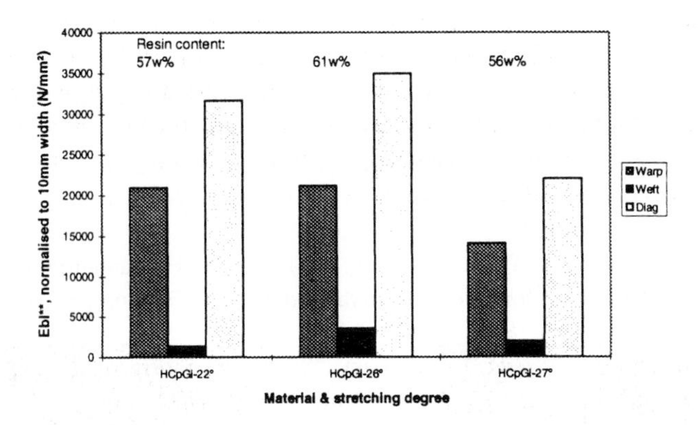

Figure 12: Flexural stiffness for a sample of 10mm wide

Stretching the fabric can also be seen as a means to vary the (areal) density of the composite which determines (partly) also the stiffness.

4. Applications

Up to now, the knitted material has not yet been used for real applications. However, some ideas already exist: light weight casing for the engines of city transport (lower inertia, ventilation capacity), crash protection in bicycle helmets (energy absorbing, ventilation), medical applications (breathability), protective parts for sporting gear, ...

5. Conclusions

The concept of 3D-knitted composites could be a solution to produce cheap, complex shaped sandwich structures. The interesting part is that deep-drawing such as used for metal sheets, can be used for mass-producing 3D-knitted composites.
The influence of geometrical and production parameters on the mechanical behaviour of the composite is being examined

6. References

1. I. Verpoest, J. Dendauw, "Mechanical properties of knitted glass fiber/epoxy resin laminates", 37th Int. SAMPE Symposium, March 9-12, 1992, pp. 369-377.
2. K. Verbrugge, J. Ivens, I. Verpoest, P. Van Der Vleuten, "Foamed 3D-fabric sandwich structures", Proc. of the 12th European SAMPE Conference, Maastricht (NL), 1991, pp. 417-425
3. I. Verpoest, J. Ivens, A.W. Van Vuure, B. Gommers, P. Vandeurzen, V. Efstratiou, D. Philips, "New developments in advanced textiles for composites", Proc. of the SAMPE Japan 4, Japan, Sept. 1995
4. A.W. Van Vuure, J. Ivens, I. Verpoest, "Sandwich fabric panels", Proc. 40th Int. SAMPE Symposium, Anaheim, USA, May 8-11, 1995

7. Acknowledgements

We would like to thank the following organisations and institutions:

- First of all, we would like to thank the Flemish Institute for the Promotion of Scientific and Technological Research in Industry (IWT) for their financial support of this research with a specialisation grant.
- This text also presents research results of the Belgian programme on Interuniversity Poles of attraction initiated by the Belgian State, Prime Minister's Office, Science Policy Programming. The scientific responsibility is assumed by its authors.

A NEW SANDWICH PANEL (SKIN IN GLASS-CARBON HYBRID FABRIC / CORE IN CARBON MAT) : MANUFACTURING AND MECHANICAL CHARACTERISTICS

M. CAVARERO, L. PRUNET, B.FERRET & D. GAY
Laboratoire de Génie Mécanique de Toulouse
50 Chemin des Maraîchers - 31077 Toulouse cedex - France

1. Introduction

The Resin Transfer Moulding process or low-pressure resin injection of preformed reinforcements is used for the manufacturing of composite material parts **[1][2]**. From a mechanical point of view, the parts obtained are between the structural composite laminates (structural parts) and the short-fibre reinforced plastics (semi-structural parts).

A previous study has dealt with the manufacture of a pressurisation valve and its moving part (the butterfly valve) for an aircraft cabin (Figure 1) **[2]**. This work enabled us to simplify the design and the mass of these parts in comparison with the metal structure. This assembly was manufactured from glass-carbon hybrid woven fabrics. Moreover, we have highlighted the difficulty in preforming fabrics for parts containing curvatures and in obtaining good adhesion between the fabrics and the metal shaft **[2]**. From these results, we have considered the possibility of the manufacturing a butterfly valve (Figure 2) with a sandwich structure made of woven skins and a mat core instead of the hybrid fabric solution.

In comparison with the original fabric design, this sandwich can ensure both the manufacture of the butterfly valve curvature and the insertion of the metal shaft.

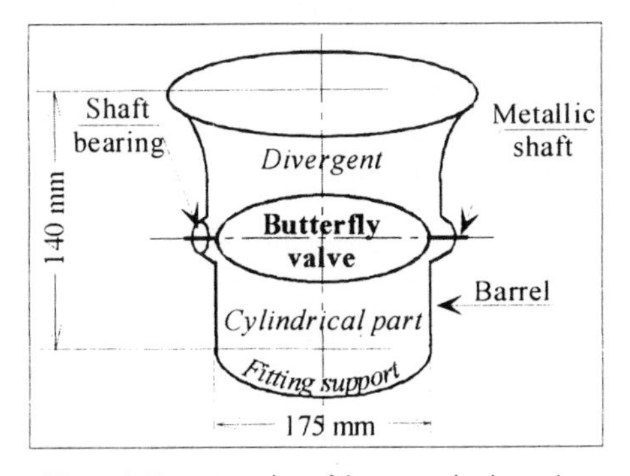

Figure 1. Representation of the pressurisation valve.

A. Vautrin (ed.), Mechanics of Sandwich Structures, 361–368.

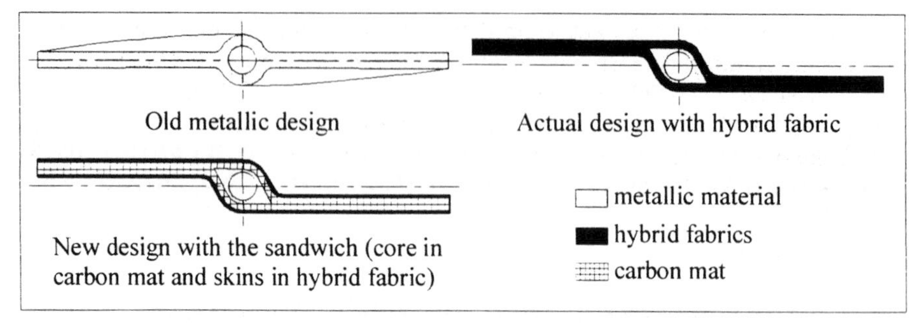

Figure 2. Evolution of the butterfly valve design.

The objective of this paper is to validate the new sandwich design of the butterfly valve. The processing feasibility of this sandwich is subsequently demonstrated. Finally, the impregnation, the flexural rigidity and the real gain sandwich mass are verified compared to the fabric design.

2. Experimental Conditions

1. The Materials

The sandwich panels have been manufactured by the R.T.M. process (Resin Transfer Moulding). It enables polymer matrix structural composite parts to be made by injecting a liquid resin into a fibrous reinforcement, which is usually preformed and placed beforehand in a mould. Polymerisation is carried out just after injection **[2][3]**.

The sandwich panels are manufactured from two skins made of glass-carbon hybrid woven fabrics and a core made of carbon mat. They are both impregnated by an epoxy resin. In most cases the resin at least, and sometimes the reinforcement, has the necessary properties for injecting according to the R.T.M. process **[4]**.

We have chosen the epoxy system Araldite LY5641-HY2954 (or RTM120) manufactured by HEXCEL. Operational curing conditions consist of three steps as catalogued in Table 1. In spite of its no-qualification for aeronautical commercial accommodation, this resin enables the sandwich material design to be validated. In the future, this panel will have to be made from a phenolic resin, in order to be commercially viable in the aerospace industry.

TABLE 1. Manufacturer's recommended curing conditions.

Injection	*Curing*	*Post-curing*
- Resin into the pressure pot at the room temperature - Mould at 65°C	120°C for 1h inside the mould	145°C for 4h in an oven

Developed by BROCHIER, the hybrid woven fabric is an epoxy-treated glass-carbon fabric Injectex (type HF-360-E 01-120) especially designed for the R.T.M.

process. This fabric facilitates resin flow and enables the manufacture of large size and/or high thickness parts and decreases the injection times.

Produced by TORAY, the carbon mat is composed of short carbon fibres (type T300) 12 mm in length, randomly oriented in the sheet. A binding agent (unsaturated polyester) maintains mat integrity and facilitates the preform processing according to appropriate temperature and pressure conditions.

2. Sandwich Definition

Design criteria prescribes a minimum thickness for the butterfly valve equal to 6mm. Consequently, we have chosen to inject 6mm thick sandwich panels, made of two skins in hybrid fabrics (thickness $\approx$ 1mm) and a carbon mat core (thickness $\approx$ 4mm).

We manufacture sandwich panels with 52% in fibre content for the skins and 20%, 30.2% and 37.2% in fibre content for the core. Consequently, the sandwiches are made up of 32.2% skin and 67.8% core. These values result from the ratio between the thickness of each components and the sandwich total thickness. The fibre content of sandwiches is therefore given by :

$$Vf = Vf_{skin} \times 0.322 + Vf_{core} \times 0.678 \quad (1)$$

where Vf is the sandwich fibre content, Vf_{skin} is the skin fibre content and Vf_{core} is the core fibre content.

From (1), we obtain 30.3%, 37.2% and 41.3% in predicted fibre content for each sandwich panels.

3. Optimising the Manufacturing Process

The manufacture of sandwich panels requires control of the resin flow in the mould in order to obtain a regular distribution of resin across the reinforcements. In order to optimise the mould filling, several parameters can be modified. Some of them, such as the admission pressure, the exit vacuum, the injection type and temperature, directly influence the time and the regularity of the impregnation. Others, such as the clamping force, the curing and the post-curing conditions, give the properties and the aspect of the finished products. The process optimisation deals with 4 parameters : reinforcement permeability, resin viscosity, injection pressure and injection method.

1. The Optimisation Parameters

1.1. THE REINFORCEMENTS PERMEABILITY

Permeability is the capacity of a porous material to absorb a fluid under a gradient pressure effect **[5]**. Darcy's law, which represents the flow of a fluid through a porous medium under a pressure gradient, defines the permeabilities **[2]**. Trevino **[6]** has compared the plane mat permeability to bi-directional reinforcement permeability. The mat (randomly oriented short fibres inside a flat sheet) has a lower plane permeability than the bi-directional tapes. But he has noticed that the transverse mat permeability is superior to the permeability of bi-directional and unidirectional tapes. Thus, before the manufacture of a sandwich panel, the filling process of the carbon mat has to be controlled as it is not easy to impregnate, unlike Injectex fabrics. Nethertheless, it is expected that the flow through the sandwich will be uniform with the aid of the high transverse permeability of the carbon mat.

1.2 THE RESIN VISCOSITY

Resin viscosity changes as a function of two parameters, temperature and degree of cure (advancement of the chemical reactions) **[7][8]**. In order to improve the impregnation and the flow across the reinforcements, resin viscosity must be low. Some viscosity measurements have enabled a slow increase of degree of cure to be maintained in order to prevent a quick raising of the viscosity and a resin polymerisation before the end of the injection. These problems will stop the resin flow.

1.3. THE INJECTION PRESSURE

According to Darcy's law **[2]**, the injection pressure directly influences the filling rate. Nevertheless, a high injection pressure could change the fibre arrangement in the mat layers. On the other hand, a very low injection pressure could reduce the flow rate. Consequently, this parameter has to be optimised in order to obtain low filling times without damaging the reinforcement characteristics.

1.4. THE INJECTION METHOD

According to the location of the injection points and the air vents it is possible to change the time and the regularity of the filling process. We present the effects of different injection methods (Figure 3) using a software of numerical flow simulation, called RTMFLOT, on a rectangular plate (300mm×300mm×2mm) made from Injectex fabrics.

Lateral point injection is a simulation with a single injection point on one of the plate's sides (Figure 3a). In order to prevent a lack of resin in the corners, the air vents are placed in each opposite corner. The different layers of colour represent the filling percentage over time. Flow is in accordance with Darcy's law since the front is elliptical.

Injection along an injection line or channel is carried out via a side feeder channel (Figure 3b). The main advantage of this type of impregnation compared with the first is the considerably reduced injection time. In fact, the resin front is obtained by leaving an empty strip in the mould which fills up with resin before the fabric is impregnated; thus, a single injection hole is needed. The same technique is used for the air vents. The air vent is therefore placed directly opposite the injection hole.

As is shown in Figure 3c, this simulation represents injection via a peripheral injection channel i.e. injection via 4 rectilinear channels surrounding the plate. Note that for this type of injection, impregnation time is even shorter than before. Air vents must be placed in the centre and elsewhere on the plate, as it is in this area that injection is finished. This solution is harder to achieve because the design and manufacture of the mould are much more complex. Moreover, the method which consists in an single injection point located at the centre of the mould, with air vents placed around the plate (Figure 3c) presents the same drawbacks as peripheral injection. In contrast, injection times are higher than the two previous methods.

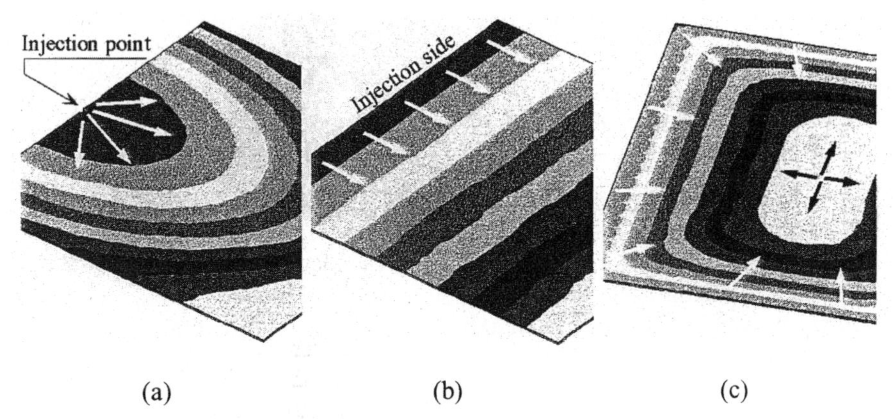

Figure 3. Different injection methods : lateral point injection simulation (a), injection along an injection line (b) and peripheral injection (white arrow) or central injection (black arrow) (c).

2. Optimising the Injection

The Injectex fabric being especially designed for the R.T.M. process, the injection problems of the sandwich result from the carbon mat. The study of the previous optimisation parameters for the manufacturing of plates made of carbon mat has highlighted several dependent characteristics :

- For a considered thickness, the permeability of the carbon mat decreases significantly as the fibre content increases. For a same fibre content, the hybrid fabric has a better flowability than the mat. Indeed, the fabric keeps its fibre arrangement despite the rise in fibre content. On the contrary, the mat has a high flow resistance due to fibre entanglement.
- The temperature increase of the resin placed inside the injection pot (50°C instead of 25°C) induces an important lowering of viscosity. Then, the injection of carbon mat plates with high fibre content has been improved. The maximum fibre content obtained is equal to 50%.
- Despite the modifications of the operational injection conditions (resin temperature and pot pressure), it is not possible to impregnate correctly high fibre content plates made of carbon mat : Vf_{maxi}=30% for an injection along a injection channel and Vf_{maxi}=50% for a peripheral injection.
- During an injection along an injection line, above an injection pressure equal to 0.2MPa, a dominant resin flow is induced along the edge of the preform.

A high flow rate gradient occurs at the flow front which induces a dry area which is impossible to fill.

These preliminary results concerning the manufacturing of carbon mat plates have enabled new operational injection conditions to be defined (Table 2). Using these injection parameters and **the peripheral injection method**, we have succeeded in manufacturing sandwich panels with different fibre contents. It is important to notice that the injection times for each sandwich are between 2 and 5 minutes.

TABLE 2. Modified curing conditions (bold font).

Injection	*Curing*	*Post-curing*
- **Resin into the pressure pot at 50°C** - **Injection pressure 0.2MPa** - Mould at 65°C	120°C for 1h inside the mould	145°C for 4h in an oven

3. Impregnation Verification

The real fibre contents for each sandwich was then checked. The standard method of resin dissolution has been used for the determination of the fibre contents of sandwich panels. They are summarised in Table 3. The difference between experimental and predicted fibre contents is attributed to the thickness change.

TABLE 3. Fibre contents of different sandwich panels.

	Sandwich		
Desired thickness (mm)		5.9	
Measured thickness (mm)	6	6.3	6
Desired fibre content (%)	30.3	37.2	41.3
Measured fibre content (%)	30.4	35.6	42

This preliminary study has established the feasibility of sandwich panels from glass-carbon hybrid fabric and carbon mat. The microscopic observations of the manufactured plates show a regular impregnation of sandwich panels (Figure 4). Excellent bonding between the fabric and the mat has taken place. There is no void and no resin-rich area at this interface.

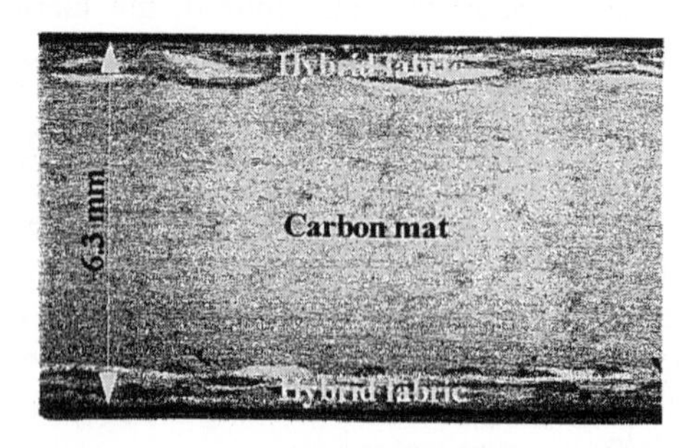

Figure 4. Micrography of the sandwich cross section.

Consequently, during the processing, a quasi-uniform resin flow exists inside

the sandwich. The high transverse permeability of carbon mat maintains a regular and rapid flow rate inside the fabric.

4. Characterisation

The mechanical characteristics of the sandwich have been determined essentially during bending tests because of real loading applied to the butterfly valve. Indeed, the butterfly valve must withstand a pressure equal to 0.134 MPa. Under this pressure, the safe deflexion must be inferior to 1mm.

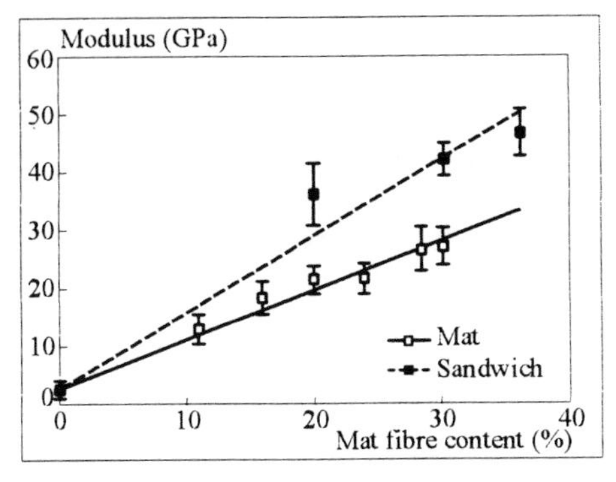

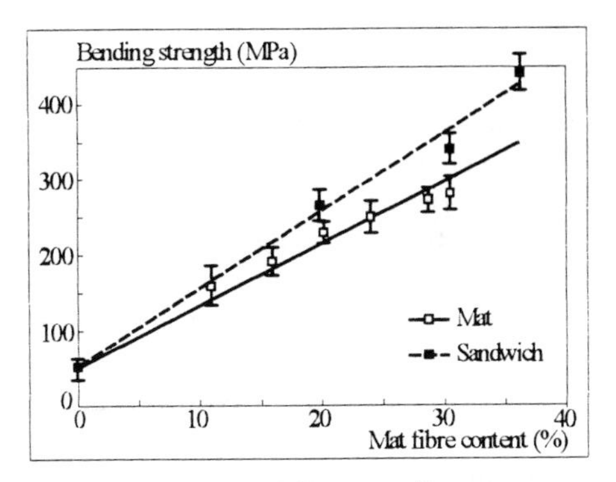

Figure 5. Moduli and bending strengths of carbon mat and sandwich plates for different mat fibre contents.

The mechanical tests are carried out on mat and sandwich samples according to French standards. These samples are made of different mat fibre contents from 0% to 36%. The flexural modulus and strength of specimens are plotted as a function of the corresponding mat fibre content in Figure 5. Notice that the results for the sandwich samples are more scattered. For a given fibre content, a high average variation (≈15%) exists between the values obtained from each tested sample. As catalogued in Table 3, the average real thickness of sandwich samples, and consequently their average fibre contents, are different when compared to the predicted values. Firstly, variations in fibre content can induce large differences in mechanical characteristics. Secondly, the resin flow may induce a significant orientation of short mat fibres inside the sandwich. In contrast to the thin mat plate (2mm in thickness), the thickness flow inside the sandwich becomes an important parameter for the distribution of short mat fibres. Lastly, the wavy interface between mat and fabric could also induce complex stress fields that influence the strength behaviour of the sandwich structure.

TABLE 4. Densities, bending strengths and moduli of sandwich panels and hybrid fabrics.

	Fabric	Sandwich *(skins / core)*		
Measured fibre content (%)	52	30.4 *(52 / 20)*	35.6 *(52 / 30.2)*	42 *(52 / 36.2)*
Density	1.62	1.35	1.39	1.43
Bending modulus (GPa)	46	36.1	42	46.6
Specific modulus (GPa)	28.4	26.7	30.2	32.6
Bending strength (MPa)	607	266	339	445
Specific strength (MPa)	375	197	244	311

The results obtained about the bending strength and modulus of the sandwich (Table 4) have been compared to the results [2] derived from the hybrid fabric solution. They are sufficiently encouraging to envisage the manufacture of semi-structural parts. Indeed, the variations between the sandwich and fabric samples are comparatively low. As is made clear by this table, on the whole, the sandwich specific modulus (modulus/density) is the best. On the contrary, the sandwich specific strengths (strength/density) are lower but remain similar to that of the fabric.

5. Conclusion

This paper has dealt with the manufacture and the mechanical characteristics of a new sandwich panel made of carbon mat core and hybrid fabric skins.

From the study of processing parameters, we have defined new injection conditions appropriate to the manufacturing of sandwich panels. They have resulted in high fibre content sandwich samples with a regular impregnation. On the basis of results obtained, recommendations for manufacturing the butterfly valve are offered. These remarks could be used in the processing of other complex parts.

The mechanical tests have shown the feasibility of the butterfly valve from this sandwich design. Indeed, compared to the hybrid fabric solution, these sandwich panels have an excellent ratio between their mechanical properties and their density. For fibre contents above 35%, the specific modulus of the sandwich increases significantly.

Despite the higher cost of carbon mat, these sandwich panels can successfully replace conventional sandwich solutions in the manufacture of a variety of structural parts.

6. References

1. Johnson, C.F. : Resin Transfer Molding., *Engineered materials handbook - Composites - ASM International* **1** (1987), 564-568.
2. Carronnier, D. : *Pièces composites structurales injectées à basse pression sur préformes. Recherche d'une méthodologie de conception couplée à la réalisation*, Ph. D. Thesis, Toulouse (France), 1995.
3. Raymer, J. : Machinery selection factors for resin transfer molding epoxies, *Composites in manufacturing* (9, 15th january 1990), 90-102.
4. Demint, T.W. & Al. : Fiber preforms and resin injection, *Engineered materials handbook - Composites - ASM International* **1** (1987), 529-532.
5. Goulley, G. & Al. : Influence de la structure de composite sur les perméabilités longitudinales et transversales des renforts en moulage R.T.M., *Composites* **3** (1993), 191-197.
6. Trevino, L. & Al. : Analysis of resin injection molding in molds with preplaced fiber mats - I: permeability and compressibility measurements, *Polymer composites* **12** (1991), 20-29.
7. Kenny, J.M. & Al. : A model for the thermal and chemorheological behavior of thermosets - I - Processing of epoxy-based composites, *Polymer engineering and science* **29** (1989), 973-983.
8. Loos, A.C. & Al. : Calculation of cure process variables during cure of graphite/epoxy composites, *Composite materials quality assurance and processing* (1983), 110-118.

We wish to thank Liebhert Aerospace (Toulouse, France), Moulage Plastique du Midi (Muret, France) and Hexcel (Montluel, France) for their financial and technical

SANDWICH PANELS FOR HIGH SPEED AIRPORT SHUTTLE

Arnt Frode Brevik
Department of Composite Engineering
ABB Offshore Technology
Nye Vakaasv. 80
1360 Nesbru
NORWAY

This paper present the sandwich train front for Gardermoen Airport shuttle in Norway. Design requirements for the structure and the materials are described, and also the test program involving high speed impact tests and tests of fire properties. Impact resistance and fire properties is given special attention.

Introduction

In October 1998 the new main airport (Gardermoen Airport) for Norway opens. Transport of passengers to and from the capital Oslo will be by 16 express airport shuttle trains. The trains will cover the 48 km stretch in 19 minutes

The shuttle train will have a maximum speed of approximately 200 km/h. Two separate parallel tracks are made, and the trains will have approximately 30% of their time in tunnels. The parallel tracks and the time in tunnel is important with regard to loads from train pass and fire respectively. The tracks are not separated by tunnel walls and train pass in tunnels will induce high pressure loads on the fronts.

In 1995 former Maritime Seanor, now part of ABB Offshore Technology was asked to participate Adtranz Norway, to develop the composite train front. Experience from composite engineering, involving high impact loads, was the main reason why former Maritime Seanor was chosen as a development partner.

The development phase was finished in summer 1996, but variations, especially regarding geometry, have continued until production started in February 1997. Figure 1 shows the first front produced. A total of 32 fronts with ploughs are to be delivered by ABB Offshore Technology to Adtranz Norway within first quarter of 1998.

A. Vautrin (ed.), Mechanics of Sandwich Structures, 369–377.

Design Requirements

The design requirements for the fronts are based on design basis (ref./1/) agreed upon by the parties in the development group and from the main contractor NSB (Norwegian Rail). The requirements for the front are based on function, loads, material and safety factor specifications:

FUNCTIONAL:

* Interface with steel base
* Ease of installation and removal
* Weak-link for crash-zone
* Interface with equipment's
* 25 years of design life
* Pressure tight cabin
* Low weight

LOADS:

* Minimum and maximum operating temperature: -40°C/+50 °C
* Impact requirements according to BR 566
* Fire requirements for inner surface NT FIRE 004 class 1
* Fire requirements cross-section: B15 acc. to NS3904 (ISO834)
* Pressure from train pass and tunnels
* Combinations of loads.

MATERIAL:

* PVC is not allowed
* Material containing halogens is not allowed
* Sound isolation
* Thermal isolation
* Surface finish ready for painting.

SAFETY FACTORS:

* Core material: 2.5
* Laminate: 5.0

The safety factors include material factors and load factors.

Materials And Process Selection

MATERIALS SELECTION

Based on the design requirements a selection of materials were evaluated. The composite engineer has a wide range of materials and material compositions available and it is necessary to reduce the amount early in the selection process. The selected materials were evaluated against the most important requirements applied for the material components and ranked. Table 1 shows an example of the evaluation of resin materials.

TABLE 1. Evaluation table of resins

Resin Type	Fire Ret.Properties	Smoke Ret.Properties	Gen.Mech. Properties	Cost
FireRet. Polyester	**	**+	*	*
Fire Ret. Epoxy	***	***	**	***
Phenolic	***+	***+	*	*

*	Low	*+	Low to medium
**	Medium	**+	Medium to High
***	High	***+	High to extra high

The above ranking can be discussed, but generally one can get a fire retarding polyester with either good fire retarding properties or good smoke retarding properties. The fire retarding epoxies have general good overall properties. Phenolic resins have both very good fire and smoke retarding properties. The above ranking is based on information from resin suppliers.

The fire properties of a laminate is also very much influenced of the reinforcing material and the fibre content. Glass is inorganic and does not burn, while carbon burns at ca. 600°C.

Early in the evaluation process it was decided to base the material composition on a sandwich construction. This material construction gives low weight, thermal and noise insulation, very good impact performance and also isolate good against fire and smoke. The rest of the materials (fibre materials, reinforcement type, core materials) were evaluated in a similar way as table 1, but with slightly different evaluation criteria's. The material evaluation ended with the following selection:

Resin:	Phenolic resin from BP Chemicals
Fibre:	Stitched glass fabric from Devold AMT
Core:	PEI (PolyEtherImide) from Airex AG

The choice of materials reflect the most important and superior requirement; safety.

PROCESS SELECTION

With respect to composite materials, material and process selection is closely linked and the process was evaluated in a similar way as the materials and simultaneously.

Evaluated processes where wet hand lay-up with vacumbag, vacumbagged pre-preg and infusion. Due to the challenges in materials and schedule, only wet hand lay-up and pre-preg where evaluated in the beginning of the project. Based on the evaluation which also considered the potential producers capability wet hand lay-up with vacumbag was chosen.

Now, also the resin infusion method SCRIMP (Seeman Composite Resin Infusion Moulding Process) has been evaluated and tested with good results and the process will be changed to SCRIMP by May/June 97.

Test Program:

A key subject to every composite project is the test programs. New combinations of resin, fibre, core and processes where properties must be evaluated and documented and also as a part of the general quality control. The two most major and important tests were the impact test and the fire test.

IMPACT TEST SPECIFICATION

The impact requirements was based on British Rail standard 566 (ref./2/) for front windows. The BR 566 specifies resistance to a sharp cornered hollow cube of 70-75 mm side having a mass of 0.9 kg and travelling at a speed of 350 km/h. (maximum speed + 100 km/h)

Test temperature is 10°C or lower and the test shall be met with the test specimen vertically mounted. For the front a zone around the driver was specified as an "90° angle impact" for the rest an 45° (135°) angle impact applied. The 135° angle impact is not a part of BS 566.

An additional requirement , which also is not specified in the BR566, was limitation to the damaged area. The specification given was a radius of 250 mm around the impact centre.

IMPACT TEST EQUIPMENT

Several possibilities to perform the test was evaluated, especially test with an air gun. However, we found that the cheapest and most flexible test method was to build our

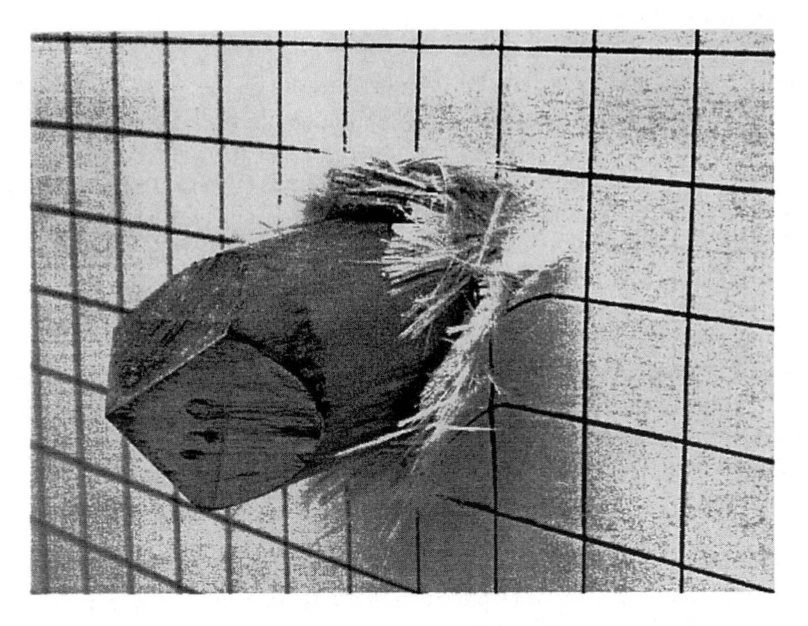

FIGURE 3 - IMPACT ON PANEL

FIGURE 4 - FIRE TEST SET UP

own powder gun. The powder gun was built in co-operation with the defence research institute in Norway (FFI).

Figure 2 shows the test set-up for an 45° (135°) angle impact. It consist of a powder gun, velocity measure, a high speed camera and the panel stand. A paper sheet was mounted 100 mm behind the test panel to detect fragments from the impact.

IMPACT TEST RESULTS

A total of 14 panels were tested. The first panel was made of reinforced polyester (not fire retarding) skins and a PVC foam core. This panel was built and tested before the ordinary test panels and used to optimise the later panels and to try out the test equipment. Figure 3 shows one of the tested panels which did not pass the test, but was very useful because all the recorded energy was absorbed in the panel. Slow speed penetration test with same specimen will be performed to see if penetration test can be used as an alternative.

In the test we ended up with the following parameters:

1 type of resin (Phenolic)
1 type of reinforcement material (glass)
2 types of lay up (biaxial and quadriaxial)
2 types of core (PolyEtherImid (PEI) and PolyMetalmid (PMI))

In addition to this the core thickness and skin thickness varied.

The panels were visual inspected after impact. Length of cuts on both sides of the sandwich were used to determine the degree of penetration. The panels were also cut in pieces to determine the area of damage (delamination etc.) and the character of the impact area (type of damage). In addition to this panel behaviour was studied on the high speed film. From the test results the following conclusion was drawn:

* The powder gun system functioned very well with small (+- 2%) velocity deviations.

* The biaxial lay-up gives better results than quadriaxial due to the shape of the impact specimen (cut more fibre at two places).

* Core with high fracture elongation gives better impact and delamination resistance.

* For this type of impact the major mechanism for energy absorption seems to be cutting of fibres, compression of core and deflection of panel.

FIGURE 1 - TRAIN FRONT

FIGURE 2 - IMPACT TEST SET UP

* For this type of panels the 45° angle impact save approximately 30 % of both core and face materials compared to a 90° angle impact. (This was not optimised and might be larger)

* Any difference between grid scoured and not grid scoured core was not discovered.

FIRE TEST

The specification for fire resistance was given as 15 minutes integrity and thermal insulation according to the NS3904 (ISO 834) test (ref./3/). The specification applies for the thinnest cross section on the front. The test requires large panels (3m by 3m) to be tested, but since our responsibility is the material cross section and not a building construction including joints, windows etc. we decided to perform the test with a 1m by 1m panel. Figure 4 shows the panel mounted in the test oven. The panel was equipped with temperature gauges on the non-exposed side to measure the temperature insulation and the oven was heated according to a standardised heating curve.

The test panel was made of glass fibre reinforced Phenolic skins and PEI foam core and was tested for 30 minutes. The panel passed the test up to 25 minutes (10 minutes more than required), when the temperature on the non exposed side become higher than specified.

Conclusion

A range of different type of composite materials have been used to produce fronts and wagons for the rail industry. The choice of materials have been based on many philosophies and specifications. Different countries operates with different fire, impact and other standards. In addition to this, parts of the rail industry are very conservative and sceptical to "plastic " materials due to lack of knowledge in this area.

Due to the large number of requirements for the front, not every requirements can be considered most important and therefore optimised. We put safety first and with that in mind we considered the impact and fire properties as most important.

By building a train front in a Phenolic reinforced sandwich construction, meeting more than the specified fire specifications and also fully meet the other requirements (like impact and strength), we have shown that even the most stringent specification, regarding fire performance, from a conservative rail industry can be met by composite materials.

By introducing processes like SCRIMP we will show the rail industry that not only the fronts, but also the rest of the train (passengers cars, etc.) can be build in composite materials.

References:

/1/ Design Basis, ABB OT Rep. no. MS/-PR/7099/RE95.001
/2/ British Rail Standard 566, 1979, British Railways Board
/3/ Test Report 846001.03 SINTEF Energy - Norwegian Fire Research Laboratory, Norway

FINITE ELEMENT AIDED DESIGN SOFTWARE FOR LAMINATED AND SANDWICH PLATES

G. EYRAUD, W.S. HAN

Materials and Mechanical Engineering Department
Ecole des Mines de Saint-Etienne
158, cours Fauriel,
42023 ST-ETIENNE CEDEX 2, FRANCE

1. Introduction

Nowadays, the finite element method is frequently used to solve the equilibrium equations and operates as a basic tool for the analysis process of structures. However, it may require a long and complex process, even for classical structures. For composite structures, the difficulties of finding pertinent and simple models for the material behaviour make finite element analysis even more difficult.

A review of existing programs for the design process shows that, in general, they are only analysis tools with no design facilities for data input or results interpretation. In structural design or optimisation, the procedures are generally iterative and require repeated studies as the structure is progressively modified. In addition the design of composite structures needs more efficient facilities in order to overcome penalties such as heavy input data, long computational output time due to material properties and the multilayer structure.

Recent reanalysis methods joined with an interactive user interface appear as one of the most realistic solutions to satisfy the designers of composite structures by countering the above drawbacks of classical finite element programs. As stated by Kane et al.[1], the term reanalysis denotes *any technique that allows for the subsequent analysis of a modified problem with less expenditure of computational resources than required to compute the response of the original problem. Generally, some information computed in the analysis of the original problem is reused in the analysis of the modified problem.*

We prove that it is possible to analyse an elastic structure with the finite element method without specialising the system of equations by taking cinematic constraints into account. A general solution can be written by using a regularised stiffness matrix.

To provide a reference software on laminated and sandwich plate design in the same

A. Vautrin (ed.), Mechanics of Sandwich Structures, 379–386.

manner as the reference book of Timoshenko and Woinowsky-Krieger [2] on homogeneous and isotropic plates, a specific program FEAD-LASP (Finite Element Aided Design for Laminated And Sandwich Plates) was developed. FEAD-LASP uses the original direct reanalysis method and provides a user friendly interface for the design of composite plates.

2. General description of FEAD-LASP Programme

Developed for laminated and sandwich plates, FEAD-LASP program permits both analysis and design of composite plates in linear elasticity [3]. Intended for the pre-design of laminated and sandwich plates, this software permits simultaneously the linear static design and analysis of composite structures with the assistance of the user interface developed on the Windows environment. This environment offers an optimal memory control, multitask facilities, and provides a high level of quality and ergonomics to the software. The present reanalysis method permits quick and consecutive modifications on the boundary conditions and loading cases at the same time, increasing the interactivity of the software.

The 16-node thick plate finite element and a pre-defined mesh technique are employed. A generalised stress-strain relation [4] is used to describe jointly the behaviour of laminated and sandwich plates taking into account shear effects:

$$\begin{Bmatrix} N \\ M \\ Q \end{Bmatrix} = \begin{bmatrix} hA^* & \frac{h^2}{2}B^* & 0 \\ \frac{h^2}{2}B^{*T} & \frac{h^3}{12}D^* & 0 \\ 0 & 0 & hG^* \end{bmatrix} \begin{Bmatrix} \varepsilon_0 \\ \kappa \\ \gamma \end{Bmatrix} \qquad (1)$$

where N, M and Q are respectively the in-plane forces, bending moments and shear forces; ε_0, κ and γ are the membrane strains, curvatures and shearing strains; A*, B*, D* and G* are the normalised in-plane, coupling, bending and shear stiffness.

In order to obtain a simple and efficient finite element model for analysis of laminate and sandwich structures, we used an equivalent material model based on the substitution of real constitutive material by a fictitious equivalent material [5]. The properties of the equivalent material are determined from laminate or sandwich properties. While this process allows use of the classical three-dimensional finite element in a slightly modified form, it avoids construction of special finite elements for the Mindlin-type laminated or sandwich plates and shells. Using elastic stiffness which vary parabolically along the thickness, the modified three dimensional element can reproduce exactly the behaviour of any composite structures, including coupling and shear effects.

A method for the calculation of the generalised shear stiffness, which does not require the so-called shear correction factors, has also been incorporated, for both sandwich and laminated plates [6, 7]. Boundary conditions are pre-defined for the standard cases (clamped and simply supported plates), but could be customised for particular cases. Uniform pressures, concentrated forces and natural weight, corresponding to the applied forces, can be customised as well.

The reanalysis method [8] permits an important gain in the computing time for simultaneous changes in cinematic constraints and applied forces. These cinematic constraints include boundary conditions, symmetry conditions, inextensibility or incompressibility constraints.

The principle of this method is based on the fact that rigid body motions cause the singularities of the classical equilibrium system of discrete elastic structures. Therefore, a general solution can be expressed prior to any assignment of displacement boundary conditions. The solution for specified load and boundary conditions can then be determined by solving an associated small-size linear system.

Many numerical methods are simultaneously used in order to reduce the computing time. Adimensional analysis showed that only 10% of the terms of the global stiffness matrix, in the adimensional form, are necessary to create this matrix. These terms, common for all the types of plates, were previously computed and stocked as data of the program. The computing time necessary for the assembly is advantageously reduced in this way.

The menu "solver" launches the first analysis and the possible ulterior reanalyses. The solving is based on the reanalysis method previously described. With this method, reanalyses under various boundary conditions and loading are made easier and quicker.

Two graphic post-treatments are dedicated to the results presentation. The first one allows the 3D visualisation of the deformed structure and gives nodal information (displacements, reaction forces, strains and stresses). The second draws the displacement, strain and stress isovalues on different surfaces of the plate.

This program is written in C language for the computation part and in C++ for the user-interface. It runs essentially on PCs. The user-interface is entirely interactive and user-friendly (multi-windows, pop-up menus, three-dimensional graphics...).

In view of the features currently provided by system environments such as Macintosh or Microsoft Windows, an interface adapted to the research field was designed, presenting similar facilities. This interface is user-friendly through the use of windows, icons, pop-up menus and dialogue boxes technique. A contextual help in the form of "Hypertext" contributes to the self-learning facility. The quasi-multitasks environment was, as much as possible, employed to keep the dialogue between the software and the user and permit a communication with others applications. Therefore, it becomes possible to quasi-simultaneously compute an analysis and to present graphic results of the previous one. The interactivity of the solver is principally obtained through the use of the reanalysis method and various specific routines developed to reduce the computation time.

To run the program satisfactorily, a graphic display and an arithmetic coprocessor are required. The minimal required free storage capacity is about 4-MByte hard disc drive and 4-MByte RAM memory. A superior configuration is preferable and will of course provide better performances.

3. Examples

To illustrate the efficiency of this software for aided design, we present two examples often used in the characterisation or modelling of the mechanical behaviours of laminated and sandwich plates.

3-1. CLAMPED SANDWICH PLATE

Let us consider a clamped square sandwich plate i) under an upward uniform pressure on the top face and ii) under a upward concentrated load at the centre of the top face. The uniform pressure is equal to 0.1 MPa and the concentrated force to 16 kN. The ratio between the thickness of the core and the total thickness is taken as 0.9 and the ration L/h is equal to 10 (Figure 1). The thickness of the core is equal to 18 mm and that of each face to 1 mm.

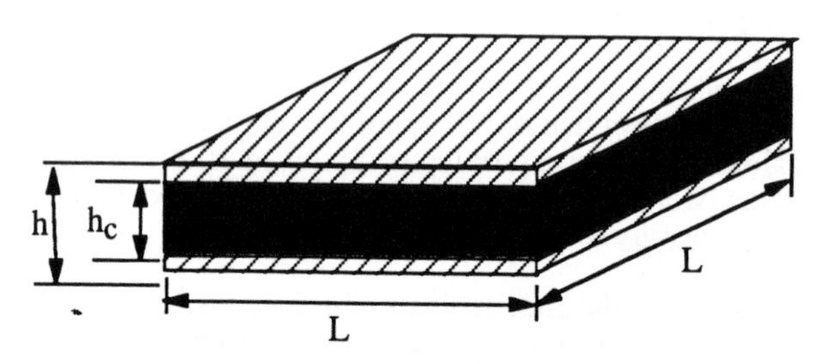

Figure 1. Square sandwich plate.

The material for the skins is an epoxy resin reinforced by unidirectional carbon fibres. Their engineering constants are given by:

$E_{11} = 175$ GPa, $E_{22} = E_{33} = 7$ GPa, $G_{12} = G_{13} = 3.5$ GPa, $G_{23} = 1.4$ GPa,
$\nu_{12} = \nu_{13} = \nu_{23} = 0.25$.

The material for the core is a honeycomb composed by epoxy resin reinforced by glass fibres. This material has the following engineering constants [9]:

$E_{11} = E_{22} = 0.28$ GPa, $E_{33} = 3$ GPa, $G_{12} = 0.112$ GPa, $G_{13} = G_{23} = 0.42$ GPa,
$\nu_{12} = 0.25$, $\nu_{13} = \nu_{23} = 0.02$.

The in-plane, in-plane-bending coupling, bending and transverse shear matrices are calculated using a special mixed variational theory developed for sandwich structures [7]. Only a quarter of the sandwich plate is considered owing to double symmetry of the problem (Figure 2). In a quadrant of the square plate, a 4x4 mesh was used for all finite elements.

In order to assess and to prove the performance of this software, the deflection of the centre of the bottom face has been compared with the different finite elements. The comparison of the central deflection values for this example is given in Table 1 for the concentrated load and Table 2 for the uniformly distributed load.

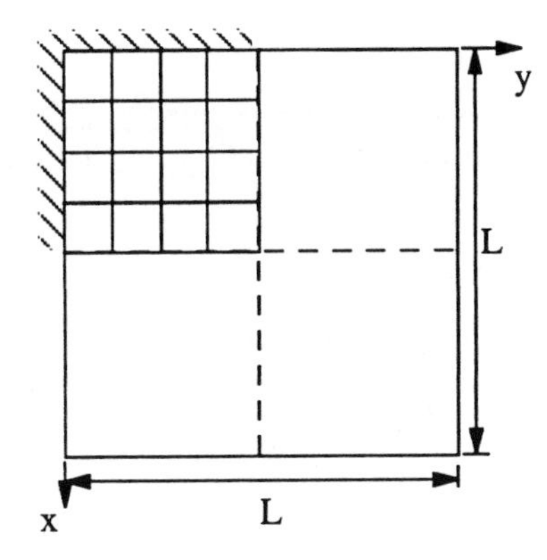

Figure 2. Finite element modelling for a quadrant of the square plate.

	DKQ	Degenerated 8-node	FEAD-LASP	3D-Quadratic	3D-Cubic
L/h = 20	1.60	3.62	3.30	3.28	3.33

Table 1 - Comparison of the central deflection of the clamped sandwich plate under concentrated load (DKQ = Discrete Kirchhoff Quadrilateral element).

	DKQ	Degenerated 8-node	FEAD-LASP	3D-Quadratic
L/H = 20	0.22	0.52	0.47	0.47

Table 2 - Comparison of the central deflection of the clamped plate under uniform loading.

The deformed shape and the displacement distributions on the top face are presented (Figures 3 and 4).

3-2. UNIDIRECTIONAL ANTICLASTIC BENDING TEST

The second example concerns an unidirectional anticlastic bending test. An unidirectional square plate in carbon epoxy is clamped at three corners and a downward concentrated force 10 N is applied at the other corner of the plate. The ratios between the length and the thickness are 10. The material properties are identical to those of the previous example for the skins. A 4x4 mesh was used for the entire plate.

One of the main advantages of FEAD-LASP is to permits changes in cinematic constraints and loading in a very simple and quick manner. To illustrate this facility, let us consider a simple design process for this test.

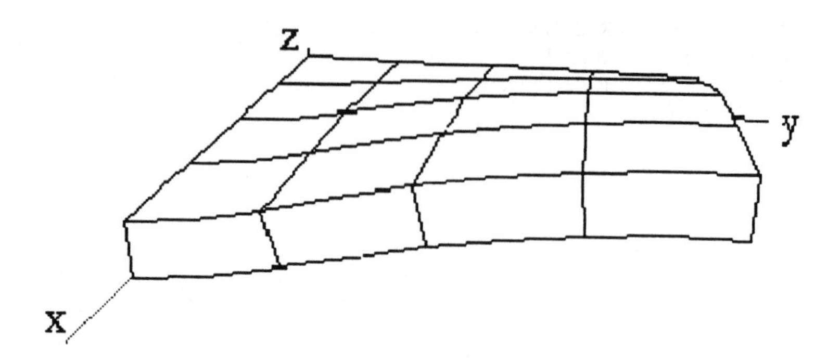

Figure 3. Deformed shape of a quadrant of the square sandwich plate.

Uniform pressure		Concentrated force	
σ_{xx} max	σ_{xx} min	σ_{xx} max	σ_{xx} min
21.24 MPa	-40.1 MPa	253.1 MPa	-157.7 MPa

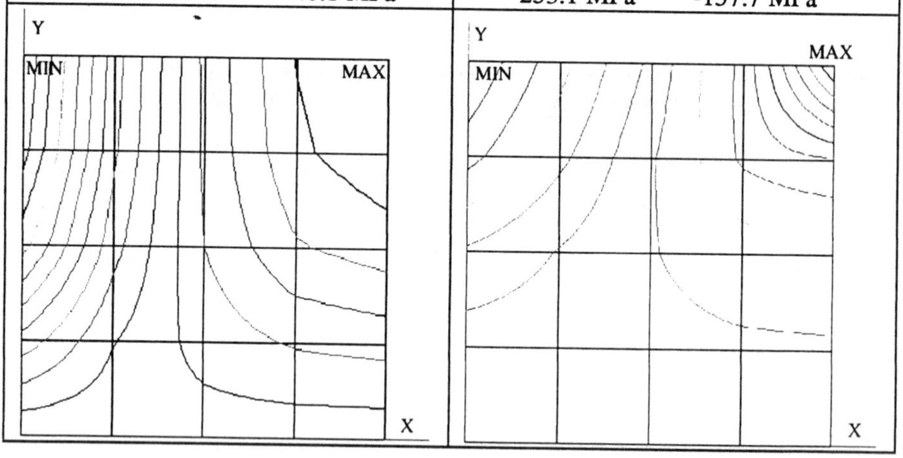

Figure 4. Stress distributions on the top face of the sandwich plate.

At first time, the fibre orientation is aligned to 0° with respect to the global axes as shown in Figure 5 (a). Then, we rotate the plate by 90° and by 45° as shown in Figure 5 (b) and (c) respectively.

Because the displacement and stress distributions of the 90° plate are the same as those of the 0° plate permuting x and y, only stress distributions on the top face for the 0° plate and 45° plate are plotted in Figure 6. The results of the 45° plate are of course better than those of the 0° plate or 90° plate.

The first computation takes about 40 seconds, but reanalyses with change on the boundary conditions and loads take only 4 seconds on the PC Pentium 133 MHz under Windows 95. So, the ratio of the computing time between the first analysis and the reanalyses is about 10.

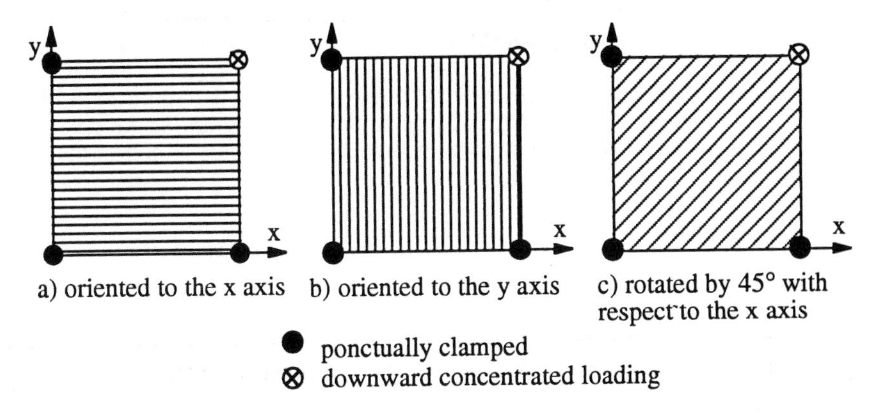

Figure 5. Anticlastic bending test with different fibre orientations.

0° plate		45° plate	
$\sigma_{xx\ max}$	$\sigma_{xx\ min}$	$\sigma_{xx\ max}$	$\sigma_{xx\ min}$
547.6 MPa	-104.3 MPa	277.2 %Pa	-36.45 MPa

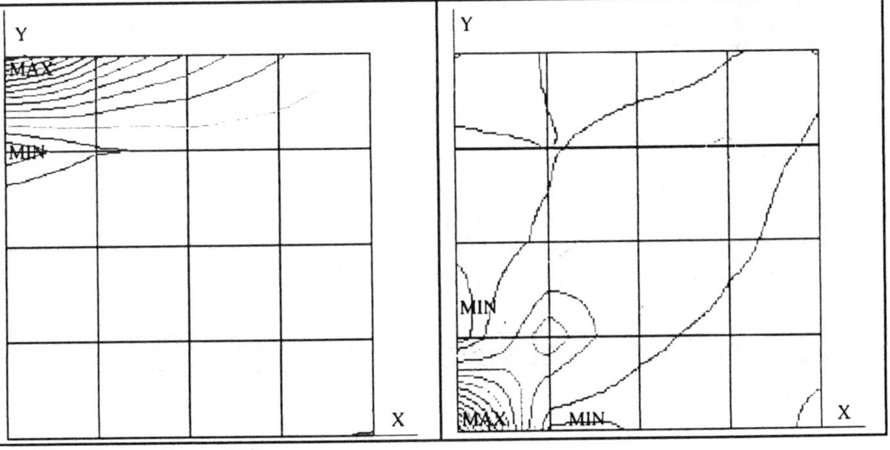

Figure 6. Stress distributions on the top face of the 0° plate and 45° plate.

4. Conclusion

A general description of FEAD-LASP program is presented with the methods of implementation and its assessment on several examples. The link between pre- and post-processors (data input and output processing) and the solution procedure by FEM to form a design process is described and the attractive features of the software, obtained by significant reductions in computing time and user friendly package, are emphasised.

The simplicity and the generality of the explicit reanalysis method incorporated in FEAD-LASP allows anticipation of a modular development of the program or the creation of new software that would consider a variety of geometrical structures or permit the use of other element types.

References

1. Kane J.H., Keshava Kumar B.L., Gallagher R.H.: Boundary-Element Iterative Reanalysis for Continuum Structures, *Journal of Engineering Mechanics* **116**, No. 10, 1990, 2293-2309.
2. Timoshenko, S. T. and Woinowsky-Krieger, S.: *Theory of plates and shells*, McGraw-Hill, 1981.
3. Eyraud, G.: *Programme par éléments finis d'aide à la conception des plaques stratifiées et sandwich basé sur une méthode de réanalyse*, Doctorate thesis, Université Claude Bernard - Lyon I, Lyon, France, 1995.
4. Tsai S.W., Hahn T.H.: *Introduction to composite materials*, Technomic Publishing, Westport, USA, 1980.
5. El Shaikh, M.S., Nor, S. and Verchery, G.: Equivalent Material for the Analysis of Laminated and Sandwich Structures, in A.R. Bunsell, C. Bathias, A. Martrenchar, D. Menkes and G. Verchery, *Advances in Composite Materials, Proceedings of the Third International Conference on Composite Materials (ICCM 3)* **2**, Pergamon Press, 1980, 1772-1782.
6. Cheikh Saad Bouh A.B.: *Evaluation des performances de divers éléments finis et des effets d'anisotropie pour les plaques composites en flexion*, Doctorate thesis, Université Claude Bernard - Lyon I, Lyon, France, 1992.
7. Pham Dang T. and Verchery G.: Théorie des plaques sandwich assurant les continuités du déplacement et de la contrainte aux interfaces, *Comptes Rendus à l'Académie des Sciences* **282**, Série A, 1976, 1039-1042.
8. Verchery, G.: Régularisation du système de l'équilibre des structures élastiques discrètes, *Comptes Rendus à l'Académie des Sciences* **311**, Série II, 1990, 585-589.
9. Whitney, J.M.: Stress analysis of thick laminated composite and sandwich plates, Journal of Composite Materials **6**, 1972, 426-440.

GENETIC ALGORITHM AND PERFORMANCE INDICES APPLIED TO OPTIMAL DESIGN OF SANDWICH STRUCTURES

D.BASSETTI (1), Y.BRECHET (1), G.HEIBERG (1),
I.LINGORSKI (2), P.PECHAMBERT(1)

(1) L.T.P.C.M. Groupe "Physique du Métal", Domaine Universitaire de Grenoble , ENSEEG, BP75, 38402 Saint Martin d'Heres, France

(2) Departments of Materials, Parks Road, Oxford University, Oxford, United Kingdom

Abstract: Materials selection methods using performance indices are applied to sandwich design. Examples of performance indices are given. Optimization method using fuzzy logic algorithm for multiple criteria selection and genetic algorithm for search strategy is presented and illustrated on two case studies.

1.Introduction

Materials and process selection is a key issue in optimal design of structures. Cost reduction and light design especially rely on optimal use of materials in order to fulfill a function (such as bearing a load or providing a thermal insulation) within a set of prerequisites (such as no fracture or limited elastic deflection) having in mind an optimization (such as weight or cost, or volume). A general systematic method to derive the combination of properties which has to be optimized has been recently developed by M.F.Ashby: the performance index method [1][2]. This method is the basis for the CMS (Cambridge Material Selector) software for materials selection [3].

This selection method aims at defining the combination of properties which has to be searched for to meet a set of requirements which can be expressed as constraints (for instance being stiff enough) , geometry (for instance a plate) , loading mode (for instance bending), and objective (for instance minimize the weight). For each set of requirements a combination of the properties describing the performance of the material can be found. The next step, i.e. multicriteria selection, can then be performed. Depending on the nature of the multicriteria selection problem, one has to deal either with a multiple constraint (for instance stiff and strong structure) or /and a multiobjective problem (for instance both light and cheap structure). Either extra information are available concerning the value of the different objectives or the loading conditions and then a completely objective multicriteria analysis can be performed using either "coupled equations" or "value analysis" [1], [4]. But in many cases , such information is lacking and for this purpose , a multicriteria selection method relying on

A. Vautrin (ed.), Mechanics of Sandwich Structures, 387–395.

fuzzy logic algorithm[5] [6] has been implemented in software for materials selection, FUZZYMAT [7].

This method however , up to now, has been applied to monolithic materials only. It has been extended to guide the design of composites [8], of polymer matrix composites [9], of glass formulation [10]. It has been applied especially on Al alloys to investigate possible competitors [11], or to help the selection of cast alloys [12]. A strong emphasis in these extensions was laid upon the need to include process parameters in design, but the selection still operates on a single material selection.

A natural extension of this method is to apply it to "structured materials". By structured materials, we mean a component whose mechanical properties are governed both by its geometry and by the materials which compose it. An example of structured material is a cable which is characterized both by the mechanical properties of the wires, but also by their geometry of their assembling. Foamed materials are also such materials whose properties can vary continuously with the relative density of the foam [13]. Sandwich structures are the simplest "structured materials" and as such they are ideal candidates to extend materials selection methods in the direction of structured materials. This possibility is especially attractive for sandwich structures since their advantages relies in the flexibility of design. It is therefore important to have an idea of which materials could be the most interesting, before starting detailed structural calculations. We will focus our attention here on this "predesign step" selecting a priori the materials and the geometry more likely to give the best solutions.

In §2, we will apply the performance index method to sandwich design. In §3, we will outline the genetic algorithm principle and the fuzzy logic algorithm used for the definition of the merit function. In §4, two case studies for a ski and a thermal insulator are presented. In §5, the selection method is applied to investigate possible applications of metallic foams in sandwiches. Limitations of the approach and indications for further development are presented in conclusion.

2.Performance indices applied to sandwich structures

The optimization of a sandwich beam for minimum weight is usually done by optimizing the geometry of the structure and the density of the foam in the core. However , a true optimization should take the constituent materials as variables of the problem. We will illustrate this point first by considering the simple case of optimizing a sandwich of prescribed stiffness and then we will give the results concerning the optimization of a sandwich of given strength. The thickness of the core and the skin are c and t respectively. b and l are the width and length of the panel. The subscript s denotes the materials of the skin, c, the core material, and * indicates the base material from which the foam is made. The objective function to be minimized is the mass M

$$M = 2\rho_S blt + \rho_c btc \qquad (1)$$

The stiffness of the sandwich has to take into account both the bending and the shear component . For a sandwich with thin skin, the stiffness deriving from both the flexural and the shear rigidity can be written [13] [14] :

$$\frac{\delta}{P}=\frac{2l^3}{B_1E_sbtc^2}+\frac{1}{B_2bcG_c} \tag{2}$$

The two constants in this equation depend on the loading and limiting conditions. For fixed materials, eliminating t from (1) using (2) and minimizing with respect to c allows one to optimize the geometry of the sandwich. The density of the foam can also be seen as a variable of the same type, i.e. geometrical. But the shear resistance of the foam is related to its density. In order to proceed in the optimization, one needs a "constitutive equation" for the foam properties. Using the classical law [13] for open cell foamed materials:

$$G_c=C_2E^*.\left(\frac{\rho_c}{\rho^*}\right)^2 \tag{3}$$

and substituting (3) in (2) allows one to calculate the optimal thickness and of the core, of the skin as well as the ideal density of the foam.

$$\left(\frac{c}{l}\right)_{opt}=4.3\left[\frac{C_2B_2}{B_1^2}\left(\frac{\rho_s}{\rho^*}\right)^2\frac{E^*}{E_s^2}.\frac{P}{b\delta}\right]^{1/5}$$

$$\left(\frac{t}{l}\right)_{opt}=0.3\left[\frac{1}{B_1(C_2B_2)^2}\left(\frac{\rho^*}{\rho_s}\right)^4\frac{1}{E_s.E^{*2}}.\frac{P^3}{(b\delta)^3}\right]^{1/5} \tag{4}$$

$$\left(\frac{\rho_c}{\rho^*}\right)_{opt}=0.6\left[\frac{B_1}{(C_2B_2)^2}\left(\frac{\rho^*}{\rho_f}\right)\frac{E_s}{E^{*3}}.\frac{P^2}{(b\delta)^2}\right]^{1/5}$$

With these optimal values of the geometrical parameters, the objective function can be expressed as a combination of parameters which depend only on the loading conditions (and thus are prescribed by the designer) and of materials properties. As we are interested in material selection, only the combination of materials properties is of interest. It can be shown that the mass is inversely proportional to a "performance index" I given by:

$$I=\frac{E_sE^{*2}}{\rho_s\rho^{*4}} \tag{5}$$

This combination gives the best choice of materials to make the lightest sandwich structure with prescribed stiffness, provided the optimal design conditions given by (4) are fulfilled. One has to notice in this expression that the material from which the core is foamed is very important. Another feature of this procedure is that the process limitations are not taken into account (there is nothing to show that the optimal foam density can indeed be obtained) and that the adhesion between the core and the skin is assumed to be perfect. Within these limitations, the procedures allows the best choice to be made. One may prefer to stop half way in the procedure and analyze the performance using existing foams. In any case the procedure gives an objective function to be maximized by the choice of two materials. The procedure we have followed here can be applied to strength design instead of stiffness design. Then, depending on the failure mode of the structure, three new performance indices may be defined. When the failure mode is skin buckling and fracture the performance index is:

$$J_1 = \frac{(\sigma_s E_s)^{1/4}}{\rho_s} \cdot \frac{E^{*1/2}}{\rho^*} \tag{6}$$

when the failure mode is foam shear and skin buckling one gets:

$$J_2 = \frac{E_s}{\sigma_s^{\,3}} \cdot \frac{E^{*2}}{\sigma^{*2} \rho^*} \tag{7}$$

when the failure mode is foam shear and skin fracture, one gets:

$$J_3 = \frac{\sigma_s}{\rho_s} \cdot \frac{\sigma^{*2}}{\rho^{*3}} \tag{8}$$

Still two questions are left unanswered. The usual selection method as implemented in the softwares allows the selection of one material, not of two. Moreover the design requirements are most of the time multiple, so that an effective performance function has to be defined. Last but not least, how can we explore the huge variety associated with the selection of two materials?

3. Fuzzy logic and genetic algorithms applied to sandwich design

Requirements in design are not only multiple, they are also somewhat undefined: for instance a trade off strategy is frequent in which one of the requirement is only partially fulfilled because the a better compromise is obtained. The fuzzy logic method we have implemented allows one to prescribe a range of requirements, instead of a strict single requirement. Moreover, the problem of multiple performance is dealt with the following way: after the set of requirements is stated, and qualitatively weighted (very important, important...etc.), possible solutions at the margin of required properties are proposed, and the qualitative answer of the user allows one to built an "aggregation function" of the requirements which accounts for the "severity" of the designer. In the present approach, the aggregation function is a product with different exponents of the individual value functions corresponding to each requirement.This kind of algorithm is derived from one developed for software guiding the choice of cars. The result is that , including both the fuzziness of the requirements and the different importance of the criteria, a "value function" can be defined which describes the set of requirement and determines for each choice of materials its respect to the requirement design. This degree of agreement is given as a 'possibility index' and a 'necessity index' which are respectively the optimistic and pessimistic evaluation of the result. This value function will be the fitness function used in the genetic algorithm procedure used for the optimization.

The optimization of the choice of materials can be done by a systematic screening of a database consisting of a subdatabase for skins, a sub database for foams (or materials to be foamed at the optimal density). The evaluation function is obtained from the aggregation of requirements after the fuzzy logic algorithm has been applied . Then a systematic screening is performed. The case study for the ski below has been done using this systematic method.

The optimization can be done using genetic algorithms. Each sandwich is coded as a skin material (itself having a set of properties listed in the database) a core material (base material to be foamed) a density of the core foam, a thickness of the skin and a thickness of the core. Due to the large number of variables (including continuum variables) genetic algorithms are an efficient way of avoiding local minima which would be poor optima of the structure, without performing a systematic screening. The initial population of 30 sandwiches undergoes a genetic algorithm which allows a probability of mutation of 0.03 and a probability of cross over of 0.6. The fitness function is the one determined by the fuzzy logic algorithm. The case study for the insulating sandwich was performed using this program.

4. Case studies of sandwich design

The software developed allows selection according to criteria built from a combination of constraints and objectives. The constraints are stiffness, strength, and thickness, the objectives are thermal conductivity, mass and materials cost. Both options (systematic screening and genetic algorithms) have been implemented.

4.1 SIMPLIFIED CASE STUDY OF A SKI

The length is of 1.76 meters, the width of 8cm, the stiffness (maximum deflection) must be at least 1000N/m and preferably above 5000 N/m, the strength has to be at least 700N, but preferably 1000N. The thickness has to be preferentially below 2cm and in any case below 2.5 cm. The weight has to be below 2kg , and the lighter the better. The result of the selection gives in the first rank sandwich structures with a bamboo skin! Of course the conditions such as the surface state or the resistance to water have not been included, so they cannot be guessed by the computer. It is always worth restating that computer aided material selection are only an indication... The next set of materials, a close second far as the global merit index is concerned, is carbon fiber polymer laminate skins with polymethacrylimide foams or polyvinylchloride foams. Then follow the same skin with a balsa core. For each set of materials the optimal geometry is calculated. Of course the inner structure of a ski is more complex. Stiffeners may be included in the foam, decoupling the protective function form the structural requirements. But the present method allows one to take into account not simply one criterion (which the performance index could do any way) but a set of requirements which can be only partially fulfilled.

4.2 SIMPLIFIED CASE STUDY OF AN INSULATING PANEL FOR A TRUCK.

The thickness must be about 60mm, the width 1200mm, the length 2500mm. For these dimensions the load is about 200 Kg /m^2. The maximum deflection allowed is 12.5 mm which amounts to a minimum stiffness of 480 N/mm. The loading in 3 point bending corresponds to B_1= 77 and B_2=8. The thermal conductivity is imposed by reference to polyurethanne foam , which is 0.023W/mK for a density of 35kg/m^3. the objective function is the mass . The cost will be considered separately since it depends mainly on the processes. The result of the selection using the genetic algorithm gives in the first rank (minimum mass) a glass fiber polymer uniply with medium density polyethylene, and a low carbon steel with polystyrenes, then carbon steels with polyethylene terephtalate (PET) is also a good solution. If the global merit index is used as the ranking criterion, pressure vessel steels with PET or Alkyds are the best solution: they are a bit heavier, but overall they have a better agreement with the ensemble of requirements. A typical result of a selection using the software coupling fuzzy logic for defining the fitness function and genetic algorithm for exploring the space of possible solutions is shown in Figure 1. The geometrical characteristics (thicknesses and foam densities) and the final properties are given for each sandwich selected.

P	Skin material	Core material	t	c	$\rho c/\rho^*$	Stiffness	Strength	Mass
96	Metal matrix composites Al-SiC	Polyvinylchlorides (PVC) rigid	1.97	49.41	0.05	482.6	20721	46.6
96	Stainless steels austenitic	Lin. Low Density Polyethylene	0.95	49.6	0.08	433.2	8177.9	51.6
96	Stainless steels ferritic	Ult. Hi. mol. Weight Polyethylene	1.03	48.33	0.17	437.2	8855.6	52.3
96	Steel. low carbon (mild)	Lin. low Density Polyethylene	0.95	49.41	0.08	427.3	8145.7	51.5
96	Steels. Carbon	Polyethylene Terephtalate PET	0.68	44.72	0.07	415	10091.3	33.4
96	Steels. Carbon	Polyethylene Terephtalate PET	0.68	44.72	0.07	415	10091.3	33.4
96	Steels. Carbon	Polyethylene Terephtalate PET	0.68	45.11	0.07	420.3	10179.4	33.4
96	Steels. Carbon	Polyethylene Terephtalate PET	0.68	45.11	0.07	420.3	10179.4	33.4
96	Steels. Medium Carbon	Magnesium alloys Wrought	0.66	45.5	0.06	472.7	11739.9	32.9
96	Steels. Medium Carbon	Lin. low Density Polyethylene	0.95	49.41	0.08	426.8	8145.7	51.3
96	Steels. Medium Carbon	Lin. low Density Polyethylene	0.95	49.6	0.08	429	8177.9	51.4
96	Steels. Medium Carbon	Lin. low Density Polyethylene	0.95	49.6	0.08	429	8177.9	51.4
96	Steels. Medium Carbon	Lin. low Density Polyethylene	0.95	49.6	0.08	429	8177.9	51.4
96	Steels. pressure vessels	Lin. low Density Polyethylene	0.95	49.41	0.08	428.1	8145.7	51.4
96	Steels. pressure vessels	Lin. low Density Polyethylene	0.95	49.6	0.08	430.2	8177.9	51.4
96	Steels. pressure vessels	Lin. low Density Polyethylene	0.99	49.6	0.08	435.9	8177.9	53.3
97	Glass fibre/ polymer. laminate	Ult. Hi. mol. Weight Polyethylene	3.53	49.41	0.19	403.9	9851.1	42.6
97	Glass fibre/ polymer. uniply	Medium Density Polyethylene	2.28	49.6	0.65	449.9	32223.1	30.3
97	Steels. Medium Carbon	Polystyrene (PS)	0.33	49.6	0.09	407.	10120.4	30.4
97	Steels. Medium Carbon	Ult. Hi. mol. Weight Polyethylene	0.66	48.14	0.22	450.9	10836.9	37.4
97	Steels. pressure vessels	Polyethylene Terephtalate PET	0.64	44.33	0.08	471.3	10949	31.7
97	Steels. pressure vessels	Polyethylene Terephtalate PET	0.64	49.51	0.08	471.3	10949	31.7
98	Glass fibre/ polymer. laminate	Ult. Hi. mol. Weight Polyethylene	3.53	46.28	0.19	408.1	24643.2	42.7
98	Glass fibre/ polymer. uniply	Polystyrene (PS)	3.14	48.04	0.05	415.8	27013.7	41.4
98	Glass fibre/ polymer. uniply	Polystyrene (PS)	3.14	48.04	0.06	433.2	8603.8	41.8
98	Glass fibre/ polymer. uniply	Ult. Hi. mol. Weight Polyethylene	3.75	48.14	0.17	424.9	10680.6	45
98	Steels. Medium Carbon	Polystyrene (PS)	0.33	49.51	0.09	413	10680.6	31
98	Steels. pressure vessels	Lin. low Density Polyethylene	0.68	48.43	0.09	473.8	11077.5	40
99	Steels. Carbon	Polyethylene Terephtalate PET	0.68	48.24	0.07	463.4	10884.5	33.5
99	Steels. pressure vessels	Polyethylene Terephtalate PET	0.8	48.24	0.06	441.5	11060.7	38.8
99	Steels. pressure vessels	Polyethylene Terephtalate PET	0.8	48.24	0.06	441.5	11060.7	38.8
100	Steels. Medium Carbon	Alkyds (ALK)	0.33	49.51	0.06	425	12047.5	35.1

Figure 1: solutions of the selection for an insulating panel listed in order of global merit (P) .

5. Possible application of new aluminium foams.

An interesting application of these methods is to consider the reverse problem [15] Let us assume that we have a potentially interesting light material, for what conditions will it be competitive with classical materials used as core in sandwiches? We have performed the exercise with the set of requirements for the ski, increasing progressively the strength constraint. When the prescribed strength is 1000N, the aluminium foams do not show up in the list. They appear in a good position as core materials when a strength above 4000N is required. Such a result indicates that, in spite of their lightness, aluminium foams in sandwiches will be interesting only for high resistance panel. A similar procedure can be applied for thermal applications.

6. Conclusions and perspectives

We have shown in this paper the applicability of materials selection methods to the optimal design of structural and thermal sandwiches. We have shown that a systematic screening of a materials database, as well as a genetic algorithm, coupled with a multicriteria evaluation based on fuzzy logic algorithm, allows an efficient evaluation of the various possible solutions, allowing one to select both materials, and geometrical characteristics. The feasibility of software guiding for materials and geometry selection for sandwiches according to a multicriteria set of requirement has been proven.
Several aspects of the problem have been however neglected and indicate the directions for future work.

The processing aspect of such sandwiches (is it possible to inject the foam? what will be the adhesion between the foam and the skin?) is likely to be an important aspect that we have not taken into account. Such a situation has already been treated in previous work on polymer matrix composites , and has been dealt with using loss factor matrices based on expertise. These processing aspects are likely to be crucial in the estimation of the economical aspects of the problem.

In order to apply the materials selection method, a basic first step of the approach is to simplify the mechanical loading. We have treated simple bending situations. More complex loading would require finite element calculations which are bound to be more complex. Interfacing a simple materials selector as this one as a "screening" stage before a complex FEM code seems a promising route toward a better integration of materials selection in the mechanical design procedure.

The simple exercise of identifying the possible applications of a metallic foam in sandwich design opens the way toward application-dedicated or materials-dedicated software. For instance, sandwich structures play an important role in civil engineering, but with very different sets of requirements: such domains are worth investigating.

Last but not least, sandwich design can be seen as the simplest case where optimization has to be performed on both a discrete database of materials and a continuum field of geometrical characteristics. We have here investigated only one way of investigating the possible solutions. Depending on the number of materials involved, on the number of criteria to be fulfilled, either the systematic approach, or the genetic algorithm approach will be the most efficient. Probably the coupling between non deterministic and deterministic methods will be the most efficient method to optimize both the materials choice and the geometry. Software which would not only select the materials and the geometry, but would point out how a slight change in the set of requirements (constraints) could allow a substantial improvement of the objective function would be extremely useful. In a way, the requirements shape the objective function (measured on the space of possible choices) just as constraints shape the thermodynamical potential of a physical system. In highly frustrated systems, a small change in the constraint can eliminate energy barriers between different states. The question is all the more important in materials selection and design problems that requirements have some intrinsic fuzziness. Optimizing sandwich structures may be a good way of exploring these concepts.

Acknowledgments

The authors want to thank Professor M.F.Ashby for enlightning discussions while this work was performed.

References

[1] M.F.Ashby (1994) *Materials selection in mechanical design*, Pergamon Press, Oxford

[2] M.F.Ashby, Y.Brechet, M.Dupeux, F.Louchet, (1996), Choix et usage des matériaux, Techniques de l'Ingénieur , Vol. Conception des produits Industriels

[3] M.F.Ashby, D.Cebon, *"Cambridge Materials Selector"*, Granta design software

[4] M.F. Ashby (1996) , Materials selection: multiple constraints and compound objectives, *Cambridge design center*, CUED/C-EDC/TR38

[5] D.Bassetti, Y.Bréchet, *"Choix des matériaux"* (1994) , Editions du CETIM, Senlis

[6] D.Bassetti, (1992) *Aides informatisées au choix des matériaux* Diplôme d'études Approfondies, Science et génie des matériaux, INPG

[7] D.Bassetti, (1996) "Fuzzymat" materials selector, B&I

[8] M.Ashby, Acta Materialia, (1995), **4**, 5-17

[9] P.Péchambert, B.Bassetti, Y.Bréchet, L.Salvo (1996) Computer aided materials selection in composites, ICCM, 283-288, Institute of Metals Publisher, London

[10] D.Bassetti, Y.Brechet (1997) "Aide au développement des compositions de verres, Saint Gobain Internal report.

[11] Y.Bréchet, D.Bassetti, P.Péchambert (1996), Rational procedures for materials and process selection in mechanical design applied to aluminium alloys ICAA5, Materials science forum, **217-222**, 121-132

[12] D.Bassetti, Y.Bréchet (1997) Sélection des alliages de moulage d'Aluminium, Pechiney Internal report

[13] M.Ashby, L.Gibson, (1988) *Cellular solids, Structures and properties*, Pergamon Press, Oxford

[14] H.G.Allen, (1969) *Analysis and design of structural sandwich panels*, Pergamon Press, Oxford

[15] G.Heiberg, (1996) Structural study and investigation of possible applications for aluminium foams, Diplôme d'Etudes Approfondies, Science et génie des matériaux, INPG

CONSTRUCTION METHODS FOR BIG AND HEAVY LOADED FIBRE REINFORCED COMPOSITE SANDWICH STRUCTURES DEMONSTRATED ON A SES HULL

RAMONA WALLAT, ANDREAS EISENHUT,
GERHARD ZIEGMANN,
Swiss Federal Institut of Technology, Zurich

1. Introduction

Surface Effect Ships (SES) are widely considered as the new water transportation system of the future. SES are catamarans (ships with slim thin hulls). By applying overpressure in the area between the hulls the ship is lifted (air cushion) and thus the hydrodynamic resistance is decreased significantly resulting in enhanced cruising speed. SES designed today are either produced with composites based on glassfibres and thus limited in length (30-40m) or in aluminium structures, like the Stena HSS, not capable to reach reasonable weight reductions.

Experience in the shipyards today is mainly restricted to wet laminating sandwich structures with glassfibres/unsaturated polyester resins, using balsa wood or PVC-foam as core material.

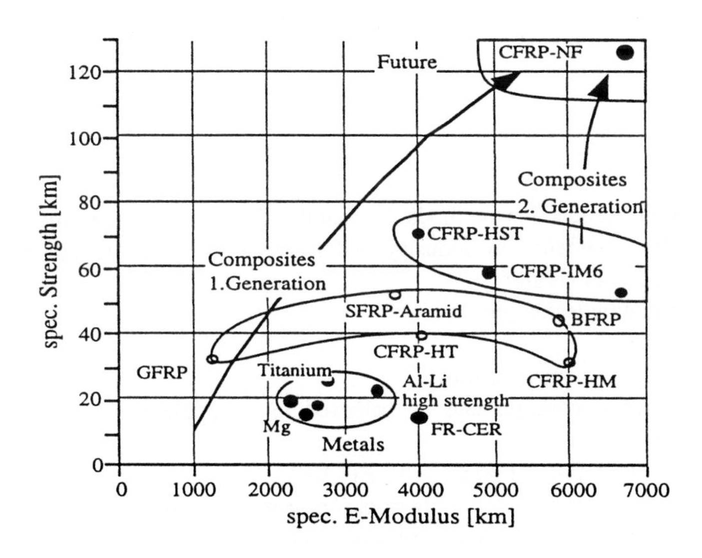

Fig. 1.1: Specific strength versus specific stiffness

A. Vautrin (ed.), Mechanics of Sandwich Structures, 397–406.

As Fig. 1.1 shows, the specific stiffness of glassfibres is below those values for metals. When using carbon fibres, specific strength and stiffness are higher as with metals. This means that advanced composite materials could lead to new improved solutions for light ship structures as big SES in particular. But up to now there is no experience how to design and produce a structure with the dimension of 160 m length and 35 m width in advanced materials. This paper describes the first steps of the design and production evolution process.

2. Evolution of Production Process

The typical cross-section (1/2) in the midship area of a SES is shown in Fig. 2.1:

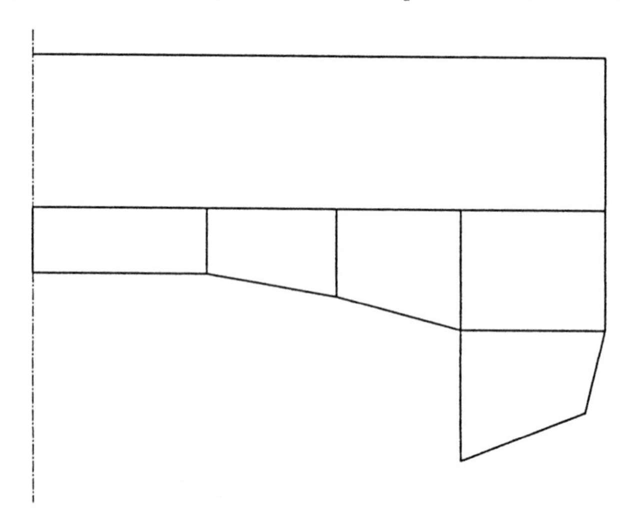

Fig.2.1: Hull section

The evaluation of the production methods is mainly done considering this midship section. Due to the experience in industry and research institutes today the most promising production methods are:

- Continuous laminating technique
- Prefabricated panels
- Modular prefabricated tubular sections

Each of these production concepts can be achieved in principle with the following processing methods:

- wet laminating with thermoset resins (UP, EP, VE)
- prepreg technique (today restricted to thermoset prepreg systems), mainly EP-systems
- RFI (Resin Film Infusion)-technique with low viscous thermoset resin systems (UP, EP, VE)

The combination of the principles mentioned above leads to the solution tree shown in Fig. 2.2

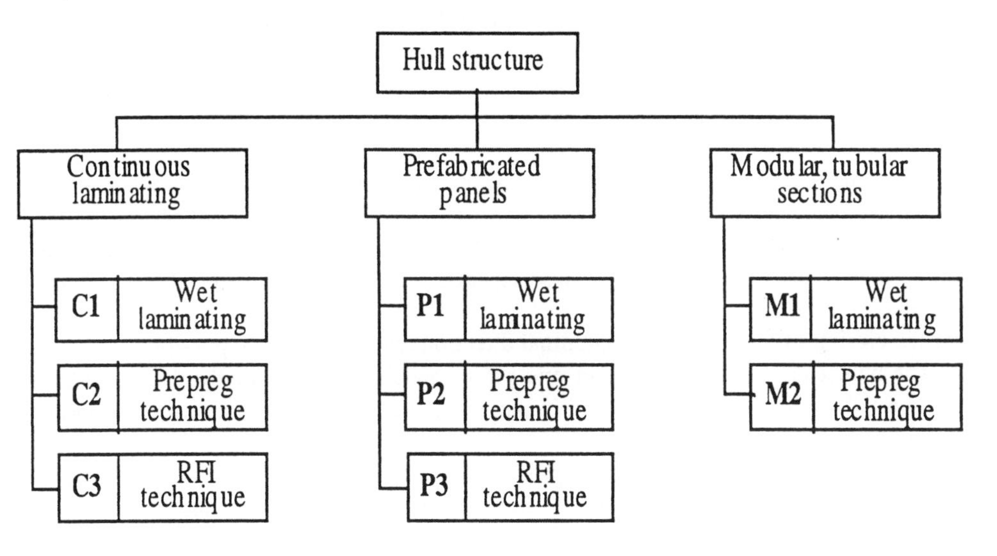

Fig.2.2: Solution tree

2.1.CONTINUOUS LAMINATING

Today ship structures up to 70 m are produced in one step using wet laminating technique. The dimensions of the envisaged SES do not allow such a concept. Therefore the idea was created to build up a moveable tool of about 10 m length with the shape of the cross section of the midship. After laminating and curing of one cross-section the tool is pushed forward to fabricate the next section with an overlap in the surface laminates.
For the stern and the bow of the hull the mould will be modified i.e. by adding elements to create the corresponding shape.
The continuous laminating technique can be applied as wet laminating, prepreg technique as well as RFI-technology.

2.2 PREFABRICATED PANELS

From the aircraft industry the production of flat or slightly curved big panels is wellknown since many years - either as a sandwich construction or a solid laminate with integrated stiffners. The aim for the construction of a ship hull is to produce flat or sligthly curved sandwich panels in a discontinuous way and to join these panels in a second step to big structures like hulls, upperdecks, bulkheads etc. Besides the fabrication of these panels one main focus has to be the joining technique with such a concept.

In principle the panel production can also be done with wet laminating technique, prepreg and RFI-production. Most experience is available today in the prepreg production method, coming from the aircraft industry [1], resulting in structures with a high and reproducible fibre content in the skins and a high reproducibility in the laminate tolerances.

2.3 MODULAR TUBULAR SECTIONS

In transportation industry the filament winding technique is used today for producing big tubular sandwich sections in one shot [3,4] very successfully. The concept of transfering this technique to the shipbuilding industry is to split the structure and the substructure into simple tubes which can be produced separately and than can be joined together by special developed joining techniques.
For the closing of the tubes in longitudinal direction two solutions are possible:

- The bulkheads are closed walls in one piece, prefabricated separately. Each module tube will be fixed to the bulkhead by applying wet laminated connections between bulkheads and tubes.
- For each module a separate cover is produced as a solid laminate or a sandwich. This concept allows the production of flexible tube lengths regarding the design requirements of the structure.

Today the filament or tape winding technique is wellknown with online wet impregnation of the textile structure [3,4]. This technique can be modified very easily by using prepreg tapes and thus eliminating the wet impregnation process.

2.4 SELECTION/VALUATION CRITERIA FOR THE PRODUCTION METHOD

To achieve production methods which can be handled for those big structures in a shipyard, 16 different criteria were selected very carefully regarding aspects like

- handling of the material
- shop life
- quality aspects
- required skill of the coworkers
- potential of automation etc.

These criteria then were weighted due to their importance for production of the ship structure in accordance to known design systematics [5].
The summary of this valuation process is shown in Fig. 2.3 including the different selected production methods and all selected criteria.
According to this evaluation the prefabrication of the panels with prepreg technique is the most favoured process.

Weighting / Production methods	Handling of Raw Material, Potential for Atuomation / Integration (6)		Laminating Quality (10)		Curing Characteristics (2)		Costs, Tooling, Necessary Investments (8)		Production Steps (4)		Quality Assurance (6)		Risks on Environment, Waste Management (2)		Required Skills of Coworkers (4)		Risks of Process Degree of Innovation (10)		Total Value	Ranking
	Scaled Value	Weighted Value	Scaled Value	Weighted Value	Scaled Value	Weighted Value	Scaled Value	Weighted Value	Scaled Value	Weighted Value	Scaled Value	Weighted Value	Scaled Value	Weighted Value	Scaled Value	Weighted Value	Scaled Value	Weighted Value		
Cont. lam. (Prepreg)	4.00	24.0	4.00	40.0	1.00	2.0	2.50	20.0	4.00	16.0	4.00	24.0	2.80	5.6	2.80	11.2	4.00	40.0	183	2
Cont. lam. (RFI)	1.00	6.0	2.00	20.0	2.50	5.0	2.50	20.0	1.00	4.0	1.00	6.0	4.00	8.0	1.00	4.0	2.50	25.0	98	5
Prefab. Pan. (Prepreg)	3.39	20.3	4.00	40.0	4.00	8.0	4.00	32.0	4.00	16.0	4.00	24.0	3.40	6.8	4.00	16.0	4.00	40.0	203	1
Prefab. Pan. (RFI)	3.10	18.6	3.00	30.0	4.00	8.0	4.00	32.0	3.57	14.3	3.25	19.5	4.00	8.0	2.80	11.2	2.50	25.0	167	3
Modular (Wet lam.)	2.80	16.8	1.00	10.0	4.00	8.0	1.00	8.0	2.29	9.2	1.00	6.0	1.00	2.0	2.20	8.8	1.00	10.0	79	6
Modular (Prepreg)	4.00	24.0	4.00	40.0	1.00	2.0	1.00	8.0	4.00	16.0	4.00	24.0	2.80	5.6	2.80	11.2	2.50	25.0	156	4

Fig. 2.3 Valuation of the production methods

3. Joining Techniques

In parallel to the selection of production techniques different joining principles were evaluated regarding the different production philosophies.

3.1 PRINCIPLE JOINING SOLUTIONS

The structure of the shiphull and the substructure like bulkheads, car decks, passenger decks etc. is requiring different kinds of solutions for several wall structures coming together (Fig. 3.1)

- Inplane joint within the structure of the continuous or discontinuous produced structure surface.
- T-joint regarding the connection of the hull with either bulkhead, car deck etc.
- X-joint in the area where the hull structure and two walls of the substructure are coming together

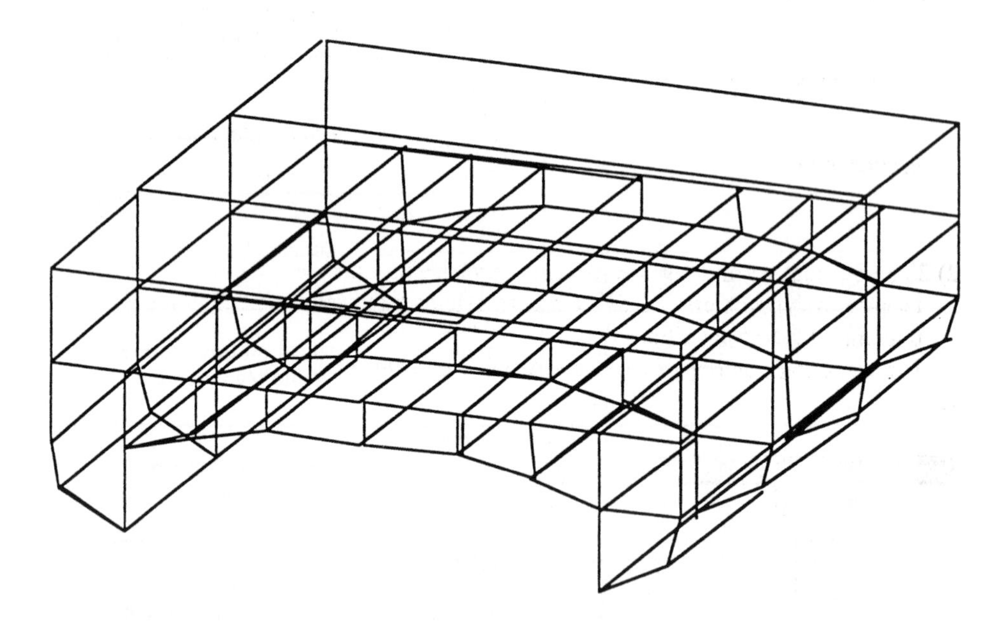

Fig. 3.1: Part of the ship hull - a summary of Inplane-, T- and X-joints

In parallel the different production techniques are influencing very heavily the selection of the joining solutions. All possible solutions were described and carefully evaluated due to their advantages and disadvantages regarding the different production techniques. Following this philosophy about 20 principles were evaluated and modified for the different structural areas described in Fig. 3.1.

The selection of the most successful joints was split into two steps:

1) Principles of joints were described with all their pros and cons due to characteristics like peeling stress stiffening effect, buckling, fatigue etc.:

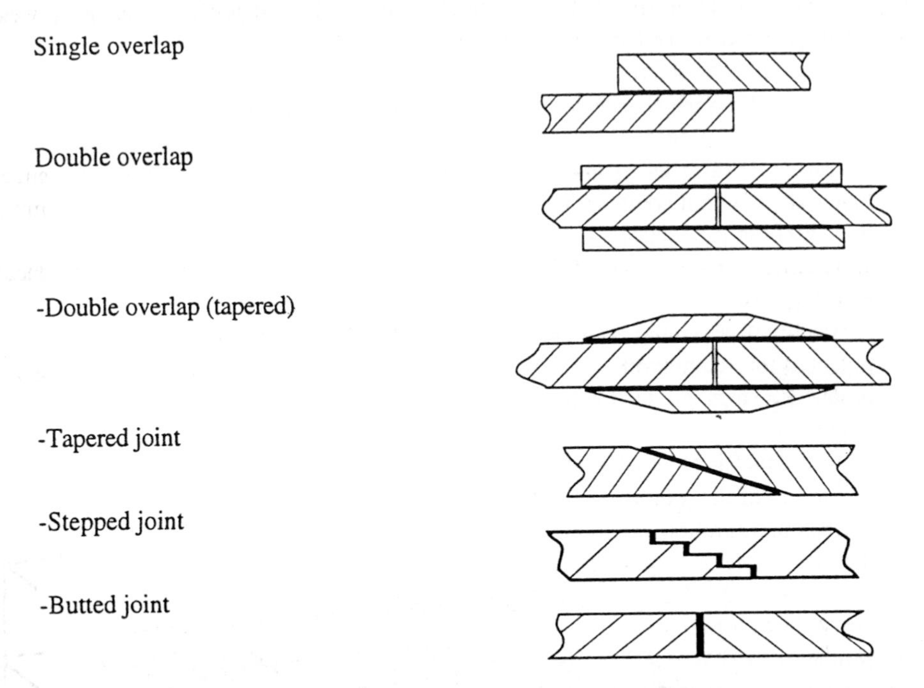

2) 20 basic joining solutions were judged very carefully and selected due to criteria like manufacturing problems related to the principle structural solutions, quality aspects, tolerance, loading conditions etc..
In the end five solutions were selected which are shown below:

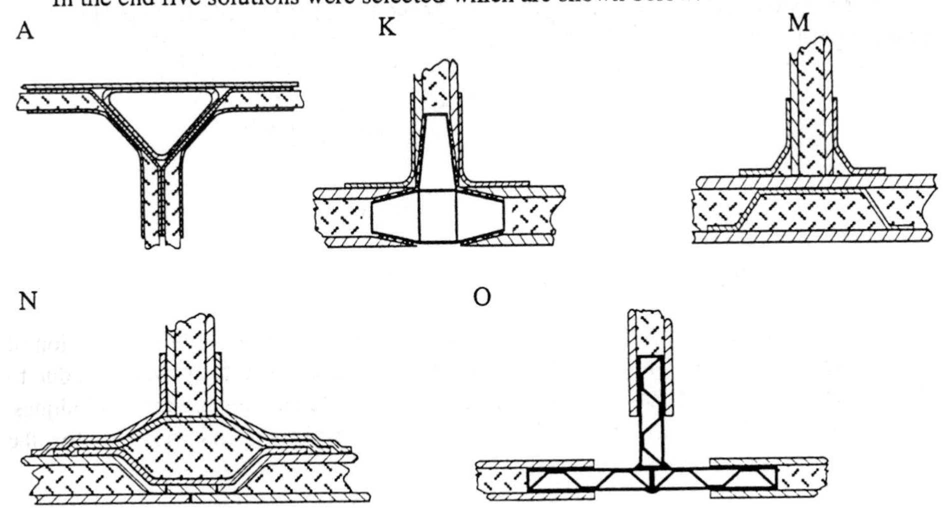

Fig. 3.2.: Construction methods for big and heavy loaded fibre reinforced composite sandwich structures demonstrated on a SES hull. R. Wallet, A. Eisenhut, G. Ziegmann

3.2 SELECTION CRITERIA

In the next step these five solutions are characterised in detail and advantages and disadvantages are described under the aspects

- Requirements/characteristics before joining
  - -requirements on panel ends dependent on prefabricated joint elements
  - -potential for prefabrication, maximum dimensions of joints/ prefabricated elements
  - -tolerances of surfaces/ geometry of joining panels
  - -materials, material combinations
  - -Sensitivity to production failures
- Requirements/characteristics during joining
  - -rigs, fixtures, handling, positioning requirements
  - -sensitivity to assembly failures, skill of coworkers
  - -potential for automation
  - -production steps, production process related to needed materials
  - -positioning
- Requirements/characteristics after joining
  - -variability of the joining concept
  - -corrosion, migration, environmental problems, water tightness
  - -thermal problems, due to fire regulations
  - -possibility of NDT of the joined area
  - -possibilities for repair

In a similar method as described in chapter 2 these joining characteristics were weighted regarding all the aspects mentioned above, especially observing the flexibility of the technique due to combining two, three or four panels.
In addition investigations were done on the stress and strength behaviour of the selected joints (in plane, out-of-plane characteristics, buckling, peeling etc.).
The combination of these ranking methods is shown in Fig. 3.3a-c, selecting *M* as the best solution.

Weighting	Materials, Material combinations Corrosion, migration		Requirements on panel closures		Potential for pre-fabrication, maximum dimensions		Requirements on surfaces, requirements on tolerances		Sensivity due to production failures		Rigs, fixures, handling of prefabricated parts		Potential for automation		Sensivity due to assembly failures		Required production steps for joining		Variability of joining concept		Possibility of non-destructive testing after joining		Possibility of repair		Thermal problems		Total value	Normalized results
	6		9		8		10		5		2		2		5		7		10		4		3		1			
	Scaled Value	Weighted Value	Scaled Value	Weighted Value	Scaled Value	Weighted Value	Scaled Value	Weighted Value	Scaled Value	Weighted Value	Scaled Value	Weighted Value	Scaled Value	Weighted Value	Scaled Value	Weighted Value	Scaled Value	Weighted Value	Scaled Value	Weighted Value	Scaled Value	Weighted Value	Scaled Value	Weighted Value	Scaled Value	Weighted Value		
Solution A	4.00	24.0	1.00	9.0	1.00	8.0	1.00	10.0	2.13	10.6	1.00	2.0	1.00	2.0	3.25	16.3	2.50	17.5	4.00	40.0	2.00	8.0	4.00	12.0	3.40	3.4	163	0.57
Solution K	3.70	22.2	3.00	27.0	3.40	27.2	1.00	10.0	1.00	5.0	3.00	6.0	2.00	4.0	4.00	20.0	2.00	14.0	4.00	40.0	2.00	8.0	4.00	12.0	2.20	2.2	198	0.69
Solution M	4.00	24.0	4.00	36.0	4.00	32.0	4.00	40.0	3.25	16.3	4.00	8.0	1.00	2.0	4.00	20.0	3.00	21.0	1.00	10.0	2.00	8.0	4.00	12.0	4.00	4.0	233	0.81
Solution N	4.00	24.0	3.00	27.0	2.80	22.4	3.40	34.0	3.25	16.3	3.33	6.7	1.00	2.0	1.00	5.0	1.00	7.0	2.50	25.0	1.00	4.0	4.00	12.0	4.00	4.0	189	0.66
Solution O	1.00	6.0	4.00	36.0	2.50	20.0	3.10	31.0	4.00	20.0	3.67	7.3	4.00	8.0	4.00	20.0	4.00	28.0	4.00	40.0	4.00	16.0	1.00	3.0	1.00	1.0	236	0.82
(Ideal Solution)	4.00	24.0	4.00	36.0	4.00	32.0	4.00	40.0	4.00	20.0	4.00	8.0	4.00	8.0	4.00	20.0	4.00	28.0	4.00	40.0	4.00	16.0	4.00	12.0	4.00	4.0	(288)	(1.00)

Fig.3.3a: Valuation of advantages/ disadvantages before, during and after joining

	Peeling stress	Stiffening effect	Buckling	In-plane tension	In-plane compression	In-plane shear	In-plane bending (up)	In-plane bending (down)	Out-of-plane shear	Out-of-plane tension	Out-of-plane compression	Flange bending	Flange shear	Total value	Normalized results
Solution A	3.00	4.00	1.00	1.00	1.00	4.00	3.25	2.00	4.00	4.00	1.00	4.00	4.00	36	0.70
Solution K	3.00	3.00	4.00	2.00	1.00	1.00	1.00	1.00	1.00	4.00	1.00	2.00	2.50	27	0.51
Solution M	4.00	2.00	4.00	4.00	4.00	4.00	4.00	4.00	4.00	1.00	4.00	1.00	1.00	41	0.79
Solution N	2.00	4.00	2.50	3.00	2.50	2.50	1.75	2.00	2.00	1.00	2.50	1.00	2.50	29	0.56
Solution O	1.00	1.00	2.50	2.00	1.00	1.00	2.50	2.00	1.00	1.00	1.00	2.00	1.00	19	0.37
(Ideal Solution)	4.00	4.00	4.00	4.00	4.00	4.00	4.00	4.00	4.00	4.00	4.00	4.00	4.00	(52)	(1.00)

Fig.3.3b: Valuation of stress and strength behaviour of finished joints

	Normalized results of Group 1	Normalized results of Group 2	Total Value	Final Ranking
Weighting	2	1		
Solution A	0.57	0.70	0.61	5
Solution K	0.69	0.51	0.63	4
Solution M	0.81	0.79	0.80	1
Solution N	0.66	0.56	0.63	4
Solution O	0.82	0.37	0.67	2
(Ideal Solution)	(1.00)	(1.00)	(1.00)	

Fig.3.3c: Final ranking of the joining solutions

4. Conclusion

The evaluation method described in [5] has led to the best solution regarding all the influencing parameters known today. Based on these results the joints are now under construction and will be tested within this year. The results should confirm the assumption described in this paper.
Applying this methodology it should be possible to judge if such a big SES can be built in composites.

5. References

1. Flemming, Ziegmann, Roth (1995/96) *Faserverbundbauweisen, 1. Fasern und Matrices, 2. Halbzeuge und Bauweisen,* Springer Verlag
2. Wallat, Eisenhut (1996) *Innovative Structural Solutions and Joints* MATSTRUTSES Report TEC/31/013-15/004-006/11
3. Anderegg (1995) *Passive Funktionsbauweisen im Schienenfahrzeugbau,* Dissertation ETH Zürich
4. Romagna (1997) *Neue Strategien in der Faserwickeltechnik,* Dissertation ETH Zürich
5. Flemming, Breiing (1993) *Theorie und Methoden des Konstruierens,* Springer Verlag

SANDWICH PLATES WITH "THROUGH-THE-THICKNESS" AND "FULLY POTTED" INSERTS

O. T. THOMSEN
Institute of Mechanical Engineering, Aalborg University, Pontoppidanstræde 101, DK-9220 Aalborg East, Denmark

Abstract

A high-order sandwich plate theory, which includes the transverse flexibility of the core, is introduced for the analysis of sandwich plates with hard points in the form of inserts. Special attention is focused on the problem of sandwich plates with inserts of the "through-the-thickness" and "fully potted" types. Numerical results obtained for sandwich plates with "through-the-thickness" and "fully potted" inserts subjected to out-of-plane loading are presented, and it is shown that the mechanical response for the latter insert type generally exhibits the most severe stress concentrations. The paper is concluded with the specification of a few guidelines for design.

1. Introduction

Advanced structural sandwich elements are used extensively for lightweight structures for spacecraft, aircraft and other applications. The introduction of loads into such elements is often accomplished using "potted" inserts, which can be of the "through-the-thickness", "fully potted" or "partially potted" type (see Figure 1) [1]. For all insert types the ideal load transfer mechanism is disturbed close to the inserts, where the face sheets tend to bend locally about their own middle surface rather than about the middle surface of the sandwich panel. This results in severe local stress concentrations.

It is generally recognised, that "through-the-thickness" inserts are superior to "fully potted" inserts with respect to load carrying capability. The primary reason for this is that the face sheets are forced to deflect together in sandwich plates with "through-the-thickness" inserts, whereas this is not the case for sandwich plates with "fully potted" inserts. However, it is not possible to distinguish between the two insert types using classical "antiplane" sandwich theories as summarised in Refs. 2-4. This is because of the fundamental assumption adopted in classical "antiplane" sandwich theories, which states that the sandwich plate thickness remains constant during deformation.

A. Vautrin (ed.), Mechanics of Sandwich Structures, 407–414.

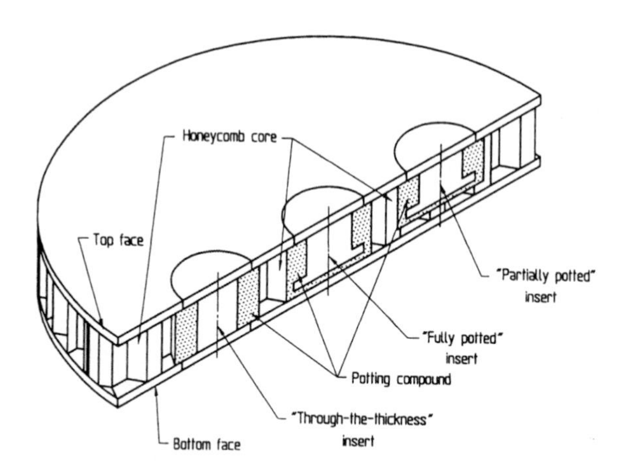

Figure 1. Potted insert types typically used for structural sandwich panels.

The importance of including the transverse core flexibility in the study of local bending phenomena has been pointed out in several investigations [5-10]. The present paper addresses the problem of approximate analysis of sandwich plates with "through-the-thickness" and "fully potted" inserts using a high-order sandwich theory, which accounts for the transverse flexibility of the core material.

2. Mathematical formulation

The high-order sandwich plate formulation includes separate descriptions of the elastic responses of the two face sheets and separate description of the elastic response of the core material. The theory can be used for approximate analysis of sandwich plates with "through-the-thickness" and "fully potted" inserts subjected to arbitrary external loading conditions. The present paper, however, only presents results obtained for sandwich plates with inserts subjected to axisymmetric out-of-plane loading. The problem is formulated by adapting and extending the principles behind the high-order theory developed for sandwich beams in Refs. 7-9 to circular sandwich plates. Elaborate details about the formulation can be found Refs. 10-12.

2.1 FORMULATION OF GOVERNING EQUATIONS

Figure 2 defines the constituent parts, the geometry and the possible external load cases for sandwich plates with "through-the-thickness" and "fully potted" inserts.

As suggested in Refs. (7)-(9), the core material is described as a transversely isotropic solid only possessing stiffness in the through-the-thickness direction. Thus, there can be no transfer of in-plane stresses in the core. Setting up the equilibrium conditions, and using the kinematic and constitutive relations for the core yields the following equations describing the core stress and displacement fields in terms of the transverse core coordinate z_c (z_c is measured from the core midsurface):

$$\tau_{rz}(r,\theta,z_c) = \tau_{rz}(r,\theta), \quad \tau_{\theta z}(r,\theta,z_c) = \tau_{\theta z}(r,\theta)$$

$$\sigma_z(r,\theta,z_c) = \frac{E_c}{c}\{w^1 - w^2\} - \left\{\tau_{rz,r} + \frac{1}{r}\tau_{rz} + \frac{1}{r}\tau_{\theta z,\theta}\right\} z_c$$

$$w_c(r,\theta,z_c) = w^1 + \frac{\{w^1 - w^2\}}{c}\left\{z_c - \frac{c}{2}\right\} - \frac{1}{2E_c}\left\{\tau_{rz,r} + \frac{1}{r}\tau_{rz} + \frac{1}{r}\tau_{\theta z,\theta}\right\}\left\{z_c^2 - \frac{c^2}{4}\right\}$$

$$u_c(r,\theta,z_c) = u_{0r}^1 - \frac{\beta_r^1}{2}\left\{f_1 - \frac{z_c^2}{c} - z_c + \frac{3c}{4}\right\} - \frac{\beta_r^2}{2}\left\{\frac{z_c^2}{c} - z_c + \frac{c}{4}\right\} + \frac{\tau_{rz}}{G_c}\left\{z_c - \frac{c}{2}\right\} \tag{1}$$

$$+ \frac{1}{2E_c}\left\{\tau_{rz,rr} + \frac{1}{r}\tau_{rz,r} - \frac{1}{r^2}\tau_{rz} + \frac{1}{r}\tau_{\theta z,\theta r} - \frac{1}{r^2}\tau_{\theta z,\theta}\right\}\left\{\frac{z_c^3}{3} - \frac{c^2 z_z}{4} + \frac{c^3}{12}\right\}$$

$$v_c(r,\theta,z_c) = u_{0\theta}^1 - \frac{\beta_\theta^1}{2}\left\{f_1 - \frac{z_c^2}{c} - z_c + \frac{3c}{4}\right\} - \frac{\beta_\theta^2}{2}\left\{\frac{z_c^2}{c} - z_c + \frac{c}{4}\right\} + \frac{\tau_{\theta z}}{G_c}\left\{z_c - \frac{c}{2}\right\}$$

$$+ \frac{1}{2rE_c}\left\{\tau_{rz,r\theta} + \frac{1}{r}\tau_{rz,\theta} + \frac{1}{r}\tau_{\theta z,\theta\theta}\right\}\left\{\frac{z_c^3}{3} - \frac{c^2 z_z}{4} + \frac{c^3}{12}\right\}$$

In Eqs. (1) E_c, G_c are the core elastic constants, σ_c is the core transverse normal stress, τ_{rz} and $\tau_{\theta z}$ are the core shear stresses, u_c, v_c and w_c are the radial, circumferential and transverse core displacements, u_{0r}^i, $u_{0\theta}^i$ and w^i $(i=1,2)$ are the radial, circumferential and transverse displacements of the faces, and r, θ are the radial and circumferential coordinates.

The core material is coupled with the face sheets by requiring continuity of the displacement and stress fields across the core/face sheet interfaces.

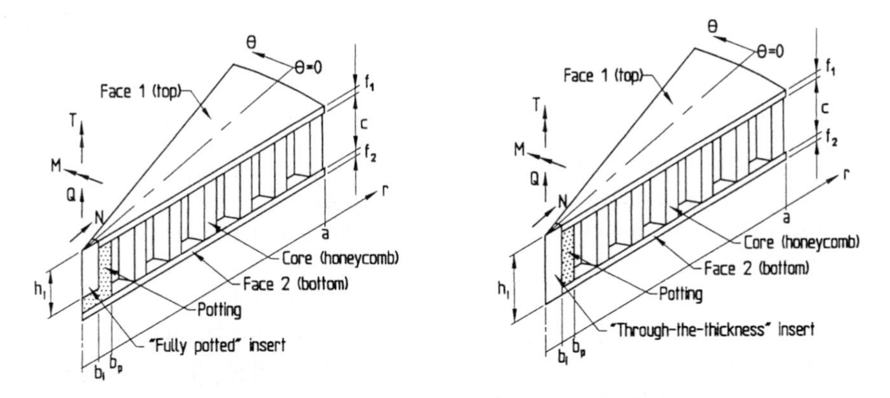

Figure 2. Illustration of sandwich plates with "fully potted" and "through-the-thickness" inserts subjected to arbitrary loading conditions.

The face sheets are modelled using a Mindlin-Reissner type of plate theory, which allows inclusion of out-of-plane shearing effects. Formulating the equilibrium, kinematic and constitutive equations for the top and bottom faces, and combining these with the "core equations" (1) together with the continuity requirements, yields a 24th order set of governing partial differential equations. Consequently, the governing equations can be reduced to 24 first order partial differential equations with 24 unknowns. If the 24 unknowns (fundamental variables) are those quantities that appear in the natural boundary conditions at an edge r=constant, then the boundary value problem can be stated completely in terms of these variables:

$$\{y(r,\theta)\} = \{u^1_{0r}, u^1_{0\theta}, w^1, \beta^1_r, \beta^1_\theta, N^1_r, N^1_{r\theta}, M^1_r, M^1_{r\theta}, Q^1_r, \tau_{rz}, q_r, \tau_{\theta z}, q_\theta, \\ u^2_{0r}, u^2_{0\theta}, w^2, \beta^2_r, \beta^2_\theta, N^2_r, N^2_{r\theta}, M^2_r, M^2_{r\theta}, Q^2_r\} \tag{2}$$

where β_r^i, β_θ^i are the midsurface rotations, N_r^i, $N_{r\theta}^i$ are the in-plane stress resultants, M_r^i, $M_{r\theta}^i$ are the moment resultants, and Q_r^i are the radial transverse shear stress resultants ($i=1,2$). In Eqn. (2) two new "core variables" q_r, q_θ have been introduced

$$q_r(r,\theta) = \tau_{rz,r}, \quad q_\theta(r,\theta) = \tau_{r\theta,r} \tag{3}$$

Adopting the matrix of fundamental variables $\{y(r,\theta)\}$, the governing equations can be reduced to the form

$$\{y(r,\theta)\}_{,r} = \Psi\left(r, \theta, \{y\}, \{y\}_{,\theta}, \{y\}_{,\theta\theta}, \ldots\right) \tag{4}$$

where Ψ denotes *24* linear functions in $\{y(r,\theta)\}$ and its derivatives with respect to θ.

The dependency of the θ-coordinate is eliminated by Fourier series expansion of the fundamental variables.

2.2 STATEMENT OF BOUNDARY CONDITIONS

2.2.1 *"Through-the-thickness" inserts*

With reference to Figure 2 the imposed boundary conditions for the "through-the-thickness" insert case are generally derived from the following assumptions:

$r=b_i$: The insert is considered as an infinitely rigid body to which the face sheets and the potting material are rigidly connected.

$r=b_p$: Continuity of the variables across the potting/honeycomb interface.

$r=a$: The face sheet and honeycomb core midsurfaces are simply supported.

The boundary conditions at $r=b_i$, $r=b_p$ and $r=a$ are stated by specifying linear combinations of the fundamental variables.

2.2.2 *"Fully potted" inserts*

The insert is modelled as a thick top face sheet (plate), even though this is a rather crude approximation. The system equations contain elements of the type r^{-1}, r^{-2} and r^{-3}, and the system equations therefore show singular behaviour for $r \rightarrow 0$. In order to overcome this the problem is rephrased slightly, now assuming that the plate centre defined by a small radius $r=b_0$ (imaginary "inner rim") is removed. If b_0 is sufficiently small, this hardly influences the solution away from the plate centre. The boundary conditions are derived from the following assumptions:

$r=b_0$: Bottom face (face 2): Free edge conditions are imposed.
Potting (core) material: Free edge conditions are imposed.
Top face (face 1): The external load is applied to the insert as a uniform surface load applied on the insert surface.

$r=b_i$: Continuity of fundamental variables, except for u_{0r}^{1} and τ_{rz}.

$r=b_p$: Continuity of fundamental variables across the potting/honeycomb interface.

$r=a$: The face sheet and honeycomb core midsurfaces are simply supported.

The boundary conditions at $r=b_0$, $r=b_i$, $r=b_p$ and $r=a$ are stated by specifying linear combinations of the fundamental variables.

2.3 NUMERICAL SOLUTION

The governing equations together with the boundary conditions, constitutes a boundary value problem, which was solved numerically using the "multi-segment method" of integration [10-12]. The method is based on a transformation of the boundary value problem into a series of interconnected initial value problems. The insert/sandwich plate configuration is divided into a finite number of segments, and the solution within each segment is derived by direct integration. Fulfilment of the boundary conditions, and continuity of the solution vectors across the segment separation points, is ensured by solving a set of linear algebraic equations.

3. Examples

3.1 GEOMETRY, MATERIAL PROPERTIES AND LOADING CONDITIONS

Two cases of sandwich plates with "through-the-thickness" and "fully potted" inserts subjected to axisymmetric out-of-plane compressive have been analysed. Both examples are based on a symmetric insert/sandwich plate system defined by:

Geometry: b_i=7.0 mm, b_p=10.0 mm, a=60.0 mm, c=20.0 mm, $f_1=f_2$=0.2 mm.
"Through-the-thickness" insert: $h_i=f_1+c+f_2$=20.4 mm.
"Fully potted" insert: h_i=9.0 mm, b_0=0.1 mm (imaginary "inner rim").
Face 1: Aluminium face sheet; E_1=71.5 GPa; ν_1=0.3.
Face 2: As face 1 (symmetric sandwich plate).
Insert: Same material properties as the face sheets.
Potting: Bulk epoxy polymer, E_p=2.5 GPa, G_p=0.93 GPa.
Honeycomb: HEXCEL Al-honeycomb 3/16"-5056-0.0007", E_h=310 MPa, $G_h \approx (G_L+G_W)/2$=138 MPa.
External loading: Q=- 1 kN (compressive out-of-plane force).

3.2 EXAMPLE 1: "THROUGH-THE-THICKNESS" INSERT

The external load Q=-1 kN has been applied by assuming a transverse rigid body motion of the insert, and by prescribing that the resultant of the face sheet transverse shear stress resultants and the core shear stresses equilibrates the applied external load.

Fig. 3 shows the out-of-plane deflections of the face sheets (w^1, w^2), and the core midsurface ($w_c(z_c=0)$). $r \leq b_i=7$ mm corresponds to the insert, $b_i=7$ mm$<r \leq b_p=10$ mm corresponds to the potting region, and $r>10$ mm corresponds to the core region.

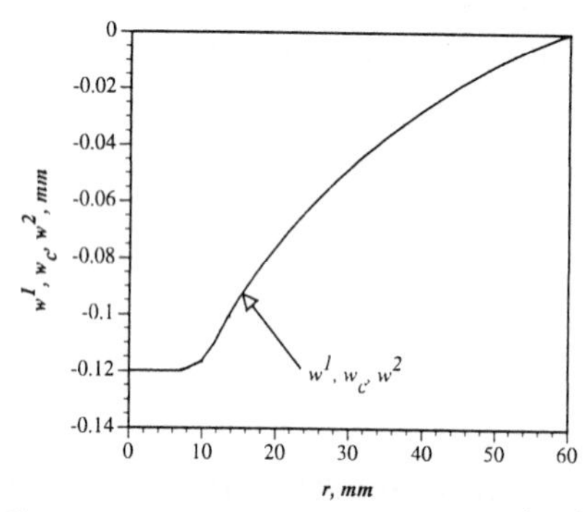

Figure 3. Out-of-plane displacements w^1, w^2, $w_c(z_c=0)$ "Through-the-thickness" insert. Q=-1 kN.

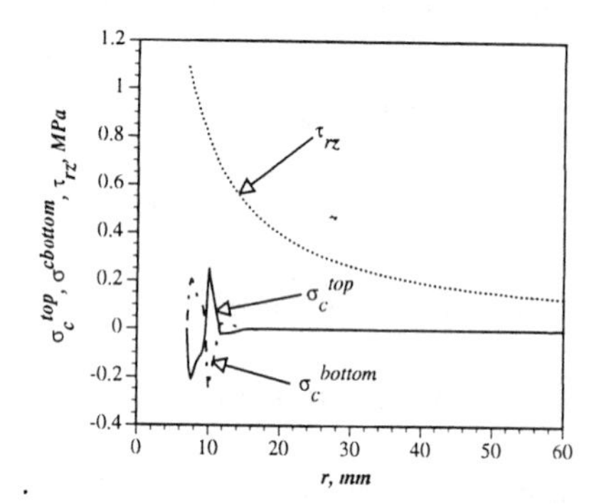

Figure 4. Core stress components τ_{rz}, σ_c^{top}, σ_c^{bottom}. "Through-the-thickness" insert. Q=-1 kN.

Fig. 4 shows the radial distribution of core stresses in the potting and core materials. σ_c is given at the interface between face sheet 1 and the potting/honeycomb (σ_c^{top}), and at the interface between face sheet 2 and the potting honeycomb (σ_c^{bottom}). Fig. 4 also shows the distribution of the core shear stress component τ_{rz}.

3.3 EXAMPLE 2: "FULLY POTTED" INSERT

For the "fully potted" insert case the external load Q=-1.0 kN has been applied as a uniform out-of-plane surface load distributed over the top surface of the insert.

Fig. 5 shows the out-of-plane deflections of the face sheets (w^1, w^2), and the core midsurface ($w_c(z_c=0)$). From Fig. 5 it is seen that w^1, w^2 and $w_c(z_c=0)$ are nearly identical for $r>b_p=10$ mm, while this is not the case for $r<b_p$. It is seen that the insert (i.e., the top face sheet for $r \leq b_i$) deflects out-of-plane as a rigid body (or nearly so).

Comparison of Fig. 5 with Fig. 3 reveals that distinctly different deflectional characteristics are observed for $r<b_p$, since the face sheets and the core have been forced to deflect the same amount at the insert/sandwich plate intersection (at $r=b_i=7$ mm) in the "through-the-thickness" insert case.

Fig. 6 shows the radial distribution of stresses in the potting and core materials. Considering at first the σ_c-distribution, it is observed that the most severe transverse normal stresses are located in the potting compound ($r<b_p=10$ mm). Comparing the "through-the-thickness" and "fully potted" insert cases it is seen, that the shear stress distribution features are almost identical, whereas the occurrence and severity of

transverse normal stresses differ considerably. Thus, the peak value of σ_c is about 25 times larger for the "fully potted" than for the "through-the-thickness" insert case.

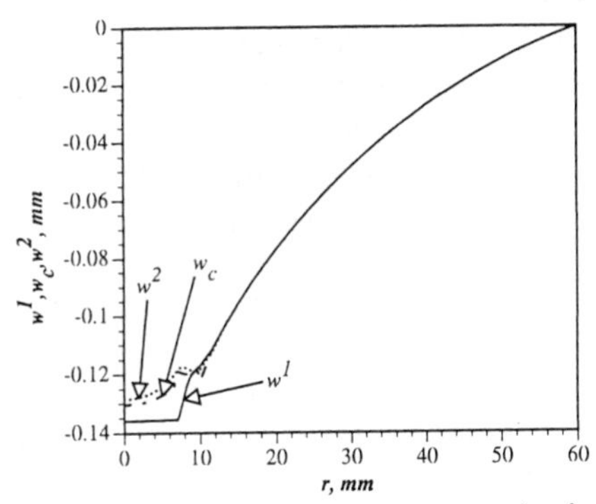

Figure 5. Out-of-plane displacements w^1, w^2, $w_c(z_c=0)$. "Fully potted" insert. Q=-1 kN.

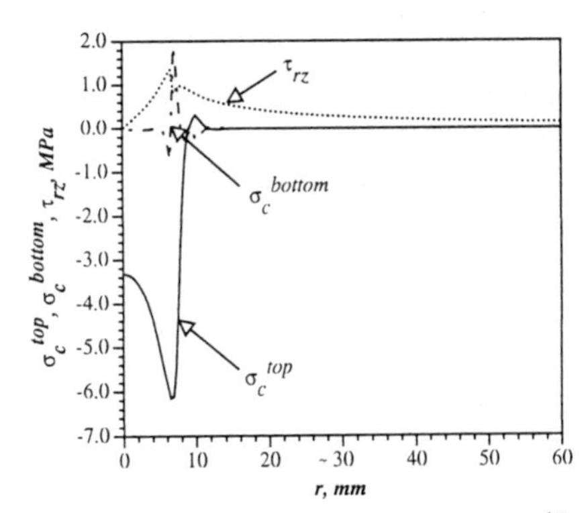

Figure 6. Core stress components τ_{rz}, σ_c^{top}, σ_c^{bottom}. "Fully potted" insert. Q=-1 kN.

4. Design guidelines

It is possible to formulate a number of simple guidelines for the design of sandwich plates with "through-the-thickness" and "fully potted". The design rules given here comply very well with established design practices [1]. The following guidelines for design can be specified for sandwich plates with "through-the-thickness" and "fully potted" inserts:

1. If possible, the radial extension of the potting compound should at least be $0.5b_i$.
2. If possible, the ratio of the potting stiffness to the honeycomb stiffness, E_p/E_h, should be chosen to $E_p/E_h \approx 3$-4.
3. It is generally recommended to avoid external bending moment. This can be achieved by using groups of inserts loaded in the out-of-plane direction.
4. If severe external loads are to be introduced, a strong advise is given to use "through-the-thickness" inserts instead of "fully potted" inserts.
5. The capability of face sheets to resist the peak stresses adjacent to the insert can be improved by reinforcing the face sheets locally.
6. As significant stress concentrations in the potting material, as well as in the bonds between the core material and face sheets, are unavoidable, the potting and bond line materials should possess long elongation to failure capability.

5. Conclusions

A high-order sandwich plate theory has presented and adapted for the analysis of sandwich plates with "through-the-thickness" and "fully potted" inserts. The high-order sandwich plate theory includes the following features: separate descriptions of the face sheets, separate description of the core ("potting" and "honeycomb"), and finally general specification of the external loading. Different formulations of the boundary conditions for the "through-the-thickness" and "fully potted" insert cases has been derived. The solution for specific insert/sandwich plate configurations has been accomplished using the "multi-segment-method of integration".

Examples studies have demonstrated that the stress concentrations induced in sandwich plates with "fully potted" inserts are much more severe, than is the case for sandwich plates with inserts of the "through-the-thickness" type. The reason for this is that the face sheets are effectively forced to deflect together for "through-the-thickness" inserts. This is not the case for "fully potted" inserts, where the inserts are only attached directly to the upper face sheet, thus effectively preventing proper collaboration between the face sheets. Thus, the suggested approximate method of analysis of sandwich plates with "potted" inserts, has proven to be capable of explaining the result known from practice, that sandwich plates with "through-the-thickness" inserts are superior to sandwich plates with "fully potted" inserts with respect to load carrying capability.

The numerical solution procedures developed for the analysis of sandwich plates with "through-the-thickness" and "fully potted" inserts (and "partially potted" inserts) will be implemented in the new composites analysis and design software package ESAComp [13], which is presently being developed for the European Space Agency.

6. References

1. *Insert Design Handbook*, European Space Agency, ESA PSS-03-1202, Issue 1, 1987.
2. Plantema, F.J.: *Sandwich Construction*, John Wiley & Sons, New York, 1966.
3. Allen, H.G.: *Analysis and Design of Structural Sandwich Panels*, Pergamon Press, Oxford, 1969.
4. Zenkert, D.: *An Introduction to Sandwich Construction*, Chameleon Press Ltd., London, 1995.
5. Thomsen, O.T., Rits, W., Eaton, D.C.G. and Brown, S.: Ply drop-off effects in CFRP/honeycomb sandwich panels - theory, *Composites Science & Technology* **56** (1996) 407-422.
6. Thomsen, O.T., Rits, W., Eaton, D.C.G., Dupont, O. and Queekers, P.: Ply drop-off effects in CFRP/honeycomb sandwich panels - experimental results, *Composites Science & Technology* **56** (1996) 425-427.
7. Frostig, Y., Baruch, M., Vilnai, O., Sheinman, I.: High-order theory for sandwich beam bending with transversely flexible core, *Journal of ASCE, EM Division* **118**(5) (May 1992) 1026-1043.
8. Frostig, Y.: On stress concentration in the bending of sandwich beams with transversely flexible core, *Composite Structures* **24** (1993) 161-169.
9. Frostig, Y. and Shenhar, Y.: High-order bending of sandwich beams with a transversely flexible core and unsymmetrical laminated composite skins, *Composites Engineering* **5** (1995) 405-414.
10. Thomsen, O.T.: *Analysis of Sandwich Plates with Through-the-Thickness Inserts Using a Higher-Order Sandwich Plate Theory*, ESA/ESTEC Report EWP-1807, European Space Agency, The Netherlands, 1994.
11. Thomsen, O.T.: *Analysis of Sandwich Plates with Fully Potted Inserts Using a Higher-Order Sandwich Plate Theory*, ESA/ESTEC Report EWP-1827, European Space Agency, The Netherlands, 1995.
12. Thomsen, O.T. and Rits, W.: Analysis and design of sandwich plates with inserts: a higher-order sandwich plate theory approach, *submitted*.
13. Saarela, O., Palanterä, M., Häberle, J. and Klein, M.: ESAComp: a powerful tool for the analysis and design of composite materials, *Proceedings of the International Symposium on Advanced Materials for Lightweight Structures*, ESA-WPP-070, ESTEC, Noordwijk, The Netherlands (1994) 161-169.

FAILURE MODES IN SANDWICH T-JOINTS

EFSTATHIOS E. THEOTOKOGLOU
Department of Engineering Science, Section of Mechanics
The National Technical University of Athens
GR-15773, Athens, Greece

Abstract

Sandwich T-joints are attractive structural members in high speed marine vehicles. Such joints can fail in several ways under static loading. In this paper we develop equations describing the load at which failure occurs for each possible failure mode for a sandwich beam with a brittle face and a core material that yields plastically. The critical loads have been compared with those from the numerical investigation of a typical GRP/PVC sandwich T-joint under pull out loads.

1. Introduction

Structural members made up of two stiff strong faces separated by a light-weight core are known as sandwich panels. The separation of the faces by the core increases the moment of inertia of the panel with little increase in weight, producing an efficient structure for resisting bending and buckling loads [1-3]. Because of this, they are often used in applications where minimizing the weight of the panel is critical. For example, sandwich panels are now commonly used in vehicle components, ranging from the motor blades of helicopters to the exterior panelling of light-weight rapid transit rail vehicles and to high speed marine vehicles.

In applications where the weight of the sandwich panel is the most important parameter, as in aerospace components, the core is usually made of an aluminium or paper-resin honeycomb, as these give the lightest structure. However, in applications in which the demand for low weight is accompanied by other demands such as low thermal conductivity or low costs, polymeric foam cores (usually, PVC (polyvinylchloride) or polyurethane foam) are usually preferred, which easily deform under compression or tension, but offer good resistance to shear forces [4,5].

In our study, a sandwich beam with PVC-core and GRP laminates is treated. This sandwich offers good mechanical properties and good chemical resistance.

Polymeric foam-cored sandwich panels can fail in several different ways under beam loading [6,7], each of which is characterized by a different failure equation. In addition, it is very difficult to handle the out of plane

A. Vautrin (ed.), Mechanics of Sandwich Structures, 415–422.

behaviour of sandwich T-joints [8-12]. The main purpose of this paper is to study the critical failure mode in sandwich panels, as a first step in optimizing the design of a sandwich T-joint for a given strength.

At first, the different failure modes which may take place in a sandwich panel under beam loading are analyzed and equations describing the failure loads for each mode of failure are developed. The proposed equations are applied to the case of sandwich T-joints loaded by symmetrical loads in order to predict the ultimate loads at which failure occurs. These critical loads are compared with those determined in the experimental investigation of a typical bulkhead-to-hull GRP/PVC sandwich T-joint under pull out forces [8,12].

Finally, a finite element analysis is applied to the T-joint and the results for the stresses are compared with those from the proposed failure equations.

2. Failure Modes of a Sandwich Panel

A typical sandwich beam is shown in Figure 1. The faces are considered of the same material and they have a Young's modulus E_f, the core has a density ρ_c, a Young's modulus E_c and a shear modulus G_c. The thickness of each face is $t_i(i=1,2)$ and that of the core is c. The beam has a width b and a span L.

A sandwich panel under beam loading may fail mainly by tensile or compressive failure of the faces, by wrinkling of the compressive face, by shear failure of the core, though compressive or tensile failure is also possible, by debonding between the core and the faces. Considering each failure mode in turn we have.

Tensile or compressive failure of the faces occurs when the normal stresses in the faces exceeds the strength of the face material in tension or compression respectively. So failure initiates when

$$\sigma_f = \sigma_{fy} \tag{1}$$

where, σ_{fy} is the strength of the face material in tension or compression.

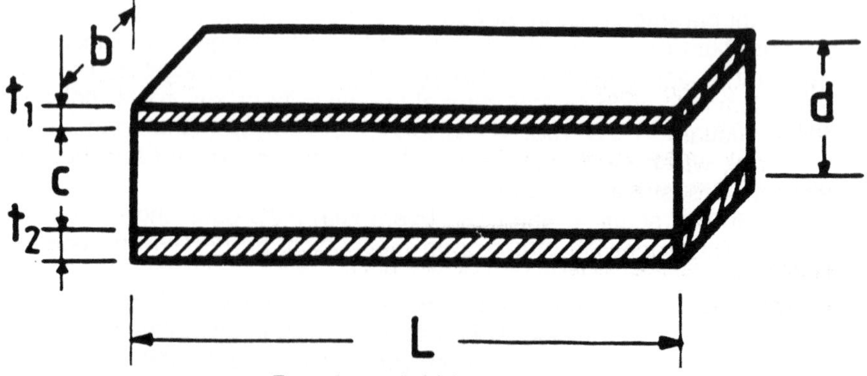

Figure 1. A sandwich beam.

Taking into consideration that the stresses in the faces may be approximated by

$$\sigma_f \cong \frac{M}{btd} \tag{2}$$

in case where the thickness of the faces is negligible compared to that of the core, we obtain

$$M_{fy} \cong \sigma_{fy}(btd) \tag{3}$$

where, M_{fy} is the moment of the beam which causes tensile or compressive failure of the faces.

The compression face of a sandwich beam may fail by local buckling or wrinkling. Face wrinkling occurs when the compressive face of the beam reaches the local instability stresss, given by [1]

$$\sigma_{fw} = \frac{3E_f^{1/3}E_c^{2/3}}{\left(12(3-v_c)^2(1+v_c)^2\right)^{1/3}} \tag{4}$$

where, v_c the Poisson's ratio of the core. Taking into consideration relation (2), we obtain

$$M_{fw} = \frac{3E_f^{1/3}E_c^{2/3}(btd)}{\left(12(3-v_c)^2(1+v_c)^2\right)^{1/3}} \tag{5}$$

where, M_{fw} is the moment of the beam at which the compression face wrinkles

Core yield in shear occurs when the maximum shear stress in the core reaches the shear yield strength of the foamed core material. So failure initiates when

$$max\,\tau_c = \tau_c^* \tag{6}$$

where, τ_c^* is the shear yield strength of the foamed core material and

$$max\,\tau_c = \left(\left(\frac{\sigma_c}{2}\right)^2 + \tau_c^2\right)^{1/2} \tag{7}$$

with σ_c, τ_c, the maximum normal stress and the mean shear stress in the core respectively.

We may have core yield in tension or compression too. In this case failure initiates when

$$max\,\sigma_c = \sigma_c^* \tag{8}$$

where, σ_c^* is the tensile or compressive yield strength of the foamed core material and

$$max\,\sigma_c = \frac{\sigma_c}{2} + \left(\left(\frac{\sigma_c}{2} \right)^2 + \tau_c^2 \right)^{1/2} \tag{9}$$

The bond between the faces and the core is the most difficult of the mechanisms to be analyzed. Fracture is usually initiated by a crack in the adhesive between the core and the face. The crack is subjected to shear stresses as well as to axial stresses parallel to it. Moreover, the stress field around the crack is complicated by the different moduli of the face, the adhesive and the core. This problem will be pursued in a subsequent study.

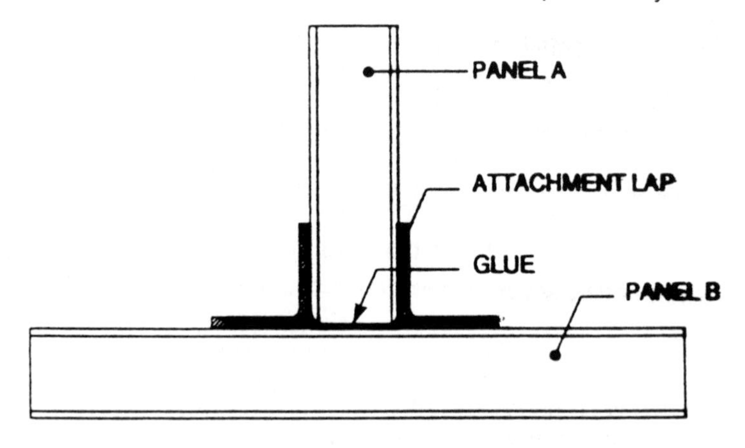

Figure 2. Connection between two sandwich beams.

3. Material Properties and Failure Modes in Sandwich T-Joints

One typical joint geometry is shown in Figure 2. The joint considered in our study is similar to that used in [8,10,12]. The T-joint is exposed to a tension force normal to the flange (Fig.3).

The face sheets of the T-joints consist of two layers of $800\,g/m^2$ each woven roven E-glass with fibers in ± 45 directions $[\pm 45 \pm 45]$ and $150\,g/m^2$ Chopped Strand Mat (CSM) on top and on bottom of the laminate. One layer of standard Al-reinforcement consists of $400\,g/m^2$ woven roven E-glass with fibers in ± 45 directions and $100\,g/m^2$ CSM on top of the layer. Their mechanical properties are given in Tables 1 [8]. The core material used is PVC-foam, Divinycell H100 with mechanical properties given in Table 2 and the glue material is Crestomer 1152 PA (Table 3).

From the experiments, force-displacement characteristics are obtained for different attachment configurations [8,12]. Two major failure modes are observed:

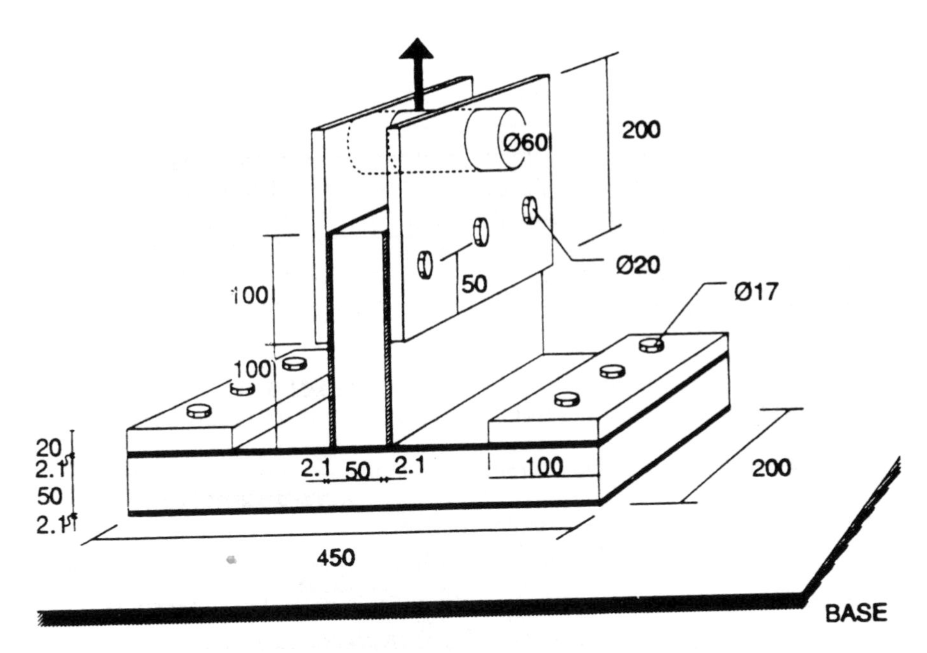

Figure 3. Test set up under pull-out force.

(I) Failure between panel laminates the glue and the attachment lap.
(II) Shear fracture of the core in panel B.

TABLE 1. Mechanical properties of E-Glass polyester laminate [8]

Longitudinal/Transverse Young's modulus (MPa)	10000/10000
Shear modulus (MPa)	6200
Longitudinal/Transverse tensile strength (MPa)	95/95
Longitudinal/Transverse compressive strength (MPa)	123/123
Shear strength (MPa)	93

TABLE 2. Mechanical properties of Divinycell PVC H100 at 20°C

Density (kg/m^3):	102
E-modulus compression, ASTM D1621 (MPa)	130
Ultimate compressive strength, ASTM D1621 (MPa)	1.7
E-modulus tension, ASTM D1623 (MPa)	105
Ultimate tensile strength, ASTM D1623 (MPa)	2.8
Shear modulus at 20°C, ASTM C273 (MPa)	42
Ultimate shear strength, ASTM C273 (MPa)	1.5

TABLE 3. Typical Properties of Crestomer 1152 PA

Property	Value
Ultimate tensile strength (MPa)	26
Elongation at break (%)	100
Initial tensile modulus (MPa)	500
Yield stress of 7% strain (MPa)	17

4. Analysis of the Problem

The load P (Fig. 3) will be carried by the laminates of panel A (Fig. 2) and will be transmitted locally into the panel B. The panel B is considered as a clamped sandwich beam loaded by two symmetrical loads $\frac{P}{2}$. The panel B may fail under the failure modes described in Section 2.

The critical load P_{fy} in order to have tensile or compressive failure of the faces, taking into consideration relation (3), is given by

$$P_{fy} \cong 2\sigma_{fy} \frac{btdL}{a^2} \tag{10}$$

The critical load P_{fw} in order to have face wrikling (relation (5)), is given by

$$P_{fw} \cong \frac{6E_f^{1/3} E_c^{2/3} btdL}{\left(12(3-v_c)^2(1+v_c)^2\right)^{1/3} a^2} \tag{11}$$

The critical load in order to have core yield in shear (relation (7)), is given by:

$$P_{cs} \cong \frac{2\tau_c^* bd}{\left(\left(\frac{a^2c}{2Ltd}\frac{E_c}{E_f}\right)^2 + 1\right)^{1/2}} \tag{12}$$

and the critical load in order to have core yield in tension or compression (relation (9)), is given by:

$$P_{cy} \cong \frac{2\sigma_c^* bd}{\frac{a^2c}{2tdL}\left(\frac{E_c}{E_f}\right) + \left(\left(\frac{a^2c}{2Ltd}\frac{E_c}{E_f}\right)^2 + 1\right)^{1/2}} \tag{13}$$

Substituting into relations (10)-(13) the dimensions of the panel B (Fig.3) and material data from Tables 1,2, we obtain

$$\left.\begin{aligned} &P_{fyt} \cong 82(KN), \quad &&P_{fyc} \cong 106(KN) \\ &P_{fw} \cong 238(KN), \quad &&P_{cs} \cong 31(KN) \\ &P_{cyt} \cong 52(KN), \quad &&P_{cyc} \cong 31(KN) \end{aligned}\right\} \tag{14}$$

where, P_{fyt}, P_{fyc} are the critical loads in order to have tensile or compressive failure of the faces respectively; and P_{cyt}, P_{cyc}, are the critical loads in order to have core yield in tension or compression respectively.

From the critical loads it is observed that the core yield in shear and in compression take place first.

5. Finite Element Analysis

The sandwich T-joint considered (Fig.3) is analyzed using the Finite Element Method (FEM) considering a plane model of the joint. The finite element analysis is performed with the use of the general purpose FEM program ANSYS [13]. Elastic plastic material behaviour for the core and the glue materials and large deflection analysis for the T-joint have been accounted for. The finite element mesh consists of triangular and rectangular elements and it has been already used in [10,12].

The analysis is carried out for the load level $31KN$ where the core starts yielding in shear and in compression (Eqs. 14). From the analysis the highest values of the maximum principal, minimum principal and shear stresses in the core and in the laminates of panel B are given in Table 4.

TABLE 4. Highest values of principal and shear stresses in Panel B

	Maximum principal stress (MPa)	Minimum principal stress (MPa)	Maximum Shear Stress (MPa)
Core	2.65	-1.53	-1.55
Laminate	40	-66	-13

It is observed that the highest values of the stresses in the laminates of panel B are lower than the strength limits (Table 1). On the contrary, the highest shear stress in the core material exceeds the shear strength at a relatively large area of panel B and the highest values of the maximum principal and minimum principal stresses are very close to the strength limits (Table 2). But the highest values for the maximum principal and the minimum principal stresses occur at a very small area where the loads carried by the laminates of panel A are transmitted locally into panel B and at small areas near the ends of panel B where the boundary conditions are imposed.

6. Conclusions

From the failure modes considered, the experimental investigation [8,12] and the FEA we have:

(I) The core of panel B yields mainly in shear. Increasing the load, the foam continues to yield until final fracture produces the diagonal cracks.

(II) The skin laminates of the panels don't fail by face yielding or face wrinkling.

The failure equations accurately describe the loads at failure for face failure, face wrinkling, core yield in tension or compression and core yield in shear.

Our results are in good agreement with those in [8,12] where only core failure of panel B is observed when there is a proper glue filling between the attachment lap and the panel laminates. Hence, if there is a proper glue filling the lap thickness and width aren't of great importance and the core failure is the dominant failure mode that takes place in a T-joint under a pull-out load.

7. References

1. Plantema, F.J. (1966) *Sandwich Construction,* Vol.3 of Airplane Missile and Spacecraft Structures, John Wiley and Sons, Inc..
2. Allen, H.G. (1969) *Analysis and Design of Structural Sandwich Panels,* Pergamon, Oxford.
3. Smith, C.S. (1990) *Design of Marine Structures in Composite Materials,* Elsevier Applied Science, London.
4. Maiti, S.K., Ashby, M.F. and Gibson, L.J. (1984) Fracture toughness of brittle cellular solids, *Scripta Metallurgica* **18**, 213-217.
5. Gibson, L. and Ashby, M.F. (1988) *Cellular Solids-Structures and Properties,* Pergamon Press.
6. Hall, D.J. and Robson, B.L. (1984) A Review of the design and materials evaluation programme for the GRP/Foam sandwich composite hull of the RAN minehunter, *Composites* **15**, 266-276.
7. Olsson, K.A. and Lonno, A. (1987) Testing procedure for core materials, K.A. Olssom and R.P. Reichard (eds.), *Proceedings of the First International Conference on Sandwich Construction,* EMAS, U.K., pp. 293-318.
8. Fagerheim, G., Moan T. and Taby, J. (1989) Experimental determination of the ultimate strength of GRP/PVC.
9. Shenoi, R.A. and Violette, F.L.M. (1990) A Study of structural composite tee joints in small boats, *Journal of Composite Materials* **24**, 644-666.
10. Theotokoglou, E.E. and Moan, T. (1991) Stress Analysis of a GRP/PVC sandwich T-joint, in S.W. Tsai and G.S. Springer (eds.), *Proceedings of the Eight International Conference on Composite Materials (ICCM/8),* Sampe, USA, **1**, pp. 9.C.1-9-C.10.
11. Kildegaard, C., (1993) Structural design of T-joints in Marine FRP-sandwich structures, Doctoral thesis, Aalborg University, Denmark, Special Report No.2.
12. Theotokoglou E.E. and Moan T. (1996) Experimental and numerical study of composite T-joints, *Journal of Composite Materials* **30**, 190-209.
13. Users Manual for ANSYS (1990), Swanson Analysis Systems, Houston, PA, USA.

LIST OF PARTICIPANTS

Prof. Allen Howard G.
University of Southampton
Dept of Civil and Environmental Engineer
Highfield
S017 1BJ SOUTHAMPTON
ANGLETERRE

Dr. Benjeddou Ayech
Conservatoire National des Arts et Métie
CNAM
2 rue Conté
75003 PARIS
FRANCE

M. Bistac S.
Institut de Chimie des Surfaces et Inter
15 rue Jean Starcky
2488
68057 MULHOUSE CEDEX
FRANCE

M. Bourgeois Stéphane
ESM2/IMT/LMA CRNS
IMT-ESM2
Technopôle de Chateau Gombert
13451 MARSEILLE Cedex 20
FRANCE

M. Boussuge M.
E.N.S.M.P.
Centre des Matériaux
87
EVRY 91003
FRANCE

Ms. Bozhevolnaya Elena
Aalborg University
Institute of Mechanical Engineering
Pontoppidanstraede 101
9220 AALBORG OUEST
DANEMARK

Prof. Brechet Yves
LTPCM
Domaine Universitaire
BP 75
38402 ST MARTIN D'HERES CEDEX
FRANCE

Mr. Brevik Arnt Frode
Maritime Seanor AS
PO Box 24
N-1360 NESBRU
NORVEGE

M. Cartraud Patrice
Ecole Centrale de Nantes
Laboratoire Mécanique et Matériaux
1, rue de la Noë
44072 NANTES CEDEX 03
FRANCE

M. Castanié Bruno
Université Paul Sabatier
Laboratoire de Génie Mécanique de Toulou
Bat 3PN
31062 TOULOUSE Cedex
FRANCE

M. César de Sà J.
Universidade do Porto
Faculdade de Engenharia
Rua dos Bragas
4099 PORTO CODEX
PORTUGAL

M. Collombet Francis
ENSAM
LAMEF
Esplanade des Arts et Métiers
33405 TALENCE CEDEX
FRANCE

Dr Davies Peter
Ifremer
DITI/GO/MM
Centre de Brest
BP 70
29280 PLOUZANE
FRANCE

M. Dufort Laurent
EMSE
SMS/Mécanique et Matériaux
158 cours Fauriel
42023 SAINT ETIENNE CEDEX 2
FRANCE

Ms. Dupuits Marie-dominique
Centre des Matériaux P.M. FOURT E.N.S.M.
P.M FOURT E.N.S.M.P.
BP 87
91003 EVRY Cedex
FRANCE

Dr. Evseev Evgeny
Russian State University of Technology
MATI
AP. 139. 15-a Suvorova Str.
KALININGRAD (MOSCOU)

Prof. Frostig Y.
Aalborg University
Pontoppidanstraede 101
9220 AALBORG OUEST
DANEMARK

Prof. Grediac Michel
EMSE
SMS/Mecanique et Matériaux
158 cours Fauriel
42023 SAINT ETIENNE CEDEX 2
FRANCE

Dr. Guedes Jose Miranda
IDMEC/IST
Av. Rovisco PAIS 1
1096 LISBOA Codex
PORTUGAL

M. Guedra-Degeorges D.
Aérospatiale
Joint Research Center Louis Blériot
12. rue Louis Pasteur
BP 76
92152 SURESNES CEDEX
FRANCE

M. Han Woo-Suck
EMSE
Mécanique et Matériaux
158 cours Fauriel
42023 SAINT ETIENNE CEDEX 2
FRANCE

Mr. Hellbratt Sven-Erik
Karlskrona varvet AB
S 37182 Karlskrona
SUEDE

Dr. Hilaire Bruno
Centre de Recherches et d'Etudes d'Arcue
16 bis Avenue Prieur de la Côte d'or
94114 ARCUEIL Cedex
FRANCE

Mr. Hildebrand Martin
VTT Manufacturing Technology
Maritime and Mechanical Engineering
PO Box1705
02044 VTT

Prof. Horgan Cornelius o.
University of Virginia
Olsson Hall
VA 22903 CHARLOTTES VILLE
U.S.A.

M. Ioualalen K.
IUT Paul Sabatier
Labo de Génie Mécanique
50 chemin des Maraîchers
31077 TOULOUSE CEDEX
FRANCE

Mr. Judawisastra Hermawan
KU-Leuven
Deprt Metallurgy and Materials Engineeri
De Croylaan 2
3001 HEVERLEE LEUVEN
BELGIQUE

Dr. Karama Moussa
Ecole Nationale d'Ingénieurs de Tarbes
Chemin Azereix
BP 1629
65000 TARBES
FRANCE

Dr. Karczmarzyk Stanislaw
Warsaw University of Technology
Institute of Machine Design Fundamentals
Narbutta 84
02-524 WARSZAWA
POLOGNE

M. Kneip Jean-Christophe
ISAT
49, rue Mademoiselle Bourgeois
58000 NEVERS
FRANCE

Dr. Limam Ali
INSA LYON
URGC Structures
20, avenue Albert Einstein Bât. 502
69621 VILLEURBANNE CEDEX
FRANCE

Prof. Maier Martin
Institut für Verbundwerkstoffe
Erwin-Schroedinger Strasse
67663 KAISERSLAUTERN
ALLEMAGNE

Mr. Mäkinen Kjell-Erik
Karlskrona varvet AB
S 37182 Karlskrona
SUEDE

M. Manet Vincent
EMSE
SMS/Mécanique et Matériaux
158 cours Fauriel
42023 SAINT ETIENNE CEDEX 2
FRANCE

Mr. MARSOLEK Jens
RWTH Aachen
Institüt für Leichtbau
Wüllenerstrasse 7
52062 AACHEN
ALLEMAGNE

M. Mendes Ferreira A.J.
Universidade do Porto
Faculdade de Engenharia
Rua dos Bragas
4099 PORTO CODEX
PORTUGAL

Prof. Meyer-Piening H.R.
Institut für Leichtbau und Sei
LEC
ETH Zürich ZURICH
SUISSE

Dr. Moreno Alain
Ministère de la Défense
Centre d'Etudes de Gramat
46500 GRAMAT
FRANCE

Dr. Mussot-Hoinard Geneviève
Université de Metz
Labor de physique et mécanique des matér
Ile du Saulcy
57045 METZ CEDEX 1
FRANCE

Mr. Olsson Robin
The Aeronautical Research Institute
FFA
POBox11021
S-161 11 BROMMA
SUEDE

Prof. Olsson Karl-Axel
Kungliga Tekniska Högskolan
Institutionen för Lättkonstruktioner
Teknikringen 8
10044 STOCKHOLM
SUEDE

M. Parewyck S.
ROYAL MILITARY ACADEMY
3 rue de la Renaissance
1000 BRUXELLES
BELGIQUE

Ms. Peeters Inge
Free University Brussels (VUB)
VUB-TW-STRU Pleinlaan 2
1050 BRUSSELS
BELGIQUE

Mr. Perche Niclolas
Centre de Recherches et d'Etudes d'Arcue
16 bis Avenue Prieur de la Côte d'or
94114 ARCUEIL Cedex
FRANCE

Mr. Pflug Jochen
Katholieke Universiteit Leuven
MTM
De Croylaan 2
B - 3001 LEUVEN
BELGIQUE

Mr. Philips Dirk
Katholieke Universiteit Leuven
De Croylaan 2
B - 3001 LEUVEN
BELGIQUE

Mr. Polit Olivier
Université Paris X
IUT - GMP
1 Chemin Desvallières
92410 Ville d'Avray
FRANCE

M. Reimerdes H.G.
IFL-RWTH
AACHEN
ALLEMAGNE

Ms. Remillat Chrystel
Ecole Centrale de Lyon
Laboratoire de Tribologie et Dynamique d
36. avenue Guy de Collongues
69131 ECULLY CEDEX
FRANCE

Prof. Skvortsov Vitali
Marine Technical University of St Peters
Department of Strength of Materials
101 Leninski Prospect
198262 ST PETERSBURG
RUSSIE

Mme Tathi Banho
Renault
60303
860 quai de Stalingrad
92109 BOULOGNE BILLANCOURT

Prof. Theotokoglou Efstathios
National Technical University of Athens
Sofouli 33
N.S 17122 ATHENS
GRECE

M. Thevenet P.
Aérospatiale
Joint Research Center Louis Blériot
12. rue Louis Pasteur
BP 76
92152 SURESNES CEDEX
FRANCE

Dr. Thomsen Ole T.
Aalborg University
Inst. of Mechanical Engineering
Pontoppidanstraede 101
9220 AALBORG OUEST
DANEMARK

Prof. Torres Marques Antonio
INEGI
Faculdade de Engenharia
Rua do Barroco. 174
4465 S. MAMEDE DE INFESTA LECA BALIO
PORTUGAL

Prof. Touratier Maurice
ENSAM
Laboratoire de Mécanique des Structures
21 rue Pinel
75013 PARIS
FRANCE

Mr. van Straalen IJsbrand J.
TNO Building and Construction Research
Dept. of Structural Eng.
Lange Kleiweg 5, Rijswijk
Box 49
2600 AA DELFT
HOLLANDE

Dr. Vannucci P.
Università di Pisa
Istituto di Scienza delle Costruzioni
Via Diotisalvi, 2
56126 PISA
ITALIE

Prof. Vautrin Alain
EMSE
SMS/Mécanique et Matériaux
158 cours Fauriel
42023 SAINT ETIENNE CEDEX 2
FRANCE

Mr. CARLSON Leif
University of Florida
Compagnie CFAU
BOCA RATON, FL-33431
USA

Ms. CARRONNIER Delphine
Aérospatiale Toulouse
A/BTE/CC/RTE
316, route de Bayonne
31060 TOULOUSE CEDEX 04
FRANCE

Mr. CARRUTHERS Joe
Advanced Railway Research Centre
The University of Sheffield, Regent Court
30 regent street
S1 4DA SHEFFIELD
GRANDE BRETAGNE

M. Ing. GOFFART David
Plastic Omnium
Rue Castellion
BP 3010
01103 OYONNAX CEDEX
FRANCE

Mr. Ing. HAGENAUER Klaus
University line, division of technical mechanics
Altenbergerstr. 69
A-4040 LINE
AUSTRALIE

Dr. HAYMAN Brian
Det Norske Veritas
N-1322 HOVIK
NORWAY

Mr. NIGLIA Francesco
DIAS - Politecnico di Torino
Corso Duce deglu Abruzzi 24
10129 TORINO
ITALIE

Mr. LONNO Anders

Mr. MEULENBERG Marc
Hunter Douglas Construction Elements
PO Box 128
g350 AC LEEK
HOLLANDE

Dr. PIERRON Fabrice
Ecole des Mines de Saint-Etienne
Département Mécanique et Matériaux
158, cours Fauriel
42023 SAINT-ETIENNE CEDEX 2
FRANCE

Mr. ROUQUAND Alain
Centre d'Etudes de Gramat
46500 GRAMAT
FRANCE

Dr. ROZENBAUM E.D.
Hunter Douglas
PO Box 128
g350 AC LEEK
HOLLANDE

Dr. Yves SURREL
Ecole des Mines de Saint-Etienne
Département Mécanique et Matériaux
158, cours Fauriel
42023 SAINT-ETIENNE CEDEX 2
FRANCE

Mr. UYTTERHAEGHE Luc
Plastic Omnium
Rue Castellion
BP 3010
01103 OYONNAX CEDEX
FRANCE

Prof. VERCHERY Georges
ISAT/LMRT
49, rue Mademoiselle Bourgeois
BP 31
58027 NEVERS
FRANCE

Mr. WEIBLEN Frank
Swiss Federal Institute of Technology
Composites Laboratory (ETH Zürich)
Wagistrasse 13
CH -8952 SCHLIEREN
SUISSE